T0391247

Lecture Notes in Electrical Engineering

Volume 260

For further volumes:
http://www.springer.com/series/7818

Yueh-Min Huang · Han-Chieh Chao
Der-Jiunn Deng · James J. (Jong Hyuk) Park
Editors

Advanced Technologies, Embedded and Multimedia for Human-centric Computing

HumanCom and EMC 2013

Volume II

 Springer

Editors
Yueh-Min Huang
ES Department
National Cheng Kung University
Tainan
Taiwan, R.O.C.

Han-Chieh Chao
Institute of Computer Science
  and Information Engineering
National Ilan University
Ilan City
Taiwan, R.O.C.

Der-Jiunn Deng
Department of Computer Science and
  Information Engineering
National Changhua University of Education
Changhua City
Taiwan, R.O.C.

James J. (Jong Hyuk) Park
Department of Computer Science and
  Engineering
Seoul University of Science and
  Technology (SeoulTech)
Seoul
Republic of Korea (South Korea)

ISSN 1876-1100   ISSN 1876-1119 (electronic)
ISBN 978-94-007-7261-8   ISBN 978-94-007-7262-5 (eBook)
DOI 10.1007/978-94-007-7262-5
Springer Dordrecht Heidelberg New York London

Library of Congress Control Number: 2013943942

Printed on acid-free paper

Springer is part of Springer Science+Business Media (www.springer.com)

Contents

Part VI   Embedded Computing

Part XI Virtual Reality for Medical Applications

**Part XIII   All-IP Platforms, Services and Internet
            of Things in Future**

Part XIV Networking and Applications

Part VIII
Smart System

Dynamic Migration Technology Platform in the Cloud Computer Forensics Applied Research

Lijuan Yang and Shuping Yang

Abstract This paper presents a new cloud computing environment, computer forensics, and migration of a virtual machine image file, the virtual machine images from cloud computing platform migration to the local forensics environment analysis, the cloud computing platform, electronic evidence and the performance and application of experimental analysis comparing the two migration strategy.

Keywords Dynamic migration · Cloud computing · Computer forensics · Virtual machine

Introduction

Computer Forensics Analysis of the computer technology, computer crime analysis to identify criminals and computer evidence, computer intrusion and criminal evidence acquisition, preservation, analysis and production, and proceedings accordingly, is an invasion the computer system for scanning and crack invasion reconstruction process. Businesses and individual users can achieve the free sharing of information by a massive repository of cloud computing, forensic technology effectively find the necessary means to establish whether violations, but the traditional paper-based forensics longer meet the cloud computing service model, forensic work to become an important issue in the cloud computing environment.

Use of data migration technology to cloud computing environment for forensic work possible, by Forensics modeling of cloud computing, cloud computing platform as the system into multiple virtual body, running virtual machine instance

L. Yang (✉) · S. Yang
School of Computer Engineering, Shenzhen PolyTechnic, Shenzhen 518055, China
e-mail: yylljj928@szpt.edu.cn

Y.-M. Huang et al. (eds.), *Advanced Technologies, Embedded and Multimedia for Human-centric Computing*, Lecture Notes in Electrical Engineering 260, DOI: 10.1007/978-94-007-7262-5_71, © Springer Science+Business Media Dordrecht 2014

as the object of forensic analysis, virtual machine instances in the virtualization software layer security, load the virtual machine image file in the localized system for forensic analysis, the use of a separate division of the temporary image file partition as the exchange of information between the image file and localization system places the virtual machine image file can be loaded correctly, the scene forensics work under the cloud computing environment [1].

Computer Forensic Model of the Cloud Platform

Cloud computing platform provider of computing resources for effective resource allocation and access control, and provide end users with a variety of cloud services, virtualization technology of distributed computer resources are managed and integrated into the virtual host resources available to the user cloud platform virtualization features enable users do not have to care about the location of the real host, maintenance and fault tolerance, to help the user to get out from the computer hardware and software resources management burden.

Computer Forensics by function can be defined as five levels, namely the discovery layer, evidence of the fixed layer, abstraction layer of evidence, evidence analysis layer and evidence presentation layer, the division of the five levels together make up the hierarchical model of computer forensics. The forensic model is divided into basic service layer, the mirror migration evidence for layer, the evidentiary regulators, as shown in Fig. 1.

Basic service layer corresponds to the layer of the discovery of evidence. This layer is the place of cloud computing services, a variety of cloud computing services through the basic service layer are classified, organized into mutual contact multiple cloud computing services sequence, the sequence of these services constitute a virtual machine instance. Distinction between cloud computing services, forensics center of gravity shifted to the virtual machine instance, while avoiding the complexity of the network and the dispersion of the cloud structure to address the relevance of the evidence [2].

The mirror migration evidence for layer corresponding evidence abstraction layer is a major component of cloud computing forensic model, the software layer divided according to the mutual isolation of virtual machine instances, the use of the virtual machine instances interfere with each other, clearly forensics object and forensics range.

Forensics evidence regulators correspond to the fixed layer, scheduling and monitoring the mirror evidence abstraction layer every migration operation, and record the migration process to ensure that the migration process of the Strategies and execution control and to generate regulatory process reported Reports, to ensure the reliability of forensic data and legal integrity.

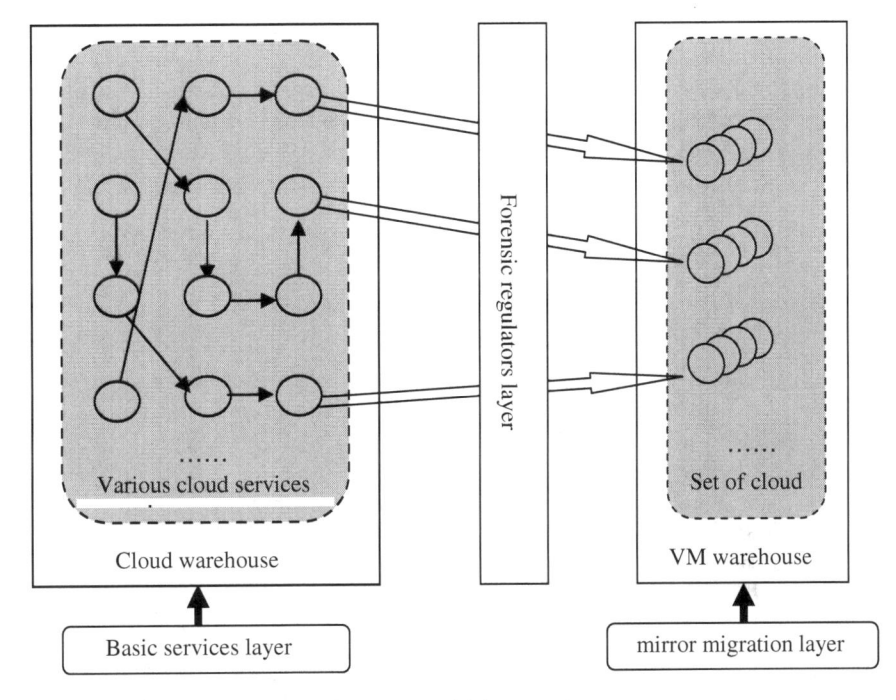

Fig. 1 Cloud platform computer forensics model framework

Dynamic Migration Technology

In the cloud computing platform built model of computer forensics, migration hardware configuration and system migration, live migration technology, the use of the operating system. Migration process needs to be saved and remodeling of the operating system, including the process of identification, information reconstruction of the file system, memory migration, network connection information monitoring, as well as many aspects of the concrete realization of registry information reconstruction. And through combined and feedback mechanisms, design data migration system architecture, data migration process will not affect the entire cloud computing platform service request. The different effects of different migration strategies of response systems, these migration strategies can be adjusted according to the working status of the migration [3].

Hardware Abstraction Layer Virtual Machine Technology

Hardware Abstraction Layer virtual machine technology is the use of an intermediate layer to isolate code running environment, virtual machine monitor layer to run on the operating system provides a variety of customer hardware response, mapping the physical device. The virtual machine monitor layer is able to create a copy of the system isolated from each other, each copy of the system shared physical device are isolated from each other, are not adversely affected, can play to the highest utilization of the physical device, the virtual machine copy of the system is concerned, this computer forensics analysis object of concern [4].

Preservation and Reconstruction of Virtual Machine Migration Process

Can achieve data migration of virtual machine instances in a virtual machine monitor layer, the virtual machine instance is saved as an image file, to facilitate the latter part of the environmental restoration and forensic analysis. The migrated data comprises a variety of sensitive data [5]:

- The page table entries and directory entries from a writable mode switch to read-only mode;
- Virtual machine's thread priority of the control module, switch from high-level to low-level, lower privileges;
- Virtualization software layer to migrate virtual machine scheduling data;
- Virtualization software layer calls to the underlying hardware of the virtual machine you want to migrate.

The entire virtual machine migration process under the state protection of sensitive data, the need to have four key modules, Is the process ID information module, memory mapping information module, the network connection module, file system information Module, in order to complete preservation of these data, to facilitate the full migration of the virtual machine instance, these four related The key part of the remodeling and save information the entire virtualization software layer module structure shown in Fig. 2 [6].

The Two Modes of Conversion of the Migration Process

Cloud computing platform virtualization software layer module add, the definition of a unified virtual communication interface, can be achieved in two modes of action, namely the ordinary cloud computing service model and cloud services migration patterns [7].

In normal mode, the cloud computing platform to provide normal cloud computing services, response to the application of the service request. Migration patterns, the virtualization software layer as multiple modules to ensure the consistency of the state of the operating system to load and record the current process state of the virtual machine system, the memory mapping, network connection and

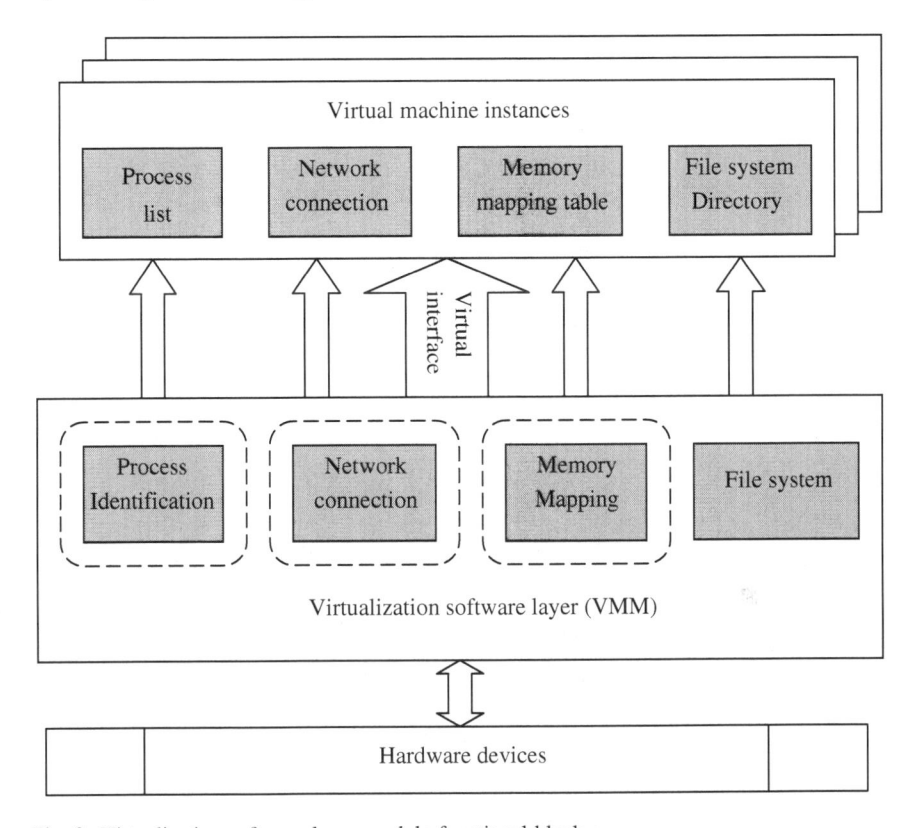

Fig. 2 Virtualization software layer module functional block

file system modifications change, Preparing Virtual Machines the mirrored data migration [8]. Data the migration scenarios migration task adaptive scheduling algorithm flow as shown in Fig. 3, the specific implementation process [9]:

Pl: The workload threshold value read from the migration task queue the migration task requests, set up to monitor the sampling period T and storage components $W_{\max}$, Counter $K = 1$, Take the initial migration rate $Migrate\_rate(0) = 0$;

P2: According to the migration strategy *MIGRATE_POLICY*, if the migration strategy is *PRIOR_LEVEL*[0], confers on the $Migrate\_rate(k)$ value is *MAX_RATE*, no rate limiting, turn P4, cycle $(k - 1) * T$ to $K * T$. The time during which a load feedback $W_f = Feedback\_load(t)$, calculate the actual load and the load threshold difference $E(k) = p \times W_{\max} - W_f$, among $P(0 < P < 1)$ is reserve coefficient.

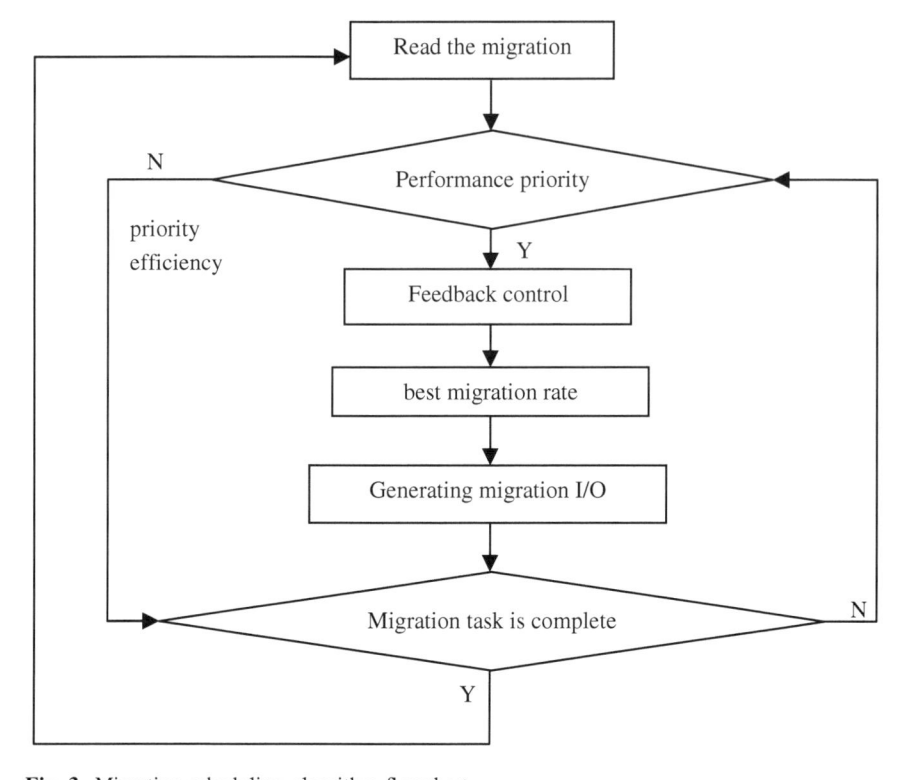

Fig. 3 Migration scheduling algorithm flowchart

P3: The difference obtained by the previous step, calculate the optimum value of the rate of migration of the next cycle:: $Migrate\_rate(k) = Migrate\_rate(k-1) + a \times E(k)$, K is the adjustment coefficient.

P4: $Migrate\_rate(k)$ send data migration generated migrate I/O, this procedure stored in the member receiving the sampling of the monitoring module, and based on the member state to modify or maintain the current task migration strategy.

P5: If this migration task is completed, turn P1, otherwise the counter K value plus 1, turn P2.

Experiment and Analysis

The Linux platform using the Xen open source virtualization tool to build a cloud computing platform, configure the virtualization monitoring layer, run the virtual machine instances, complete cloud computing platform features, and then use the testing instrument for the Linux platform, by analyzing the dynamic migration process virtual converted to the migration from the service-state mode of monitoring layer state mode, the system state transition time and performance comparison, draw mode changes the impact of cloud computing services platform, followed by the use of different data migration strategy, different migration strategy process in the amount of time the performance impact of cloud computing platform, the conclusion that the mirror migration method of performance evaluation [10, 11].

1. Performance test

Virtualization software layer needs to correspond to a virtual machine system for a recording and reconstruction, when the platform mode conversion from the normal mode to the migration mode takes considerable platform resources, and therefore significantly extend the run time of several test procedures, fork program as it relates to the large number of page table updates, adds to the reconstruction of platform information resource overhead.

2. Time Test

The global file system module using virtualization software layer traversal completed to measure the time course of the mode converter. Switch between normal mode and migration patterns of time longer than the switching time between migration mode and normal mode, migration mode and normal mode conversion services in the Xen test environment under quite small.

3. Data migration process virtual machine performance impact

The data migration process, the performance of the entire cloud computing platform is quite large, because the migration takes up bandwidth, resulting in customer service response is too slow.

4. The performance test results of different migration strategies

The efficiency priority migration strategy for the 1/0 response still have a considerable impact, due to the migration process need to ensure that the normal response of cloud computing services in a complete image file migration process is complete. Takes a large number of 1/0 channel.

The performance priority migration strategy, the migration process duration becomes longer, from the beginning lasted until 3005, than giving priority to efficiency strategy takes a long time, but did not bring the platform I/O response time Much impact.

Different from the experimental data can be concluded that, during the migration process can use different migration strategies, the use of giving priority to efficiency or service priority principle to divide, making migration Cheng Duiyun computing services impact as small as possible, but also in the cloud computing services when the system is idle, the migration of virtual machine images.

References

1. Clark C, Fraser K, Hand S et al (2005) Live migration of virtual machines. In: Proceedings of the 2nd international conference on networked systems design and implementation, Berkeley, CA, USA
2. Wood T (2007) Blaek-box and Gray-box strategies for virtual machine migration. In: Proceedings of the 4th international conference on networked systems design and implementation, IEEE Press
3. LaVelle C, Konrad A (2007) FriendlyRoboCoPy: A GUI to RoboCoPy for computer forensic investigators. Digital Invest 4:16–23
4. Arnold J (2008) Ksplice: An automatic system for rebootless kernel security updates. Ph.D. thesis, Massachusetts Institute of Technology
5. Zhou G, Cao Q, Mai Y (2011) Forensic analysis using migration in cloud computing environment. In: 2nd International conference on intelligent transportation systems and intelligent computing, Suzhou, China
6. Fasheng, based on resource management end-system traffic shaping algorithm (2007). Comput Eng Appl (18):117–119
7. Grossman RL, Yunhong Gu (2009) On the varieties of clouds for data intensive computing. IEEE Data Eng Bull 32(1):44–50
8. Tu WW, Zhang J, Zhang X (2006) Grade allocation token parameter of traffic shaping algorithm. Comput Appl (9):2175–2177
9. Ladan-Mozes E, Shavit N (2008) An optimistic approach to lock-free FIFO queues. Distrib Comput 20(5):323–341
10. Yadav AK, Tomar R, Kumar D, Gupta H (2012) Security and privacy concerns in cloud computing. Comput Sci Softw Eng 2(5)
11. Valenzuela JL (2004) A hierarchical token bucket algorithm to enhance QoS in IEEE 802.11: proposal, implementation and evaluation. In: IEEE 60th vehicular technology conference, 2004. VTC2004-Fall

Research on Parameters of Affinity Propagation Clustering

Bin Gui and Xiaoping Yang

Abstract The affinity propagation clustering is a new clustering algorithm. The volatility is introduced to measure the degree of the numerical oscillations. The research focuses on two main parameters of affinity propagation: preference and damping factor, and considers their relation with the numerical oscillating and volatility, and we find that the volatility can be reduced by increasing the damping factor or preference, which provides the basis for eliminating the numerical oscillating.

Keywords Affinity propagation · Damping factor · Preference · Volatility

Introduction

Affinity propagation (AP) is a new clustering algorithm published in Science magazine and proposed by Frey and Dueck [1]. Its advantage is that it finds clusters with much lower error than other methods, and it does so in less than one-hundredth the amount of time [1].Although results on small data sets (900 ≤points) demonstrate that vertex substitution heuristic (VSH) is competitive with AP, VSH is prohibitively slow for moderate-to-large problems, whereas AP is much faster and could achieve lower error [2].

There are some scholars who have done some improvements on AP algorithm. Literature [3, 4] proposed a new message transmission algorithm soft constraint

B. Gui (✉)
School of Information, Remin University of China, Beijing, China
e-mail: guibin_163@163.com

X. Yang
School of Computer Science and Technology, Huaiyin Normal University, Huaian, China
e-mail: yang@ruc.edu.cn

Y.-M. Huang et al. (eds.), *Advanced Technologies, Embedded and Multimedia for Human-centric Computing*, Lecture Notes in Electrical Engineering 260, DOI: 10.1007/978-94-007-7262-5_72, © Springer Science+Business Media Dordrecht 2014

condition who allowed a single cluster exemplar did not represented itself in the search and optimization process of the cluster exemplar. Kaijun Wang proposed an algorithm that could adjust adaptively the damping factors and preferences to eliminate oscillations and find the optimal clustering result [5]. Xiao Yu proposed a semi-supervised clustering method based on affinity propagation, it can produce good clustering results for the datasets with complex cluster structures by using the prior known labeled data or pairwise constraints to adjust the similarity matrix [6]. Although there are some research for AP, no one does overall research on AP.

Affinity Propagation Clustering

AP works based on similarities between pairs of data points (Euclidean distance), each similarity is set to negative squared error:

For points x_i and x_k, $s(i, k) = -||x_i - x_k||^2$.Take the negative in order to facilitate the calculation, the greater the value, the higher similarity. These similarity can be symmetric, i.e., $s(i, k) = s(i, k)$; they can also be asymmetric. i.e., $s(i, k) \neq s(i, k)$. The similarity between the N data points is composed of $N \times N$ similarity matrix S.Rather than requiring that the number of clusters be prespecified, AP takes as input a real number $s(k, k)$ for each data point k so that data points with larger values of $s(k, k)$ are more likely to be chosen as exemplars. These values are referred to as "preferences". The number of identified exemplars (number of clusters) is influenced by the values of the input preferences. If a prior, all data points are equally suitable as exemplars, the preferences should be set to a common value this value can be varied to produce different numbers of clusters. The shared value could be the median of the input similarities (resulting in a moderate number of clusters) or their minimum (resulting in a small number of clusters). There are two kinds of message exchanged between data points: responsibility and availability. The "responsibility" r(i, k), sent from data point i to candidate exemplar point k, reflects the accumulate evidence for how well-suited point k is to serve as the exemplar for point i, taking into account other potential exemplars for point i. The "availability" a(i, k),sent from candidate exemplar point k to point i, reflects the accumulated evidence for how appropriate it would be for point i to choose point k as its exemplar, taking into account the support from other points that point k should be an exemplar, and the support should be positive.

AP searches for clusters through an iterative process. The iteration may be terminated after a fixed number of iterations, after the changes in the messages fall below a threshold, or after the local decisions stay constant for some number of iterations. As affinity propagation is an iterative method, a damping factor $\lambda(0 \leq \lambda < 1)$ is introduced to reserve some information about the old values of messages to avoid numerical oscillations that arise in some circumstances. Each message is set λ times its value of the previous iteration plus $1 - \lambda$ times its prescribed updated value. The process of affinity propagation clustering is described as follow:

INPUT: data points $/x_i, i = 1, 2, \ldots, n$
OUTPUT: K cluster centers

1. Initialization: set $r(x_i, x_j) = 0$ and $a(x_i, x_j) = 0$ for all x_i and x_j. Compute similarity to generate the similarity matrix.

$$p(k) = s(k, k) = \underset{i \neq k}{median} \, (s(i, k)) \tag{1}$$

2. Responsibility updates:

$$r(i, k) \leftarrow s(i, k) - \max_{k' \neq k} (a(i, k') + s(i, k')) \tag{2}$$

$$r(k,k) = p(k) - \max_{j \neq k} \{ a(k,j) + s(k,j) \} \tag{3}$$

3. Availability updates:

$$a(i, k) \leftarrow \min\{0, r(k, k) + \sum_{i' \notin \{i,k\}} \max(0, r(i', k))\} \tag{4}$$

$$a(k, k) \leftarrow \sum_{i' \neq k} \max(0, r(i', k)) \tag{5}$$

4. Iterate (2) and (3) until a fixed number of iterations, or the changes in the messages fall below a threshold, or the local decisions stay constant for some number of iterations.
5. if $/ r(k, k) + a(k, k) > 0$, then data point k is the center.
6. Data points are assigned to the corresponding cluster center End.

The above algorithm would produce numerical oscillations. The numerical oscillation is the number of cluster centers generated in the iterative process continue to fluctuate. AP adjusts the damping factors to eliminate oscillations, each message is set to lam $(0 \leq lam < 1)$ times its value from the previous iteration plus $1 - lam$ times its prescribed updated value. The formula used for this method is the following:

$$R_i = (1 - lam) * R_i + lam * R_{i-1} \tag{6}$$

$$A_i = (1 - lam) * A_i + lam * A_{i-1} \tag{7}$$

According to Eqs. (6) and (7), we can see that R_i and A_i are influenced by the lam in each iteration. When lam is small, R_i and A_i are very different from R_{i-1} and A_{i-1}, and when lam is large, R_i and A_i are very close to R_{i-1} and A_{i-1}.

Research on Affinity Propagation Parameters

In this section, we will study the AP parameters based on experiment result. The experimental data is randomly generated in order to make the conclusions more catholicity. The experimental platform is Matlab 2009a. The hardware environment is Intel CPU PD T4500 2.3G, main memory is 2G.

AP algorithm will produce numerical oscillation in the process of iteration. it is usually too hard for AP to converge. So, it is very important to study to eliminate the numerical oscillation. In order to better reflect the degree of oscillation in the clustering process, we introduced a volatility index. Volatility is an index to measure the fluctuation degree of the underlying assets price, and it is usually expressed by standard deviation. The essence of numerical oscillation is the clustering number varying with time, Therefore, it is appropriate to use volatility to measure the degree of it.

Although AP need not require the number of clusters be prespecified, the final clustering result can not be known exactly. We usually wish it to produce wanted clustering number in practice. It is important to study the relation between clustering number and related parameters in forecasting the AP the clustering number. When the similarity matrix is determined, there are some parameters like preference, damping factor, maxits which could be adjusted in the run-time of AP Algorithm

Maxits Parameter

Maxits means maximum number of iterations for AP. Like convits, We are also more concerned about its relations to the running time and the clustering number. Set N = 100, lam = 0.5, convicts = 5, p = median (s). The relations are shown in Fig. 1.

It can be seen from Fig. 1 that the running time and the clustering number was little influenced by the maxits. Therefore, in order to ensure adequate iteration for AP,we recommend the value of maxits may be set higher.

Fig. 1 Maxits's relations to the running time and the clustering number

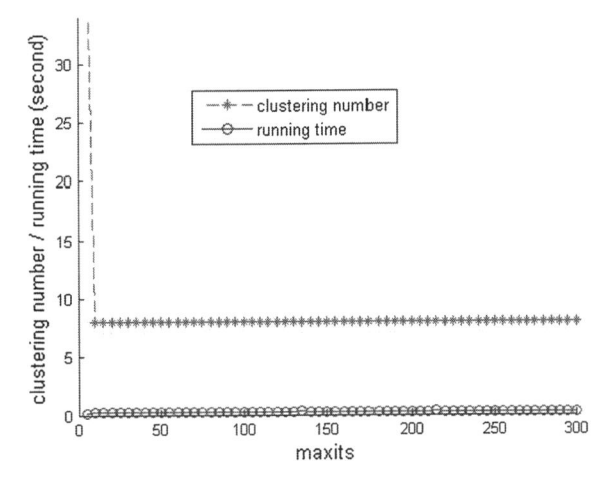

Preference Parameter

As the above points out, preference is a very important parameter for AP. It determines the clustering number, it also exercises an crucial influence over AP convergence rate. So, it is necessary to study its relations to the numerical oscillation, running time, and the clustering number in details. Set $N = 100$, lam = 0.5, convicts = 50, maxits = 50000. The relations between preference and the running time, and the relations between preference and the clustering number are shown in Fig. 2.

The horizontal axis shows preference varies from median(s)*2^15 to median(s)*2^ − 15. In order to make the conclusions more universal, convicts and maxits are set relatively larger values according to above conclusion. It can be seen

Fig. 2 Preference's relations to the running time and the clustering number

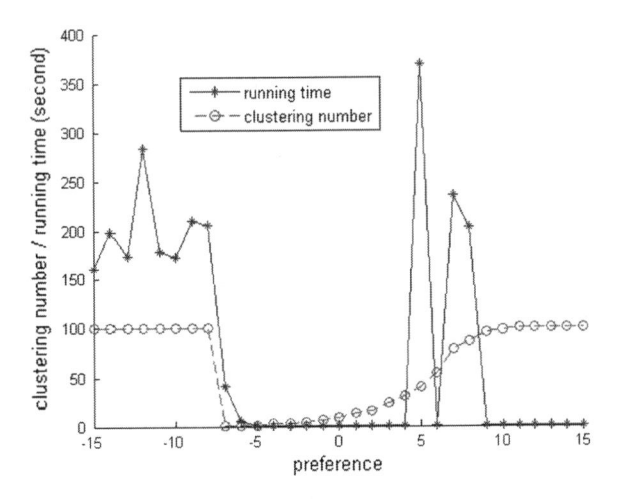

Fig. 3 The relation between preference and volatility

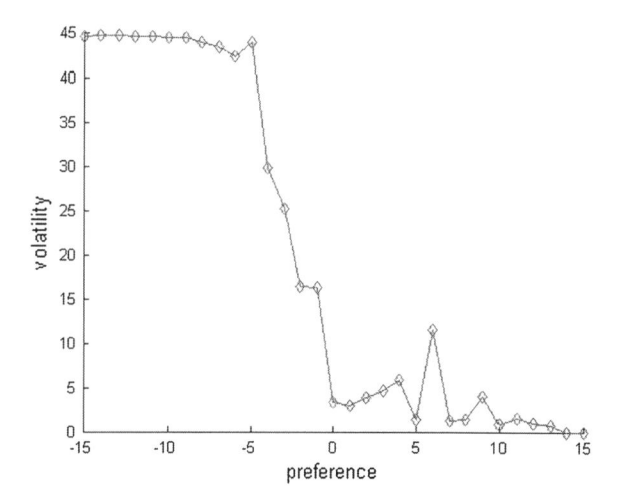

from Fig. 3 that the clustering number varies from 1 to N when the preference is in a certain range; When p = median (s), the clustering number is slightly less than $\sqrt{N}$; It will be difficult to converge and more clustering number, if the preference value too high or too low; Increase the algorithm scale and repeated runs on the certain dataset, it can still come to these conclusions. According to the above conclusions, we can get the idea for Re–K-AP, re-runs AP until the specified clustering number is generated. Of course, there are other ways for K-AP, Zhang proposed an idea for K-AP by introducing a constraint in the process of message passing, was more effective than k-medoids w.r.t. the distortion minimization and higher clustering purity [7]. Our experiments shows that K-AP preserves the clustering quality as Re–K-AP in terms of the distortion, and it has negligible increase of computational cost compared to Re–K-AP.

The relation between preference and volatility is shown in Fig. 3.

The horizontal axis means preference varies from median(s)*2^−15 to median(s)*2^15. It can be seen from Fig. 3 that the volatility declines with increasing the value of preference on the whole. Therefore, the degree of numerical oscillation can be reduced by enlarging the value of preference. Combined with Fig. 3, We find that it does not mean AP must converge when volatility declines. Literature [5] pointed out that numerical oscillation could be escaped by decreasing the preference. According to our conclusions, it may be right under certain circumstance, but it is more like wrong, because numerical oscillation will increase. Through a lot of studies and experiments, it shows that better experiment result can be gained when p is in the range: median(s)* 2^−5 ~ median(s)* 2^5.

Fig. 4 Damp factor'
relations to the running time
and the clustering number

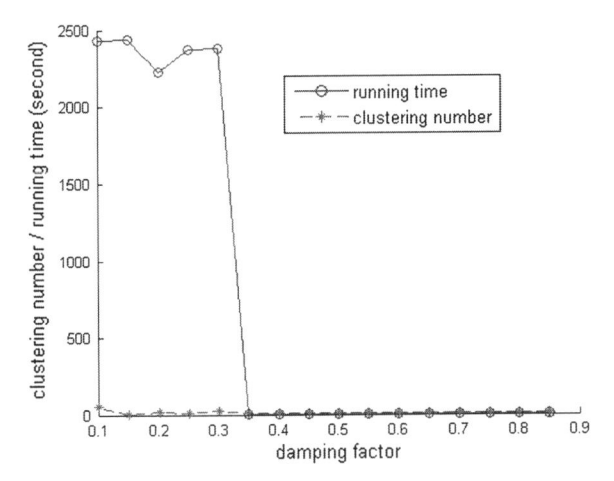

Damping Factor Parameter

Like preference, damping factor is also an vital parameter. According to the above
conclusions, the damping factor can be used to eliminate the numerical oscillation.
We are also more concerned about its relations to numerical oscillation, the running time, and the clustering number. Set $N = 100$, convicts $= 50$, maxits $= 50,000$, $p = $ median(s).The relation between the damping factor and running
time,and the relation between damping factor and clustering number are shown in
Fig. 4.

It can be seen from Fig. 4 that the clustering number is relatively steady when
the damping factor increases from 0.3 to 0.85, and the running time is also steady

Fig. 5 The relation between
damping factor and volatility

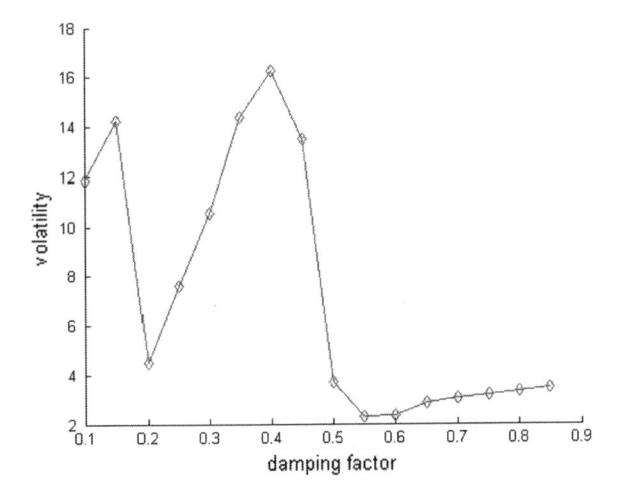

when the damping factor is greater then 0.35. Therefore, damping factor has a little effect on the clustering number and running time.

The relation between damping factor and volatility is shown in Fig. 5.

It can be seen from Fig. 5 that the volatility declines with enlarging the value of damping factor on the whole. Therefore, the degree of numerical oscillation can be reduced by enlarging the value of damping factor. Through a lot of studies and experiments, the damping factor should be set a large value from the beginning and subject to/$0.5 \leq \lambda < 1$.

Conclusions

The main parameters of AP are discussed in this paper. Because of the complex using circumstance, our conclusions is only for reference. Of course, It may further develop the studies on some parameters, for example, how to generate the specified clustering number quickly according to the relation between preference and clustering number? how to combine and balance the preference and damping factor to get the best result? and so on. All of these will be researched in the future.

Acknowledgments. This research was supported by the grants from the Natural Science Foundation of China (No. 71271209); Huaiyin Normal University Youth Talents Support Project (NO. 11HSQNZ18).

References

1. Frey BJ, Dueck D (2007) Clustering by passing messages between data points. Science 315(5814):972–976
2. Frey BJ, Dueck D (2008) Response to comment on "clustering by passing messages between data points". Science 319(5864):726
3. Leone M, Sumedha S, Weigt M (2007) Clustering by soft-constraint affinity propagation: applications to gene-expression data. Bioinformatics 23(20):2708–2715
4. Sumedha ML, Weigt M (2008) Unsupervised and semi-supervised clustering by message passing: Soft-constrain affinity propagation. Eur Phys J B 66:125–135
5. Wang K, Zhang J, Li D, Zhang X, Guo T (2007) Adaptive affinity propagation clustering. J Acta Automatica Sinica, 33(12): 1242–1246, (In Chinese)
6. Yu X, Yu J (2008) Semi-supervised clustering based on affinity propagation algorithm. J Software, 19(11):2803–2813, (In Chinese)
7. Zhang X, Wang W, Nørvåg K, Sebag M (2010) K-AP: generating specified K clusters by efficient affinity propagation. ICDM 2010: 1187–1192

Ant Colony Algorithm and its Application in QoS Routing with Multiple Constraints

H. E. Huilin and Y. I. Fazhen

Abstract In the modern communication network, QoS routing optimization problem acts as one of the most important types of discrete optimization problem which could normally be solved by heuristic algorithm. And Ant Colony Algorithm as a new heuristic optimization algorithm shows good performance in solving complex optimization problems. In this paper, Ant Colony Algorithm is presented to solve the QoS unicast routing problem under the constraints of bandwidth and delay, using the mechanism that ants are able to find the optimal path through pheromone and joining the heuristic strategy. The simulation results show that this algorithm can quickly find the routing that meets the constraints of time delay and bandwidth with minimum cost and minimum time delay.

Keywords Ant colony algorithm · QoS routing · Cost · Delay · Dijkstra algorithm

H. E. Huilin (✉) · Y. I. Fazhen
School of Electronic and Information Engineering, Beijing Jiaotong University,
Beijing 100044, China
e-mail: 10271037@bjtu.edu.cn

Y. I. Fazhen
e-mail: fzhyi@bjtu.edu.cn

Y.-M. Huang et al. (eds.), *Advanced Technologies, Embedded and Multimedia for Human-centric Computing*, Lecture Notes in Electrical Engineering 260, DOI: 10.1007/978-94-007-7262-5_73, © Springer Science+Business Media Dordrecht 2014

The Research and Exploration of the Development Trend of Cloud Computing

Shuping Yang and Lijuan Yang

Abstract Cloud computing is praised as the third-time IT revolution following the transformation of the personal computer, and the Internet change, its future trend of development will greatly impact on the existing IT concept that will change the business models and and people's way of life and work. This paper do the summary analysis and research for present global cloud computing, put forward the development measures and the problem to solve about China's cloud computing.

Keywords Cloud computing · Moble internet · SME

Introduction

In the era of cloud computing, companies no longer need to self-built data centers, since the group IT team to maintain and manage the system, enterprises need IT services via the Internet by specialized companies, like running water or electricity.

According to the prediction of international authoritative institutions, in 2014 global cloud computing output will reach 1,700 billion dollars [1]. BRICS cloud computing investment and research in 2012 has been far behind the United States, Canada and other developed countries, at present, the development of cloud computing has entered the stage of the campaign of, in China, If we do not increase investment and research of cloud computing, we are be again behind the developed countries in Europe and the United States after the Internet.

S. Yang (✉) · L. Yang (✉)
School of Computer Engineering, Shenzhen Polytechnic, Shenzhen 518055, China
e-mail: ysping@szpt.edu.cn

L. Yang
e-mail: Yylljj928@szpt.edu.cn

Y.-M. Huang et al. (eds.), *Advanced Technologies, Embedded and Multimedia for Human-centric Computing*, Lecture Notes in Electrical Engineering 260, DOI: 10.1007/978-94-007-7262-5_74, © Springer Science+Business Media Dordrecht 2014

People have a certain inertia and dependence using technologies or products, everyone has the inertia you want to re-capture the market is not easy.

What is Cloud Computing?

The concept of cloud computing is Google first proposed in 2006, 2007 IBM technical white paper also mentioned cloud computing, since the industry began an extensive discussion and research for cloud computing [2]. A time, on overwhelming the concept of cloud computing, so as SaaS, IaaS, PaaS, and so all of a sudden people into the "cloud", "fog".

In fact, Google and Baidu's search engine is the earliest of cloud computing, search engine to the majority of individuals and enterprises to provide information query function, the sharing of information resources, is cloud computing; Sina's microblogging, Tencent's micro-letters, etc., for the users to provide a platform for communication, sharing of platforms and services, it is cloud computing, Jinshan network fast disk, etc., are cloud computing technology. This is what we see around the touch of cloud computing. In fact, we are already living in the "cloud".

So, what is cloud computing? Although the definition given we are not the same, but the core idea is the same. Simply put, cloud computing is to optimize computing resources on the Internet integration, unified management and scheduling, constitute a powerful computing pool to provide services to the user on demand.

The cloud computing system architecture is shown in Fig. 1.

Type of Cloud Computing

Cloud computing application range is divided into public cloud, private and hybrid clouds [3].

- The public cloud is to provide services for the majority of consumers or businesses via the Internet, it is a new business model using cloud services to provide services to end users.
- Private cloud is the enterprise cloud technology to build enterprise network applications within the enterprise information platform, more emphasis on efficiency and cost advantages of cloud applications within the enterprise, as well as to the internal IT management.
- Hybrid cloud companies face different applications use to hire an external public cloud service or self-built cloud platform.

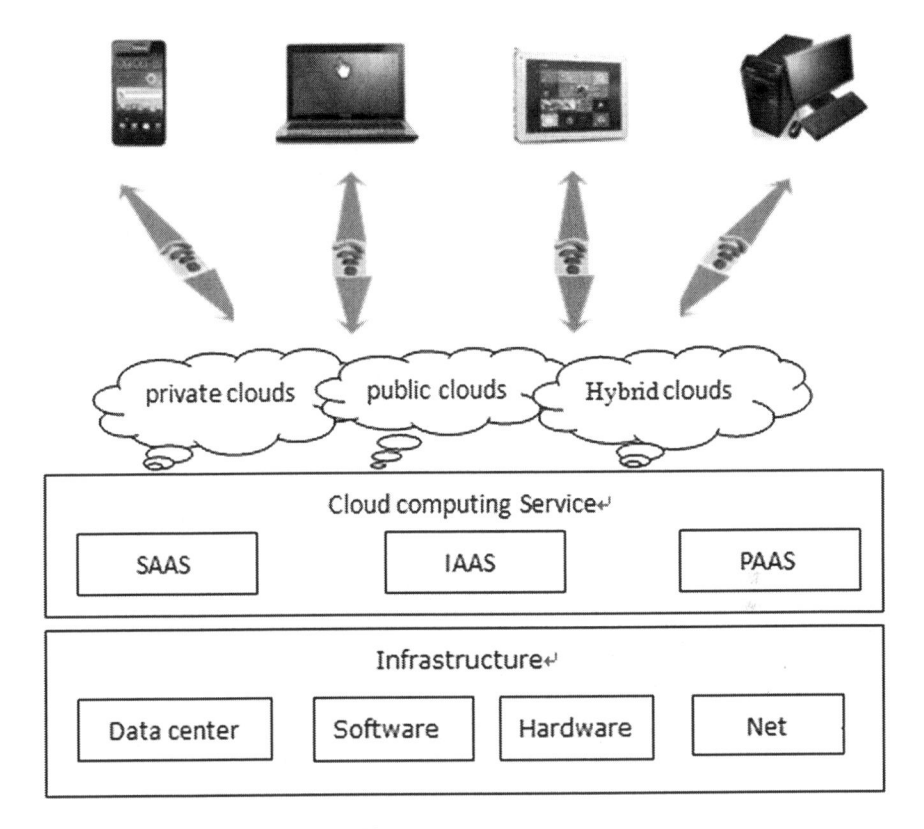

Fig. 1 Cloud computing system architecture

Services and Key Technology of Cloud Computing

Application structure in terms of, in order from the underlying hardware to the upper application, cloud computing offers three levels of service [3].

- IaaS:Infrastructure as a Service, infrastructure cloud.
- Paas:Platform as a Service,platform cloud.
- SaaS:software as a service,application cloud.

Involved in cloud computing technology [3]: Data Storage Technology, Data Management Technology, programming model, cloud security etc.

Features of Cloud Computing

Cloud computing is to continuously improve the processing capability of the "cloud", to reduce the burden of the user terminal handling, and finally to the user terminal simplified to a simple input and output devices, and can enjoy a "cloud" of the powerful computing capabilities demand. Cloud computing has the following characteristics [3]:

- scalability, flexibility and on-demand services and billing. Cloud services with on demand resource allocation and charges.
- extremely cheap. Is the internal private clouds or external public cloud, due to the centralization of the data center scale, economies of scale led to the rapid decline in costs and expenses.
- The reliability and versatility. Cloud computing can provide more reliability than the local computer. At the same time, cloud computing has the versatility with a "cloud" can support different applications run.
- large-scale. Google "cloud" has more than 100 million servers, declared objective is the development of more than 10 million units, Amazon, IBM, Microsoft and other "cloud" also hundreds of thousands of enterprise private cloud scale is generally within a few hundreds to thousands of servers. Larger centralized data center industry trends.
- high efficiency. The industry estimates the cloud computing implementation of the current IT resource utilization can be increased by 5–7 times.

How Cloud Computing from "Clouds" into "Spring Rain"?

Now, cloud computing technology has been steadily through the early "most difficult" stage of development, many industries also fully aware of the benefits of cloud computing, more and more companies have adopted or integrated some cloud computing technology to the business, which means that they benefit a lot from the cloud.

The saying goes, "your spring such as oil", cloud computing from the "cloud" becomes priceless "spring rain", it has the inevitability of technology and market.

Mobile Internet to Promote Cloud Computing Priceless

Cloud Computing Center of the investment of resources and energy consumption and waste is amazing, the initial proposed not to be optimistic. However, the widespread use of smart phones, tablet PCs cloud computing leaps and bounds, it can be said mobile Internet cloud computing priceless. The rapid expansion of the user base so that cloud computing offers free platform to become possible, open platform Baidu cloud, computing, storage, and use of network resources is no longer charges are freely available to the majority of small and medium-sized enterprises and application developers to attract massive subscribers and to generate a great deal of network traffic, then flow returns to achieve profitable platform.

Market research firm IDC survey forecast, the 2013 iphones and ipads market will grow by 20 %. Among them, the screen is smaller than an 8-inch mini ipads accounted for 60 % of the tablet market. Mobile manufacturers will break the present status, or the rise or decline [4].

SaaS Applications to Flourish

At present, SaaS is the most mature, the most famous and most widely used cloud computing model. In the past few years, has paid off handsomely for many companies to bring the rapidly expanding enterprise, such as Salesforce, Amazon and other. Today, cloud computing service objects are small and medium-sized enterprises from around the world gradually extended to the government, research, education, medical and other fields, and large enterprises.

2013 SME SaaS market booming. According to Gartner research, in 2012 the global SaaS applications market has reached $12 billion, in 2015 to more than 20 billion U.S. dollars. Although the main consumers of SaaS applications or Europe and the United States, Japan and South Korea, Australia and New Zealand and other developed countries and regions, but the growth of the emerging markets, including China, should not be underestimated. IDC survey shows that in 2013 global IT spending will be up to $2.1 trillion [5], the golden age of cloud computing upcoming.

Cloud Computing Trends

- mobile cloud will be more brilliant

Mobile Internet to a large customer base, the flow its development offers great potential. Commissioned a slightly more complex due to the restrictions on the capacity of smart phones, cloud computing has some limitations, advantages and disadvantages, and commissioned the task is relatively simple, relatively easy to implement and manage SME service providers and high enthusiasm, to facilitate the formation of a huge market. In addition, the development of smart phone hardware is also important part of the mobile Internet, I believe that in the near future, smart phones will be more capacity.

- small-scale industry cloud colorful

We will provide a small cloud computing platform cloud services in specialized industry applications and services known as the small-scale industry. Small industry cloud computing platforms to specialized areas, professional services, fast and efficient and has broad prospects.

- cloud security issues highlighted

Cloud computing as a system service center under attack is inevitable, the consequences are disastrous. Increased protection measures to improve the efficiency of security is bound to affect the security of cloud even affect the development of cloud computing, governments must attach great importance, sincere cooperation, security technology is important, but the fight against cybercrime must be ruthlessly expected to cloud security technology in there will be a $10 billion market in the next five years [6].

- cloud integration imperative

Cloud computing is still in its infancy, the relevant technical standards, cloud agreements are still inadequate perfect, cloud computing integration, large IT companies will acquire smaller cloud computing companies, systems and norm-setting is imperative. IDC believes that in the next 20 months, the cost of acquisition of SaaS will be more than 25 billion U.S. dollars [4].

- Data rapid market growth

IDC said that the big data market will be an annual growth rate of 40 %. 2012 big data market size of approximately $ 5 billion, will double in 2013, will reach $ 53 billion in 2017. This will cloud computing bring considerable profit margins [4].

China's Cloud Computing How to Achieve the Bend to Overtake?

China's cloud computing research and investment is relatively backward, you want to bend to overtake the need to rely on innovation. As the saying goes "non-profits can not afford early, the development of new technologies ultimately rely on the enterprise. If there is no profit, the firm is not dry.

- cloud computing has the advantage. At present, China's number of Internet users is about more than 300 million mobile phone users reached more than 600 million, the market potential is huge, which means the companies to develop sufficient power.
- American experience detours, open platform for Internet sharing, and energy conservation can be more optimized solution.
- construction of smart city for cloud computing to provide a broad market prospect. In 2012, the mobile Internet, cloud computing, Internet of Things rapid development, further promote economic and social informatization process. Than 320 cities across the country to invest 30 billion yuan in building smart city [7].

- China's three major telecom operators can be greater as. China's huge market potential for the three major telecom operators to provide a broader prospects and kinetic energy, telecommunications operators in the traditional line services can provide cloud computing services platform.
- Baidu, Tencent, Huawei, ZTE, wave and other large domestic IT companies have been established in the field of cloud computing cloud computing service platform, will be to lead and promote the development of China's cloud computing.

Conclusion

Cloud computing has moved from concept to the market and a huge market driven by the interests of governments and companies are competing invested heavily. China should develop countermeasures as soon as possible, otherwise it will again in the wave of cloud computing in a backward position, the domestic Baidu, Tencent, Huawei, Inspur and other large enterprises should seize the opportunity to turn overtaking, the three major telecom operators should also be to actively participate in the competition of cloud computing.

References

1. Gao W, He B, Li J (2012) The development and analysis of the global cloud computing, telecommunications network technology 2(2), 31–34
2. Zhou S, Su C (2012) Cloud computing technology research. Sci Technol Inf 24 (4)
3. Zeng X (2012) The concept of cloud computing applications, wireless internet technology, (2), 32–33
4. Predict 2013 nine technology trend of mobile will break the status quo. http://www.csdn.net/article/2012-12-04/2812492
5. Across the tide SaaS. http://www.programmer.com.cn/15835/
6. Network security overhead climbed to $10 billion after five years. http://net.chinabyte.com/230/11774230.shtml
7. 320 of city of our country investment wisdom city in 2012. http://www.cnscn.com.cn/policy/show-htm-itemid-538.html

The Research of Intelligent Storage System Based on UHF RFID

Yong Lu, Zhao Wu and Zeng-Mo Gao

Abstract Warehousing is an important link of the enterprise operation. Usually, traditional management of warehousing has many defects and low efficiency. We need to change its management mode for improving the efficiency of logistics. Currently, most of warehousing and logistics take barcode as an intelligent way on management. In this way, the intelligence of the system has greatly improved than before, but it still need a lot of manpower and material resources into warehousing and logistics. This thesis designed an automatic identification system which bases on UHF-RFID and embedded control technology. The system can identify goods wirelessly, and realize automated management of warehousing goods.

Keywords RFID · Warehousing system

Background

RFID, as a new generation of automated data collection tool, has been widely used in many fields and gradually began to be used in warehouse management. Compared with the barcode which has been widely used in warehousing, RFID has higher efficiency, longer read distance and large capacity for storage. With economic development and the advancement of technology, the application of RFID in warehousing and logistics will become more and more common.

Y. Lu (✉) · Z. Wu · Z.-M. Gao
School of Electronic and Information Engineering, Beijing Jiaotong University,
Beijing, China
e-mail: 11120003@bjtu.edu.cn

Y. Lu
e-mail: ylu@bjtu.edu.cn

Z. Wu
e-mail: 10120017@bjtu.edu.cn

Y.-M. Huang et al. (eds.), *Advanced Technologies, Embedded and Multimedia for Human-centric Computing*, Lecture Notes in Electrical Engineering 260, DOI: 10.1007/978-94-007-7262-5_75, © Springer Science+Business Media Dordrecht 2014

Introduction of RFID

The Structure of Passive RFID

Passive RFID system generally consists of two parts, electronic labels and readers. Usually some specific format information stores in the electronic label. In actual applications, electronic label attached to the surface or inside of the object to be identified. When the object with electronic label falls into the area where the reader can connect, the reader automatically read out the information contained in the electronic label with non-contact manner.

In the field of UHF RFID [1], the mainstream protocol is ISO/IEC 18000-6C and EPC G1C2 [2]. In both protocols, the communication between readers to label is based on the PIE encoding. All communications should be started with previously synchronized code or frame synchronization. The reader takes DSB-ASK, SSB-ASK or PR-ASK modulation scheme for communication. The label should be able to demodulate the above-mentioned three types of modulation. The encoding of data signals are mainly use the FM0 baseband coding and Miller subcarrier modulation.

Algorithm for Anti-Collision in RFID

During the multi-target communication, multiple readers or multiple labels take up a communication channel to send data at the same time. As a result, collision and the interference between devices will occur. In this condition, communication absolutely fails. Taking the above situation into consideration, an algorithm for anti-collision is of great significance for a warehousing system based on UHF RFID.

The algorithm for anti-collision based on ALOHA is suitable for our system because it has been widely used in many fields and easy to implement [3]. ALOHA can be divided into pure ALOHA, slotted ALOHA, frame slotted ALOHA and dynamic frame slotted ALOHA.

Combining the rules of the EPC Gen2 protocol label, this design make a definite improvement on the traditional frame slot ALOHA anti-collision algorithm. The improved algorithm similar to the dynamic frame slotted ALOHA algorithm. But the concept of a frame is replaced by the identification cycle which is the time between two Query commend. In this algorithm, the number of slots in each identification cycle can adaptively change According to the label identification availability. In addition, every identification cycle can stops at any time based on the adaptive and then start another identification cycle. It replaces the mechanism that the number of slots changes after the identification cycle end. This algorithm has good performance on throughput and stability and is applicable to the warehousing system based on RFID.

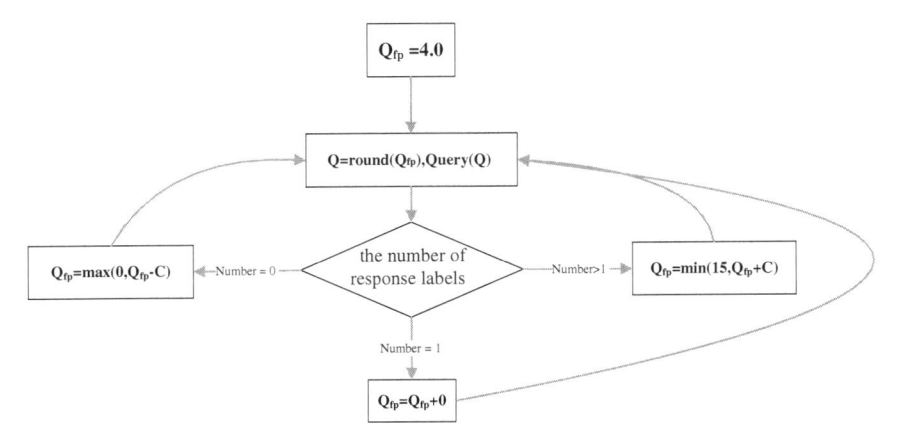

Fig. 1 Anti-collision a algorithm

In this algorithm reader sent EPC protocol command Query [4] to define the number of time slots contained in each piece of identification cycle. Query command sends a parameter Q to the label in RF field. After receiving the Q value, the slot counter in respective label will generate a random number within the range of 0–2Q-1 and then whenever the reader sends QueryRep command to the label, random number of the label is decremented by one automatically. To the random number down to 0, the label will return the RN16 handle and the identification is successful. The collision appears when more than one label's random number reaches to 0. The unidentified labels enter the sleep state and wait for getting a new random number in next identification cycle. Based on the number of unidentified labels, Q value can be made adaptive. This thesis set 4 as the initial value of Q and the algorithm processes shown in the Fig. 1.

After simulated in MATLAB, the algorithm shows a good performance. When the number of labels is under 30, the throughput rate increases from 0.27 to 0.35 with the increase in the number of labels; when the number of labels is more than 30 and less than 500, the throughput rate is floating around 0.35.

Program Design

Hardware Structure

The design goal of intelligent warehousing recognition system RFID reader [5] is to accurately identify the goods affixed with EPC labels through the working table at a certain speed and transmit the label information to the host computer. According to the label content, host computer perform specific statistical functions

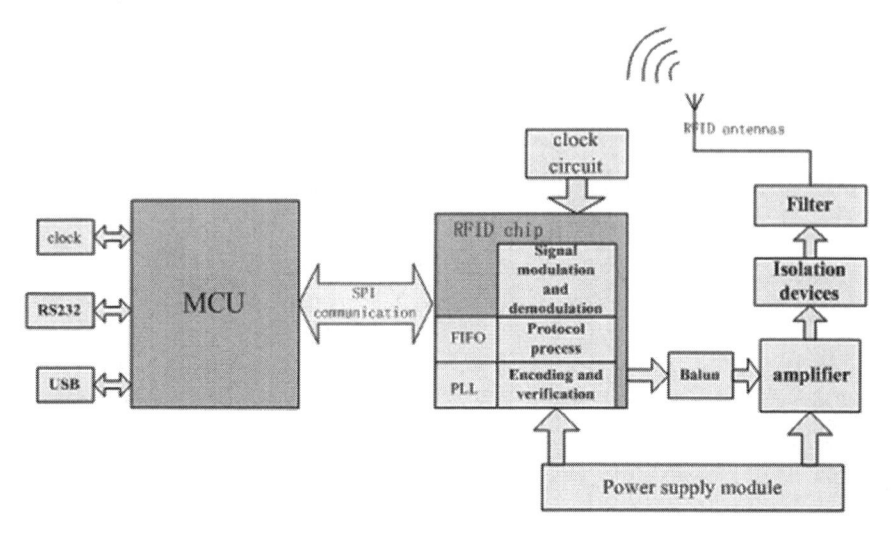

Fig. 2 Hardware system

and update the database. The following figure shows the structure of the hardware of the RFID reader.

The function of the RFID reader:

Launch command to search for labels; Detected electronic labels and generate RF continuous carrier; Provide energy to the labels to communicate, modulation, demodulation; read, write and encrypt the label information.

The hardware part includes a RFID chip, controller, signal isolation circuit, power supply circuit, a voltage -controlled oscillator and clock circuit. Figure 2 shows the structure.

In this design, control part and RF part coexist in one circuit board. FL2440 ARM9 development kit is selected to be control part of the reader. The control part and RF part is connected via SPI communication.

The design choices AS3992 as the RFID chip. AS3992 provide a downlink communication rate up to 640 kbps and add DRM process.

Software Structure

The structure of RFID system software part is shown below. The entire software part is mainly divided into the application interface module, the configuration control module, the protocol processing module and the data storage module (Fig. 3).

Function of each module.

The application interface module: This part provides the connection between protocol layers and the host computer application layer. Operation from PC passed

Fig. 3 Software system

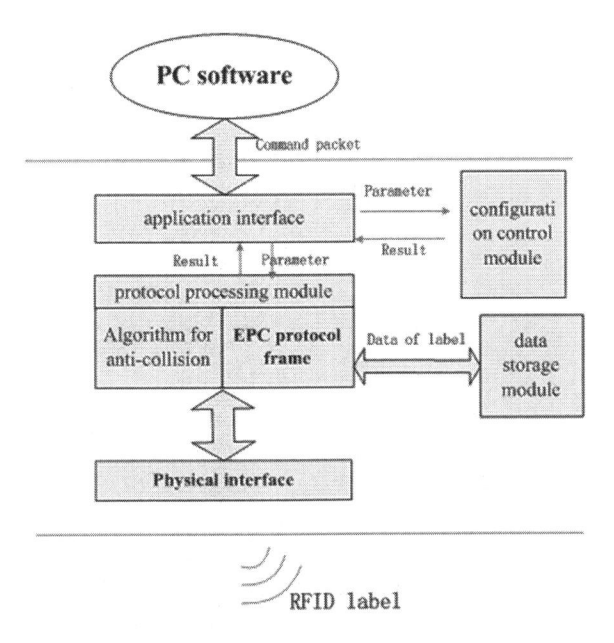

to the protocol layer through the application interface module. After get information from the packet from PC, the protocol processing module completes the corresponding task.

The configuration control module: The entire RFID systems program is burned into the ARM9 processor by CodeWarrior and then the ARM control module will do control and communicate with AS3992. According to the different needs of the user, the reader configured for different environments. Moreover, software design automatic mode for reader. ARM control module get configuration parameters from application interface module. When configuration finish, result will send to PC.

The protocol processing module: This part is the core module of the reader software system and mainly deal with RFID protocol, the related anti-collision algorithm. The reader uses a different protocol processing module can match different RF protocols to operate different protocol version label. This module exchange data with others and the identification work complete in this part.

The data storage module: This module mainly used for the storage of temporary label data. Because the speed of reading label is faster than outputting speed, the reader cannot guarantee outputting data timely when numbers of labels has been read. So system needs to store the data and output at spare time. Storage management obtains each label input and read record. The module cooperates with the corresponding memory interface to form a complete storage.

The RF label identification software implementation process.

RFID software process starts with creating connection between the reader and PC. The reader receives command from PC and completes corresponding work

Fig. 4 Software
implementation process

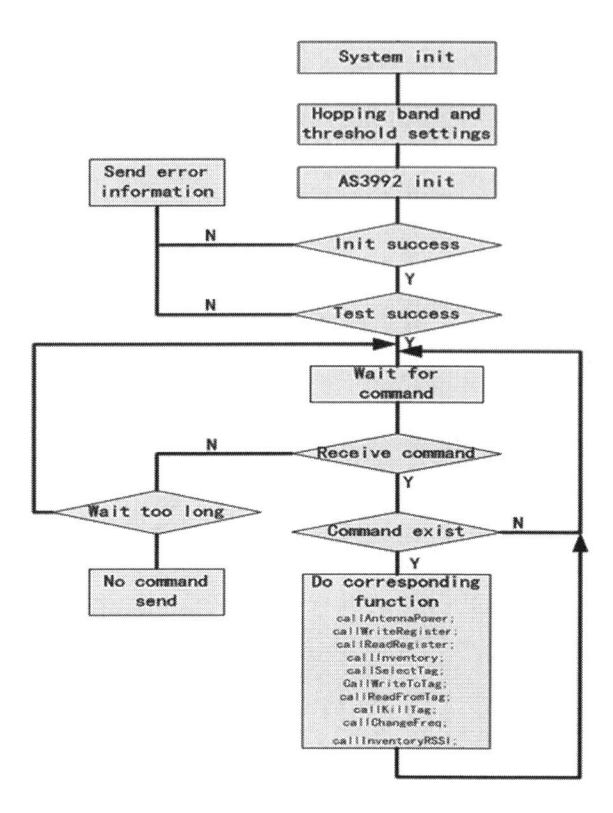

such as read information from label and write information to label. As the core of
the reader, ARM coordinates each module and makes the EPC protocol work. The
system works as the following Fig. 4.

Test and Conclusion

Author puts 30 RFID labels at positive direction to the antenna and makes the
distance between labels and antenna different.

First test is to confirm the limit distance that the reader can indentify whole
labels successfully. The conclusion is: when the distance is less than 4.9 m, all of
the label can be identified and when the distance is longer than 4.9 m, the identification rate decrease with the increase of the distance.

Second test is to confirm the limit distance that the reader can write information
to labels successfully. The conclusion is: when the distance is less than 2.0 m, all
of the label can receive the new data from the reader and when the distance is
longer than 2.0 m, the information in some labels cannot be changed.

Third test is to test the identification speed. The conclusion is: when the label in the normal distance, the time of recognize all 30 label is less than 1 s.

The result shows that the RFID identification system can basically meet the requirements of the automated identification of the goods in real life.

Prospect

With the in-depth development of the Internet of Things, intelligent storage will no longer be confined to the closed environment of automated identification. In the future, the information of the goods will form a database. The reader can connect to the database and download or update the information. That is more convenient and makes warehousing intelligent.

References

1. Lee J (2007) A UHF mobile RFID reader IC with self-leakage canceller. Radio Freq Integ Cir (RFIC) Symposium. IEEE, 273–276
2. International Standard Organization radio frequency identification for item management-part6: parameters for air interface communications at 860–960 MHz 2004
3. Li M, Qian ZH, Zhang X, Wang Y-J (2011) Slot-predicting based ALOHA algorithm for RFID anti-collision. J Commun 32 (12):43–50
4. International Standard (2005) ISO-IEC_CD 18000-6C. 1
5. Zhang X-P, Zhu Y-l, Luo H (2005) Design of UHF RFID interrogator. Chinese J Elec Dev 28(3):542–545

An Efficient Detecting Mechanism for Cross-Site Script Attacks in the Cloud

Wei Kan, Tsu-Yang Wu, Tao Han, Chun-Wei Lin, Chien-Ming Chen and Jeng-Shyang Pan

Abstract Cloud computing is one of the most prospect technologies due to its flexibility and low-cost usage. Several security issues in the cloud are raised by researchers. Cross-site script (XSS) attack is one of the most threats in the Internet. In the past, there are many literatures for detecting XSS attacks were proposed. Unfortunately, fewer studies focus on the detection of XSS attacks in the cloud. In this paper, we propose a mechanism to detect XSS attacks in cloud environments. The framework is also presented. In particular, our mechanism is not need to modify browsers and applications. We demonstrate our mechanism has higher accuracy rate and lower impact on performance of applications in the experiment. It sufficiently shows our mechanism is suitable for real-time detection in XSS attacks for cloud environments.

Keywords XSS attack · Cloud computing · Detection · Real-time

W. Kan · T.-Y. Wu (✉) · T. Han · C.-W. Lin · C.-M. Chen · J.-S. Pan
Innovative Information Industry Research Center, Shenzhen Graduate School, Harbin Institute of Technology, Shenzhen 518055, China
e-mail: wutsuyang@gmail.com

W. Kan
e-mail: hbrkanwei@163.com

T. Han
e-mail: 284763655@qq.com

C.-W. Lin
e-mail: jerrylin@ieee.org

C.-M. Chen
e-mail: chienming.taiwan@gmail.com

J.-S. Pan
e-mail: jengshyangpan@gmail.com

C.-W. Lin · C.-M. Chen · J.-S. Pan
Shenzhen Key Laboratory of Internet Information Collaboration, Shenzhen 518055, China

Y.-M. Huang et al. (eds.), *Advanced Technologies, Embedded and Multimedia for Human-centric Computing*, Lecture Notes in Electrical Engineering 260, DOI: 10.1007/978-94-007-7262-5_76, © Springer Science+Business Media Dordrecht 2014

Introduction

Nowadays, cloud computing is obviously one of the most prospect technologies due to its flexibility and low-cost usage. Different from the traditional Internet environments, enterprises can exempt from building expensive infrastructure by the cloud techniques. Hence, many IT companies deploy their web applications into the cloud environments. As we all know, the cloud environments are based on the Internet. Hence, the threats in the traditional Internet environments still exist in the cloud environments.

Cross-site script (XSS) attack is one of the most threats in the Internet. It can grab the user's privacy information and leads other attacks such as fishing, SQL injection, and DDoS. This attack is caused by the script language embed into web applications. In general, the web applications are adopted HTML language, script language, hyperlinks, and other languages to provide resources operating and interaction between the client and the server. However, these languages and methods lead web applications are vulnerable to XSS attacks.

Typically, the XSS attack contains the three attacks models: (1) Reflected XSS, (2) Stored XSS, and (3) Dom-based XSS. Up to now, several approaches were proposed to prevent XSS attack such as static analysis [1–3], black-white list [4], taint and flow analysis [5–8], string injection [9–12], machine learning [13–16], rewriting [17, 18]. For the static analysis approach, it addresses XSS attacks by means of static source analyses such as Jovanovic et al. [1] presented a method called alias analysis for PHP language in 2006 and Wassermann et al. [2] proposed the method which combines tainted information flow with string analysis in 2008. For the back-white list issue, it uses back and while list between the client and server to prevent XSS attacks. In 2007, Jim et al. [4] present a method called BEEP. In BEEP, scripts written by developer are stored in whitelist and malicious scripts are stored in blacklist. Before browser executes JavaScript codes, it will check the black and white lists. For the taint and flow analysis approach, it addresses XSS attacks by tracking the sensitive data such as Vogt et al.'s method [5]. For the string injection issue, Shahriar and Zulkernine [11] proposed a method by inserting comment tags containing rules into HTML page to restrict the content in 2011. For the rewriting approach, it addresses XSS attacks by inserting some tokens into the request and the response messages as Cookie and Headers. In 2011, Putthacharoen and Bunyatnoparat [18] put forward a dynamic cookies rewriting method. Recently, Shar and Tan [19] proposed a detecting method which can automatically audit and remove the XSS attacks.

With the fist growing of cloud computing, the research on the protection of application system in cloud platform becomes an important issue. XSS attack is one of the most threats in traditional application system. Hence, the protection of XSS attack should be considered in the cloud. However, the traditional detection methods for XSS attacks are not suitable in the cloud environments because most methods need to modify applications source codes or browsers. Meanwhile, most methods also cannot provide real-time protecting. In this paper, we first define the framework and related notions of XSS attacks in cloud environments. A concrete detecting mechanism is then proposed. Our mechanism can solve the mentioned disadvantages of traditional detecting methods of XSS attacks. In the experiments, our mechanism has a higher precision rate and a low impact on the performance of the application systems in the cloud.

The remainder of this paper is organized as follows. In section Framework, we propose the framework of detecting mechanism. A concrete detecting mechanism is presented in section Concrete Mechanism. In section Experiments and Comparisons, we describe the experiments and the conclusions are given in section Conclusions.

Framework

In this section, we present a sketch framework which is depicted in Fig. 1. Our framework consists of three entities: User, App Server, and XSS Detection Server as follows.

(1) **User**: The user can play the following three roles: app user, app administrator, and platform administrator. We assume the user visits App Server in first time which is not malicious. This assumption is called *F-secure*.

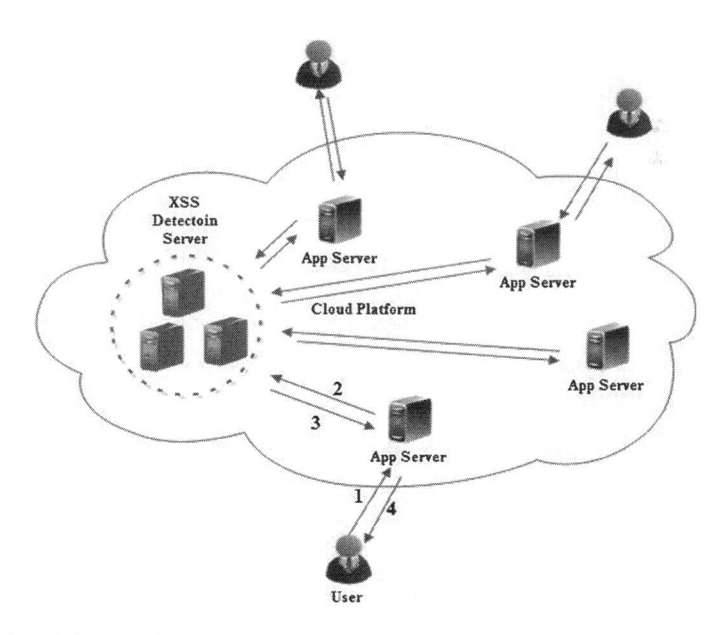

Fig. 1 Sketch framework

(2) **XSS Detection Server** (**XDS**): XDS is responsible to dispose the request messages sent by user and checks the messages whether they are malicious script languages.

(3) **App Server**: In App Server, there is a filter (named XFD) which is responsible to gather the request messages sent by user and then sends the gathered messages to XDS. Meanwhile, we define the following two rules for app in App Server.

> **Rule 1**: *HTML pages should not include comment tags.*
> **Rule 2**: *Script tags should not be dynamically generated in HTML pages.*
> The flowchart of our framework contains the following four steps:

Step 1: User sends the request messages to App Server.

Step 2: XFD collects the request messages sent by user and sends them to XDS.

Step 3: After receiving the messages from XFD, XDS verifies them whether they are malicious script languages. Then, XDS outputs the result and sends to XFD. We use the symbol NO_XSS is denoted by the request messages are secure and the symbol HAVE_XSS is denoted by the request messages are malicious script languages.

Step 4: Upon receiving the result form XDS, App Server provides the service for the user if the result is *NO_XSS*. Otherwise, App Server terminates the connection and responds a warning to the user.

Concrete Mechanism

Parsing Phase

Here, we first define a new data structure called *levelgroup* which is used to represent the structure of HTML pages. Then, we propose an algorithm named *FP* algorithm which is used to transform HTML pages into *levelgroup*. By the *F-secure* assumption, we can obtain an initial *levelgroup* in which the elements are not attacked.

Definition 1 Let *levelgroup* be a set which contains $n + 1$ subsets G_i for $i = 0, 1, \ldots, n$, where each G_i is an ordered set. Here, each element of G_i is consists of tag name and attribute.

For convenience to describe *FP* algorithm, we define *Page* by HTML page visited by user and pseudo code of *FP* algorithm is proposed as follows:

Algorithm 1: FP algorithm

Input: *Page*
Output: *levelgroup*
(1) *PStructure* ← obtaining the structure of *Page*;
(2) *levelgroup* ← defining a map<key, G>, where key denotes the index of G;
(3) **For** node *n* in *PStructure* using depth-first method **do**
   i ← calculating the corresponding level of *n*;
   If *levelgroup* contains key *i* **then**
     G_i ← getting value from *levelgroup* by key *i*;
     Storing *n* into G_i;
   Else
     G_i ← defining a subset of *levelgroup*;
     Storing *n* into G_i;
     Putting entity(i, G_i) into *levelgroup*;
(4) Return *levelgroup*;

Identifying Phase

Here, we first define some functions as follows:

Definition 2 (*H function*) *H* function is defined by $H(G^{old}, G^{new})$. It responsible to calculate added and deleted elements between the two subsets of *Levelgroup*. Then, it outputs two lists $List_{add}$ and $List_{delete}$. Formally,

$$\text{add } E^{new} \text{ into } List_{add} \text{ and add } E^{old} \text{ into } List_{delete}$$

Definition 3 (*C function*) *C* function is defined as $C(G^{old}, G^{new})$. It is responsible to check the name and attributes of elements in the same position of two subsets G^{old} and G^{new}. Then, it outputs the corresponding results *HAVE_XSS*, *NO_XSS*, or *UNCERTAIN*. The formal definition is described below. Note that E_j is a *j*th element in subgroup *G* and is denoted by the number of attributes of element E_j. Formally,

If ∃ outputting *HAVE_XSS*
If outputting *NO_XSS* for $\in G^{new}$ and $\in G^{old}$
Else adding into $List_{UNCERTAIN}$ Outputting *UNCERTAIN*

Definition 4 (*F Function*) *F* function is defined by $F(List_{add}, List_{delete}, G^{old}, G^{new})$. It is responsible to verify the elements of G_{new} whether they contain malicious scripts. Then, it outputs the corresponding results *HAVE_XSS*, *NO_XSS*, or *UNCERTAIN*. Formally,

If ∃ $E \in List_{add}$ contains malicious scripts (by **Rule1** and **Rule2**) outputting

HAVA_XSS

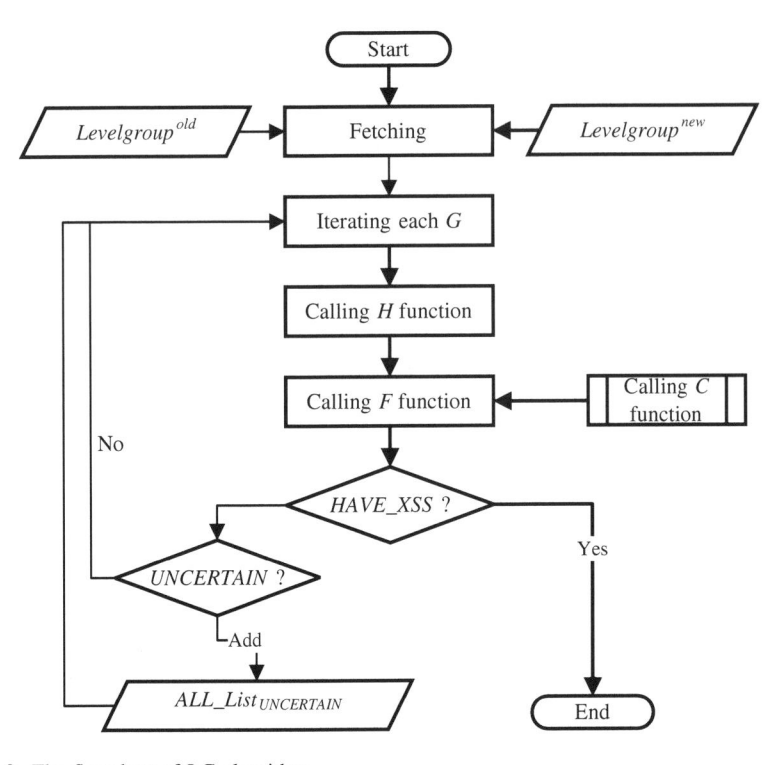

Fig. 2 The flowchart of LC algorithm

Else if $List_{delete} \neq \emptyset$ outputting NO_XSS

Else if $List_{add} \neq \emptyset$, adding $List_{add}$ into $List_{UNCERTAIN}$ and outputting $UNCERTAIN$

Else call the C function with (G^{old}, G^{new})

Here, we propose an algorithm named LC algorithm which is used to identify $Page$ whether it suffered from XSS attacks. The pseudo code of LC Algorithm and the flowchart in Fig. 2 are presented as follows:

Algorithm 2: LC algorithm

Input: *Page*

Output: *HAVE_XSS* or *NO_XSS*

(1) Calculating $levelgroup^{new}$ of *Page*;

(2) Fetching $levelgroup^{old}$ of *Page*;

(3) Defining a list, $ALL\_List_{UNCERTAIN}$;

(4) **For each** $\in levelgroup^{new}$ and $\in levelgroup^{old}$ **do**

 Calling H function with (,);

 Calling F function with $(List_{add}, List_{delete}, ,)$;

  If F function returns *HAVE_XSS*, **then** breaking ;

  Else if F function return *NO_XSS*, **then** continue;

  Else if F function return *UNCERTAIN*, adding $List_{UNCERTAIN}$ into $ALL\_List_{UNCERTAIN}$, **then** continue;

(5) **If** $ALL\_List_{UNCERTAIN}$ is not empty, calling the *CSE* algorithm;

 Else return *NO_XSS*;

Verifying Phase

Here, we propose an algorithm named *CSE* algorithm which is used to verify the list *ALL_List$_{UNCERTAIN}$* by rules.

Algorithm 3: CSE algorithm

INPUT: *ALL_List$_{UNCERTAIN}$*, application ID
OUTPUT: *HAVE_XSS* or *NO_XSS*
(1) Fetching the rules by ID;
(2) Transferring the rules to automata;
(3) Iterating all the elements in *ALL_List$_{UNCERTAIN}$*;
      Running the automata with the element;
      If the result of automata is *HAVE_XSS* **then** breaking and returns *HAVE_XSS*;
      Else continue;
(4) Returning *NO_XSS*;

Experiments and Comparisons

The experiment environment is simulated by a standard PC, where the processor Intel(R) Core(TM) i5-2400 3.10 GHz, the RAM is 8 GB, and the operating system is Windows 7. Meanwhile, we use VMware Workstation 7 to customize three virtual PCs to be the XSS Detection Server, one virtual PC to be the App Server. Here, we design an application (app_xss) which suffered from XSS attack.

In Table 1, we demonstrate the accuracy rate of our proposed mechanism with 32,560 malicious URLs [20]. The pre-treatment of URLs are consisted of following steps:

- Decoding and deleting invalid parameters in URLs.
- Changing parameter names and encoding URLs.

The details of experiment are described in the following steps:

1. We honestly visit app_xss in App Server.
2. We fetch disposed URLs to request app_xss in App Server.
3. XDS verifies them and outputs the accuracy rate as shown in Table 1.

In Fig. 3, we demonstrate the accuracy rate of our proposed mechanism with 26 websites such as Sina.com, QQ.com, ifeng.com, etc. It is easy to see that our mechanism provides an insignificant performance overhead of applications. Thus, our mechanism is suitable for real-time detection in XSS attacks.

Table 1 The accuracy rate of our mechanism

Total URLs	Malicious URLs	None malicious URLs	Error URLs
32,560	32,195 (99.18 %)	216 (0.67 %)	49 (0.15 %)

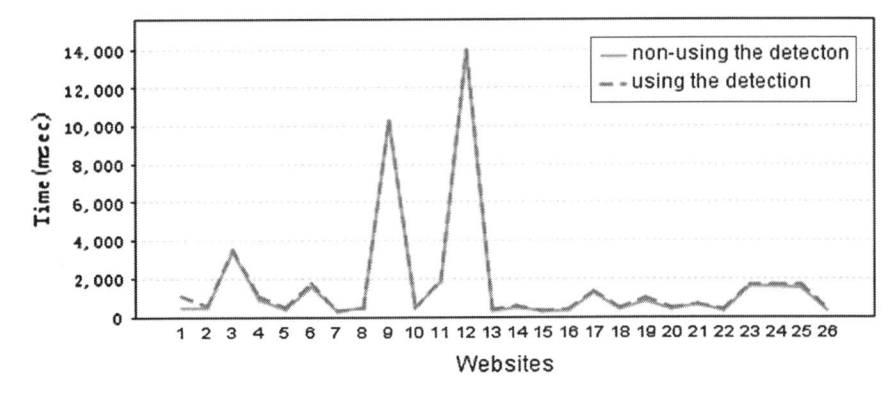

Fig. 3 The efficiency of our mechanism

Table 2 Comparing with other detection methods

Method	BSC	MSC	MB
Static analysis [1–3]	Yes	No	No
Black-white list [4]	Yes	Yes	Yes
Taint and flow analysis [5–8]	Yes	No	No
String injection [9–12]	Yes	Yes	Yes
Machine learning [13–16]	No	No	No
Rewriting [17, 18]	No	No	Yes
Our method	No	No	No

In Table 2, we list the comparison between some of previous XSS detecting methods in terms of whether modifying browsers (MB), browsing source codes (BSC), and modifying source codes (MSC). For instance, Black-white list method needs MB to insert black-white list parsing codes into the JavaScript engine of browsers. Meanwhile, this method needs BSC and MSC to insert black and white list into HITML pages. Obviously, our method is based on the structure of HTML pages such that we need not to do extra operation for the languages. Thus, it sufficient demonstrates the advantage of our mechanism.

Conclusions

In this paper, we have proposed a detecting mechanism of XSS attacks and defined the corresponding framework in the cloud. Our mechanism is not need to modify browsers and applications. The experiment results are demonstrated our method has higher accuracy rate and lower impact on performance of applications.

Acknowledgments The authors thank the referees for their valuable comments and constructive suggestions. This research was partially supported by Shenzhen peacock project, China under contract No. KQC201109020055A and Shenzhen Strategic Emerging Industries Program, China under Grants No. ZDSY20120613125016389 China.

References

1. Jovanovic N, Kruegel K, Kirda E (2006) Precise alias analysis for static detection of web application vulnerabilities. In: 2006 workshop on programming languages and analysis for security. ACM press, New York, pp 27–36
2. Wassermann G, Su z (2008) Static detection of cross-site scripting vulnerabilities. In: 30th international conference on software engineering. IEEE press, New York, pp 171–180
3. Zhang XH, Wang ZJ (2010) A static analysis tool for detecting web application injection vulnerabilities for asp program. In: 2nd international conference on e-business and information system security. IEEE press, New York, pp 1–5
4. Jim T, Swamy N, Hicks M (2007) Defeating script injection attacks with browser-enforced embedded policies. In: 16th international conference on World Wide Web. ACM press, New York, pp 601–610
5. Vogt P, Nentwich F, Jovanovic N, Kirda E, Christopher K, Vigna G (2007) Cross-site scripting prevention with dynamic data tainting and static analysis. In: international symposium on network and distributed system security. IEEE press, New York, pp 201–210
6. Lam MS, Martin M, Whaley J (2008) Securing web applications with static and dynamic information flow tracking. In: 2008 ACM SIGPLAN symposium on evaluation and semantics-based program manipulation. ACM press, New York, pp 3–12
7. Zhang Q, Chen H, Sun J (2010) An execution-flow based method for detecting cross-site scripting attacks. In: 2nd international conference on software engineering and data mining. IEEE press, New York, pp 160–165
8. Guarnieri S, Pistoia M, Tripp O, Dolby J, Teihet S, Berg R (2011) Saving the World Wide Web from vulnerable JavaScript. In: 11th international symposium on software testing and analysis. ACM press, New York, pp 177–187
9. Gundy M, Chen H (2009) Noncespaces: using randomization to enforce information flow tracking and thwart cross-site scripting attacks. In: International symposium on network and distributed system security. IEEE press, New York, pp 123–130
10. Johns M, Engelmann B, Posegga J (2011) S2XS2: a server side approach to automatically detect XSS attacks. In: International conference on computer security applications. IEEE press, New York, pp 335–344
11. Shahriar H, Zulkernine M (2009) Injecting comments to detect JavaScript code injection attacks. In: 35th international conference on computer software and applications. IEEE press, New York, pp 104–109
12. Wurzinger P, Platzer C, Ludl C, Kirda E, Kruegel C (2009) SWAP: mitigating XSS attacks using a reverse proxy. In: 2009 ICSE workshop on software engineering for secure systems. IEEE press, New York, pp 33–39
13. Komiya R, Paik I, Hisada M (2011) Classification of malicious web code by machine learning. In: 3rd international conference on awareness science and technology. IEEE press, New York, pp 406–411
14. Choi J, Kim H, Choi C, Kim Pk (2011) Efficient malicious code detection using n-gram analysis and SVM. In: 14th international conference on network-based information systems. IEEE press, New York, pp 618–621
15. Nunan AE, Souto E, Santos EMD, Feitosa E (2012) Automatic classification of cross-site scripting in web pages using document-based and URL-based features. In: 2012 IEEE symposium on computers and communications. IEEE press, New York, pp 702–707

16. Shar LK, Tan HBK (2012) Mining input sanitization patterns for predicting SQL injection and cross site scripting vulnerabilities. In: 2012 ICSE international conference on software engineering. IEEE press, New York, pp 1293–1296
17. Iha G, Doi H (2009) An implementation of the binding mechanism in the web browser for preventing XSS attacks. In: international conference on availability, reliability and security. IEEE press, New York, pp 996–971
18. Putthacharoen R, Bunyatnoparat P (2011) Protecting cookies from cross site script attacks using dynamic cookies rewriting technique. In: international conference on advanced communication technology. IEEE press, New York, pp 1090–1094
19. Shar LK, Tan HBK (2012) Auditing the XSS defence features implemented in web application programs. IET Software 6(4):377–390
20. XSS Attacks Information. http://www.xssed.com

A Novel Clustering Based Collaborative Filtering Recommendation System Algorithm

Qi Wang, Wei Cao and Yun Liu

Abstract Traditional collaborative filtering algorithms compute the similarity of items or users according to a user-item rating matrix. However, traditional collaborative filtering algorithms face very severe data sparsity, which causes a discount of the performance of recommendation. In this paper, we proposed an improved clustering based collaborative filtering algorithm for dealing with data sparsity. We first clustered the users set into k clusters using K-means algorithm. Then we presented a formula to estimate those absent ratings in the user-item rating matrix and acquired a high density matrix. After that, we use the new rating matrix to calculate the similarity of items and predict the ratings of a target user on items which have not been rated and recommend Top-N items to the target user. We also implemented experiments and demonstrated that our proposed algorithm has better accuracy than traditional collaborative filtering algorithms.

Keywords Collaborative filtering · Recommendation systems · K-means algorithm · Data sparsity

Q. Wang · Y. Liu (✉)
School of Communication and Information Engineering, Beijing Jiaotong University, Beijing 100044, China
e-mail: liuyun@bjtu.edu.cn

Q. Wang
e-mail: 12125042@bjtu.edu.cn

Q. Wang · Y. Liu
Key Laboratory of Communication and Information Systems, Beijing Municipal Commission of Education, Beijing Jiaotong University, Beijing 100044, China

W. Cao
China Information Technology Security Evaluation Center, Beijing, China
e-mail: caow@itsec.gov.cn

Y.-M. Huang et al. (eds.), *Advanced Technologies, Embedded and Multimedia for Human-centric Computing*, Lecture Notes in Electrical Engineering 260, DOI: 10.1007/978-94-007-7262-5_77, © Springer Science+Business Media Dordrecht 2014

Introduction

Recommendation systems can automatically recommend to users what they might be interested in. Usually we divide recommendation system algorithms into content-based algorithms [1, 2] and collaborative filtering algorithms [3–6]. Content-based algorithms recommend to users items which are similar to what users have already bought or rated by analyzing the features of users or items. These algorithms can solve the problem called "cold start" and also won't face the challenge of data sparsity because they don't depend on the rating matrix. But they have a serious drawback that they can't deal with pictures, video, music and other products difficult to be analyzed and extracted features from. On the contrary, collaborative filtering algorithms utilize a user-item rating matrix to calculate the similarity between users or items and then predict those items which have not been rated or bought depending on the ratings of neighbors which have high similarity with the target users. However, the number of items which each user has bought is usually less than 1 % of the total number of items in a site, which causes severe data sparsity and a decrease of the performance.

In this paper, we proposed an improved clustering based collaborative filtering algorithm for dealing with data sparsity. We combined K-means algorithms and a formula dealing with data sparsity of the user-item matrix. After that we implemented some experiments and it was shown that our proposed algorithms have a better performance than the traditional algorithms.

The Proposed Algorithm

Traditional collaborative filtering algorithms [4] create a user-item rating matrix and calculate similarity between users or items, but very often this matrix is sparsely populated leading to poor coverage of the recommendation space and ultimately limiting recommendation effectiveness. Our proposed method combines K-means algorithm [7] and a formula which can assign an estimation rating to an unrated item, so we can get a high-density matrix and resolve the data sparsity.

User Clustering

The first step is user clustering, and clustering is a preliminary step for the subsequent step to gather those similar users. In this paper, we use K-means algorithm to cluster our user set in the user-item rating matrix. Furthermore, we use Euclidean distance to represent the distance between users. Let a centroid is denoted by $U_{mid} = (r_{01}, r_{02}, \ldots, r_{0n})$ and user i can be denoted by $U_i = (r_{i1}, r_{i2}, \ldots, r_{in})$, where r_{mn} states the rating of user m on item n, then the distance between the two users is given by,

Fig. 1 K-means clustering
algorithm

```
Input: k original user centroids
1.for each user vector U
2.      for the k^th centroid C
3.          distance[k]=Euclidean(U,C);
4.      end
5.      find the shortest distance—distance[i]
6.      assign user U to cluster i
7.end
```

$$D = \sqrt{\sum_{j=1}^{n} (r_{ij} - r_{oj})^2} \tag{1}$$

Figure 1 shows the process of the K-means algorithm in our recommendation system and Fig. 2 shows the result of the K-means algorithm.

Constructing the Rating Matrix

Assuming there are M users and N items in the rating matrix, we define the sparsity level of the matrix as $1 -$ the number of ratings/M*N. Usually the sparsity level is very high, so in order to resolve this problem, we proposed a formula to calculate an estimation rating for an absent rating. The estimation rating of user c on item s, $R_{c,s}$ is computed as,

$$R_{c,s} = \bar{R}_c + \frac{1}{|U|} \sum_{\hat{c} \in U} (R_{\hat{c},s} - \bar{R}_{\hat{c}}) \tag{2}$$

where $\bar{R}_c$ is the average rating of the user c, U is the set of users that belong to the same cluster which is formed through the K-means algorithm and moreover have rated item s. Because $R_{c,s}$ is computed by the users in one same cluster and users belonging to one same cluster have higher similarity with each other, the estimation rating is more accurate. In addition, the rating scales of different users are varied, so in order to eliminate this inaccuracy, we let a rating subtract the average rating of a user and utilize the difference to calculate the predicting rating. Apparently, this method can eliminate data sparsity.

Similarity Computation

Similarity computation is the most important step for collaborative filtering algorithms. There are three different ways to compute the similarity between items. They are cosine-based similarity, Person correlation-based similarity and adjusted cosine similarity respectively as shown in Eqs. (3)–(5). We define the set

	Item2	Item3	...	...	Item i-1	Item i
U1	3					4
U345	3				5	
U143		3			2	4
...	...	...	...	...	...	...
Ui			5	6	6	
Uj		4	4	5	5	
Um			3		4	1
Un			2	3	5	5

Fig. 2 The result of clustering

of users that have rated both item i and item j as U, and the average rating of user u is denoted by $\bar{R}_u$. $\bar{R}_i$ and $\bar{R}_j$ denote the average rating of item i and item j respectively.

- Cosine-based Similarity

$$sim\,(i,j) = \cos(\vec{i},\vec{j}) = \frac{\vec{i}\cdot\vec{j}}{||\vec{i}||_2 * ||\vec{j}||_2} \tag{3}$$

- Person Correlation-based Similarity

$$sim\,(i,j) = \frac{\sum_{u\in U}(R_{u,i} - \bar{R}_i)(R_{u,j} - \bar{R}_j)}{\sqrt{\sum_{u\in U}(R_{u,i} - \bar{R}_i)^2}\sqrt{\sum_{u\in U}(R_{u,j} - \bar{R}_j)^2}} \tag{4}$$

- Adjusted Cosine Similarity

$$sim\,(i,j) = \frac{\sum_{u\in U}(R_{u,i} - \bar{R}_u)(R_{u,j} - \bar{R}_u)}{\sqrt{\sum_{u\in U}(R_{u,i} - \bar{R}_u)^2}\sqrt{\sum_{u\in U}(R_{u,j} - \bar{R}_u)^2}} \tag{5}$$

In the paper [3], Sarwar et al. have demonstrated that adjusted cosine similarity performs best among them in the recommendation system through their experiments. So, in this paper, we will use adjusted cosine similarity as the similarity computation method.

Prediction and Recommendation

After the former similarity computation, we will get a $N * N$ similarity matrix, where N represents the total number of items. The last step is predicting and recommendation. Firstly, we predict the items which have not been bought or rated by the target user, after that we recommend the top-N items to the target user basing the predicting. We use weighted sum to calculate the predicting ratings. Let the target user is u and the unrated item is i. $N = \{$all the items that have high similarity with i and have been rated by user u in the mean time$\}$.The predicting rating of user u on the item i is denoted by the Eq. (6).

$$R'_{u,i} = \frac{\sum_N (sim_{i,N} * R_{u,N})}{\sum_N (|sim_{i,N}|)} \tag{6}$$

Experimental Evaluation

Our data set is from the Movielens which is a web-based research recommender system. The data set includes 100,000 ratings of 943 users on 1,682 items and each user has rated 20 or more movies. The data set is divided into a training set and a test set. The 80 % of the data is used as the training set and the rest 20 % is used as the test set. All our experiments were implemented at Matlab7.0 and run in a PC with Intel Core processor having a speed of 2.2 GHz and 2 GB of RAM.

Metric

We use Mean Absolute Error (MAE) as our evaluation metric, and the MAE is defined as,

$$MAE = \frac{1}{N} \sum_{i=1}^{N} |P_i - R_i| \tag{7}$$

where N states the total number of predicting ratings in the test set, P_i is the ith predicting rating, and R_i is the ith actual rating in the test set.

Experimental Results

First, we implemented an experiment to see the performance of recommendation with the increasing of the density of rating matrix. We let the density of rating

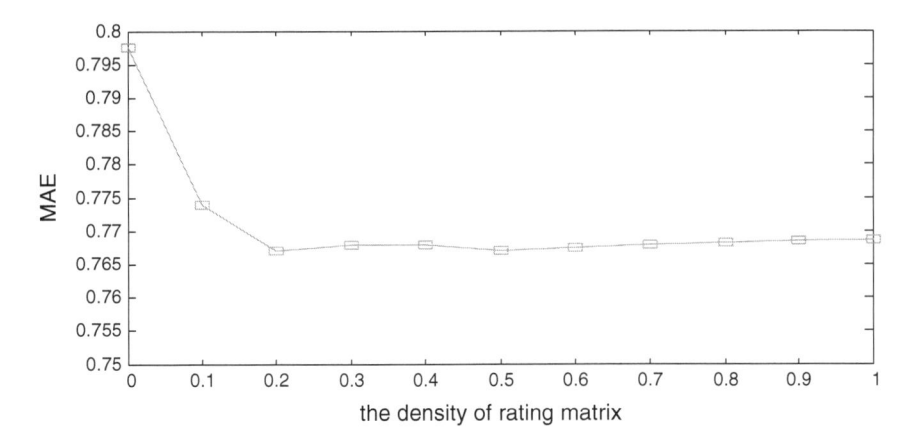

Fig. 3 Performance of the algorithm under different data sparsity

matrix increase at the speed of 10 % and then computed their MAE of predicting ratings. In addition, in this experiment, we set the parameter k = 30 in K-means algorithm and we didn't design an extra experiment to determine the optimal k. The results are shown in Fig. 3. From the Fig. 3 we can observe that with the increasing of the density of rating matrix the MAE of predicting decreases, which means the performance of recommendation getting better. Furthermore, the performance of recommendation improves as we increase the density of rating matrix from 0 to 20 %, after that the curve tends to be flat. With the increasing of the density of rating matrix, the computation also will increase rapidly, so we select the 20 % as the optimal choice of the density. On the other hand, the optimal value of sparsity level is 80 %.

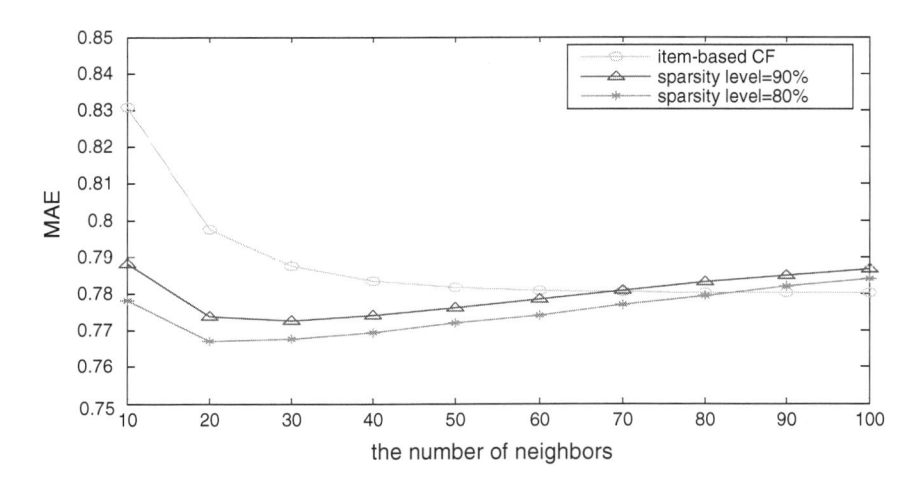

Fig. 4 Performance of the item-based algorithm and our proposed algorithm under the different number of neighbors

Then, we performed an experiment where we varied the number of neighbors and compared the results of the proposed algorithm with the traditional item-based collaborative filtering algorithm. The results are shown in Fig. 4. There are three curves in the figure which represent the normal item-based algorithm, the proposed algorithm where sparsity level equals 90 %, and the proposed algorithm where sparsity level equals 80 %. From the figure, we can observe that our proposed algorithm performs better than the item-based collaborative filtering when the number of neighbors is from 10 to 70, and our proposed algorithm can acquire the minimum MAE when the sparsity level = 80 % and the neighbors = 20.

Conclusion

In this paper, we proposed a new CF recommendation algorithm which introduces the K-means algorithm and estimation ratings to resolve data sparsity. By employing the experiments we observe the performance of our proposed algorithm is much better than the traditional CF algorithm and we also find when de sparsity level = 80 % and the number of neighbors = 20 the MAE of predicting ratings is minimum and the performance is the best. On the other hand, our proposed algorithm resolves data sparsity and acquires a more accurate recommendation. In addition, we can implement more experiments to determine the optimal number k of clusters in the K-means algorithm.

Acknowledgments This work has been supported by the National Natural Science Foundation of China under Grant 61172072, 61271308, the Beijing Natural Science Foundation under Grant 4112045, the Research Fund for the Doctoral Program of Higher Education of China under Grant W11C100030, the Beijing Science and Technology Program under Grant Z121100000312024.

References

1. Balabnovic M, Shoham Y (1997) Fab: content-based, collaborative recommendation. Comm ACM 40(3):66–72
2. Mooney RJ, Bennett PN, Roy L (1998) Book recommending using text categorization with extracted information. In: Procedings of recommender systems papers from 1998 workshop, technical report WS-98-08
3. Linden G, Smith B, York J (2003) Amazon. com recommendations: Item-to-item collaborative filtering. IEEE Internet Comput 7(1):76–80
4. Sarwar B, Karypis G, Konstan J et al (2001) Item-based collaborative filtering recommendation algorithms. In: Proceedings of 10th international WWW conference, Hong Kong, pp 1–5
5. Deshpande M, Karypis G (2004) Item-based top-N recommendation algorithms. ACM Trans Inf Syst 22(1):143–177

6. Koren Y (2010) Factor in the neighbors: scalable and accurate collaborative filtering. ACM Trans Knowl Disc Data 4:1–24
7. Kim KJ, Ahn HC (2008) A recommender system using GA K-means clustering in an online shopping market. Expert Syst Appl 34:1200–1209

A Secure and Flexible Data Aggregation Framework for Smart Grid

Lun-Pin Yuan, Bing-Zhe He, Chang-Shiun Liu and Hung-Min Sun

Abstract Smart grids are electrical grids that take advantage of information and communication technologies to achieve energy-efficiency, automation and reliability. Smart grids include renewable energy, electrical vehicles, phasor measurement unit (PMU) and advanced metering infrastructure system (AMI) etc. The system's availability can be achieved via data aggregation technique by reducing the overhead of networks. However, since smart grids have become more popular in recent years, many researches have been done on the security issue of the smart grid such as confidentiality, integrity and availability. For these security issues, many researchers adopt secure data aggregation algorithms to protect the data transmission and to reduce the overhead of networks. In this paper, we propose a secure data aggregation framework, which provides multi-level security for different kinds of applications.

Keywords Smart grid · Data aggregation

L.-P. Yuan · B.-Z. He · C.-S. Liu · H.-M. Sun (✉)
Department of Computer Science, National Tsing Hua University, Hsinchu 300, Taiwan, Republic of China
e-mail: hmsun@cs.nthu.edu.tw

L.-P. Yuan
e-mail: lunpin@is.cs.nthu.edu.tw

B.-Z. He
e-mail: ckshejrho@is.cs.nthu.edu.tw

C.-S. Liu
e-mail: monkey10020@is.cs.nthu.edu.tw

Y.-M. Huang et al. (eds.), *Advanced Technologies, Embedded and Multimedia for Human-centric Computing*, Lecture Notes in Electrical Engineering 260, DOI: 10.1007/978-94-007-7262-5_78, © Springer Science+Business Media Dordrecht 2014

Introduction

Smart grid is the next generation of electricity gird which can help us to monitor and to manage energy usage so that we can accomplish energy conservation and carbon reduction. In a smart grid, meters will report the usage of reading periodically to the energy producer via wireless or power line communication (PLC). In addition, smart grids are required to be self-healing and to protect customers' privacy. Nowadays, there are many research issues [3, 8, 9] need to be discussed in smart grids such as security and network issues.

The network overhead in the smart grid is a serious problem, which needs to be addressed carefully. The metering messages are able to waste the bandwidth of the network due to the increasing requests of sending similar messages. To resolve above problem, many researchers apply secure data aggregation to reduce the overhead in smart grids. The data aggregation algorithms can be divided into end-to-end based and hop-by-hop based algorithms according to whether the aggregator can obtain the messages or not. In 2010, Li et al. [4] proposed a secure data aggregation scheme which is based on the homomorphic encryption. Later, many researchers proposed the end-to-end based data aggregation schemes for smart grids. For instance, in 2011, the Elster group [1] provides a proposal for privacy enhancing technology implementation on smart grids. Later, Lu et al. [7] proposed efficient and privacy-preserving aggregation (EPPA) scheme for smart grid communication. Besides the above works, Kamto et al. [2] proposed an aggregation protocol for advanced metering infrastructure (AMI) system, which support both end-to-end model and hop-by-hop model. Furthermore, Li et al. [11] studied the signature issues on smart grids in 2012.

Although homomorphic cryptosystems can protect customers' privacy, generating metering packages has expensive electricity costs which need to be taken into account. Sometimes, the real-time property is more important than the security property. As an illustration, a phasor measurement unit (PMU) submits the data to the server, and the server should be able to analyze the data in a very short period of time such as 30 ms [5]. In this particular case, a lightweight aggregation approach needs to be designed to satisfy real-time requirement. In this paper, we propose a secure and flexible data aggregation framework, which can satisfy different requirements in smart grids such as security property and real-time property. The proposed framework consists of end-to-end aggregation and hop-by-hop aggregation. Compared to hop-by-hop based approach, the end-to-end based approach can achieve more security since the aggregator cannot learn the plaintext in the aggregation phase. On the other hand, the hop-by-hop based approach can do the aggregation faster than the end-to-end based approach because of using symmetric encryption.

The rest of the paper is organized as follows: In the section Framework, we describe our framework in detail. Then, we analyze the proposed framework in section Analysis. Finally, we summarize our results in section Conclusion.

Framework

In this section, the workflow of our framework is described. The proposed framework consists of four parts: secure end-to-end data aggregation, signature aggregation, secure hop-by-hop data aggregation, and MAC aggregation. Figure 1 shows the architecture of the proposed framework.

The secure end-to-end data aggregation permits the nodes to gather information (i.e. power consumption) from other nodes without disclosing their privacy. Additionally, we propose a signature aggregation approach which is compatible with the secure end-to-end data aggregation. Therefore, the overheads can be minimized while delivering security services. Besides, since all real-time information (i.e. PMU information) should be aggregated by a faster approach, we propose a secure hop-by-hop data aggregation and a MAC aggregation approach.

Secure End-to-End Data Aggregation

In this approach, one node (either a meter or a sensor) generates a data to be sent to a server. Data from nodes have to be aggregated in order to minimize the overheads. The data are aggregated into one result and sent in cipher to an aggregator using homomorphic encryption with server's public key. Once an aggregator receives the data, the aggregator can apply operations (i.e. summation or multiplication) without knowing server's private key. After calculation, the aggregator sends the aggregated ciphertext to another aggregator, until reaching a control server.

For instance, we can aggregate power consumption information by summing them up together. Therefore, we may use EC-Elgamal encryption [10], which is an additive homomorphic encryption, so that the secure data aggregation can be done without decryption.

EC-Elgamal encryption mainly consists of the following four parts: KeyGen, Encrypt, Aggregate, and Decrypt.

- *KeyGen*(τ): Given a security parameter τ, outputs an elliptic curve $E(F_q)$, where F_q is a finite field, and q is a large prime which is related to τ. Points on the

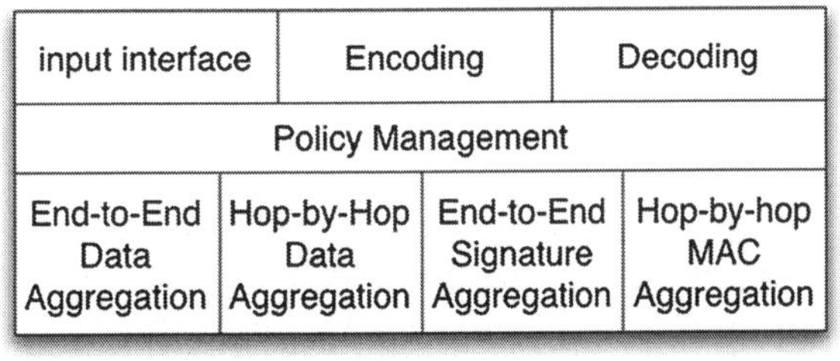

Fig. 1 The architecture of the proposed framework

input interface	Encoding		Decoding
Policy Management			
End-to-End Data Aggregation	Hop-by-Hop Data Aggregation	End-to-End Signature Aggregation	Hop-by-hop MAC Aggregation

elliptic curve $E(F_q)$ form an additive cyclic group G_1, and $ord(G_1) = n$. Choose a generator $G \in G_1$, $n * G = \infty$. Randomly choose a server's private key $Priv = x$, where $x \in Z_q^*$. Calculate the server's public key $Pub = Y = x * G$. Finally publish (E, G, q, n) as system parameters.

- $Encrypt(M, Pub)$: Given a message M to be encrypted, where $M \in [1, \ldots, q - 1]$, and given the server's public key Pub and system parameters (E, G, q, n). One can calculate the ciphertext C by applying the following steps:

1. Choose a random number $k \in Z_n^*$.
2. Ciphertext $C = (R, S) = (k * G, M * G + k * Pub)$.

- $Aggregate(C_1, C_2)$: Given two (or more) ciphertexts C_1 and C_2, the output of aggregating ciphertexst can be calculated by applying the following step:

$$C' = C_1 + C_2 = (R_1 + R_2, S_1 + S_2)$$
$$= ((k_1 + k_2) * G, (M_1 + M_2) * G + (k_1 + k_2) * Pub)$$

- $Decrypt(C, Priv)$: Given a ciphertext C and the server's private key Priv, one can calculate the aggregated plaintext as the equation, $M * G = -Priv * R + S$. The consistency is shown in the following equation:

$$M * G = -Priv * R + S = -x * \left(\sum_i k_i * G\right) + \left(\sum_i M_i * G + \sum_i k_i * Pub\right)$$
$$= -x * \left(\sum_i k_i * G\right) + \left(\sum_i M_i * G + \sum_i k_i(x * G)\right)$$
$$= \sum_i M_i * G = M * G$$

Signature Aggregation

In this subsection, we show an example of signature aggregation, which is a simple short signature based on bilinear pairings. These kinds of signatures are shorter than traditional ones. Furthermore, signatures can be aggregated and verified in batch processing.

In this approach, every node must sign the message with its private key. Once an aggregator receives messages, the aggregator can either verify them one by one, or verify them in once by applying batch verification process.

In addition, in some particular cases, a node is directly connected to another node instead of an aggregator. And the fact that these nodes have less computational power than an aggregator, all in all affects the execution time. For instance, a group of meters are connected to an aggregator through another meter. In these situations, it is impractical to perform secure data aggregation over a meter. However, it can perform signature aggregation to reduce overheads.

In this example, by using EC-Elgamal encryption, we use short signature, which is based on bilinear pairings, so that the number of system parameters on curve can be reduced.

The simple short signature is generally consists of five parts: KeyGen, Sign, Verify, Aggregation, and Batch–Varify.

- *KeyGen*(τ): This part is quite similar to the *KeyGen* part described in the previous subsection. Therefore, if a proper τ is given, the number of system parameters on curve may not be increased. In Addition, choose a secure hash function $H : \{0, 1\}^* \to h$, $h \in G$. Each user U_i, should choose a random number $x_i \in Z_q^*$ as his private key *Priv*$_i$, and calculate *Pub*$_i = x_i * G$ as his public key.
- *Sign*(M, *Priv*$_i$): Given M and *Priv*$_i$, where M is a message to be signed and *Priv*$_i$ is the private key of user U_i.. The result of signing a signature σ can be calculated by the following steps:

1. Calculate the message digest $h_i = H(M)$.
2. Signature $\sigma = Priv_i * h_i$.

- *Verify*(M, σ, *Pub*$_i$): Given M, σ, and *Pub*$_i$, where M is the message, σ is the signature to be verified, and *Pub*$_i$ is the public key of user U_i. Determine the validity of the signature by the following steps.
1. Calculate the message digest $h_i = H(M)$.
2. Determine whether $e(\sigma, G) = e(h, Pub_i)$ is equal or not. If the equation holds, then the message M is acceptable. The consistency of a valid signature is shown as follows:

$$e(\sigma, G) = e(Priv_i * h_i, G) = e(h_i, Priv_i * G) = e(h_i, Pub_i)$$

- *Aggregate*(σ_1, σ_2): Given two (or more) signatures σ_1 and σ_2, the output of aggregating signatures can be calculated as follows:

$$\sigma' = \sigma_1 + \sigma_2 = (Priv_1 * h_1) + (Priv_2 * h_2)$$

- *Batch–Verify*(σ, $\{M_i\}$, $\{Pub_i\}$): Given σ, $\{M_i\}$, and $\{Pub_i\}$, where σ is an aggregated signature, $\{M_i\}$ is a list of messages, and $\{Pub_i\}$ is a list of public keys of users. Determine the validity of the signatures as follows:

1. Calculate every message digests $h_i = H(M)$.
2. Determine whether $e(\sigma, G) = \prod_i e(h_i, Pub_i)$ is true or not. If it is true, then accept the messages. The consistency of valid signatures is shown as follows:

$$e(\sigma, G) = e(\sum_i (Priv_i * h_i), G) = \prod_i e(Priv_i * h_i, G)$$
$$= \prod_i e(h_i, Priv_i * G) = \prod_i e(h_i, Pub_i)$$

Secure Hop-by-Hop Data Aggregation

The previous approach cannot satisfy real-time communication. Some metering messages (i.e. PMU information) should be transmitted in real time (e.g., less than 30 ms). Security and performance are in a trade-off. Therefore, we propose a secure hop-by-hop data aggregation, which is faster than the secure end-to-end data aggregation but less secure, and MAC aggregation, which is also faster than the signature aggregation approach.

In this approach, a node sends the data to a server. The data from nodes have to be aggregated in order to minimize the overheads. The node encrypts the data by symmetric encryption, and then sends the ciphertext to an aggregator. Once an aggregator receives packets, the aggregator first decrypts the packets, and then the aggregator can apply operations (i.e. calculate sum, average, deviation). After encrypting the results, the aggregator then sends the ciphertext to a server through other aggregators.

We use Advanced Encryption Standard (AES) [6] in this particular example. Let $sk_{A_1 n_1}$ denote the session key between an aggregator A_1 and a node n_1.

- n_1 encrypts a message M_1 using AES: $C_1 = E_{sk_{A_1 n_1}}(M_1)$.
- Once A_1 receives i packets C_i, A_1 decrypts ciphertext C_i to its corresponding message: $M_i = D_{sk_{A_1 n_1}}(C_i)$.
- Afterwards, A_1 can use the plaintexts obtained from the previous step to apply operations, such as calculating sum, average, and deviation. Then A_1 stores the result into M_A.
- A_1 encrypts the result message M_A using AES: $C_A = E_{sk_{A_2 A_1}}(M_A)$, where $sk_{A_2 A_1}$ is a session key between A_1 and another aggregator A_2.
- A_2 can repeat step 2 to step 4 until reaching a control server.

MAC Aggregation

In this subsection, we propose an aggregated message authentication code (MAC) method for data integrity. As mentioned before, every node must sign his messages. However, using homomorphic signature is not practical when dealing with transmitting data real-timely. Therefore, we propose a MAC aggregation approach.

In this approach, every node must send their message along with a MAC. The MAC function takes a shared secret key and an arbitrary-length message as inputs. Applying the same session key that is used in secure hop-by-hop data aggregation is also possible.

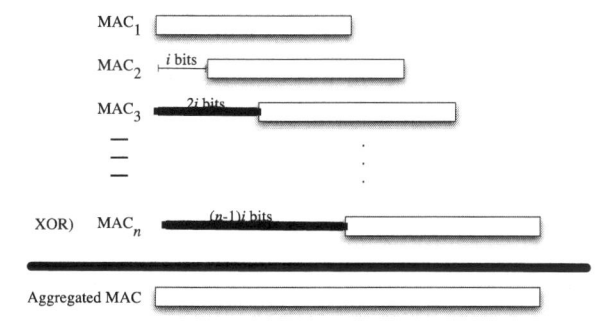

Fig. 2 An example of an aggregated MAC

Once an aggregator receives packets, the aggregator can either verify them one by one, or verify them in once by aggregating MACs.

In some particular cases, a node is directly connected to another node instead of an aggregator. For instance, a group of sensors are connected to an aggregator through another sensor. In these situations, it is impractical to perform secure data aggregation over a sensor. However, it can perform MAC aggregation to reduce overheads. An example of aggregated MAC is shown in Fig. 2.

- A node n_1 calculates MAC_1 by performing MAC algorithm with M and $sk_{A_1 n_1}$.
- An aggregator A_1 can aggregate MAC_1 and MAC_2 by performing the following steps. Basically, we are aware of the order of MACs by applying bitwise shifts as shown in Fig. 1. As a result, in every step of verification process, we know which MAC_i fails. Therefore, we can reduce the time to be spent for re-verification. The aggregation method is described briefly in the following steps:

1. Shift MAC_2 by i bits.
2. Performing exclusive-or (XOR) on MAC_1 and shifted MAC_2.

- A_1 can verify MACs by performing the same MAC algorithm with same inputs to check whether the outputs are equal or not.

Analysis

In this section, we compare our framework with related work in order to show the strength of our scheme.

Examine the Proposed Framework

As mentioned in the previous parts, our framework consists of two main parts: end-to-end based algorithm and hop-by-hop based algorithm. The end-to-end model consists of data aggregation and signature aggregation whereas the hop-by-hop model includes data aggregation and message authentication code (MAC) aggregation instead.

The hop-by-hop model is more efficient but less secure than the end-to-end model, while the hop-by-hop model uses symmetric encryption, and as a result, encryption/decryption phase will be executed faster. The fact that the aggregator can access the users' data, thus, the aggregator must be fully trusted.

On the other hand, the end-to-end model is based on public key cryptosystems where the aggregator can aggregate data without knowing the plaintext even under the honest- but-curious model. Therefore, the end-to-end based can provide more security for users' privacy than the hop-by-hop model. The end-to-end based approach has more computational costs because of inheriting properties from public key cryptosystems.

In order to satisfy data integrity and non-repudiation properties, our framework offered signature aggregation for end-to-end model and message authentication code aggregation for hop-by-hop model.

Compare Our Framework with Previous Works

In [2], Kamto et al. established a protocol for advanced metering infrastructure system (AMI). This protocol assumes that the relay gateways are fully trusted. The relay gateway can decrypt and collect the data from several meters into a single data to be sent to the next substation. But their protocol is insecure under the honest-but-curious model because the relay gateways are able to decrypt the data.

Li et al. [4] designed a secure data aggregation protocol using Paillier cryptosystem to encrypt data. In this protocol, each meter will send the data in ciphertext to the parent node. By using homomorphic encryption, the parent node can aggregate all the ciphertexts from other nodes into one ciphertext to be sent to the collector. This protocol assumes that all smart meters follow the honest-but-curious model. But this protocol is not suitable in some real-time required environment because the end-to-end model requires more computational cost.

Lu et al. (2012) [4] provided the efficient and privacy-preserving aggregation (EPPA) scheme which is constructed under the end-to-end model for smart grid communications. This scheme addressed a multidimensional data aggregation approach based on the homomorphic Paillier cryptosystem. Comparing with the traditional methods, EPPA can reduce computational costs and improve the throughput of communication. Moreover, EPPA satisfies the real-time high-frequency data collection requirements in smart grid communications. Above all, the EPPA scheme can recover the aggregated data into individual data.

Elster group proposed the privacy enhancing technologies (PETs) [1] for the smart grid. These technologies employ the homomorphic cryptographic technique for aggregating data from several nodes. Because these technologies are for the end-to-end model, PETs are not compatible with some real-time systems.

Table 1 Comparison between our framework and previous works

	[2]	[4]	[7]	[1]	Our scheme
End-to-end data aggregation	√	√	√	√	√
End-to-end signature aggregation			√		√
Hop-by-hop data aggregation	√				√
Hop-by-hop MAC aggregation					√
Honest-but-curious model		√	√	√	√
Real-time	Δ		Δ		√

Δ is partially offers the benefit

In this paper, we present a framework that gives a variety of configurations according to different requirements. We can apply the end-to-end model when we need more security, since it is more secure than the hop-by-hop model. On the other hand, for some real-time system, such as PMU system, where the data are delivered in less than 30 ms, we can choose the hop-by-hop model.

We summarize the comparisons of our scheme based on the result of Table 1. As shown in Table 1, all of them can achieve the end-to-end data aggregation. The honest-but-curious model is achieved by all except [2]. However, [4] and Elster [1] cannot provide data integrity and non-repudiation services while our framework can use signature aggregation to deliver these services. Although EPPA [7] scheme can work with many options that listed in the table, it is not appropriate for the PMU system where real-time data transmission is required. By adopting the hop-by-hop model, our framework ensures real-time transmission. Compared with the previous studies, our framework is more flexible in most situations than the others.

Conclusion

In this paper, we propose a secure and flexible data aggregation framework for smart grids. Our framework consists of end-to-end based and hop-by-hop based aggregation approaches. The end-to-end based approach is more secure but less efficient since the aggregator cannot learn any plaintext from all collected data. For this reason, we also design the hop-by-hop based approach which is suitable for real-time requirements. However, the hop-by-hop based approach is less secure than the end-to-end based approach. In addition, in order to resolve data integrity and non-repudiation, we also construct an end-to-end based signature algorithm and a hop-by-hop based MAC aggregation. Consequently, our framework is more flexible in most situations, compared to the previous works.

Acknowledgments The authors would like to thank anonymous reviewers for their valuable comments and suggestions that certainly led to improvements of this paper. This research was partially supported by the National Science Council of the Republic of China under Contract NSC-101-2221-E-007-026-MY3 and NSC-100-2628-E-007-018-MY3. The corresponding author is Professor Hung-Min Sun.

References

1. Elsters proposal for privacy enhancing technology implementation. http://www.elster.com/en/privacy-enhancing-technologies-for-the-smart-grid
2. Kamto J, Qian L, Fuller J, Attia J, Qian Y (2012) Key distribution and management for power aggregation and accountability in advance metering infrastructure. In: Smart grid communications (SmartGridComm), 2012 IEEE international conference
3. Khurana H, Hadley M, Lu N, Frincke D (2010) Smart-grid security issues. Secur Priv IEEE 8(1):81–85
4. Li F, Luo B, Liu P (2010) Secure information aggregation for smart grids using homomorphic encryption. In: Smart grid communications (SmartGridComm), 2010 first IEEE international conference, pp 327–332
11. Li F, Luo B (2012) Preserving data integrity for smart grid data aggregation. In: Smart grid communications (SmartGridComm), IEEE third international conference, pp 366–371
5. Lin H, Deng Y, Shukla S, Throp J, Mili L (2012) Cyber security impacts on all-pmu state estimator: a case study on co-simulation platform GECO. In: Smart grid communications (SmartGridComm), 2012 IEEE international conference. pp 587–592
6. Deamen J, Rijmen V (1998) AES proposal: Rijndael. In: First advanced encryption standard (AES) conference
7. Lu R, Liang X, Li X, Lin X, Shen X (2012) EPPA: an efficient and privacy-preserving aggregation scheme for secure smart grid communications. Parallel Distrib Syst IEEE Trans 23(9):1621–1631
8. Metke A, Ekl R (2010) Security technology for smart grid networks. Smart Grid IEEE Trans 1(1):99–107
9. Parikh P, Kanabar M, Sidhu T (2010) Opportunities and challenges of wireless communication technologies for smart grid applications. In: Power and energy society general meeting, 2010 IEEE. pp 1–7
10. Rabah K (2005) Elliptic curve elgamal encryption and signature schemes. Inf Technol J 4(3):299–306

Data Integrity Checking for iSCSI with Dm-verity

Rui Zhou, Zhu Ai, Jun Hu, Qun Liu, Qingguo Zhou, Xuan Wang, Hai Jiang and Kuan-Ching Li

Abstract With the ever increasing popularity of web service and e-commerce, there is a high demand on data storage. Because of the development of Internet infrastructure and the low cost of deployment, implementing storage over IP has become a trend. For the utilization of network storage, one important issue is the way to achieve data integrity. Usually, the application of Internet Small Computer System Interface (iSCSI), which is a kind of network storage technology, is to store some read-only or important data remotely. I/O requests, which may cause data loss and data error, are frequent in a traditional distributed network storage system like

R. Zhou · Z. Ai · J. Hu · Q. Liu · Q. Zhou (✉)
School of Information Science and Engineering, Lanzhou University Lanzhou, Lanzhou, People's Republic of China
e-mail: zhouqg@lzu.edu.cn

R. Zhou
e-mail: zr@lzu.edu.cn

Z. Ai
e-mail: aiz11@lzu.edu.cn

J. Hu
e-mail: huj11@lzu.edu.cn

Q. Liu
e-mail: liuq11@lzu.edu.cn

X. Wang
School of Science, Lanzhou University of Technology, Lanzhou, People's Republic of China
e-mail: wangxuan2010@lut.cn

H. Jiang
Dept. of Computer Science, Arkansas State University, Arkansas, USA
e-mail: hjiang@astate.edu

K.-C. Li
Dept. of Computer Science and Information Engineering (CSIE), Providence University, Taichung, Taiwan
e-mail: kuancli@pu.edu.tw

Y.-M. Huang et al. (eds.), *Advanced Technologies, Embedded and Multimedia for Human-centric Computing*, Lecture Notes in Electrical Engineering 260, DOI: 10.1007/978-94-007-7262-5_79, © Springer Science+Business Media Dordrecht 2014

iSCSI. In this paper, the data integrity of iSCSI is analyzed and Dm-verity mechanism is utilized to provide read-only transparent integrity checking for iSCSI, which could avert data loss and data error, increasing overall system reliability.

Keywords Dm-verity · iSCSI · Data integrity

Introduction

Dm-verity [1] stands for *device-mapper verity*, which aims to provide read-only transparent integrity checking of block devices. It was originally developed by Google Chromium OS team and introduced later in Linux kernel 3.4.0. Dm-verity is based on device-mapper [2] used to verify the integrity of the root filesystem on boot and supported in many applications, such as LVM, RAID, and multi-path. The core component of dm-verity mechanism is a cryptographic hash tree, in which the leaf nodes of the tree store data blocks, so the hash nodes of the intermediary nodes are calculated based on all of its child nodes and hash function. When some data blocks are accessed, related hash nodes will be verified. In case one of hash nodes fails in the verification, the access will be denied. Therefore, dm-verity could ensure the integrity of data blocks. iSCSI, is an IP based network storage standard for linking data storage facilities. Also like IP-SAN (Storage Area Networks), iSCSI is a standard to encapsulate one SCSI command into a TCP/IP (Ethernet) packet, while users can access the storage system with commodity IP devices only. There are some security issues in iSCSI, since it is a type of network storage and data are transferred by network. In this paper, dm-verity is applied to iSCSI, arming it with an additional layer of security. iSCSI provides two disks, one for data disk of dm-verity, and another one for hash disk of dm-verity. Once the performance is estimated, the research shows that it pays to cost verification overhead to improve the security of iSCSI.

The paper is organized as follows. Section Related Work introduces related work and analyzes the past work of iSCSI security. Section Design and Implementation describes the ways to deploy dm-verity into iSCSI, while section Performance Evaluation evaluates the reading performance of iSCSI with dm-verity embedded, and finally some conclusions are discussed in section Conclusions.

Related Work

There exist limitations in security in present iSCSI, based on the way it works. Mainly they can be categorized as [3]:

1. Active attack (modifying/deleting data, inserting illusive data),
2. Passive attack (listening to Internet lines, data analysis),

A lot of research on storage security has been done to deal with attacks from the above two categories, such as identification and protection of TCP/IP package. As identification, connection authentication is the iSCSI way to determine trustworthiness via CHAP, SRP, SPKM-1 or SPKM-2. This method only protects the identification of target and initiator. On the other hand, in data encryption schemes, two alternatives are considered, IPSecurity Protocol (IPSec) and Secure Socket Layer (SSL). iSCSI uses Cyclic Redundancy Check (CRC) digest to solve the security problems when the 32-bit check word algorithm can ensure end to end checking. iSCSI can have digests for iSCSI headers and data. Header digest can ensure correct operation and data placement, whereas data digest can ensure that data is unmodified throughout network path. Moreover, the performance of iSCSI with CRC has been evaluated in [4].

In particular, some previous works have put forward schemes to ensure the security of network storage. A framework of storage security was proposed in [5]. Analyzing the framework shows that encrypt-on-disk systems not only are more secure but also provide better performance than encrypt-on-wire systems. Other frameworks about iSCSI security have also been brought. For example, a security encryption storage system named ANGLE, which contains two major parts—the Key Management System (KMS) and the Encryption Engine (E-Engine) in [6]. Related transfer protocols have also been target of researches, as analysis of the iSCSI protocol presented in [7], enabling accessing SCSI I/O devices over an IP network.

A number of recent storage security works have been focusing on protecting communication between servers and clients in a distrustful networked world. In particular, the focus is on data integrity: preventing unauthorized modification of data or commands, replaying of requests and modification of requests in transit. Some of these systems further address the issue of privacy, or confidentiality of data transfer: preventing the leaking of data in transit by snooping on the network. Those typical solutions could solve the security problems of identification, provide data integrity to transferred data and evaluate the security of iSCSI, but there are no good measures to make sure whether the important and read-only data of iSCSI application is unmodified or not when being visited. Therefore, it is introduced in this paper a new mechanism, called dm-verity, to protect iSCSI data from being polluted, and mainly fighting against active attacks.

Design and Implementation

Dm-verity, aimed to data integrity protection, has been applied to Chromebooks to protect OS information from being modified. Based on this mechanism, we consider applying it to iSCSI for data integrity.

The iSCSI environment configured with dm-verity is depicted in Fig. 1. It is build up with two targets (target1 and target2) and one initiator. The terminals are connected by a switch and both of the targets will map a disk into the initiator, so

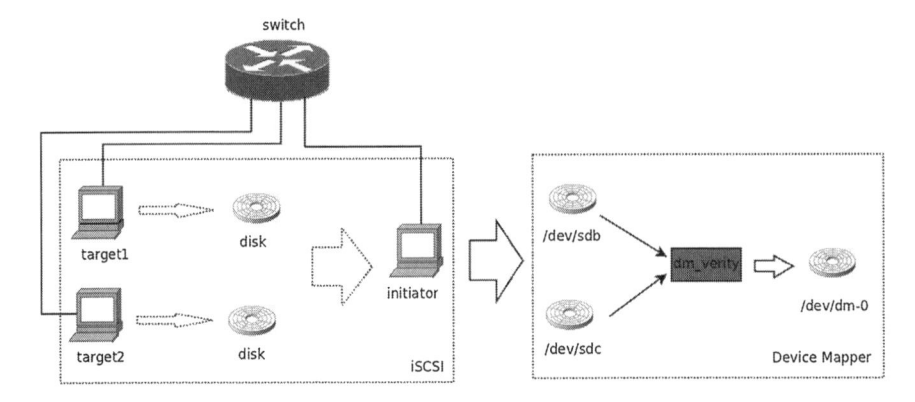

Fig. 1 Experimental system setup

the initiator would have two disks, */dev/sdb* and */dev/sdc*. Then dm-verity mechanism is utilized to configure the integrity checking environment with the disks in initiator.

The configuration information of three computers is as follows:

target1: Ubuntu OS 10.04, Linux kernel 3.0.0, CPU: AMD, 64-bit, 2.2 GHz
target2: Ubuntu OS 10.04, Linux kernel 2.6.35, CPU: AMD, 64-bit, 2.2 GHz
initiator: Ubuntu OS 11.10, Linux kernel 3.4.0, CPU: AMD, 64-bit, 2.2 GHz

iSCSI Configuration

iSCSI technology is very mature and the process of configuration is easy and straightforward. In the experiments, both of the two targets provide two disks for the initiator.

Dm-verity Over iSCSI Configuration [8]

There are four steps to configure dm-verity over iSCSI:

Step 1. Loading related module and preparing for the disks

We insert the *dm-verity.ko* module into kernel. It will be available once the kernel is re-compiled with dm-verity enabled. Also a data disk and a hash disk should be specified. We regard */dev/sdb1* (from target1) as the data disk and */dev/sdc1* (from target2) as the hash disk.

Step 2. Preparing hash device (formatting)

All of the following steps are taken in the initiator. We format the hash device by *veritysetup* command, and then get the root hash value, salt value and other information. After that a hash tree will be generated. The root hash value is the root node of the hash tree.

Step 3. Activating verity data device

Once data device is activated by *veritysetup*, the mapping device will be created. Then there could be a mapped device named */dev/dm-0* from data disk. The most difference between */dev/dm-0* and normal disk is that, the sector capacity of */dev/dm-0* is 4,096 bytes, which is the same as a data block of most system, but the normal size is 512 bytes, and *dm-0* is read-only.

Step 4. Verification

We verify all data block by checking the integrity of block devices through related commands. This process will take several minutes until all nodes of the hash tree are checked. Integrity checking for any specific block does not happen until that block is read, and other data blocks will not be verified. If some blocks to be read are polluted, it will fail to read. Once some blocks that are to be read are correct, while other blocks are broken, the verification will succeed.

Performance Evaluation

In order to evaluate reading performance (writing performance is not available because the disk is read-only) of iSCSI with dm-verity, we take two experiments. First, we evaluate iSCSI without dm-verity. In this experiment, we just set up the iSCSI environment with two targets and one initiator. The formatted and partitioned */dev/sdb* storage device is the mapper device from one target. Finally, we test the reading capability of */dev/sdb1* by *hdparm* tool. Second, we evaluate reading performance of iSCSI with dm-verity. After configuring the iSCSI with dm-verity mechanism, we would get a virtual disk named */dev/dm-0*, and test the reading capability of */dev/dm-0* in the same method.

We carry out three groups of tests, for each of which the capacity of data disk and hash disk is 2, 4 and 8 GB. The testing results are shown as Fig. 2, in which horizontal axis stands for the times of testing, while vertical axis stands for the throughput of reading. The average throughput without verification are 11.0788, 9.3048 and 10.6691 MB/s in Fig. 2a, b and c respectively. Meanwhile, the average throughputs with verification are 11.042, 9.155, and 10.3053 MB/s, so the average performance overhead are 0.33, 1.61 and 3.41 % respectively based on the three-condition set, which implies a very small performance loss with application of dm-verity.

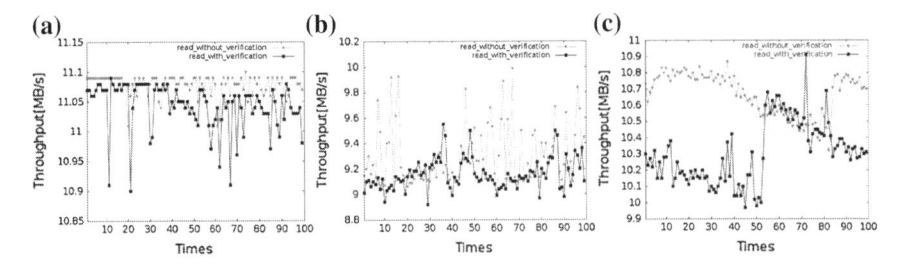

Fig. 2 Reading performance (the capacity of each disk is 2G in (**a**), 4G in (**b**) and 8G in (**c**))

Fig. 3 Verification time

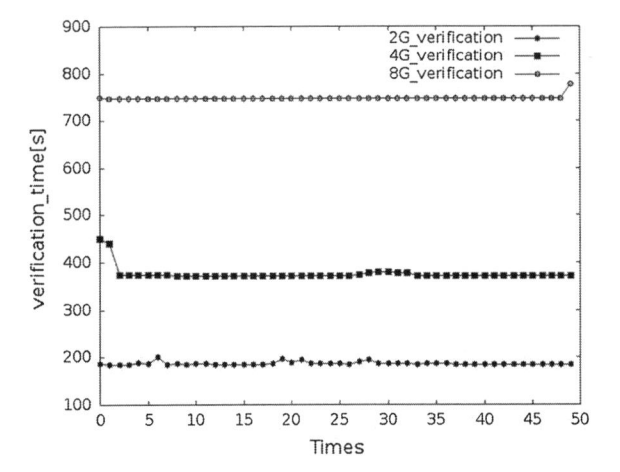

The overhead results from the process of verification costing extra time. Considering the improvement of security, this overhead is acceptable.

Additionally, the bigger the capacity of the disk is, the more time of verification will cost and the larger their margin of read throughput will be. In order to evaluate the relation between verification time and capacity of disk, we carry out another three group experiments—the capacity of data disks also are 2, 4 and 8 GB. The process costs some time, since it must verify from every leaf nodes to root node to assure each hash node of the tree to be verified. This verification could test whether all the data are correct, and testing results are depicted in Fig. 3.

In Fig. 3, horizontal axis stands for the times of verification, and vertical axis stands for verification time. The verification time depends on the size of the data, hash algorithm and network speed [9], and the results are almost stable. In tests performed, the hash algorithm is SHA256. The verification time of 2 GB disk is about 180 s, and the verification speed is about 11.37 MB/s. The verification time of 4 GB disk is about 370 s, and the verification speed is about 11.03 MB/s. The verification time of 8 GB disk is about 750 s, and the verification speed is about 10.9 MB/s. We can see the verification speed is similar to reading throughput.

Conclusions

This paper analyzes the data integrity of iSCSI and utilizes dm-verity to ensure the data integrity of iSCSI. It is an efficient way to protect important data from being polluted. Experiments have shown that application of dm-verity mechanism to iSCSI is feasible with tiny and ignorable performance loss. As some significant data are visited, it can protect the data from being modified and ensuring that callers obtain correct data. Dm-verity for iSCSI not only improves data integrity of iSCSI, but also eases the data management of iSCSI.

Acknowledgments We would like to thank the anonymous reviewers for their constructive comments. This work is supported in part by National Natural Science Foundation of China under Grant No. 60973137, Program for New Century Excellent Talents in University under Grant No. NCET-12-0250, Gansu Sci.&Tech. Program under Grant No. 1104GKCA049, 1204GKCA061 and 1212RJYA003, The Fundamental Research Funds for the Central Universities under Grant No. lzujbky-2012-44, lzujbky-2013-k05, lzujbky-2013-43 and lzujbky-2013-44, Google Research Awards, Google Faculty Award, China and National Science Council (NSC), Taiwan, under grant NSC101-2221-E-126-002.

References

1. Corbet J (2011) dm-verity overview: http://lwn.net/Articles/459420/ LWN: 9/19
2. Device-mapper Resource Page: http://sourcewre.org/dm/
3. Kaufman C, Perlman R, Speciner M (2002) Network security: private communication in a public world, 2nd edn. Prentice Hall, New Jersey
4. Wendt J, Thaler P, Satran J, Shimony I, Makhervaks V (2010) iSCSI-R data integrity: http://www.research.ibm.com/haifa/satran/ips/iscsi-r-data-integrity-v1d.pdf
5. Riedel E, Kallahalla M, Swaminathan R (2002) A framework for evaluating storage system security. In: The 1st USENIX Conference on File and Storage Technologies, pp 15–30, USENIX Association, Berkeley, CA
6. Di CY, Li KC, Hung JC, Yu Q, Zhou R, Hung CH, Zhou QG (2013) A case of security encryption storage system based on SAN environments: intelligent technologies and engineering systems lecture notes in electrical engineering, vol 234. Springer, Heidelberg, pp 27–32
7. Meth KZ, Satran J (2003) Design of the iSCSI protocol. In: The 20th IEEE/11th NASA Goddard Conference on Mass Storage Systems and Technologies (MSS'03), pp 116–122, IEEE Computer Society, Washington, DC
8. Linux Unified Key Setup (2012) http://code.google.com/p/cryptsetup/wiki/DMVerity 6/12
9. Wu GF, Wu HH, Chang L (2011) On performance optimization of iSCSI SAN. In: 3rd International Conference On Information Technology and Computer Science (ITCS 2011), pp 430–433, ASME, New York

Opportunistic Admission and Scheduling of Remote Processes in Large Scale Distributed Systems

Susmit Bagchi

Abstract The large scale loosely-coupled distributed systems such as, grid and cloud computing systems employ opportunistic execution mechanism of remote processes in order to utilize computing resources of idle nodes. The opportunistic admission and scheduling of remote processes at a node need to balance the enhanced resource utilization and the performance of local processes at the node. This paper proposes the design and implementation of a novel Admission Control and Scheduling (ACS) algorithm for opportunistic execution of remote processes in a distributed system based on online estimation method. The experimental results illustrate that the algorithm can schedule the CPU-bound and IO-bound remote processes without degrading overall performance of a node. The CPU-utilization and memory-utilization of a node are enhanced by 26.65 and 24.5 % respectively on the average without degrading the performance of local processes executing at the node.

Keywords Distributed systems · Opportunistic scheduling · Online estimation

Introduction

The large scale loosely-coupled distributed computing systems such as, grid computing and cloud computing systems are gaining attention as next generation computing platforms due to the availability of computing machines as well as Internet at lower cost [1, 2]. The large scale distributed computing systems can be viewed as the virtual supercomputer infrastructure offering location transparent execution and faster response time [1, 3–6]. However, the statistical observations

S. Bagchi (✉)
Department of Informatics, Gyeongsang National University, Jinju, South Korea
e-mail: susmitbagchi@yahoo.co.uk

Y.-M. Huang et al. (eds.), *Advanced Technologies, Embedded and Multimedia for Human-centric Computing*, Lecture Notes in Electrical Engineering 260, DOI: 10.1007/978-94-007-7262-5_80, © Springer Science+Business Media Dordrecht 2014

about distributed resource utilizations have revealed that 50 ∼ 70 % of computing resources remain idle or unutilized on the average in the large scale grid and cluster computing systems [7–10]. In the large scale distributed computing systems, the scheduling of the remote processes remains an important research issue [2, 3, 11–13]. The global scheduling algorithms are often employed in grid computing to enhance throughput, to scale up the global resource utilization and to manage idle nodes [7, 8]. However, the global scheduling of processes does not consider the dynamics of available resources at individual nodes. Due to this reason, the opportunistic scheduling of processes tends to degrade the performance of existing local processes at a node. The local scheduling of remote processes in large scale distributed computing systems has received less attention [7, 9]. It has been proposed that scheduling of processes should be done at two levels such as, at local cluster level and at grid level [3, 14]. Thus, admission control of remote processes and local scheduling of the remote processes at a node are the two important research areas with the aim to maximize the resource utilization of a node while maintaining the throughput of the local processes. This paper proposes the novel Admission Control and Scheduling (ACS) algorithm based on online estimation of available resources and the process-load in a system. The proposed architecture utilizes the kernel address-space virtual devices and user-space application to create a remote process execution framework in a node in the distributed systems. The goals of the proposed design framework are, (1) *enhancing the utilization of idle resources in a node* and, (2) *maintaining load-balance between local as well as remote processes and system resources of a node preventing the system from thrashing or resource-starvation*. Rest of the paper is organized as follows. Section Related Work describes related work. Section Resource Estimation and Algorithm Design depicts the design model and description of the algorithm. Section Experimental Evaluations presents implementation and experimental evaluation. Section Conclusion concludes the paper.

Related Work

The wide array of computing applications employ large scale loosely-coupled distributed computing systems. The examples of such systems are grid, cluster and cloud computing systems [1, 2, 8, 15]. In the grid computing systems, the global scheduling of jobs in the batches is one of the main research challenges. The scheduling of jobs in such large scale distributed systems can be broadly classified into three classes namely, periodic resource allocation based scheduling, opportunistic scheduling and, reliability-based scheduling. Furthermore, the reliability-based scheduling algorithms can be classified into two broad groups such as, hierarchical scheduling and, genetic algorithm based scheduling. In the Periodic Resource Reallocation (PRR) based global scheduling technique, the resource broker maps the jobs to the available computing nodes in multiple stages [1]. The periodic reinforcing and reclamation algorithm schedules the jobs at the nodes to minimize the execution time of the jobs at each stage [1]. However, the algorithm is prone to frequent process migration through checkpointing, which is an expensive operation. Moreover, the migration of processes between heterogeneous systems would degrade the system performance. On the other hand, the opportunistic scheduling model aims to utilize the resources of idle nodes in a distributed system [7, 8, 16]. The dilation of response time of short jobs is reduced by opportunistic scheduling (OPS) algorithm proposed in [8]. The OPS algorithm is an augmented form of Up-Down algorithm [7]. The algorithm reduces the mean slowdown of processes in a system by rotating the jobs in a batch within a queue [8]. However, the algorithm is based on the assumptions that, local processes at a node have exponential inter-arrival time and, the distribution of local process-load in a system is hyper-exponential in nature [8]. These are fairly rigid assumptions in case of real-life execution environments. If the arrival rate of processes varies significantly then, the performance of the opportunistic algorithm degrades quickly with increasing remote process-load. The Condor distributed system employs Up-Down algorithm to implement fair access to remote resources for executing processes [7]. The performance of the system is further improved by introducing Multilevel Queue Opportunistic (MOF) policy [16]. However, the policy is dependent on workload model, where the processing load estimation assumes hyper-exponential distribution.

The hierarchical reliability-driven (HRD) job scheduling algorithms have gained research attention in recent time. The hierarchical reliability-driven scheduling algorithm employs two stages to schedule the jobs in a grid based on the distributed network reliability model [2, 5]. The first stage scheduling decision is based on local schedulers controlling a subset of nodes and the second stage of scheduling is done by global scheduler located at a central server [2]. The hierarchical scheme aims to enhance system reliability through distributed decision making in a computing system. The main two disadvantages of the hierarchical scheme are employment of non-preemptive scheduling model by the local scheduler and, possibility of single-point of failure due to centralized global scheduler system. In another approach, the reliability-aware genetic algorithm is proposed, where the transmission time and waiting time in resource queue are taken into consideration while making scheduling decisions [15]. The sampling model employed in the genetic algorithm is stochastic in nature in order to speed up the convergence. The stochastic universal sampling enhances the performance of the algorithm over the other genetic algorithms [15]. However, the algorithm is computationally expensive. The master–slave architecture is often employed for the realization of grid scheduling of jobs in batches. A multilevel hybrid scheduling algorithm is proposed based on master–slave model [3]. The algorithm uses two different schemes such as, single queue model and dual queue model of task scheduling. Although the algorithm enhances performances however, one of the major disadvantages of the algorithm is the centralized master–slave architectural model, which is prone to the single point of failure.

Resource Estimation and Algorithm Design

The main two computing resources required by any process are CPU and main memory, where the availability of both can vary randomly in time. The online estimation of the computing resources in a system is carried out based on the time-windows having fixed quanta $\varepsilon > 0$ and having total memory M, a constant. Let, P(t) is the total number of processes existing in a node in the distributed computing system at time t, whereas n(t) and p(t) represent the remote process load (RPL) and local process load (LPL) in the node at time t, respectively. Let, the instantaneous memory-load of the node is denoted by function m(t) and the CPU-load is denoted by function c(t), where both are dimensionless. Hence, the instantaneous ratio of number of remote processes and local processes including the administrator process (i.e. the process which admits the remote processes in a node) can be given by,

$$y(t) = [n(t) + 1]/[p(t) + 1] \tag{1}$$

The ratio of number of local processes including the administrator process and total number of processes at time t is given by,

$$z(t) = [p(t) + 1]/P(t) \tag{2}$$

Accordingly, the memory-load at time t is given by, m(t) = u(t)/M, where u(t) is the amount of used memory at time t. Thus, the process-load at a node is computed as,

$$c(t) = [n(t) + 1]/p(t) \tag{3}$$

From Eqs. (1) and (3),

$$c(t) = [y(t)(p(t) + 1)]/[P(t) - y(t)(p(t) + 1)] \tag{4}$$

From Eqs. (2) and (4),

$$c(t) = [y(t)(p(t) + 1)]/[P(t)z(t) - 1] \tag{5}$$

Now, from Eqs. (4) and (5), it can be further derived as,

$$P(t) = [(y(t)(p(t) + 1)) - 1]/[1 - z(t)] \tag{6}$$

It is evident from Eq. (6) that, it is a balancing equation representing the dynamics of process-load on a CPU of a system. Thus, dynamics of the co-variation of the instantaneous memory-load and process-load in a system at time t is computed by,

$$h(t) = m(t) - c(t) \tag{7}$$

Fig. 1 Dynamics of cross-over distribution for monotonically increasing RPL

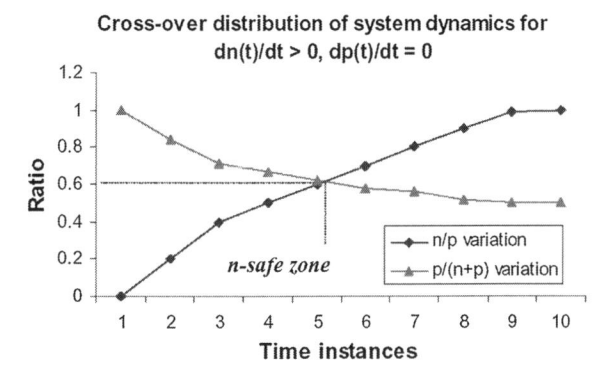

The difference equation for any time-window $\varepsilon = t_2 - t_1$ can be computed from equation (7) as,

$$\Delta h(\varepsilon) = h(t_2) - h(t_1) = [m(t_2) - m(t_1)] - [c(t_2) - c(t_1)] \qquad (8)$$

Hence, for the sufficiently small ε, $lim_{\varepsilon \to 0}(\Delta h(\varepsilon)/\varepsilon) = lim_{\Delta t \to 0}(\Delta h(\Delta t)/\Delta t) = d/dt(h(t))$, where $\Delta t = t_2 - t_1$. This indicates that Eq. (8) computes the dynamics of resource-load in a system as the differential on continuous time plane for sufficiently small quanta. However, for larger quanta, Eq. (8) transforms into difference equation computing the dynamics of resource-load variations in the discrete time plane. In the real-life systems, the LPL and RPL can vary randomly in time. The cross-over distribution of monotonically increasing RPL with respect to time t $(dn(t)/dt > 0)$ along with static LPL $(dp(t)/dt = 0)$ is illustrated in Fig. 1. It is evident that, on the left side of cross-over point (n-safe zone), the resource utilization by remote processes is ensured without creating the possibility of thrashing. On the other hand, if LPL increases monotonically in time along with static RPL, then mainly the local processes utilize the system resources on the right side of the cross-over point (p-safe zone) as illustrated in Fig. 2. Thus, a computing node requires the balance of the admixture of load by keeping local processes on upper right side of the cross-over point and remote processes on the lower left side of the cross-over point on their respective curves.

Designing the Algorithm

The ACS algorithm is comprised of two logical components namely, the online periodic estimation and the scheduler. The estimator periodically makes online estimation of system resources available to the local as well as remote processes. On the other hand, the scheduling algorithm takes the estimated parameters as input and controls the admission as well as scheduling of the remote processes in a node. The pseudo-code representation of the ACS algorithm is illustrated in Fig. 3,

Fig. 2 Dynamics of cross-over distribution for monotonically increasing LPL

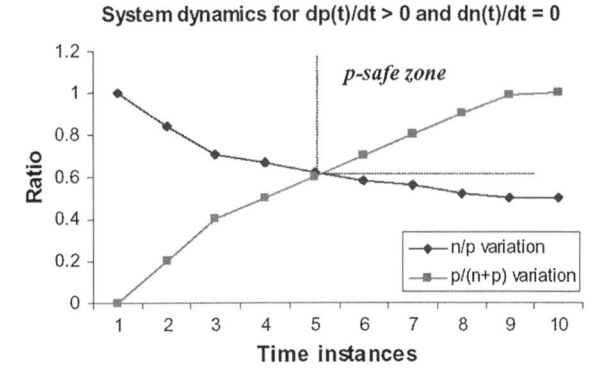

where `delta_h1`, `delta_h2` and `delta_h3` represent estimated values of Eq. (8) in three time-windows.

The algorithm does not assume any fixed number of local processes in a node and, it controls the admission of remote processes based on the instantaneous number of local processes in a node at any point of time. One of the aims of the algorithm is to restrict the possibility of creation of monopoly of remote processes in a system leading to starvation of the existing local processes. On the other hand, the algorithm aims to enhance the overall resource utilization in a system by admitting the remote processes following the controlled proliferation technique. According to the resource estimation model, 50 % of instantaneously available main memory (i.e. in n-safe zone below the cross-over point) is kept free as upper bound (`ram2` in Fig. 3) in order to avoid memory-overload and thrashing due to sharp rise of local processes in a short duration of time. However, the ratio of number of remote processes and local processes (`f1` in Fig. 3) is kept near to cross-over point at 0.45 at any point of time in order to control the load-balancing in a node. On the contrary, if the ratio of number of local process and total number of processes in a system (`f2` in Fig. 3) falls below 70 % (i.e. in p-safe zone above the cross-over point), the algorithm stops admission of remote processes into the system to achieve load-balancing. If any of the aforesaid three conditions are satisfied, then the algorithm temporarily stops admitting remote processes in a node by changing the flag (line 1). If the estimated values of co-variance of resource-load in a system are non-negative and system is lightly loaded, then remote processes are loaded one by one from the queue of processes (line 2). Initially, all the newly loaded remote processes are allocated zero priority and the remote processes run following the preemptive scheduling policy. However, if the system is on the process of getting lightly-loaded due to completion of some of the local or remote processes, then the blocking of the remote-process admission into the node is cleared by setting the flag (line 3). The `set_scheduling_param` procedure in Fig. 3 changes the scheduling parameters of the corresponding process as mentioned in the argument. In order to allow greater time-slicing between local and remote processes in a node, the remote processes running in non-

Variables:
Integer i = 0, policy, priority, flag = 0;
Float f1, f2, ram2, delta_h1, delta_h2, delta_h3;

Procedure scheduler (ram2, f1, f2, delta_h1, delta_h2, delta_h3)

```
    {
1.  if ((ram2 < 0.5) || (f1 > 0.45) || (f2 < 0.7)) flag = – 1;
2.  if ((delta_h1 >= 0.0) && (delta_h2 >= 0.0) && (delta_h3 >= 0.0) && (flag != –1))
        load_and_start_remote_proc (priority = 0; policy = preemptive);

3.  if ((delta_h1 < 0.0) || (delta_h2 < 0.0) || (delta_h3 < 0.0)) {
        flag = –1;
        for (i = get_index (list_of_all_remote_procs_loaded)) {
            policy = get_scheduling_param (i);
            if (policy != preemptive) {set_scheduling_param (i, policy = preemptive);
            break; }
        }
    }

4.  else if ((delta_h2 >= 0.0) && (delta_h3 >= 0.0)) {
        for (i = get_index (list_of_all_remote_procs_loaded)) {
        if (i != 0){ policy = get_scheduling_param (i);
            if (policy == preemptive) {
                set_scheduling_param (i, priority = 50, policy = round_robin);
                break;}
            }
        }
    }

5.  if ((delta_h1 >= 0.0) && (delta_h2 >= 0.0) &&
        (delta_h3 >= 0.0) && (flag == –1)) flag = 0;

6.  else flag = 0;
    }
```

Fig. 3 Pseudo-code of the admission control and scheduling algorithm

preemptive mode are changed to preemptive mode of execution. On the other hand, if the system continues to be in lightly loaded state in three consecutive time-window frames, then the priorities of remote processes are increased to 50 one by one and the scheduling policy is changed to round-robin in order to speed up the processing (line 4). Lastly, if the system had entered into highly-loaded stage previously and currently is in lightly-loaded stage in three consecutive time-window frames, then the blocking of the admission of remote process into a node is cleared by resetting the flag (line 5).

Experimental Evaluations

The implementation of the algorithm and other software components are made in Linux 2.6 kernel using C language. The implementation is comprised of one user-space daemon and one kernel-space module having secured communication channel between them. Experiments are carried out on the platforms having dual core Pentium processor equipped with 4 GB RAM and 250 GB disk drive connected to 100 Mbps wired network. The list of concurrent local process load (LPL) and, multiple instances of user applications are illustrated in Table 1, where the processes are characterized as IO-bound and CPU-bound depending on their nature. The experimental observations are made at a node in four phases such as, Phase I: RPL < 50, Phase II: 50 < RPL < 100, Phase III: 100 < RPL < 150 and, Phase IV: Stable system. A system is called stable when ACS algorithm temporarily stops admission of remote processes depending upon the dynamics of resource-load in the system and waits for some of the executing processes to complete execution before admitting any remote processes further in the system.

Table 1 The distribution of LPL and RPL in a system

Process load	Admission	Type	Locality
• System processes	Pre-existing for concurrent execution	• CPU and IO bound	LPL
• Adobe photoshop		• CPU-bound and memory-loading	
Processes with multiple instances:		• CPU-bound and memory-loading	
• Multimedia editor			
• Spawning application		• CPU-bound	
• Media streaming		• IO-bound	
Type I (multiple instances):	From queue for concurrent execution	Type I:	RPL
• Video encoder and streaming		CPU-bound and memory-loading	
• DSP simulator		Type II:	
• Big-data encryption algorithm		IO-bound and memory-loading	
• Continuous-fork application			
Type II (multiple instances):			
• Repeated file IO application			
• Streaming server application			
• Memory-grabber application			

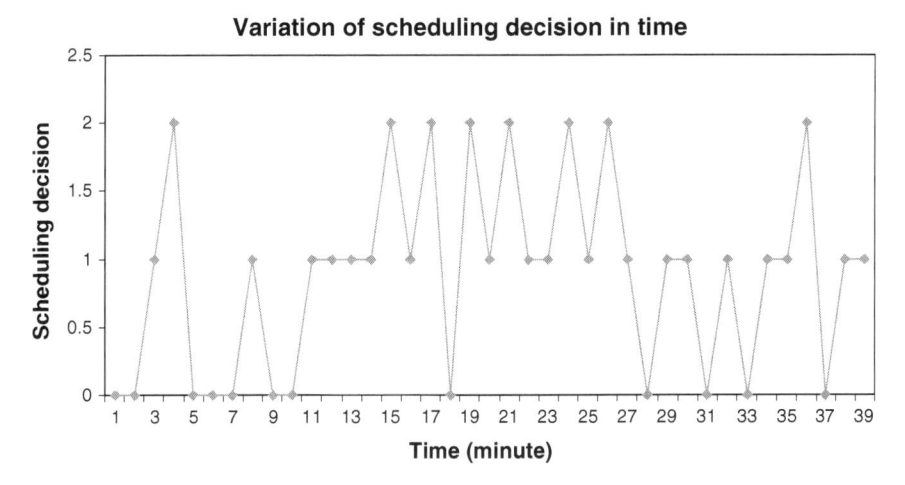

Fig. 4 Snapshot of variations of scheduling decisions for RPL <50

Phase I

The snapshot of the variations of admission and scheduling decisions taken by ACS algorithm is illustrated in Fig. 4, where on y-axis numerical value 0 denotes the reduction of priority and changing the scheduling policy of remote processes, 1 denotes the admission of remote processes and, 2 denotes the increment of priority and changing the scheduling policy of remote processes. The variation of average % utilization of the system resources by LPL and RPL is illustrated in Fig. 5.

It is evident from Fig. 4 that, in Phase I, initially a node behaves as highly unstable system due to repeated admission of remote processes in the system,

Fig. 5 Snapshot of variation of %utilization of CPUs and RAM for RPL <50

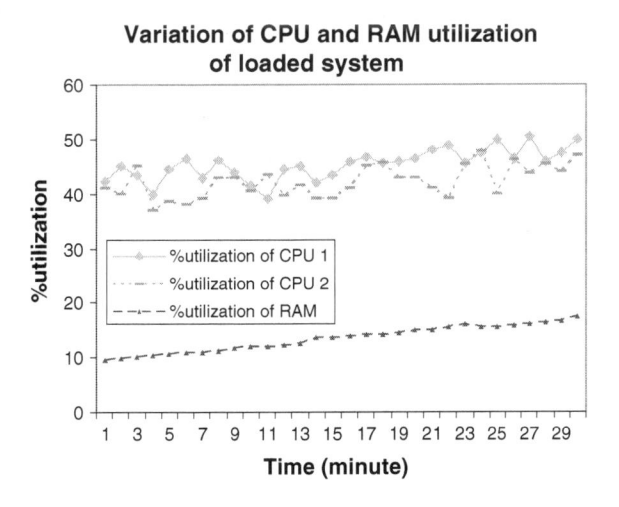

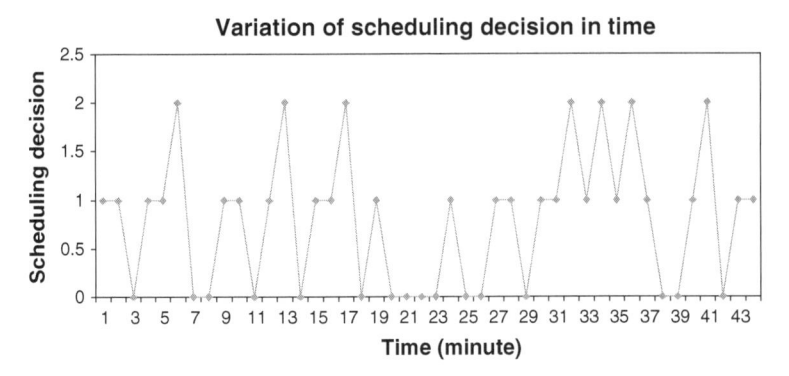

Fig. 6 Snapshot of time variation of scheduling decision for 50 < RPL < 100

whereas the resulted utilization of main memory of the system increases mono-tonically. However, occasionally, the ACS algorithm reduces the priority of remote processes in order to maintain the resource-demand and performance of the local processes. In this phase, on the average CPU 1 and CPU 2 are utilized up to 50 and 46.7 %, respectively. However, the main memory utilization is 21.4 % on the average.

Phase II

The snapshot of the variations of admission and scheduling decisions taken by ACS algorithm in this phase is illustrated in Fig. 6, where on y-axis numerical value 0 denotes the decrement of priority and changing scheduling policy of remote processes, 1 denotes the admission of remote processes and, 2 denotes the

Fig. 7 Snapshot of variation of %utilization of CPUs and RAM for 50 < RPL < 100

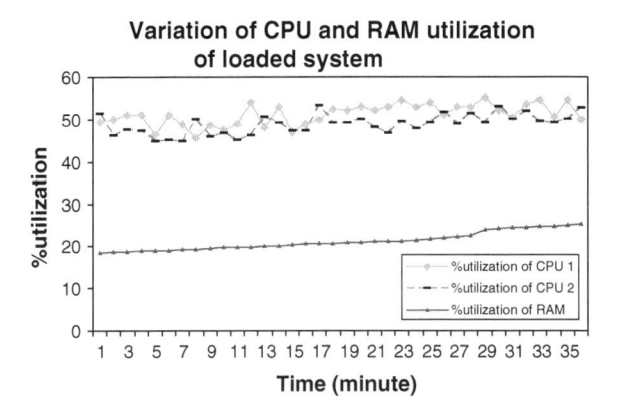

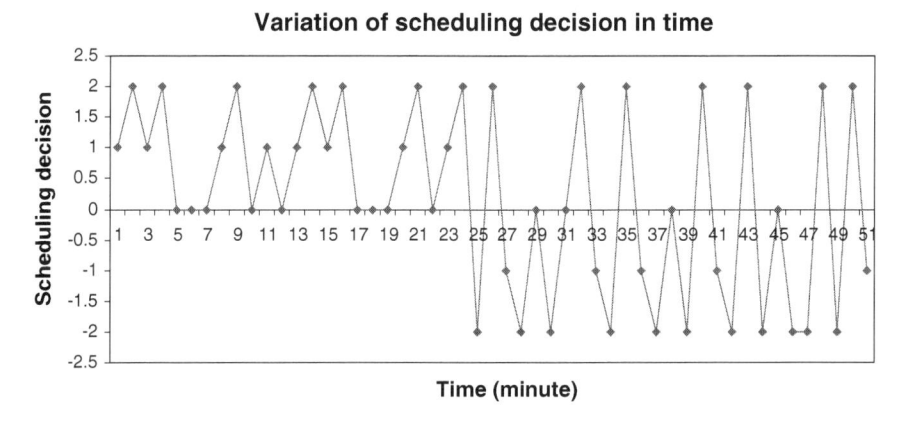

Fig. 8 Snapshot of time variation of scheduling decision for $100 < \text{RPL} < 150$

increment of priority and changing scheduling policy of remote processes. The average % utilization of the system resources by LPL and RPL in Phase II is illustrated in Fig. 7. In this phase, Fig. 6 illustrates that, due to enhanced processing-load and resource-demand on the system, the kernel is reclaiming the idle resources (such as, fragmented memory pages) and making them available to LPL as well as RPL executing in the system. This phenomenon induces the frequent transitions between two extreme scheduling decisions as depicted in Fig. 6.

However, according to Fig. 7, the overall utilization of resources is monotonically increased. In this phase, on the average CPU 1 and CPU 2 are utilized up to 51.5 and 45.3 %, respectively. However, the main memory utilization is increased to 22.7 % on the average.

Phase III

The snapshot of the variations of admission and scheduling decisions taken by ACS algorithm in this phase is illustrated in Fig. 8, where on y-axis numerical value 0 denotes the reduction of priority and changing scheduling policy of remote processes, 1 denotes the admission of remote processes, 2 denotes the increment of priority and changing scheduling policy of remote processes, -1 denotes the readmission of remote processes and, -2 denotes stopping admission of remote processes. The average % utilization of the system resources by LPL and RPL in Phase III is illustrated in Fig. 9. According to Fig. 8, through the kernel activation the resource-availability is increased and the ACS algorithm rapidly admits remote processes into the system while increasing scheduling priorities for faster execution of remote processes. However, due to variations of LPL, the ACS algorithm later stops admitting any RPL further when the scheduling decision makes frequent transitions between two extremes in the highly-loaded system in Phase III.

Fig. 9 Snapshot of variation of %utilization of CPUs and RAM for $100 < RPL < 150$

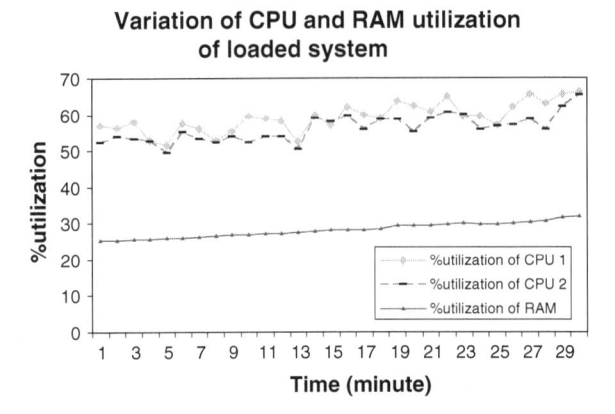

In this phase, utilization of CPUs has increased to 78.6 and 73.4 % on the average, whereas the main memory utilization is increased to 40.5 % on the average scale.

Phase IV

This is the phase of temporary stability, where the ACS algorithm temporarily stops any admission of remote processes into the system allowing the LPL and existing RPL to execute enhancing resource utilization without causing memory-overloading and thrashing. The screenshot of resource monitor of the system in

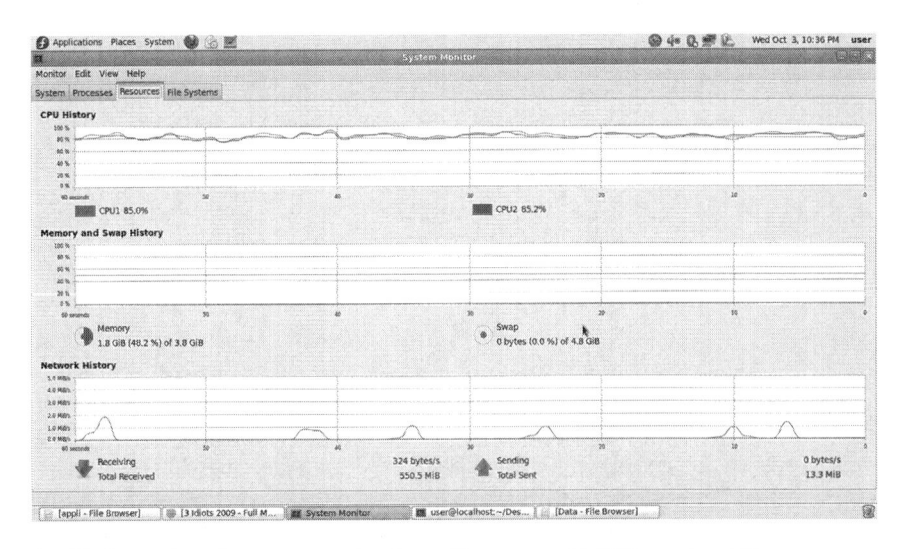

Fig. 10 Screenshot of resource monitor for stable system with RPL equal to 150

maximally-loaded stable phase is illustrated in Fig. 10, where RPL is equal to 150 after 3 h of execution.

The CPU utilization is enhanced to 85 and 85.2 % at maximum for CPU 1 and CPU 2, respectively. The main memory utilization is increased up to 48.2 %. The performances of local processes remain unaffected.

Conclusion

The global scheduling algorithms are employed to determine the location of execution of remote processes in large scale distributed computing systems. However, the admission control and scheduling of remote processes locally at the nodes are required to maintain the load-balance between local processes and remote processes, while enhancing resource utilization of a node. The proposed lazy invocation based ACS algorithm implements opportunistic admission and scheduling of remote processes in a node. The algorithm enhances the over all CPU and memory utilizations up to 85.2 and 48.2 % during the phase of temporary stability. The average value of resource utilization is increased by 26.65 and 24.5 % for CPU and memory, respectively, without affecting the performance of the local processes. The algorithm is easy to realize and computationally inexpensive.

References

1. Lin CC, Shih CW (2008) An efficient scheduling algorithm for grid computing with periodical resource reallocation. In: 8th IEEE International Conference on Computer and Information Technology, IEEE
2. Tang X, Li K, Qiu M, Sha EHM (2011) A hierarchical reliability-driven scheduling algorithm in grid systems. J Parallel Distribut Comput 72(4):525–535
3. Shah SNM, Mahmood AKB, Oxley A (2010) Analysis and evaluation of grid scheduling algorithms using real workload traces. In: ACM MEDES'10, ACM
4. Wieczoreka M et al (2009) Towards a general model of the multi-criteria workflow scheduling on the grid. Future Generation Computer Systems Journal, Vol. 25, No. 3, Elsevier (2009)
5. Fan H, Sun X (2010) A multi-state reliability evaluation model for P2P networks. Reliab Eng Syst Saf J 95(4): 402–411
6. Fernandez-Baca D (1989) Allocating modules to processors in a distributed system. IEEE Transact Softw Eng 15(11): 1427–1436
7. Litzkow MJ, Livny M, Mutka MW (1988) Condor: a hunter of idle workstations. In: 8th IEEE International Conference on Distributed Computing Systems, IEEE
8. Ghare GD, Leutenegger ST (2000) Improving small job response time for opportunistic scheduling. In: 8th International Symposium on Modeling, Analysis and Simulation of Computer and Telecommunication Systems, IEEE
9. Mutka M, Livny M (1991) The available capacity of a privately owned workstation environment. Perform Eval J 12(4), 269–284

10. Tchernykh A et al (2009) Idle regulation in non-clairvoyant scheduling of parallel jobs. Discrete Appl Math J 157(2): 364–376
11. Livny M, Basney J, Raman R, Tannenbaum T (1997) Mechanisms for high throughput computing. SPEEDUP J 11(1)
12. Zhihong X, Xiangdan H, Jizhou S (2003) Ant algorithm-based task scheduling in grid computing. In: CCECE'03, IEEE
13. Kiran N, Maheswaran V, Shyam M, Narayanasamy P (2007) A novel task replica based resource scheduling algorithm in grid computing. In: The 14th HiPC Conference
14. Li H (2009) Workload dynamics on clusters and grids. J Supercomputing 47(1): 1–20
15. Abdulal W, Ramachandran S (2011) Reliability-aware genetic scheduling algorithm in grid environment. In: International Conference on Communication Systems and Network Technologies, IEEE
16. Abawajy JH (2002) Job scheduling policy for high throughput computing environments. In: 9th International Conference on Parallel and Distributed Systems, IEEE

Feasible Life Extension Concept for Aged Tactical Simulation System by HLA Architecture Design

Lin Hui and Kuei Min Wang

Abstract The military owned simulation systems are facing legacy situation causing sky-high maintenance cost and unstable condition. It would be costly and risky for retiring them without considering life extension alternative. The objective of this research is to provide a feasible concept for migrating the structure of aged Tactical Training Simulation (TTS) system to High Level Architecture (HLA), which not only can retain the original feature but also offer an extra function of interoperability. This research offers a solution for military legacy simulation systems and paves the way for creating the synthetic battlefield environment that military can benefit its training, planning, acquisition and resource allocation.

Keywords HLA · Interoperability · PDU · RTI · Simulation · TTS

Introduction

The original Tactical Training Simulation (TTS) system has been built for over decades being not only uneasy to maintain but also very hard to have the merit of interoperability and reusability for this synthetic environment. The causes of that are the out of date of aging hardware and software system for being uneasy maintained which is even costly, imbedded models in the system with its limitation in function expansion which includes new forces add in and exercises size

L. Hui
Department of Innovative Information and Technology, Tamkang University, Tamsui, Taiwan, Republic of China
e-mail: amar0627@gmail.com

K. M. Wang (✉)
Department of Information Management, Shih Chien University, Taipei, Taiwan, Republic of China
e-mail: willymarkov@gmail.com

Y.-M. Huang et al. (eds.), *Advanced Technologies, Embedded and Multimedia for Human-centric Computing*, Lecture Notes in Electrical Engineering 260, DOI: 10.1007/978-94-007-7262-5_81, © Springer Science+Business Media Dordrecht 2014

expanding, as well as unable to interact with other simulation systems which is one of the most important abilities in the joint exercise simulation environment.

This paper proposes a way to migrate aging simulation system to the interoperability-based High Level Architecture (HLA). In addition to the hardware that still in use in the system, the function of interoperability for meeting the reality of live (real people with real equipment in the field), virtual (real people with simulated equipment) and constructive (simulated people using simulated equipment) in simulation world is also a major concern.

The latest trend with high fidelity synthetic environment, based on HLA support trainings and exercises with virtual forces, has replaced the most of live forces in the field that not only can cut down the military training cost hugely but also reduce the probable casualties in the unexpected incidents.

Based upon the migration strategy of legacy system [1], study of TTS has been carried out in sequence as follows: trace TTS framework for understanding its original design concept and function of modules; the second is to check the way of message interchanged in the system; the third is to analyze HLA framework for the purpose of transforming it into the TTS. The purpose of the first two sequences is to restore TTS original technical and functional features for upgrades or migration concerns. The third is mainly for interoperability concern dressing TTS with interaction capability.

The upgraded TTS will have a new system procedure according to the six services of HLA Run Time Infrastructure (RTI). In the process, the primitive TTS military functions would be all retained, other than that the more flexible and convenient feature will be presented with the function of add-in, interoperability and reusability.

Literature Review

The aging Tactical Training Simulation system has fatal defects from the perspective from this decade, which is very hard to keep it operational when any system-malfunctions, such as expanded size of exercises and the new military assets that designed and setup in the system, occurred. The U.S.A. Army's Enhanced REmoted Target System (ERETS) has the operational problems caused by over the system life cycle that system components had no longer been available or repaired [2]. Most of the Tactical Training Simulation systems are unique and independent from each other due to they came from different vendors under without standards set by government authorities that make these systems become lower performing value, decomposability, obsolescence and deterioration, after years in service and it is called the legacy system [3].

U.S. Department of Defense (DoD) regards the migration of legacy systems as critical issues for there are so many systems have been in use by military for so many years. The legacy systems produced by various companies with different software that most of the systems are required to be upgraded or migrated to new hardware and software platforms. Therefore, "DoD Legacy System Migration Guidelines" is issued for being as an aid in the migration. There are 10 guidelines in total for risk-reduction while in the process of the migration of legacy systems. Guideline #6 indicates: Make software architecture a primary reengineering consideration, and define that a methodical evaluation of the software architectures of the legacy and target systems should be a driving factor in the development of the reengineering technical approach [1, 4, 5].

Well-known protocols for this paradigm include Distributed Interactive Simulation (DIS) [6], Aggregate Level Simulation Protocol (ALSP) [7], and the High Level Architecture (HLA) [8]. The HLA is an ameliorative technique that combines the operational concepts of DIS and ALSP, as well as other distributed simulation issues, into a standardized infrastructure. The technical innovation of HLA allows a radical upward scaling of the number of distributed simulation nodes to participate in a single scenario.

The HLA is a project initiated by the U.S. DoD to support interoperability among geographically distributed simulators. It later became an international standard, IEEE 1516, in 2001 [8]. The HLA defines an infrastructure to support the reusability and interoperability among heterogeneous simulations. To achieve this goal, the HLA defines a time management service for a simulation system to coordinate its execution pace with others. The time management service defines the synchronization mechanism to ensure causal order consistency of the event sequence among distributed simulation nodes. In HLA terminology, each distributed node of a simulation is called a federate. Furthermore, a federation is a simulation environment that consists of a set of federates. Depending upon the simulating model, a federate can use the time-stepped or event driven time advancing technique [9, 10].

Concept of TTS Operation

After the current TTS operation being examined, the documented concept of fundamental operation can be classified into five steps that they are Pre-setup before system initialization, initialization, communication link establishment, message transmission and termination.

The exercise is initiated after the system completion of its initialization. There are a series of weapon deployments, detection message transmitting etc. are followed for being set up. When an exercise is about to finish, the 'end' button can be chosen on the exercise screen of each computer. Then the system transmits finish message to server for the logout in the client end computer. The training simulated exercise flow chart is as Fig. 1.

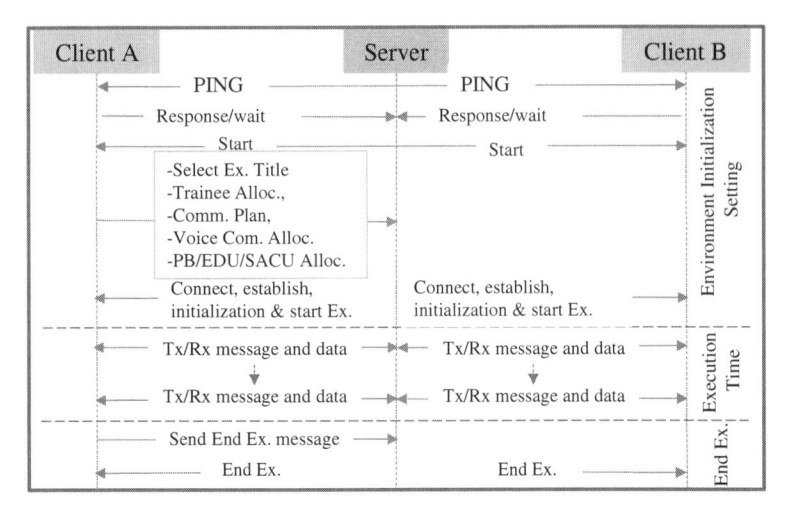

Fig. 1 TTS system flow chart

Upgrades Analysis

TTS and HLA architectures are analyzed for developing the concept of TTS migration that will follow the upgrades on the communication and application (APP).

HLA Architecture

There are advantages of HLA framework: first, it provides an interface for all computers through network that increases the function of interaction among users and data requesters (including system designer and analysts). Second, the modeling and simulation environment is under the open system framework allowing adopting reusable components to construct simulation that will maximize the utility rate and increase the flexibility. Third, HLA framework can integrate all simulators and simulation systems, which can be homogeneous or heterogeneous type that makes the expansion of the simulation range possible. Forth, the RTI mechanism reduces the quantity of data transmission in network. The data transmission for original TTS was in Protocol Data Unit (PDU) format, but PDU format is complicated and most of PDUs were defined for military purpose that would be a bottleneck for TTS with a very limited simulation performance. Also in the procedure of data transmission, due to PDU format, lots of unchanged data are still transmitted in packet by broadcasting that will waste bandwidth hugely. Fifth, the new system will be modulated for making it more convenient to have the following functions such as add in, modify and delete that can increase system flexibility. Sixth, DIS has time

Fig. 2 HLA architecture

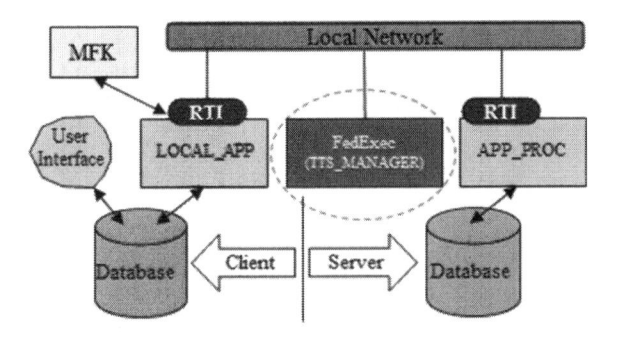

managing problem that is prone to make time disorder in data receiving but with complete time managing functional services, HLA can illuminate the time disordering problem.

The HLA-based architecture is as Fig. 2. The comparison between TTS_MANAGER and FedExec is as Table 1.

Upgrade Communication and APP Module

Two major issues are essential with top priority to concern in the process of migration toward HLA architecture. The first concern is the modification of the communication that can be done by using RTI to take charge of all interchanged message. The advantage is to allow tactical module concentrating on its tactics design and mission planning without the distraction of taking care of data transfer event. The second concern would be the modification of APP on both of local computer and main computer for their Local_APP and APP_PROC, respectively. For current APP modules, they will be improved by having additional functions on addition, modification and deletion under HLA architecture that allows the system to edit all of the weapon systems, equipment and platforms base on the users' need.

Fig. 3 depicts the contrast of APP module in TTS and HLA. In TSS APP module, the send-out from client A to client B must by server, therefore some APPs would appear at both of main computer and local computer, such as Dynamics, MSG_OUT, EX_MNG and Detection, etc. In fact, the function of APPs are not the same at both main and local computer, e.g. at main computer, the tactical APP is simply responsible for receiving and send-out the message to the clients but it's not the case at local computer that would make mistakes that is ambiguous.

Table 1 TTS_MANAGER versus FedExec

TTS_Manager in TTS	FedExec in HLA
• Checking the status of all joined computers	• Managing the joining and leaving federate in federation
• Handling additional joining computer	• Helps Message transmission among federates in their federation
• Message transmission	• Restored function in federation management
• Exercise recovery	• RTI's time management

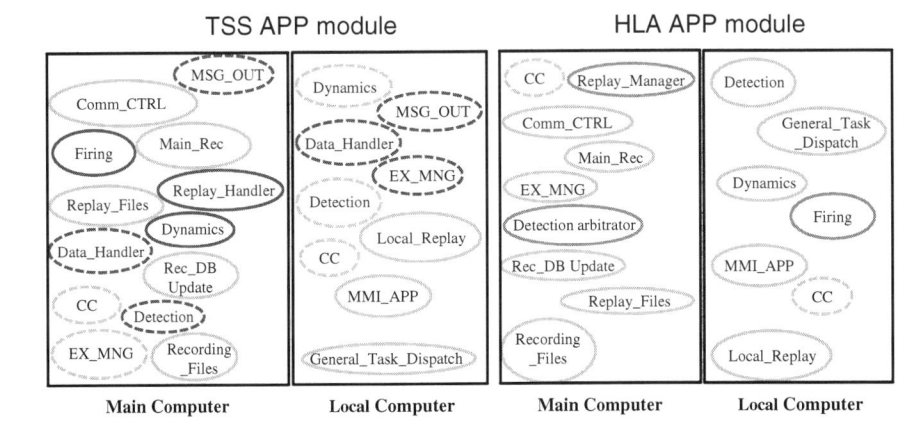

Fig. 3 APP in TTS and HLA (*dash line* indicates the same APP at both computer)

Conclusion

Most of military simulation systems which ranged from tactical simulations to technical simulators have been facing legacy problem and lack of interoperable capability. To bridge the current legacy TTS to HLA based concept under the considerations of life extension with interoperability, cost-effectiveness and risk, analyzing TTS by tracing codes thoroughly for restoring its original feature is substantial, from where the way of migration become clearer and the logic is developed with less difficult.

The serious issues encountered in migrating TTS to be under HLA architecture are repetitive sending message over the network and the ambiguous function emerged with APP module. This paper presents the solution with using RTI to handle all the in and out message instead of server, and simplify APP module with clear function in order to avoid the ambiguity.

Contribution of this paper is to provide a way solving not only the issue of life extension with the required more powerful additional functions but also the problem of making a legacy simulation system interoperable. With the migrated system being part of the military's synthetic environment, it become possible for having live, virtual and constructive (LVC) exercise tightly integrated that multiplier on training which would beyond uses' expectation. Moreover, from the Ministry of National Defense perspective, the LVC environment can also benefit the issues of military planning, acquisition and doctrine development.

References

1. Samuel ARB (2007) Legacy system: migration strategy. Technowave, Inc., pp 4–5
2. Smith J, Todd J, Kahl R (2009) Training range modernization: new technology on old infrastructure. Interservice/Industry Training, Simulation, and Education Conference (I/ITSEC). Orlando, Floria: National Training and Simulation Association, pp 1–13
3. Fronckowiak D (2008) Extending the life of legacy software systems: a decision process model. In: Proceedings of Student-Faculty Research Day. New York City: Pace University, D1.2–D1.6
4. Bass L, Clements P, Kazman R (2003) Software architecture in practice, 2nd edn. Addison-Wesley Professional, MA
5. Bergey J, Smith D, Weiderman N (1999) DoD Legacy System Migration Guidelines. Carnegie Mellon Software Engineering Institute, Pittsburgh, pp 2–12
6. IEEE, Std-1278.1 (1995) IEEE standard for distributed interactive. Institute of Electrical and Electronics Engineers, Inc., New York
7. Wilson AL, Weatherly RM (1994) The aggregate level simulation protocol: an evolving system. In: Proceedings of the Winter Simulation Conference, IEEE. pp 781–787
8. IEEE, Std. -1516.1 (2000) IEEE standard for modeling and simulation (M&S), high level architecture (HLA)—federate interface specification. New York: IEEE, p.467
9. Huang J (2007) Design of an online decision-making support system for joint training simulation. JDMS, pp 43–53
10. Löfstrand B (2006) HLA Standard and Certification. Pitch Technologies, Rome

Mobile Agents for CPS in Intelligent Transportation Systems

Yingying Wang, Hehua Yan, Jiafu Wan and Keliang Zhou

Abstract Recently, cyber-physical systems (CPS) have emerged as a promising direction to enrich the interactions between physical and virtual worlds. Because of the large-scale features of CPS, mobile agents (MA) technology can promote the performance of CPS. In this article, we first introduce the concept and characteristics of CPS, MA, and intelligent transportation system (ITS). Then, we propose the structure of intelligent transportation CPS (ITCPS). On this basis, giving the case of mobile agents for ITCPS, we exploit a mobile agent by three levels (node level, task level, and combined task level) to reduce the information redundancy and communication overhead. Finally, we in brief outline the technical challenges for ITCPS.

Keywords Adaptive mobile agents · Cyber-physical systems · Intelligent transportation system

Introduction

In recent years, along with the rapid development and matures of Machine-to-Machine (M2M) technology, wireless sensor technology, cloud computing technology, and distributed real-time control technology, Cyber-Physical Systems (CPS) integrating computing, communications, and control in one fusion system has become one of the forefront interdisciplinary areas. CPS is closely related to

Y. Wang · H. Yan (✉) · J. Wan
School of Information Engineering, Guangdong Jidian Polytechinc, Guangzhou, China
e-mail: hehua_yan@126.com

K. Zhou
College of Electrical Engineering and Automation, Jiangxi University of Science and Technology, Ganzhou, China
e-mail: nyzkl@sina.com

Y.-M. Huang et al. (eds.), *Advanced Technologies, Embedded and Multimedia for Human-centric Computing*, Lecture Notes in Electrical Engineering 260, DOI: 10.1007/978-94-007-7262-5_82, © Springer Science+Business Media Dordrecht 2014

the development of human life and society, ranging from the small nanoscale biological robots to global energy coordination and management systems and involving human infrastructure construction. Currently, CPS applications include many domains, such as intelligent transportation and security, advanced automotive systems, industrial process control, distributed robotics, environmental control, and avionics, forming the basis to build smart city and smart planet of the future [1, 2].

CPS has emerged as a promising direction to enrich human-to-human, human-to-object, and object-to-object interactions in the physical world as well as in the virtual world, and typically involves multiple dimensions of sensing data, crosses multiple sensor networks and the Internet, and aims at constructing intelligence across these domains. So, dynamic participation and departure of a sensor network is possible. CPS imposes the same requirements for a WSN, but different levels of connectivity and coverage for different WSNs [3]. Furthermore, cross-domain communications may happen quite frequently in CPS applications [4]. In CPS applications, sensing data may be collected from static and mobile sensor nodes (such as those in vehicles, smart phones, etc.,) with both controllable and uncontrollable mobility. Apparently, we need to address the basic problems about the large data traffic, heterogeneous network communication, and real-time multitask in CPS applications.

A mobile agent (MA) is a composition of computer software and data which is able to migrate from one machine to another autonomously and continue its execution on the destination machine, with the feature of autonomy, social ability, learning, and most importantly, mobility [5–8]. The feasibility of using mobile agents in CPS is reflected in the following aspects:

- CPS exists the great data traffic, while MA can move to the machine to calculate, reducing the network load.
- CPS is composed by a plurality of heterogeneous network, and MA is conducive to parallel processing to be executed on multiple heterogeneous network machines.
- CPS autonomously adapts to changes in the physical environment, and MA can react according to the state of the machine environment. For example, by the load condition of the current machine, MA decides whether to move to another machine.
- CPS meets the reliability requirements, and MA is tolerant to network faults and able to operate without an active connection between client and server.
- CPS includes a variety of applications, and the use of MA may reduce the deployment of applications and enhance portability, consequently maintenance is more flexible.

Intelligent transportation system (ITS) is the effective integration of advanced information technology, communications technology, sensor technology, control technology and computer technology, over a large distance area and in all directions. ITS is a real-time, accurate, efficient and integrated transport and management system [9–11]. The ITS research is in full swing in the United States, Europe,

Japan, with amazing scale of development and alarming rate. Studies have shown that the use of CPS technology is feasible to achieve intelligent transportation system.

Intelligent Transportation CPS

Intelligent transportation CPS (ITCPS) needs to coordinate the different particle size of manned and unmanned equipment. To build reasonable CPS traffic architecture is the basic work of the future transportation system. The ITCPS is a typical CPS system, and has the typical characteristics of CPS, such as high flexibility, security, stability, reliability, efficiency, and seamless coupling. As Fig. 1 shows, the architecture of ITCPS can be divided into the following three layers [12–15]:

- Physical layer: In ITCPS, car, road equipment is no longer just a simple mechanical device, and will be embedded in a large number of sensors, calculators, and controllers. There are a lot of smart traffic devices distributed in the environment, such as smart cars, smart traffic lights, smart roads, smart bridge and so on, directly interacting with the physical environment. These transport equipment having sensing, computing and control functions constitute the physical layer of the ITCPS.
- Network layer: Network layer including satellite communications, base station communications, and other means of communications. All devices of the physical layer share the communication information through the network layer.

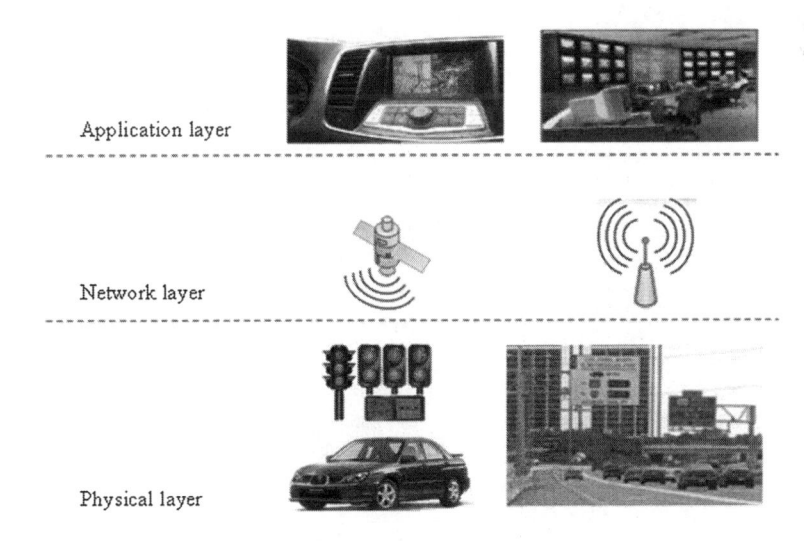

Fig. 1 A simple architecture of ITCPS

A smart car not only obtains surrounding environmental information through its own sensors, but also gets geographical environment information sensed by other vehicle sensors through the network layer, in order to facilitate the path forecast planning. Shielding the heterogeneity of the physical layer, the network layer seamlessly connects to share resource network-based for the application layer, and provide plug-and-play service for the user with the fully transparent way.

- Application layer: The application layer is the application of the user-oriented services, such as Smart car software, centralized management platform for traffic management department. Smart does not mean to exclude people, but in order to better serving people. The application layer is the part interacting with people directly; it needs to consider the people-oriented issues of design, implementation and confirmation, such as comfort, availability and correctness.

Case Study: Mobile Agents for ITCPS

In ITCPS, vehicles are furnished with on-board equipment (OBE) which includes computing, positioning, sensing, and wireless communication devises. As such, these vehicles are able to communicate with other equipped vehicles as well as with similarly instrumented roadside equipment (RSE). Thus, these components form vehicular ad-hoc networks (VANET), within which connected vehicles will be able to "talk" to each other. Belonging to the wireless sensor network, the link bandwidth of VANET is usually much lower than the wired network. However, ITCPS has frequent tasks, and the great data traffic is likely to exceed the capacity of wireless networks.

A mobile agent is a process that can transport its state from one environment to another, with its data intact, and be capable of performing appropriately in the new environment. MAs execute parallel processing on multiple heterogeneous network machines. MA's actions are dependent on the state of the machine environment. Thus, MA may reduce the deployment of applications and enhance portability, and maintenance is more flexible. In conclusion, MA has the mobility, autonomy, collaboration and intelligence. Therefore, the use of the Mobile Agent technology in VANET can effectively reduce redundant information and communication overhead, thereby reducing network load.

As Fig. 2 shows, the RSE produces MA to perform specific tasks as a server. The MA should adjust its own behaviors depending on quality of service needs (e.g., data delivery latency) and the network characteristics to increase network lifetime while still meeting those quality of service needs. The information contained in a MA packet is shown in Table 1. A MA is dedicated to reduce the information redundancy and communication overhead in three levels so as to prolong VANET lifetime [16–23].

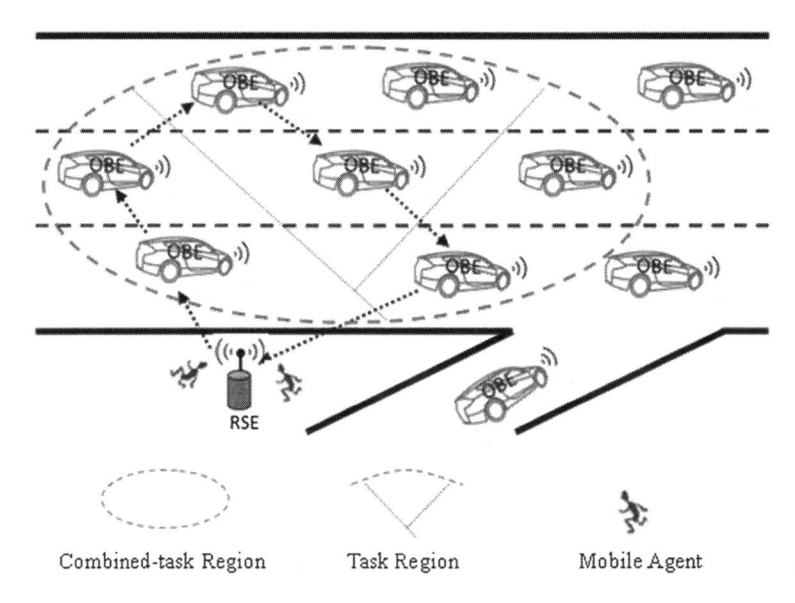

Combined-task Region　　　　　Task Region　　　　　Mobile Agent

Fig. 2 A simple architecture of MA based VANET

Table 1 MA structure

Fixed attributes	Variable attributes	Payload
RSE_ID	NextOBESrc	Processing code
MA_SeqNum	NextHop	Accumulated data and result
FirstOBESrc	ToRSEFlag	
LastOBESrc	OBESrcList	
RoundIdx		
LastRoundFlag		

- Node Level: Application Redundancy Eliminating by local Processing.

In Fig. 2, the RSE can assign the processing code (behavior) of MA to the target OBE, based on the requirement of a specific application (e.g., measuring road surface conditions). The MA in OBE processes the local raw data as requested by the application, only to transmit, extract and send relevant information. So the capability enables a reduction in the amount of data transmission.

- Task Level: Spatial Redundancy Elimination by Data Aggregation.

The MA aggregates individual sensed data when it visits each target OBE in the task region. Such as navigation, the RSE assigns a MA one-by-one to access the target OBE and aggregate individual location data to achieve precise positioning.

- Combined Task Level: Communication Overhead Saved by Multiple Tasks' Data Concatenation.

The packet unification technique that unifies the several short data packets to one longer packet can reduce the communication overhead in combined-task level. Due to data concatenation, the duty cycle and communication overhead of intermediate OBE can be reduced so as to prolong VANET lifetime. In order to reduce the delay, multiple tasks should be executed simultaneously by replicated MAs in the combined-task region to decrease the execution time of all tasks. Since copying MAs will bring additional overhead, the MA number of the combined-task region should be carefully designed depending on application's requirements (e.g., intersection traffic safety, dangerous road traffic safety).

The MA performing tasks algorithm is as follows:

```
If RSE has new tasks then
  RSE creates MA;
  Set RSE_ID, MA_SeqNum, FirstOBESrc, LastOBESrc, LastRoundFlag,
    NextOBESrc, NextHop, ToRSEFlag, OBESrcList;
RSE dispatch MA to FirstOBESrc;
FirstOBESrc sets RoundIdx = 1;
MA processes Data and gets accumulated data result;
Do While ToRSEFlag is false
  FirstOBESrc sets LastRoundFlag;
  MA sets NextOBESrc;
  If NextOBESrc = RSE then
    MA sets ToRSEFlag true;
  elseif NextOBESrc = LastOBESrc
      MA migrates toward LastOBESrc;
      MA processes Data and gets accumulated data result;
      MA sets ToRSEFlag true;
  elseif NextOBESrc = FirstOBESrc
      MA migrates toward FirstOBESrc;
      RoundIdx = RoundIdx + 1;
      MA processes Data and gets accumulated data result;
  else
      MA migrates toward NextOBESrc;
      MA processes Data and gets accumulated data result;
  endif
End While
MA migrates back to RSE;
endif
```

Discussion and Outlook

CPS research is still at a preliminary stage, so adaptive mobile agents for CPS also face many key technical issues, such as optimal allocation of resources, tasks real-time assurance, security control, standards development [21, 24–26]. Below, we in brief explain the technical challenges:

- Optimal allocation of resources: Multiple tasks are executed simultaneously by replicated MAs in the combined-task region to decree the execution time of all tasks. But copying MAs will bring additional overhead. Hence we should carefully design the MA number of the combined-task region and the RSE number according to application's requirements in order to avoid waste of resources.
- Real-time capabilities: We must ensure that the real-time performance must meet the specific application requirements. However, MAs execute the operation associated with migration, communication, control and safe, along with the additional running load. Therefore need to find a solution, such as better migration technology, better communication mechanisms, and multi-agency coordination mechanism.
- Security and privacy challenges: Since sensing data is no longer owned by local devices, security and privacy issues become more critical. MAs and machines (e.g., OBE, RSE) are vulnerable to a number of threats, including MA to machine threats, MA to MA threats and machine to MA threats. Security issue has been a hot and difficult research.
- Standards development. ITCPS applications depend on many technologies across multiple industries. Consequently, the required scope of standardization is significantly greater than that of any traditional standards development.

Conclusions

In recent years, CPS has become one of the forefront interdisciplinary areas. In this article, we first analyze the concept and characteristics of CPS, MA and ITS. Next, we present the structure of ITCPS. Then, through the case of mobile agents for ITCPS, we illustrate how a mobile agent is exploited in three levels (node level, task level, and combined task level) so as to reduce the information redundancy and communication overhead. Finally, we outlined the relevant research challenges. We hope to inspire more technological development and progress for CPS applications.

Acknowledgments The authors would like to thank the National Natural Science Foundation of China (No. 61262013), the Natural Science Foundation of Guangdong Province, China (No. S2011010001155), and the High-level Talent Project for Universities, Guangdong Province, China (No. 431, YueCai Jiao 2011) for their support in this research.

References

1. Wen JR, Wu MQ, Su JF (2012) Cyber–physical system. Acta Automatica Sinica 38(4):507–515

2. Wang ZJ, Xie LL (2011) Cyber–physical systems: a survey. Acta Automatica Sinica 37(10):1157–1166
3. Wang X, Xing G, Zhang Y, Lu C, Pless R, Gill C (2003) Integrated coverage and connectivity configuration in wireless sensor networks. In: Proceedings of the 1st international conference on Embedded networked sensor systems, Los Angeles, California, USA, pp 28–39
4. Han L, Potter S, Beckett G, Pringle G, Welch S, Koo SH, Tate A (2010) FireGrid: an e-infrastructure for next-generation emergency response support. J Parallel Distrib Comput 70(11):1128–1141
5. Ranjan S, Gupta A, Basu A, Meka A, Chaturvedi A (2000) Adaptive mobile agents: modeling and a case study. In: Proceedings of 2nd workshop on distributed computing IEEE Ind CFP, WDC
6. Zhao J, Liu B (2010) An overview of mobile agent (Mogent). Micro Process 31(1):1–5
7. Chen M, Gonzalez S, Leung V (2007) Applications and design issues for mobile agents in wireless sensor networks. IEEE Wirel Commun 14(6):20–26
8. Wang R, Zhou C (2001) The study of mogent (mobile agent): an overview. Appl Res Comput 18(6):9–11
9. Beresford AR, Bacon J (2006) Intelligent transportation systems. IEEE Pervasive Comput 5(4):63–67
10. Dimitrakopoulos G, Demestichas P (2010) Intelligent transportation systems. IEEE Veh Technol Mag 5(1):77–84
11. Weiland RJ, Purser LB (2000) Intelligent transportation systems, transportation in the new millennium
12. Miller J (2008) Vehicle–to–vehicle–to–infrastructure (V2V2I) intelligent transportation system architecture. IEEE intelligent vehicles Symposium, pp. 715–720
13. Wang F (2010) Parallel control and management for intelligent transportation systems: concepts, architectures, and applications. IEEE Intell Transp Syst 11(3):630–638
14. Chen M, Wan J, Li F (2012) Machine-to-machine communications: architectures, standards, and applications. KSII Trans Internet Inf Syst 6(2):480–497
15. Li LI, Liu YA, Tang BH (2007) SNMS: an intelligent transportation system network architecture based on WSN and P2P network. J China Univ Posts Telecommun 14(1):65–70
16. Suo H, Wan J, Huang L, Zou C (2012) Issues and challenges of wireless sensor networks localization in emerging applications. In: Proceedings of 2012 international conference on computer science and electronic engineering, Hangzhou, China, pp 447–451
17. Wan J, Li D (2010) Fuzzy feedback scheduling algorithm based on output jitter in resource–constrained embedded systems. In Proceedings of international conference on challenges in environmental science and computer engineering, Wuhan, China, pp 457–460
18. Chen M, Gonzalez S, Zhang Q, Leung VC (2010) Code-centric RFID system based on software agent intelligence. IEEE Intell Syst 25(2):12–19
19. Tseng YC, Kuo SP, Lee HW, Huang CF (2004) Location tracking in a wireless sensor network by mobile agents and its data fusion strategies. Comput J 47(4):448–460
20. Zou C, Wan J, Chen M, Li D (2012) Simulation modeling of cyber-physical systems exemplified by unmanned vehicles with WSNs navigation. In Proceedings of the 7th international conference on embedded and multimedia computing technology and service, Gwangju, Korea, pp 269–275
21. Chen M, Leung V, Mao S, Kwon T (2009) Receiver-oriented load-balancing and reliable routing in wireless sensor networks. Wirel Commun Mob Comput 9(3):405–416
22. Wan J, Yan H, Suo H, Li F (2011) Advances in cyber-physical systems research. KSII Trans Internet Inf Syst 5(11):1891–1908
23. Chen M, Gonzalez S, Zhang Y, Leung V (2009) Multi-agent itinerary planning for wireless sensor networks. Qual Serv Heterogen Netw 22:584–597
24. Wu FJ, Kao YF, Tseng YC (2011) From wireless sensor networks towards cyber physical systems. Pervasive Mob Comput 7(4):397–413

25. Liu J, Wang Q, Wan J, Xiong J (2012) Towards real-time indoor localization in wireless sensor networks. In: Proceedings of 12th IEEE international conference on computer and information technology, Chengdu, China, pp 877–884
26. Yan H, Wan J, Suo H (2011) Adaptive resource management for cyber–physical systems. In: Proceedings of international conference on mechatronics and applied mechanics, HongKong, pp 747–751

Improving Spectator Sports Safety by Cyber-Physical Systems: Challenges and Solutions

Hehua Yan, Zhuohua Liu, Jiafu Wan and Keliang Zhou

Abstract With the support of wireless networking, cloud computing, and advanced control technology, cyber-physical systems (CPS) applications are gradually becoming a reality. However, different applications (e.g., spectator sports safety) are confronted with diverse challenges and issues. In this article, we innovate an autonomous dynamic spatial panoramic video surveillance system that is a typical CPS to provide effective technologies and systems to protect spectators in safety and security. We first introduce the system model and components, and then propose the methodologies for quality of service (QoS) improvement from the following aspects: distributed real-time panoramic video stitching, autonomous unmanned aerial vehicles, autonomous unmanned ground vehicles, and wireless networking. The proposed methodologies can help the system design and improve the QoS.

Keywords Cyber-physical systems · Wireless networking · Unmanned aerial vehicles · Unmanned ground vehicles · Challenges

H. Yan · Z. Liu (✉) · J. Wan
School of Electrical Engineering, Guangdong Jidian Polytechinc, Guangzhou, China
e-mail: cnhope@tom.com

H. Yan
e-mail: hehua_Yan@126.com

J. Wan
e-mail: jiafu_wan@ieee.org

K. Zhou
College of Electrical Engineering and Automation, Jiangxi University of Science and Technology, Ganzhou, China
e-mail: nyzkl@sina.com

Y.-M. Huang et al. (eds.), *Advanced Technologies, Embedded and Multimedia for Human-centric Computing*, Lecture Notes in Electrical Engineering 260, DOI: 10.1007/978-94-007-7262-5_83, © Springer Science+Business Media Dordrecht 2014

Introduction

The importance of spectator sports safety and security is often overlooked in research and education. The large sports events may face various threats. For example, recall the bombing threat in the 2004 Summer Olympic Games [1] and the Munich massacre during the 1972 Olympics [2]. Other threats can come from accidents, such as fire and gas explosions. Confronting these challenges, sports spectators are hard to be secured with the capabilities of traditional sport security management mainly relying on human labor. In our view, an interdisciplinary effort is required. This paper takes the multidisciplinary research initiative for a cyber-physical system (CPS) of dynamic spatial video surveillance to effectively protect spectators.

In recent years, the research on CPS has made much progress in prototype testbed, resource allocation, control methods, system model, etc., [3–11]. Depending on the previous research results, we innovate an autonomous dynamic spatial panoramic video surveillance CPS to provide effective technologies and systems to protect spectators in safety and security. This system consists of two primary instruments: (1) an autonomous spatial video surveillance platform consisting of heterogeneous unmanned physical objects (unmanned aerial vehicles (UAVs) and unmanned ground vehicles (UGVs)) networked by wireless communication, and (2) a distributed multi-tier hierarchical real-time panoramic video stitching solution overlaying across the spatial surveillance platform. As shown in Fig. 1, the autonomous platform is driven by surveillance missions and collaborates with the distributed video stitching to accomplish the surveillance missions. The system communication and the video are secured with confidentiality and integrity.

Fig. 1 Conceptual system

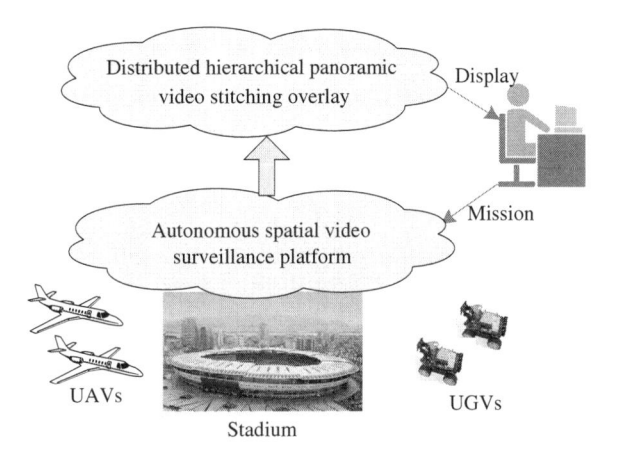

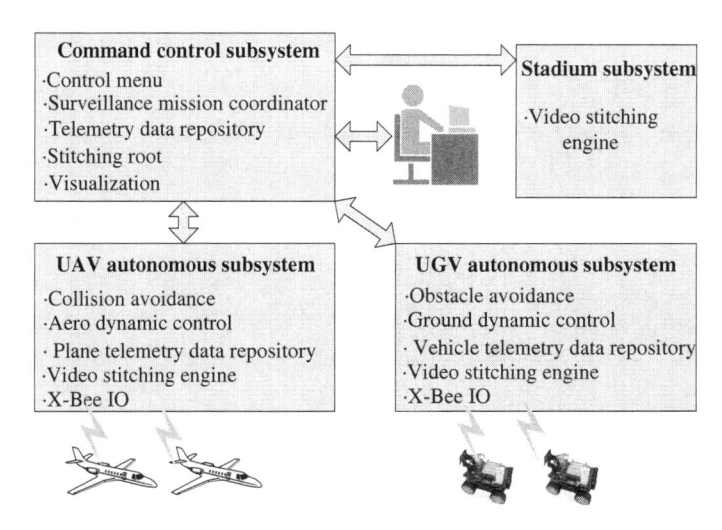

Fig. 2 System model

System Model and Components

A loosely conceptualized system model of the proposed CPS is shown in Fig. 2. The model consists of three major subsystems: command control subsystem, UAV autonomous subsystem and UGV autonomous subsystem, and one minor subsystem: stadium subsystem. Actually, the whole system is an autonomous one with three major subsystems being semi-autonomous in that:

- The command control subsystem has the top-level mission-based intelligence, but not to every detail, to coordinate the UAVs and UGVs.
- The UAV and UGV subsystems have their own low-level intelligence on autonomous coordination to accomplish their assigned missions, but they have to follow the control and missions from the command control subsystem.
- They all can revise their control during a mission based on the closed-loop feedback of physical surroundings.
- Command Control Subsystem: This subsystem provides the interface to the commander or system administrator and is the "brain" of the entire system in coordinating UAVs and UGVs subsystems to fulfill a surveillance mission. The commander initiates a surveillance mission through the *control menu*, for example, the concerned surveillance area/object is specified. The mission can be interpreted into a set of parameters such as the target Geo location, concerned objects, expected video resolution, etc. Then, the mission is delivered to the *surveillance mission coordinator* that computes the preliminary paths for the UAVs and UGVs to move based on their current physical information such as battery (power) and current Geo location from the *telemetry data repository*. The computed path data are delivered via wireless links to the UAV and UGV

autonomous subsystems for further process and control. The *telemetry data repository* polls the current physical system information of UAVs and UGVs from their subsystems. Thus, it also supports the visualization of the UAVs and UGVs on the map with Google Maps KML [12] to the commander. The *stitching root* stitches the panoramic videos from the UAV and UTV subsystems into the final panoramic real-time video and visualizes it to the commander for surveillance.

- UAV Autonomous Subsystem: This is a distributed system working across all UAVs. When the UAV subsystem receives UAV control messages with path data from the *command control subsystem*, the *collision avoidance* inspects any potential collision and makes necessary updates of the data before passing them to the *aero dynamic control* module. It keeps running with the continual updates of the latest UAV physical system information in flights. The *aero dynamic control* computes the aero dynamic parameters for UAVs and then notify all UAVs through the *X-Bee IO* that is a tri-band wireless interface on 2.4/5/60 GHz. Meanwhile, all UAVs continually update their flight physical information such as location, speed and battery to the *plane telemetry data repository* in the subsystem via the *X-Bee IO* so that the system is up-to-date of their status. In addition, the surveillance video captured by the UAVs will be hierarchically stitched by the *video stitching engine* in a distributed mode. The eventual stitched panoramic video of the UAV subsystem is pushed to the *command control subsystem* for a final stitching.
- UGV Autonomous Subsystem: This subsystem works similarly to the UAV subsystem except that the UGVs have to avoid obstacles in their paths and have different control and mechanical systems for their motions on the ground. *X-Bug IO* is also a tri-band wireless interface that exchanges the messages containing UGV physical system information or surveillance video.
- Stadium Subsystem (Immobile Camera): This subsystem refers to the traditional video surveillance system deployed in the venue or its surroundings. This subsystem does not require any mechanical control as the UAV and UGV systems do. It accepts inputs such as video resolutions from the *command control subsystem* and runs the distributed *video stitching engine* to generate the panoramic real-time video of its coverage to command control office. It might also provide wireless communication replay to the UAVs and/or UGVs when possible.

Distributed Real-Time Panoramic Video Stitching

The stitching solution currently only runs in standalone mode on a single computer that is short of stitching videos from a large number of surveillance cameras in events like sporting games. To extend the standalone stitching to a distributed real-time video stitching solution, we address a couple of challenges.

Fig. 3 Synchronization on stitching

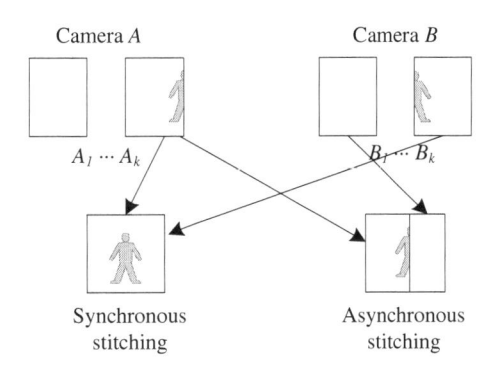

Synchronization: Synchronization between video frames to be stitched is critical to the stitching result. Stitching asynchronous video frames may lead to the displacement of objects, which is illustrated in Fig. 3 where two video streams come from neighboring surveillance cameras A and B. Suppose that an object is moving from A to B at the time of frame k. If both camera videos arrive synchronously at the stitching point, frame A_k is stitched with frame B_k and the object will appear naturally in the stitched frame. Otherwise, in an asynchronous case where frame A_k is stitched with frame B_l that arrives late, partial of the moving object disappears in the stitched result as in the right bottom on the figure. In the distributed video stitching, the synchronization of video frames from different cameras is significantly challenged even if the cameras are tightly synchronized because of

- The different delays that video streams may undergo when transmitted over the network.
- The different processing time at stitching nodes at different tiers.

We consider a delay-tolerant stitch-or-store solution to address the asynchronous issue. At stitching nodes, the video frames are verified of the synchronization in a two-step procedure before stitched. The timestamps of the frames are first inspected. When the system time of the cameras is tightly synchronized, identical timestamps hint synchronization. Otherwise, mobile objects in the overlapping coverage among frames are examined. Synchronous frames should have the same mobile object at the same place in their overlapped coverages at the same time. If the frames are synchronous, then they are stitched. If only some frames are synchronous, these frames are first stitched and stored with other asynchronous frames locally for a short period to wait for other frames synchronous to this stitching set.

Scalability: Another challenge is the scalability of the stitching due to the large number of cameras in the spatial surveillance platform. We could run the standalone video stitching solution on a central powerful server such as a high performance cluster to achieve large scale stitching. But, we have to transmit all videos across the network to the central computing server. As a result, the huge amount of data will definitely overwhelm the network and the resulted traffic congestion will lead to severely asynchronous videos. Meanwhile, the overlapping information in video frames wastes the network bandwidth. Instead of transmitting all videos to a central server for processing, we will take a strategy similar to clouding computing that pushes the computation to the fronts. This strategy leads to a distributed multi-tier hierarchical stitching architecture. The camera nodes form into clusters and each cluster has an elected cluster lead that stitches the local panoramic video from the cameras in the same cluster. Then, the cluster leads at the same tier form into higher-tier clusters and do the election and stitching again. This process repeats till reaching the stitching root, where the final panoramic video is generated. The size of a cluster depends on the computation resources of the embedded devices available at the distributed stitching nodes.

Autonomous UAVs

At this time, we already know how to fly UAVs through a set of predetermined or uploaded waypoints. The proposed CPS requires UAVs fly autonomously and collaboratively to accomplish tasks in taking and stitching videos. The flight paths are often determined by the surveillance tasks. In this paper, we may design task-driven algorithms that allow genuine autonomous flight of a large number of cooperative UAVs.

One challenge is that the *Multiplex Easystar* UAV is not appropriate for surveillance purpose due to its high speed. For surveillance, we often need to browse slowly or even stop to focus on suspicious events. For such constraints, a multi-copter (quad) is more appropriate. The DIY Drone community developed an open source auto-pilot for a quadcopter: the *ArduCopter*. The autopilot uses the *Arduino* microcontroller. Given Auburn familiarity with this platform and the versatility of the *ATTRACT-ROS* software architecture, it is easy to adapt the collision avoidance algorithms designed initially for the *Multiplex Easystar*. The maneuverability of the *ArduCopter* greatly simplifies the problem of collision avoidance in that: (1) the speed can vary, (2) it can stop, and (3) it can take sharp turns. With this architecture, *ArduCopters* can be assigned surveillance missions with physical flexibility for collisions avoidance. Another challenge is in the UAV capability. The current *ArduCopter* can fly only about 10 min. We need to increase the flight time and the payload capacity to carry a camera.

Autonomous UGVs

The UGVs in the proposed system face various challenges in the populated venues. The largest challenge comes from the mobility of human beings in sports games. A UGV has to detect and avoid any obstacles in its path. To this day, it is not a problem anymore for a robot to avoid static obstacles in indoor or outdoor environments with many-year research. However, the situation becomes extremely complicated at a sport stadium with a big crowd: people often walk around. It is challenging for a UGV to act in such an environment where obstacles are mobile. An appropriate moving path of a UGV must be dynamically computed based on the mobility of obstacles, the stadium environment and the sensibility of the UGV. New algorithms are required to provide high accuracy, effectiveness and efficiency in UGVs' autonomous driving. We may address this problem by exploiting the vector field based (VFB) guiding model to UGVs with the global knowledge from the UAVs in the sky in the spatial platform. VFB extends the potential based guiding (PBG) model that is widely adopted in obstacle avoidances for robots. In PBG, a robot divides scanned area into Z scanned lines that point to different directions. The main idea of the PBG is to repulse (or attract) the robot from or to the obstacles. However, this model relies only on the sensing information obtained by robot itself, which easily leads to a mistaken decision due to the lack of global knowledge. VFB extends the PBG model in such a way that a direction can be defined at every point of the potential surface in PBG model. Therefore, the global information discovered by the UAV system in our spatial platform can be modeled into the UGV's decision-making process. The globe information such as human density, traffic pattern, and moving speed of a swarm can be useful in computing the potential best path data. The data will then be fed into the VFB model so that UGVs can make a motion decision based on both the local and global information in their movement.

Another challenge is in the UGV mechanical capability. In the application of spectator sports safety and security, UGVs confront the complexity of terrestrial surroundings of a venue that is normally in an urban area where there are various physical obstacles such as curbs and steps. In addition, the battery of the Lunabot does not last long. In the surveillance, this is a potential issue.

Wireless Networking

Networking protocols are necessary to drive the wireless communication. We expect to use the IEEE 802.11 as the MAC/PHY protocols. However, special routing protocols are required to: (1) forward the coordination messages of the spatial platform, especially between the command control and the UAV and UGV systems and (2) deliver the surveillance videos for stitching or visualization. The routing faces the challenges of (1) tri-band wireless communication, (2) directional antennas in 60 GHz for power gain, and (3) special hierarchical topologies in distributed video stitching. Now, we have extensive expertise on multi-hop wireless network routing and design and develop a geographical routing protocol to support the communication of the coordination across the system, since the system components have GPS equipped [13, 14]. In addition, a tree-like routing protocol should be developed to support the distributed video stitching and transmission over the hierarchical multi-tier topology. Since the wireless networking is based on tri-band wireless communication, the routing protocols should consider the impact of multiple bands.

Conclusions

Recently, CPS has emerged as a promising direction to enrich the interactions between physical and virtual worlds. This paper integrates the traditional disciplines into a complex system with the clear goal of autonomous dynamic spatial panoramic video surveillance to address the challenges. In order to protect spectators in safety and security, we introduce the system model and components of CPS, and propose the methodologies for QoS improvement from the following aspects: distributed real-time panoramic video stitching, autonomous unmanned aerial vehicles, autonomous unmanned ground vehicles, and wireless networking. We have proposed several research challenges and solutions to encourage more insight into this meaningful application.

Acknowledgments The authors would like to thank the National Natural Science Foundation of China (No. 61262013), the Natural Science Foundation of Guangdong Province, China (No. S2011010001155), and the High-level Talent Project for Universities, Guangdong Province, China (No. 431, YueCaiJiao 2011) for their support in this research.

References

1. Taylor T, Toohey K (2006) Perceptions of terrorism threats at the 2004 Olympic Games: implications for sport events. J Sport Tourism 12(2):99–114
2. Wikipedia http://en.wikipedia.org/wiki/munichmassacre
3. Wan J, Chen M, Xia F, Li D, Zhou K (2013) From machine-to-machine communications towards cyber-physical systems. Comp Sci Inf Syst. doi:10.2298/CSIS120326018W
4. Chen M, Wan J, Li F (2012) Machine-to-machine communications: architectures, standards, and applications. KSII Trans Internet Inf Syst 6(2):480–497
5. Wan J, Yan H, Li D, Zhou K, Zeng L (2013) Cyber-physical systems for optimal energy management scheme of autonomous electric vehicle. Comput J. doi:10.1093/comjnl/bxt043
6. Wan J, Yan H, Suo H, Li F (2011) Advances in cyber-physical systems research. KSII Trans Internet Inf Syst 5(11):1891–1908
7. Suo H, Wan J, Huang L, Zou C (2012) Issues and challenges of wireless sensor networks localization in emerging applications. In: Proceedings of 2012 international conference on computer science and electronic engineering, Hangzhou, China, pp 447–451, March 2012
8. Zou C, Wan J, Chen M, Li D (2012) Simulation modeling of cyber-physical systems exemplified by unmanned vehicles with WSNs navigation. In: Proceedings of the 7th international conference on embedded and multimedia computing technology and service, Gwangju, Korea, pp 269–275, Sept 2012
9. Chen M, Gonzalez S, Zhang Q, Leung V (2010) Code-centric RFID system based on software agent intelligence. IEEE Intell Syst 25(2):12–19
10. Suo H, Wan J, Li D, Zou C (2012) Energy management framework designed for autonomous electric vehicle with sensor networks navigation. In: Proceedings of the 12th IEEE international conference on computer and information technology, Chengdu, China, pp 914–920, Oct 2012
11. Shi J, Wan J, Yan H, Suo H (2011) A survey of cyber-physical systems. In: Proceedings of the international conference on wireless communications and signal processing, Nanjing, China, pp 1–6, Nov 2011
12. Google Inc. http://code.google.com/apis/kml/documentation/whatiskml.html

13. Wang X, Vasilakos A, Chen M, Liu Y (2012) A survey of green mobile networks: opportunities and challenges. Mob Netw Appl 17(1):4–20
14. Chen M, Leung V, Mao S, Li M (2008) Cross-layer and path priority scheduling based real-time video communications over wireless sensor networks. In: Proceedings of IEEE vehicular technology conference, pp 2873–2877

MapReduce Application Profiler

Tzu-Chi Huang, Kuo-Chih Chu, Chui-Ming Chiu and Ce-Kuen Shieh

Abstract MapReduce is a programming model popularized by Google to process large data in clusters and has become a key technology on cloud computing nowadays. Due to the feature of simplicity, MapReduce attracts many application developers to develop related applications. However, MapReduce currently has few solutions to help application developers with the tasks of profiling their applications. In this paper, MapReduce Application Profiler (MRAP) is proposed to facilitate profiling MapReduce applications in clusters.

Keywords MapReduce · Cloud computing · MRAP · Profiler

Introduction

MapReduce [1] is a programming model popularized by Google to process large data in clusters. MapReduce has become a key technology on cloud computing to make an application easily utilize resources in computers. Today, MapReduce has been widely used by many public services such as Google Search Engine, Yahoo

T.-C. Huang (✉) · K.-C. Chu
Department of Electronic Engineering, Lunghwa University of Science and Technology,
Guihan, Taiwan, Republic of China
e-mail: tzuchi@mail.lhu.edu.tw

K.-C. Chu
e-mail: kcchu@mail.lhu.edu.tw

C.-M. Chiu · C.-K. Shieh
Institute of Computer and Communication Engineering, Department of Electrical
Engineering, National Cheng Kung University, Tainan, Taiwan, Republic of China
e-mail: cmchiu@ee.ncku.edu.tw

C.-K. Shieh
e-mail: shieh@ee.ncku.edu.tw

Y.-M. Huang et al. (eds.), *Advanced Technologies, Embedded and Multimedia*
for Human-centric Computing, Lecture Notes in Electrical Engineering 260,
DOI: 10.1007/978-94-007-7262-5_84, © Springer Science+Business Media Dordrecht 2014

Search Engine, Amazon EC2, and Facebook to deal with a huge amount of data. In the future, MapReduce definitely will have more and more related applications.

Due to the feature of simplicity, MapReduce allows application developers to easily develop applications in clusters even though they do not have many experiences on developing distributed and parallelized applications. MapReduce relies on the runtime system to automatically distribute an application over computers in clusters, provide the application with input data at runtime, and collect outputs of the application from computers in clusters. All MapReduce needs is a Map function and a Reduce function developed by application developers to implement the functionality of their applications.

Since simplicity is the main contribution of MapReduce to the development of applications on cloud computing, MapReduce should offer application developers a way to easily profile their applications as well. MapReduce should care about that application developers may have very few experiences on developing distributed and paralleled applications in clusters. Furthermore, MapReduce should care about that application developers may neither know how the runtime system serves their applications nor understand how to improve their applications in order to get a better performance. However, MapReduce currently has few solutions specifically designed to help application developers with the tasks of profiling their applications.

In this paper, MapReduce Application Profiler (MRAP) is proposed to facilitate profiling MapReduce applications in clusters. MRAP can collect runtime information about the application execution progress at runtime and save runtime information in files compatible to Microsoft Excel. Accordingly, MRAP allows application developers to quickly profile their applications, e.g., doing statistics or drawing charts about the application execution progress. MRAP provides application developers with runtime information about the application execution progress, so they can easily improve their applications to get a better performance. In this paper, MRAP demonstrates the performance metrics of Word Count in a generic runtime system as the example.

This paper is organized as follows. Section Background reviews MapReduce as the background. Section MapReduce Application Profiler (MRAP) addresses MRAP. Section Implementation introduces the MRAP implementation. Section Demonstration has a demonstration of MRAP with Word Count. Section Conclusions concludes this paper.

Background

MapReduce is a programming model relying on its runtime system to distribute an application over computers in clusters, provide the application with input data at runtime, and collect outputs of the application from the computers in clusters. For serving an application, MapReduce needs application developers to develop a Map function (a.k.a. Mapper) and a Reduce function (a.k.a. Reducer) to implement the functionality of their applications. Accordingly, MapReduce divides the application execution progress roughly into the Map phase and the Reduce phase.

At the Map phase, the runtime system reads input files from local disks of a master node, i.e., a computer responsible for controlling the entire application execution progress, or certain distributed file systems such as the Google File System (GFS) [2] or the Hadoop Distributed File System (HDFS) [3] specifically designed for cloud computing. The runtime system distributes Mappers over computers in clusters according a certain node selection algorithm, e.g., finding the idle ones to run the Mappers, and provides the Mappers with contents of input files. After Mappers process contents of input files and generate intermediate data in a format of key and value pairs, the runtime system enters the Reduce phase. At the Reduce phase, the runtime system chooses suitable computers to run Reducers and forwards the intermediate data from Mappers to Reducers. Finally, the runtime system collects outputs from Reducers as the result of the application. Because the runtime system manages Mappers and Reducers of applications at runtime on behalf of application developers, application developers can easily develop applications on cloud computing by focusing on the development of Mappers and Reducers.

Word Count [1] is a canonical application often used to explain MapReduce on cloud computing. Word Count uses a Mapper to process input data offered by the runtime system and generate a key and value pair as intermediate data for each word found in input data, e.g., "word 1". When processing intermediate data, Word Count uses a Reducer to merge the key and value pairs generated by a Mapper. Word Count may count words by summing up their values in a Reducer and outputs partial results to the runtime system. Finally, Word Count waits the runtime system to collect all partial results from Reducers to get the final result of the application.

MapReduce Application Profiler (MRAP)

Overview

MRAP has an overview in Fig. 1 where a cluster is composed of a master node and several slave nodes. MRAP uses a master node to not only do the works as the master node in other MapReduce systems but also accept runtime information reported by slave nodes. MRAP uses slave nodes to not only execute tasks dispatched by the master node, e.g., running a Mapper or Reducer, but also collect runtime information of the runtime system and the Operating System (OS) at local nodes. In this paper, MRAP refers to information about the application execution progress in a slave node as runtime information, e.g., CPU utilization, number of Mapper instance, number of Reducer instance, intermediate data size, and quantity

Fig. 1 MRAP overview

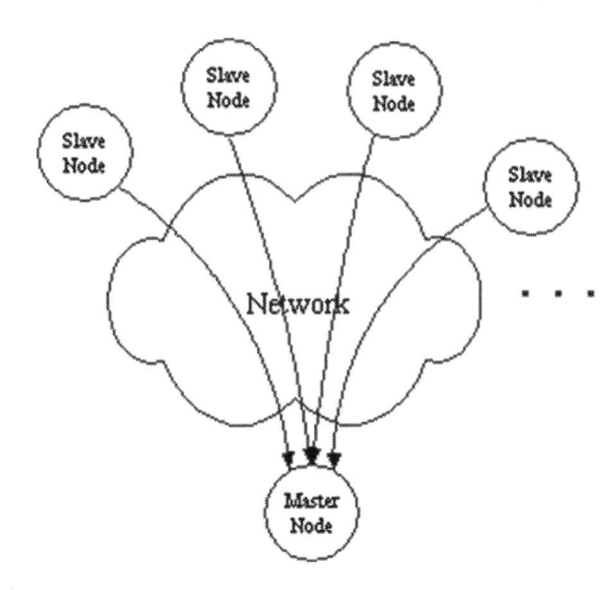

of network bandwidth consumption. At the end of the application, MRAP allows application developers to use Microsoft Excel to examine runtime information in files saved by the master node.

System Components

MRAP has components in Fig. 2. MRAP has the Runtime System to MRAP (RS-MRAP) interface between the runtime system and itself, and the MRAP to Operating System (MRAP-OS) interface between the Operating System (OS) and itself. MRAP exports the RS-MRAP interface to the runtime system and communicates with the runtime system in order to get the MapReduce-related information in the runtime system, e.g., number of Mapper instance, number of Reducer instance, and intermediate data size, because the runtime system is responsible for managing the execution of Mappers and Reducers and intermediate data at runtime. MRAP utilizes the MRAP-OS interface (usually embodied with OS calls) to communicate with the OS in order to get the OS-related information in the OS, e.g., CPU utilization and quantity of network bandwidth consumption. Through the MRAP-OS interface, MRAP also can report runtime information to the master node via the network service, collect runtime information from slave nodes via the network service, and save runtime information to files via the file system.

MRAP uses Information Reporter to periodically report runtime information of the local node to the master node via the network service in the OS. MRAP uses Information Recorder to collect runtime information from salve nodes via the

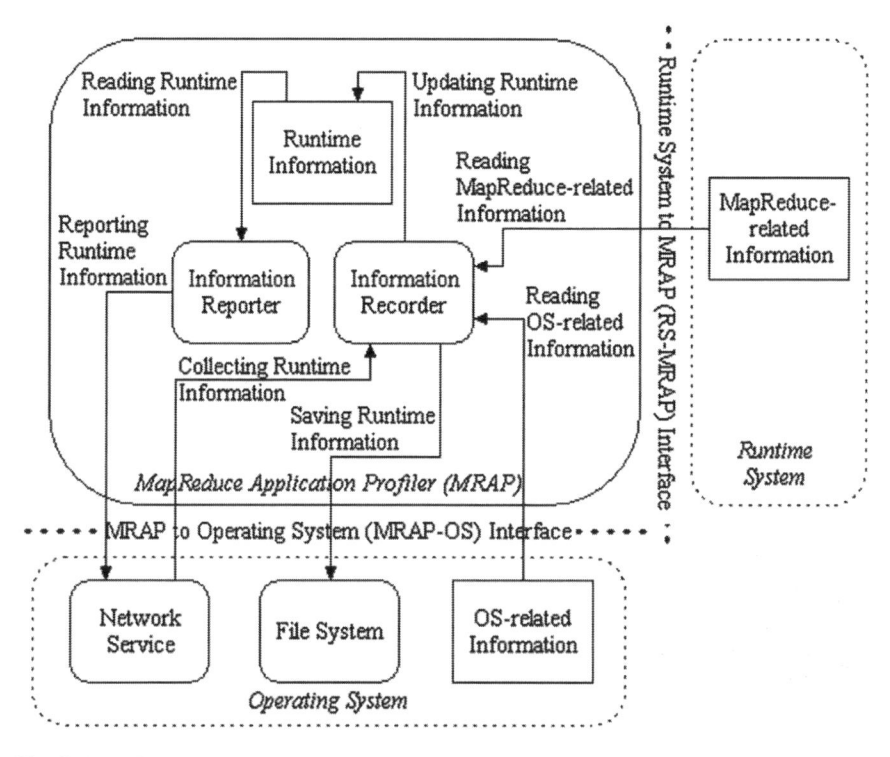

Fig. 2 MRAP components

network service in the OS, save runtime information to files via the file system, read OS-related information in the OS, and read MapReduce-related information in the runtime system. Besides, MRAP uses Information Recorder to update runtime information that is the collection of the OS-related information and the MapReduce-related information. While MRAP usually caches runtime information in the memory in slave nodes to facilitate reporting runtime information to the master node periodically, it saves runtime information to files in the master node directly once receiving runtime information reported by slave nodes from networks.

Implementation

We use the C language to implement MRAP as a library that can be linked to the runtime system. At the initial time of MRAP, we create a thread to periodically query not only the runtime system about number of Mapper instance, number of Reducer instance, and intermediate data size, but also the OS about CPU utilization and quantity of network bandwidth consumption. Next, we arrange the

MapReduce-related information and the OS-related information in a Type-Length-Value (TLV) format [4] and use the UDP socket library to send them to a configurable IP address as the master node. Currently, we set the interval to 1 s for MRAP in a slave node to periodically report runtime information to MRAR in the master node.

For a proof of concept, we implement a runtime system capable of reading input files from disks in the master node, finding the idle nodes to run Mappers and Reducers of an application, automatically feeding Mappers with input files downloaded from the master node, forwarding intermediate data files from Mappers to Reducers on demand, and collecting outputs produced by Reducers from all computers in clusters as the result of the application. Next, we link MRAP to the runtime system and make it periodically query the runtime system about the MapReduce-related information through subroutine calls. Besides, we make MRAP query the OS about the OS-related information through the Windows Management Instrumentation (WMI) APIs [5], because Windows Server 2003 is our current platform.

Demonstration

We construct a cluster with 9 computers connected to each other via Gigabit Ethernet. We setup an AMD Athlon II X4 620 CPU (2.6 GHz) and 4 GB DDR2 RAM in each of the computers. We install a runtime system and MRAP on Windows Server 2003 at each computer. We select one of the 9 computers as the master node while selecting the other computers as slave nodes. We prepare 210 input files having 64 MB in the master node in order to test Word Count, the canonical application widely used to observe MapReduce runtime systems. Accordingly, we give Word Count a total workload of 13,440 MB (i.e., 210 × 64 MB). We use MRAP to profile Word Count and Microsoft Excel to draw charts about various metrics in runtime information as the demonstration in this paper.

We show CPU utilization of all slave nodes in Fig. 3. We observe that Word Count roughly can achieve 60 % CPU utilization in all slave nodes in the runtime

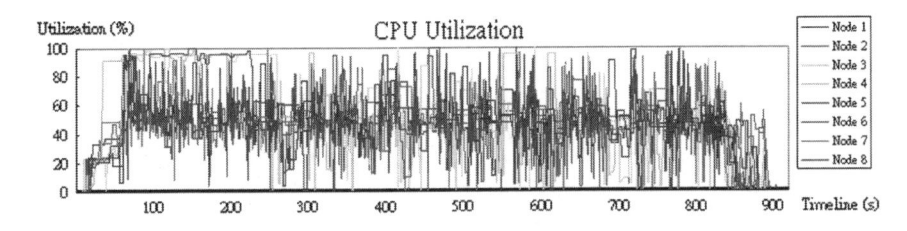

Fig. 3 CPU utilization

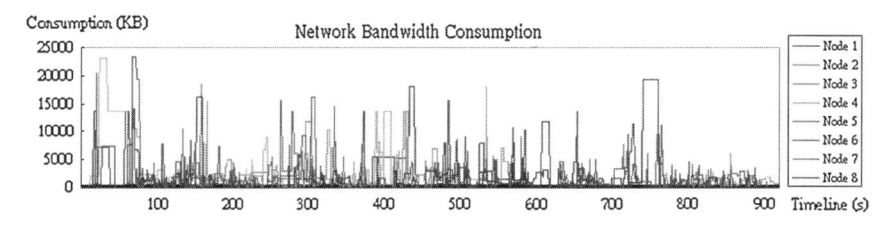

Fig. 4 Network bandwidth consumption

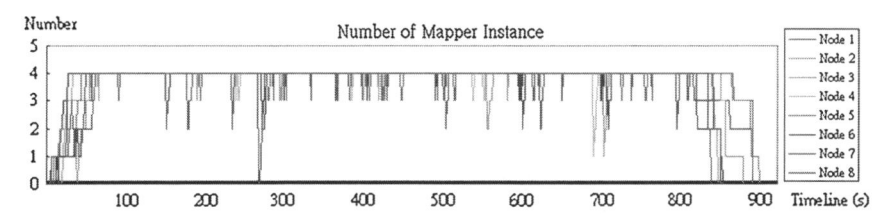

Fig. 5 Number of mapper instance

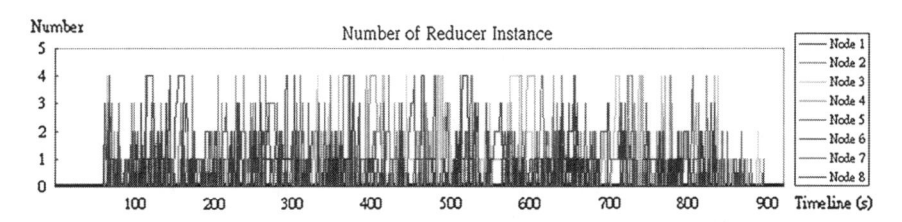

Fig. 6 Number of reducer instance

system. In other words, we use CPU utilization in runtime information offered by MRAP to know that Word Count is not an application costing much CPU time.

In Fig. 4, we observe network bandwidth consumed by each slave in the cluster. We note that network bandwidth consumption in Fig. 4 corresponds to CPU utilization in Fig. 3 because Mappers have to receive input files downloaded by the runtime system from the master node and because Reducers have to receive intermediate data forwarded by the runtime system from Mappers to Reducers before any computation can be made.

With the help of MRAP, we can clearly observe the number of Mapper instances in the runtime system. According to Fig. 5, we observe that Word Count almost has 4 Mapper instances activated in each slave node during the entire application execution progress, which implies that the runtime system can keep the application busy at processing input data with Mappers.

In Fig. 6, we observe the number of Reducer instances in the runtime system. We note that the average number of Reducer instances in each slave node is 2,

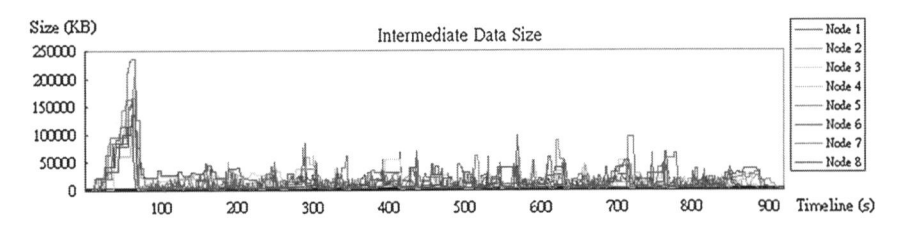

Fig. 7 Intermediate data size

which implies that the runtime system does not use a Reducer to process much intermediate data. Accordingly, we can deduce that Word Count does not generate much intermediate data in Mappers.

Finally, we observe intermediate date size of Word Count in Fig. 7. We confirm that Word Count does not generate much intermediate data, which corresponds to the observation in Fig. 6. Moreover, we note that Word Count generate much intermediate data before 60 s because the runtime system is busy at uploading input files to slave nodes and running Mappers to process input data without invoking Reducers to process intermediate data. We confirm the deduction because Fig. 6 shows that there is no Reducer instance before 60 s in the runtime system.

Conclusions

In this paper, we propose MapReduce Application Profiler (MRAP) to facilitate the profile of an application on cloud computing. We design MRAP to periodically query the runtime system and the Operating System (OS) about runtime information, the collection of the MapReduce-related information and the OS-related information. We design runtime information in MRAP to include information about CPU utilization, network bandwidth consumption, number of Mapper instance, number of Reducer instance, and intermediate data size in each slave node at runtime. We make MRAP collect and output runtime information to files compatible to Microsoft Excel, so application developers can easily profile their applications by observing the application execution progress with statistics or charts. We believe that application developers with the help of MRAP can easily improve their applications to get a better performance.

Acknowledgments We gratefully acknowledge the National Science Council of Taiwan for their support of this project under grant numbers NSC 100-2628-E-262-001-MY2 and 101-2219-E-006-005. We further offer our special thanks to the reviewers for their valuable comments and suggestions, which materially improved the quality of this paper.

References

1. Dean J, Ghemawat S (2008) MapReduce: simplified data processing on large clusters. In: Communications of the ACM, vol. 51. ACM, New York, USA
2. Ghemawat S, Gobioff H, Leung S-T (2003) The Google file system. In: ACM SIGOPS operating systems review, vol. 37. ACM, New York, USA, pp 29–43
3. Shvachko K, Kuang H, Radia S, Chansler R (2010) The Hadoop distributed file system. In: 2010 IEEE 26th Symposium on mass storage systems and technologies (MSST), IEEE Press, New York, USA, pp 1–10
4. Liu H, Zhang D (2012) A TLV-structured data naming scheme for content-oriented networking. In: 2012 IEEE international conference on communications (ICC), IEEE Press, New York, USA, pp 5822–5827
5. Jayaputera J, Poernomo I, Schmidt H (2004) Runtime verification of timing and probabilistic properties using WMI and NET. In: 30th Euro micro conference, IEEE Press, New York, USA, pp 100–106

Part IX
Cross Strait Conference on Information Science and Technology

Research on Micro-Blog Information Perception and Mining Platform

Xing Wang, Fei Xiong and Yun Liu

Abstract To predict the tendency of Micro-blog information dissemination, provide the early warning of the Internet emergencies, and contribute to the content security of micro-blog, the paper offers a platform for Micro-blog information perceiving and mining. This platform is an integration of Micro-blog data collection and processing module, topic detection and tracking module, user behavior analysis module, trend prediction module, etc. It could access and analyze micro-blog information automatically, leading a positive significance to grasp the emergencies on micro-blog. This paper puts forward methods based on the Latent Dirichlet Allocation (LDA) document clustering and hot topics prediction, which could analysis and predict the micro-blog data effectively, avoiding the problems in the traditional algorithm. Also, these methods have a higher accuracy for clustering and prediction.

Keywords Micro-blog · Data mining · Text clustering · Hot topic detection

X. Wang
China Information Technology Security Evaluation Center, Beijing, China
e-mail: wangx@itsec.gov.cn

F. Xiong (✉) · Y. Liu
School of Electronic and Information Engineering, Beijing Jiaotong University, Beijing 100044, China
e-mail: 08111029@bjtu.edu.cn

Y. Liu
e-mail: liuyun@bjtu.edu.cn

F. Xiong · Y. Liu
Key Laboratory of Communication and Information Systems, Beijing Municipal Commission of Education, Beijing Jiaotong University, Beijing 100044, China

Y.-M. Huang et al. (eds.), *Advanced Technologies, Embedded and Multimedia for Human-centric Computing*, Lecture Notes in Electrical Engineering 260, DOI: 10.1007/978-94-007-7262-5_85, © Springer Science+Business Media Dordrecht 2014

Introduction

Micro-blog, is an integrated, open Internet social networking services in the Web2.0 era. It is an system of immediate message releasing, similar to the blog. Users can use the following ways to announce its messages, such as mobile phones, instant messaging software, and external application programming interface,. The general message is less than 200 words (in practice, generally 140 words). The micro-blog makes more and more people participate in the Internet interactions. Many scholars at home and abroad, have carried out in-depth research on micro-blog, which includes the characteristics, propagation mechanism, development trends, and the social influences.

In recent years, scholars have conducted a lot of research on Internet information processing and mining. Han Ruixia introduced the characteristics and basic concepts of micro-blogging platform [1]. Kang Shulong studied the user network structure and degree distribution characteristics of Sina micro-blog [2]. Wang Rui explained empirically the relationship between the number of user's friends and the popularity level of the user [3]. In the area of text clustering and topic detection, Bruno et al. proposed a logarithmic maximum-likelihood criterion to calculate the word weights, instead of the traditional algorithm [4]. In the field of text vector extraction, Yang et al. improved the similarity formula, and presented a decay function of similarity calculation with time [5]. A lot of research methods for hot topic mining originate from Topic Detection and Tracking (TDT). TDT usually handles the document stream which dynamically changes with the news, rather than a set of texts on Internet [6]. Gabriel Piu et al. proposed a hot topic identification scheme based on free parameters, which determines the distribution of hot topics within a certain period of time by the characteristics of text [7]. In the field of user behavior and trend forecasting, Jiang Zhu extracted the features related to user's retweeting behavior, including content influence, network influence, and time influence [8]. Fei Xiong et al. analyzed the factors associated with the post propagation in online forums and combined these factors to predict the potential hot topics [9].

The recent research mainly aims at information on the Web pages, such as news and online forums. However, information on micro-blog and other social networks is more latent, and user behavior and social relationships are often heterogeneous and variable. Moreover, the text information of a message is very small. Therefore, traditional network monitoring systems based on text analysis cannot work well on micro-blog. We need to develop a content-aware and user behavior mining platform for micro-blog to provide public opinion analysis and early warning. This paper presents a micro-blog information perception and mining platform, which integrates the function of micro-blog user behavior mining, the social relationship mining, the topic detection, and so on. This work can help grasp the online emergencies, guide public opinion towards the right direction, and build a green and healthy online environment.

The remainder of the paper is organized as follows. In section Functional overview of the platform, we give the overview of the platform, and describe its functional modules. In section Research on key technology we introduce the key methods of our platform. We put forward a clustering algorithm for micro-blog short texts, and present a new method to detect hot topics in terms of ensemble learning. We conclude the paper in section Conclusions.

Functional Overview of the Platform

Our system includes the module of micro-blog data acquisition and processing, topic detection and tracking, user relationship analysis, user behavior mining, and propagation trend prediction. It can collect micro-blog information, provide information monitoring and early warning, discover potential hot topics, and detect user abnormal behavior. The functional modules of our platform are shown as Fig. 1.

1. Micro-blog data acquisition and processing

In consideration of the problems in micro-blog data acquisition and processing, this module puts forward a method for micro-blog data collection based on the

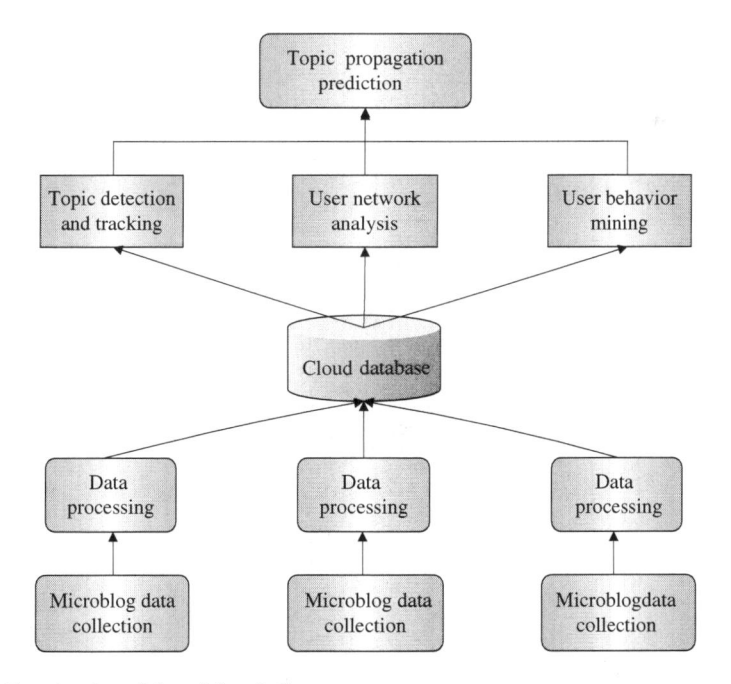

Fig. 1 Functional modules of the platform

combination of open API and page crawling and parsing technique. In the data collection module based on open API, the system obtains the data results that the API requests by parsing JSON object, and controls the query frequency of API by thread scheme. For the data outside the available scope of API, we use the web crawler to collect relevant page files by simulating the user's login process on micro-blog. By establishing different web page templates, the system can parse different categories of web pages downloaded by crawler. Therefore, our platform realizes the comprehensive and efficient data acquisition.

2. Micro-blog topic detection and tracking

This module can identify topics in the information flow of micro-blog media automatically by combining methods of information extraction and content clustering. Considering the feature of small text for micro-blog posts, this module builds a model of the topic expression and detection. The module achieves the mining of meaningful strings and the extraction of network characteristics, and can also identify different sentimental values of posts belonging to the same topic. Moreover, it can detect new and unknown topics including the event occurring first and related events, from the massive information on Internet.

3. Micro-blog user relationship analysis

The user relationship network on micro-blog is different from the traditional social network or interpersonal network. According to complex network theory, this module computes the macroscopic characteristics of network structure, and compares these characteristics in different network communities. This module can also find the heterogeneity of different virtual communities, and analyze user behavior and preferences.

4. Micro-blog user behavior mining

The module processes and analyzes user data, and finds out the potential rules of user behavior. According to the queuing theory, it researches on the interaction between users and micro-blog communities, and simulates user's behavior in the network, and reflects the characteristic distribution of user participation.

5. Micro-blog topic trend prediction

This module uses mathematics, information science, and other interdisciplinary approaches to model and forecast the diffusion process of information and evolution of public opinion. Meanwhile, in order to find general control strategy, this module can simulate the effect of different guide techniques on the opinion formation.

Research on Key Technology

Micro-Blog Short Text Clustering Based on LDA Algorithm

A topic consists of a series of related posts, containing the description of cause of the incident, the evolution, results and impact of the event. In our micro-blog perception and mining platform, micro-blog posts should be clustered into topics, and these posts about the same event are grouped in the same cluster. Only in this way, the impact of the social event on Internet can be measured, and the clustering also provides a basis for hot topic detection and evolutionary trend prediction. Topic clustering is implemented according to the similarity between the contents of posts.

However, all the micro-blog posts are short texts, which are no more than 140 words. After converting these posts to term vectors, any two posts rarely have overlap words. Therefore, the accuracy of the similarity between these posts is not high even if synonym substitution is used in the calculation. Brief contents make the cosine similarity between the posts belonging to the same topic very low. Here, we use the LDA algorithm to carry out document clustering. LDA is a generative model which treats a post as the mixing of multiple topics [10]. Each post has a different function of topic distribution, and therefore a post may belong to more than one topic. LDA does not need to compare the similarity between two posts, greatly improving the accuracy of clustering.

LDA model assumes that each word in posts is generated by topics, each topic corresponds to a word distribution function, and each post also corresponds to a topic function. The topic distribution of post i is defined as θ_i, and the initial distribution of θ_i is Dirichlet distribution for parameter α. The word distribution of topic k is φ_k, and the initial distribution of φ_k is also Dirichlet distribution for parameter β [11, 12]. With these two distribution functions, the post i can be constructed from these topics. For each word in the post, we choose a topic z_{ij} according to multinomial distribution of θ_i, and then in the light of the topic, choose a word w_{ij} following the multinomial distribution of $\varphi_{z_{ij}}$. We repeat this process until we reproduce the entire post. Therefore, if word t is assigned to topic k for n_k^t times, and then the posterior probability of words is

$$P(w|z, \varphi) = \prod_{k=1}^{K} \prod_{t=1}^{V} \left(\varphi_k^t\right)^{n_k^t} \tag{1}$$

K is the total number of topics, V is the total number of words in the lexicon. Similarly, if the topic k appears n_i^k times in the document i, then the posterior probability of topics is

$$P(z|\theta) = \prod_{i=1}^{M} \prod_{k=1}^{K} \left(\theta_i^k\right)^{n_i^k} \tag{2}$$

M is the total number of posts.

For each micro-blog post, we get its topic distribution θ. If the largest component of θ for a post is greater than 0.5, then the post is assigned to the topic corresponding to that component. If the largest component of θ is lower than 0.5, then the post is assigned to the first three topics ranking by probability of topic distribution. Before LDA clustering, we need to determine the total number of topics K based on validity index. Here, we calculate the number of topics by text databases [13]. Figure 2 illustrates the process of short text clustering for micro-blog posts.

Hot Topic Prediction Based on Ensemble Learning

The propagation extent of a micro-blog topic is closely related to the attraction of the topic itself, current active users' behavior, and the competition between the topics. During the process of micro-blog topic dissemination, there are new posts published about this topic continuously, and the most powerful mechanism to diffuse information is post retweeting based on user relationship. Thus, the unique way of micro-blog information retweeting, makes the topic propagation show different characteristics from traditional networks. The diffusion process is more

Fig. 2 Flow chart of LDA clustering for short texts

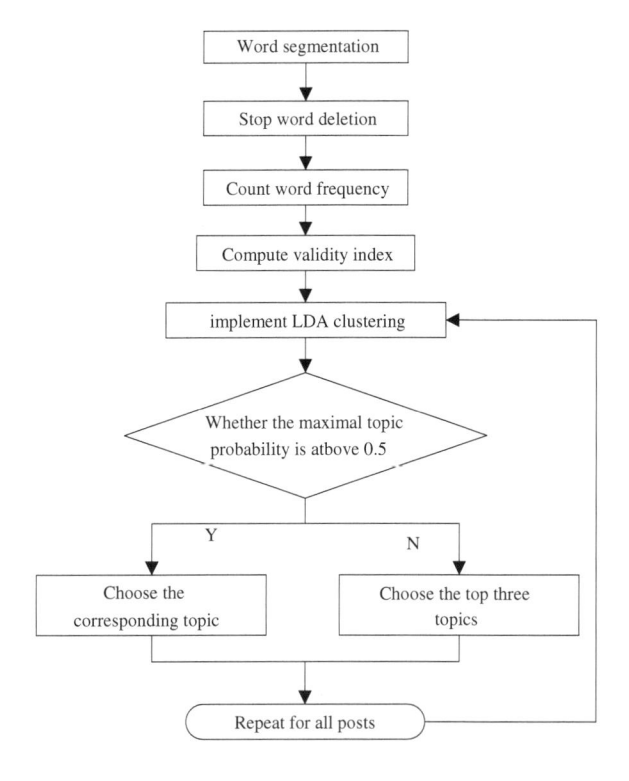

explosive and more capricious. The features of initial topic dissemination reflect the overall trend to a certain extent, and can be used as an indicator for judging whether a topic is popular. However, some topics absorb only a few users at the beginning, but after some influential users retweet it, it becomes a hot topic. The topic attracts numerous retweeting, and affects a large number of users in the network. Therefore, we need to extract multiple features of topic dissemination, and create a model to predict potential hot topics.

First, we implement short text clustering for micro-blog posts. Posts with similar contents are put into a cluster, and each cluster corresponds to a topic. Compared with topics about biology or tourism, topics of social issues are always more popular. The more popular a topic is, the more users it would attract. Generally, most topics cannot get a mass of attention, and will gradually die out, while a few topics suddenly become hot after coherent interaction. Now we have a question how to define a hot topic. There are a variety of definitions. Erzhong Zhou [14] defined hot topics by the number of related posts, the number of replies, and user participation. Zhongfeng Zhang [15] described the popularity of a topic by the total number and time evolutionary frequency of hot words that appear in a document. Here, we do not concern more about the detailed definition of hot topics, but only use the total number of participants as the popularity of a topic. If the number of participants is greater than 1,000, the topic is considered as a hot topic.

Now, according to the data in the initial δ hours of each topic, we extract the features related to topic dissemination, including the attractiveness of topic content, and user behavior. The initial data of a topic is related to its popularity. In the topic content influence, we calculate the total number of posts in the first δ hours, the total number of replies, the total times of retweeting, the number of participants, the current number of other topics, and the average frequency of retweeting. In all, we have 6 content features. User behavior also affects the propagation of topics. We can extract 5 user features, that is, the number of friends for post author, the number of followers, the average times of being retweeted for post author, the total times of being retweeted for the users who has retweeted the post, and the current number of active users.

According to the previously mentioned 11 features that are related to the popularity of posts, we use support vector machines to combine these features to generate a predictive model. The support vector machine is an optimal margin classifier, which uses a linear mapping, to find a hyperplane as the boundary that can divide the data into two classes. The support vector machine makes the data points nearest from the hyperplane in both sides, get greater spacing [16].

The larger δ is, the higher recall and accuracy of predicted results we can get. Considering the need of network monitoring, we hope to identify potential hot topics as soon as possible, so as to provide early warning of network emergencies. In practice, we usually set δ at $\delta < 8$ h.

In order to improve the performance of the model, now we use the bagging method of ensemble learning to deal with the training data set. If the number of samples in training sets D is n, the bagging method divides D into m subsets D_i,

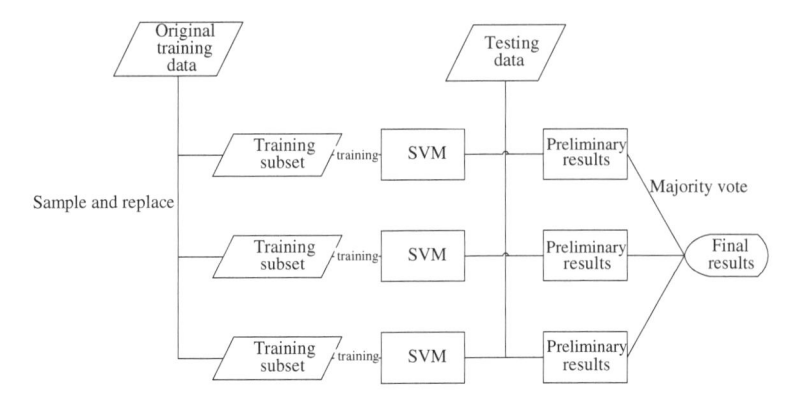

Fig. 3 Flow chart of ensemble learning process

and D_i includes n_i samples satisfying $n_i < n$ [17]. We uniformly sample data from D, and replace the elements randomly, to generate a subset D_i. We use each D_i respectively to train a model, and then use these trained support vector machines to predict new topic. Then we calculate the average or do majority vote of all the preliminary results to get our final prediction result. Using bagging algorithm to deal with the training data set can improve the robustness and the precision of the model, and avoid over fitting as well. The testing results indicate that the bagging method can, to some extent, improve the accuracy of the predicted results while keeping the recall almost unchanged. We sample data from the original training set. The original data set is divided into 3 groups, each of which is used to train a support vector machine separately. For a new topic, we use the three support vector machines to predict its popularity, and do majority vote for the 3 preliminary results. Figure 3 shows the ensemble learning process.

Conclusions

With the development of micro-blog and the explosive growth of information, the security of micro-blog content has attracted more and more attention. In recent years, the processing and mining of large amounts of data for micro-blog become a research hotspot. In this paper, we presented a micro-blog data perception and mining platform. It can automatically collect micro-blog data and analyze the data, detect hot topics and predict the diffusion trends of topics. Moreover, the platform shows a better performance on short text processing of micro-blog, and is able to predict hot topics effectively. Therefore, the system can provide early warning of online emergencies on micro-blog, and ensure the content security of micro-blog more forcefully.

Acknowledgments This work has been supported by the National Natural Science Foundation of China under Grant 61172072, 61271308, the Beijing Natural Science Foundation under Grant 4112045, the Research Fund for the Doctoral Program of Higher Education of China under Grant W11C100030, the Beijing Science and Technology Program under Grant Z121100000312024.

References

1. Han R (2010) The influence of micro-blogging on personal public participation. In: Proceeding(s) of 2010 IEEE 2nd symposium on web society, pp 615–618
2. Kang S, Zhang C (2010) Complexity research of massively micro-blogging based on human behaviors. In: Proceeding(s) of 2nd international workshop on database technology and applications, pp 1–4
3. Wang R, Jin Y (2010) An empirical study on the relationship between the followers' number and influence of micro-blogging. In: Proceeding(s) of the international conference on e-business and e-Government, pp 2014–2017
4. Pouliquen B, Steinberger R et al (2004) Multilingual and cross-lingual news topic tracking. In: Proceeding(s) of the 20th international conference on computational linguistics, pp 23–27
5. Yang Y, Pierce T, Carbonell J (1998) A study on retrospective and on-line event detection. In: Proceeding(s) of the 21st annual international ACM SIGIR conference on research and development in information retrieval, pp 28–36
6. Jin H, Schwartz R, Wall F (1999) Topic tracking for radio, TV broadcast, and newswire. In: Proceeding(s) of the DARPA broadcast news workshop, pp 199–204
7. Pui G, Fung C, Yu JX, Lu H (2005) Parameter free bursty events detection in text streams. In: Proceeding(s) of the 31st international conference on very large data bases, pp 181–192
8. Zhu J, Xiong F, Piao D, Liu Y, Zhang Y (2011) Statistically modeling the effectiveness of disaster information in social media. In: Proceeding(s) of the IEEE global humanitarian technology conference, pp 431–436
9. Xiong F, Liu Y, Zhu J, Lian J, Zhang Y (2012) Hot post prediction in BBS forums based on multifactor fusion. J Convergence Inf Technol 7(12):129–137
10. http://en.wikipedia.org/wiki/Latent_Dirichlet_allocation
11. Blei DM, Ng AY, Jordan MI (2003) Latent dirichlet allocation. J Mach Learn Res 3(4–5):993–1022
12. Farrahi K, Gatica-Perez (2011) Discovering routines from large-scale human locations using probabilistic topic models. ACM Trans Comput Logic 2(1):3
13. Can F, Ozkarahan EA (1990) Concepts and effectiveness of the cover-coefficient-based clustering methodology for text databases. ACM Trans Database Syst 15:483–517
14. Zhou E, Zhong N, Li Y (2011) Hot topic detection in professional blogs. Active Media Technol 6890:141–152
15. Zhang Z, Li Q (2011) QuestionHolic: hot topic discovery and trend analysis in community question answering systems. Expert Syst Appl 38(6):6848–6855
16. Suykens JAK, Vandewalle J (1999) Least squares support vector machine classifiers. Neural Process Lett 9(3):293–300
17. http://en.wikipedia.org/wiki/Bootstrap_aggregating

A New Algorithm for Personalized Recommendations in Community Networks

Xin Zhou, XinXiang Xing and Yun Liu

Abstract In a graph theory model, clustering is the process of division of vertices in groups, with a higher density of edges in groups than among them. In this paper, we introduce a new clustering algorithm for detecting such groups; we use it to analyze some classic social networks. The new algorithm has two distinguished features: non-binary hierarchical tree and the feature of overlapping clustering. A non-binary hierarchical tree is much smaller than the binary-trees constructed by most traditional algorithms; it clearly highlights meaningful clusters which significantly reduce further manual efforts for cluster selections. The present algorithm is tested by several bench mark data sets for which the community structure was known in advance and the results indicate that it is a sensitive and accurate algorithm for extracting community structure from social networks.

Keywords Clustering · Graph theory · Hierarchical tree · Social network

X. Zhou
China Information Technology Security Evaluation Center, Beijing, China
e-mail: zhoux@itsec.gov.cn

X. Xing · Y. Liu (✉)
School of Electronic and Information Engineering, Beijing Jiaotong University,
Beijing 100044, China
e-mail: liuyun@bjtu.edu.cn

X. Xing
e-mail: 10111028@bjtu.edu.cn

X. Xing · Y. Liu
Key Laboratory of Communication and Information Systems, Beijing Municipal
Commission of Education, Beijing Jiaotong University, Beijing 100044, China

Y.-M. Huang et al. (eds.), *Advanced Technologies, Embedded and Multimedia
for Human-centric Computing*, Lecture Notes in Electrical Engineering 260,
DOI: 10.1007/978-94-007-7262-5_86, © Springer Science+Business Media Dordrecht 2014

Introduction

Clustering is an important task for the discovery of community structures in networks. Its goal is to sort cases (people, things, events, etc.) into clusters so that the degree of association is relatively strong between members of the same cluster and relatively weak between members of different clusters. Webster [1] defines cluster analysis as "a statistical classification technique for discovering whether the individuals of a population fall into different groups by making quantitative comparisons of multiple characteristics". Various clustering algorithms have been proposed in the literature in many different scientific disciplines. Jain [2] broadly divided these algorithms into two groups: (1) hierarchical algorithm and (2) partitional algorithm. Hierarchical clustering algorithms recursively find nested clusters either in agglomerative mode or in divisive mode. The most well-known hierarchical algorithms are single-link and complete-link; in single-link hierarchical clustering, the two clusters whose two closest members have the smallest distance are merged in each step; in complete-link case, the two clusters whose merger has the smallest diameter are merged in each step. Compared to hierarchical clustering algorithms, partitional clustering algorithms find all the clusters simultaneously as a partition of the data and do not impose a hierarchical structure. The most popular and the simplest partitional algorithm is K-means [3]. Berkhin [4] listed another six classifications besides the above two main groups: (3) Grid-Based Algorithms, (4) Algorithms Based on Co-Occurrence of Categorical Data, (5) Constraint-Based Clustering, (6) Clustering Algorithms used in Machine Learning, (7) Scalable Clustering Algorithms, (8) Algorithms for High Dimensional Data.

In recent years, a growing number of clustering algorithms for categorical data have been proposed based on various centrality measures. For instance, vertex betweenness has been studied by Freeman [5] as a measure of the centrality to detect communities in a network; Girvan and Newman [6] generalize vertex betweenness centrality to edge in order to discover community structures; Frey and Dueck [7] devised a algorithm called affinity propagation, which takes as input measures of similarity between pairs of data points. Newman [8] utilizes the eigenvectors of matrices to find community structure in networks; Rosvall and Bergstrom [9] use the probability flow of random walks on a network as a proxy for information flows in the real system and decompose the network into modules. Wu and Huberman [10] propose an approach for discovering the communities based on the property of resistor networks. For reviews see Refs. [11, 12].

A social network [13, 14] is a set of people or groups each of which has connections of some kind to some or all of the others. A cluster is a collection of individuals with dense relationship internally and sparse relationships externally. Based on this criterion, we introduce a new clustering algorithm for detecting community structures. In this algorithm, individuals and their relationships are denoted by weighted graphs, then the graph density we defined gives a better quantity depict of whole correlation among individuals in a community, so that a reasonable clustering output can be presented. Compared with other algorithms, this algorithm has two important features:

1. A much smaller hierarchical tree that clearly highlight meaningful clusters.
2. Overlapping clusters.

To evaluate the effectiveness of our algorithm, we applied it to analyse some classic bench mark data sets whose clusters are already known. These data sets include Karate Club, Davis southern club women, Dolphin, Books about US politics, American College Football. The accuracy of the outputs in those classical benchmark data sets is a supporting evidence of further applicability of the new algorithm.

The rest of the paper is organized as follows. In section A new clustering algorithm, we introduce the details of the new dense subgraph clustering algorithm. In section Applications of the algorithm in social networks, we apply it to some classic social networks and compare its results with that of known clusters. Finally, a summary and conclusions are given in section Conclusion.

A New Clustering Algorithm

A graph or network is one of the most commonly used models to present real-valued relationships of a set of input items. Let $G = (V, E)$ be a graph with the vertex set V and the edge set E with weight w(e) on every edge e. Models with un-weighted graphs (the weight of every edge is set to 1) have been extensively studied in graph theory. In an un-weighted graph G, a subgraph H of G is defined as a clique if every pair of vertices of H is joined by one edge [15–17]. It is well-known that the search of maximum cliques in graphs is an NP-complete problem [18]. Therefore, it is not practical to define cliques as clusters. Furthermore, there is no appropriate definition for a clique in a weighted graph. However, in order to closely represent the nature and the real situation of the inputs in most applications (different degrees of similarity for clustering problems), we should use weighted graph models which are much more appropriate than un-weighted models. For simplification or other practical reasons, many designers of clustering algorithms may set a specific threshold, such as that any edge with weight below the threshold is deleted and the remaining ones have no associated weight. However, one may not be able to expect an accurate output since the cut-off (by threshold) may cause a loss of important information.

For a subgraph $C(|V(C)| > 1)$, we define the density of C by

$$d(C) = \frac{2\sum_{e \in E(C)} w(e)}{|V(C)|(|V(C)| - 1)} \tag{1}$$

As seen above, if $w(e) = 1$ for every edge e in C and $d(C) = 1$, then the subgraph C induces a clique. For a weighted graph, a subgraph C is called a Δ-quasi-clique if $d(C) \geq \Delta$ for some positive real number Δ.

Since clustering is a process that detects all dense subgraphs in G and construct a hierarchically nested system to illustrate their inclusion relation, a heuristic process is applied here for finding all quasi-cliques with density invarious levels. The core of the algorithm is deciding whether or not to add a vertex to an already selected dense subgraph C. For a vertex $v \notin V(C)$, we define the contribution of v to C by

$$c(v, C) = \frac{\sum_{u \in V(C)} w(uv)}{|V(C)|} \tag{2}$$

A vertex v is added into C if $c(v, C) > \alpha d(C)$ where α is a function of some user specified parameters.

Instance: G = (V, E) is a graph with edge weights $w : E(G) \mapsto R^+$.

Question: Detects Δ-quasi-cliques in G with various levels of Δ, and construct a hierarchically nested system to illustrate their inclusion relation.

Sub-Algorithm Growing(C, G):

(Grow a Community C in G)

```
while V(G) − V(C) ≠ ∅
begin
pick v ∈ V(G) − V(C) such that c(v, C) is a maximum
if  c(v,C) > αₙd(C)  then  add  v  to  C  (where  n = |V(C)|,
αₙ = 1 − 1/(2λ(n+t)), with λ ≥ 1 and
t ≥ 1 as user specified parameters)
else return
end
```

Sub-Algorithm Decompose(G, w_0):

(decompose a graph G into communities using edges with weights at least w0).

Let $E_0 = \{e \in E(G) : w(e) \geq w_0\}$

```
For each e = uv ∈ E₀ in decreasing order of w(e)
begin
if either u or v is not in any community
then
begin
create a new empty community C and add u, v in it
   Growing(C, G)
end
end
```

Sub-Algorithm Merging(G):

```
For any two communities Cᵢ and Cⱼ in G, if
```

$$|C_i \cap C_j| > \beta \min(|C_i|, |C_j|)$$

then merge C_i and C_j into a new community $C = C_i \cup C_j$ (where β is a userspecified parameter).

Contract each community to a vertex. The weight of an edge is defined by

$$w(C_i, C_j) = \frac{\sum_{e \in E_{i,j}} w(e)}{|E_{i,j}|}$$

where the set of crossing edges

$$E_{i,j} = \{v_i v_j : v_i \in C_i, v_j \in C_j, v_i \neq v_j\}$$

Main-Algorithm
(generate hierarchic clustering tree or a graph G)
While $E(G) \neq \emptyset$
begin
Choose w_0 according to some criterion
Decompose(G, w_0)
Merging(G)
Store the resulted graph to G
end
Trace the movement of each vertex and generate the hierarchic tree.

Applications of the Algorithm in Social Networks

In this section, we present a number of applications of our algorithm to some classic social networks for which the community structure is already known and compare its results with that of preceded algorithms.

Zachary's Karate Club Study

The first social network is the well-known "karate club" of Zachary [19]. He observed 34 members of a karate club over 2 years. During the course of observation, the club members split into two groups because of the disagreement between the administrator of the club and the club's instructor, the members of one group left to start their own club. Zachary constructed a simple unweighted graph to show the friendships between two members of the club, each member in the club is represented by a node, and edge is drawn if the two members are friends outside the club activities.

Davis Southern Club Women

The data of the social participation of eighteen women in "Old City" was collected by Davis et al. [20]. The data (see Table 1) is a table with 18 rows—one for each woman—and 14 columns, one for each "event" (such as a club meeting, a church supper, a card party, etc.), held during the course of a year. For the simplicity, we use number 1–18 denote the 18 women, then a matrix A is generated to record their attendances of events: A(i, j) is 1 if woman i attended social event j, and 0 otherwise. The goal of the study was to determine the clique structure according to their records of attendances. The clique membership reported by Homans [21] is as follows. Group 1:1,2,7,8,14,15,16; Group 2: 11,12,13,17,18; women not clearly belonging to either groups: 3,4,5,6,9,10.

Dolphin's Network

The next social network was constructed from observations of a bottlenose dolphin community [22–24]. There are 62 nodes and 159 edges in this network: nodes represent the dolphins, edges between nodes represent associations between dolphin pairs occurring more often than expected by chance. This network is interest because, during the course of the study, the dolphin group split into two smaller subgroups following the departure of a key member of the population. The subgroups of the actual division in Newman [8] are represented by the shapes of the vertices (see Fig. 6), the squares and circles represent the actual division of the network observed when the dolphin community split into two as a result of the departure of a keystone individual. The individual who departed is represented by the yellow triangle. The dotted line denotes the division of the network into two equal-sized groups found by the standard spectral partitioning algorithm. The solid curve represents the division found by the algorithm based on the leading eigenvector of the modularity matrix in Newman [8]. Its result corresponds quite closely to the actually split—all but 3 of the 62 dolphins are misclassified.

We have also applied our algorithm to this network, the result corresponds perfectly with the actual division. The algorithms of Girvan and Newman [25] and Newman [12] also give precisely the same result.

An Example of Overlapping

Most of clustering algorithms generate non-overlapping groups, which are useful for graph drawing but are not as good for group analysis, since real social groups are usually more complex, involving different degrees of overlap among groups. In this section, we use an example in Santamaria and Theron [26] to illustrate the overlapping feature of our algorithm.

Conclusion

In this paper, we introduce a new clustering algorithm for detecting community structure in network and use it to analyze some classic social networks. Compared with the existing algorithms, the new algorithm has two distinguished features:

1. A much smaller hierarchical trees that clearly highlight meaningful clusters. It was pointed out in SAS/STAT User's Guide [27], "there are no completely satisfactory algorithms for determining the number of population clusters for any type of cluster analysis". Hence, a relatively small hierarchical tree in an output will significantly reduce the human involvement in the final selection of clusters. In Fig. 2 and Fig. 3, one may notice that each hierarchical tree (for karate club) with 34 leafs (inputs) has 33 internal nodes. By using the new algorithm, the hierarchical tree (see Fig. 4) contains only 22 internal vertices. Similarly, a hierarchical tree for women club with 18 leafs usually have 17 internal nodes, while the tree in Fig. 5 has only 8 internal vertices.
2. The feature of overlapping clustering does reflect the complexity of our real world. One may notice the overlapping clustering feature in Fig. 4: At the right end of the graph, a small group of club members {24, 30, 33, 34, 9, 31} hold multi-memberships in three internal clusters (internal nodes on the third level, right end). The overlapping clustering is a concept that has recently received increased attention in Palla et al. [28], Pereira-Leal et al. [29], Futschik and Carlisle [30], etc. One may also notice this feature in Fig. 14: The element n3 and n7 present in two groups and the node n4 presents in three groups. According to mathematical definition in graph theory, the "hierarchical trees" in Figs. 4 and 14 are not really trees–they are hierarchical networks in which the relations of clusters are hierarchically nested.

For further research, we will consider to develop an automated value selection algorithm for each parameter. Determine a function for each parameter in terms of some structural information of the input graph, so that graphs with different structures (density, connectivity, locally or globally) will be automatically assigned proper values.

Acknowledgments This work has been supported by the National Natural Science Foundation of China under Grant 61172072, 61271308, the Beijing Natural Science Foundation under Grant 4112045, the Research Fund for the Doctoral Program of Higher Education of China under Grant W11C100030, the Beijing Science and Technology Program under Grant Z121100000312024.

References

1. Merriam-Webster Online Dictionary (2008) Cluster analysis. http://www.merriam-webster-online.com
2. Jain AK (2009) Data clustering: 50 years beyond K-means. http://dataclustering.cse.msu.edu/papers/JainDataClusteringPRL09.pdf

3. Steinhaus H (1956) Sur la division des corpmateriels en parties. Bull Acad Polon Sci C1. III IV:801–804
4. Berkhin P (2009) Survey of clustering data mining techniques, http://www.ee.ucr.edu/barth/EE242/
5. Freeman L (1977) A set of measures of centrality based upon betweenness. Sociometry 40:35–41
6. Girvan M, Newman MEJ (2002) Community structure in social and biological networks. Proc Natl Acad Sci 99(12):7821–7826
7. Frey BJ, Dueck D (2007) Clustering by passing messages between data points. Science 315:972–976
8. Newman MEJ (2006) Finding community structure in networks using the eigenvectors of matrices. Phys Rev E 74:036104
9. Rosvall M, Bergstrom CT (2008) Maps of random walks on complex networks reveal community structure. Proc Natl Acad Sci USA 105(4):1118–1123
10. Wu F, Huberman BA (2004) Finding communities in linear time: A physics approach. Eur Phys J B 38:331–338
11. Danon L, Duch J, Diaz-Guilera A, Arenas A (2005) Comparing community structure identification. J Stat Mech 2005:P09008
12. Newman MEJ (2004a) Detecting community structure in networks. Eur Phys J B 38;321–330
13. Scott J (2000) Social network analysis: a handbook, 2nd ed. Sage Publications, London
14. Wasserman S, Faust K (1994) Social network analysis. Cambridge University Press, Cambridge
15. Bondy JA, Murty USR (1976). Graph theory with applications. Macmillan, London
16. Diestel R (2005) Graph theory, Graduate texts in mathematics, vol 173, 3r edn. Springer, Heidelberg
17. West W (1996) Introduction to graph theory. Prentice Hall, Upper Saddle River,NJ
18. Gary MR, Johnson DS (1979) Computers and intractability. Freeman, NY
19. Zachary WW (1977) An information flow model for conflict and fission in small groups. J Anthropol Res 33(4):452–473
20. Davis A, Gardner BB, Gardner MR (1941) Deep South: a social anthropological study of caste and class. University of Chicago Press, Chicago
21. Homans GC (1950) The human group. Harcourt, Brace and World, New York
22. Lusseau D, Schneider K, Boisseau OJ, Haase P, Slooten E, Dawson SM (2003) The bottlenose dolphin community of doubtful sound features a large proportion of long-lasting associations. Can geographic isolation explain this unique trait? Behav Ecol Sociobiol 54:396–405
23. Lusseau D (2003) The emergent properties of a dolphin social network. Proc R Soc London B270:S186–S188
24. Lusseau D, Newman MEJ (2004) Identifying the role that individual animals play in their social network. Proc R Soc London B(Suppl.) 271:S377–S481
25. Newman MEJ, Girvan M (2004) Finding and evaluating community structure in networks. Phys Rev E 69:026113
26. Santamaria R, Theron R (2008) Overlapping clustered graphs: coauthorship networks visualization. Lect Notes Comput Sci 5166:190–199
27. SAS Institute Inc (2003) Introduction to clustering procedures, Chapter 8 of SAS/STAT User's Guide.(SAS OnlineDocTM: Version 8) http://www.math.wpi.edu/saspdf/stat/pdfidx.htm
28. Palla G, Derenyi I, Farkas I, Vicsek T (2005) Uncovering the overlapping community structure of complex networks in nature and society. Nature 435(7043):814–818
29. Pereira-Leal JB, Enright AJ, Ouzounis CA (2004) Detection of functional modules from protein interaction networks. PROTEINS: Struct Funct Bioinf 54:49–57
30. Futschik ME, Carlisle B (2005) Noise-robust soft clustering of gene expression timecourse. J Bioinf Comput Biol 3:965–988
31. Breiger RL (1974) The duality of persons and groups. Soc Forces 53(2):181–190

32. http://www-personal.umich.edu/mejn/netdata/
33. Kernighan BW, Lin S (1970) An efficient heuristic procedure for partitioning graphs. Bell Syst Tech J 49:291–307
34. Krebs V (2009) http://www.orgnet.com/
35. Newman MEJ (2004b) Fast algorithm for detecting community structure in networks. Phys Rev E 69:066133

Design and Implementation of NGDC Geospatial Metadata Management System

Li Zhang

Abstract As an important part of the geospatial framework, National Geospatial Data Center (NGDC) is a website which can provide potential users with convenient geospatial metadata query services. It will improve the development of Digital City strongly. This paper firstly analyzes the content of the geospatial metadata. It puts forward the concept of the Integrated Geospatial Metadata based on the analysis of geospatial metadata query conditions entered in the NGDC website. The NGDC geospatial metadata management system is developed according to the Integrated Geospatial Metadata.

Keywords Geospatial metadata · National geospatial data center (NGDC) · Geospatial framework · Digital city · Internet · Geographical information system (GIS)

Introduction

With the deepening of information construction, the State Bureau of Surveying and mapping geographic information made the strategic decision of building a national public service platform of geographic information in order to implement the scientific development concept [1]. In 2006, the State Bureau of Surveying and Mapping took the Digital City Geospatial Framework Construction as one of "Eleventh Five-Year" key project [2, 3]. Geospatial framework includes the policies, regulations, standards, technology, facilities, mechanisms and human resources involved with the collection, processing, exchange of geospatial data. It is the basis of the national economic and social information supporting platform and the foundation which a variety of information resources spatially reference to [4, 5].

L. Zhang (✉)
School of Computer Engineering, Shenzhen Polytechnic, Shenzhen 518055 Guangdong, China
e-mail: zhangli@szpt.edu.cn

Y.-M. Huang et al. (eds.), *Advanced Technologies, Embedded and Multimedia for Human-centric Computing*, Lecture Notes in Electrical Engineering 260, DOI: 10.1007/978-94-007-7262-5_87, © Springer Science+Business Media Dordrecht 2014

In order to meet the growing demand for geospatial data, geospatial data production in China has begun to take shape in recent years. A platform is required, which provides users with the geospatial metadata query service according to coverage, layer features, price and other factors. National Geospatial Data Center (National Geospatial Data Center, hereinafter referred to as "NGDC") will be able to provide such a platform which can complete the transition of geospatial data from data producers to data users.

Geospatial metadata is the data foundation to achieve NGDC. The geospatial metadata management system in NGDC can effectively avoid the repetitive production of geospatial data, and can achieve the spatial information resource sharing between different agencies or departments.

Geospatial Metadata and NGDC

How to help potential data users to search the specified geospatial data set more effectively, more easily from such a wide variety of geospatial data products has become an unavoidable problem. It makes the study of geospatial metadata become so important.

Geospatial Metadata refers to the description of geospatial data and information resources. In fact, it is the generalization and extraction of the attributes and spatial characteristics of geospatial data sets or products. In short, the geospatial metadata will help data users know whether the content and quality of the specified geospatial data set meets the requirements of their actual GIS applications, whether the source of geospatial data set is reliable, whether the data format is compatible with other data formats before they get or buy the geospatial product.

Actually NGDC is a geospatial information service network platform. As a bridge between geospatial data producers and data users, NGDC can provide the data producers with a unified geospatial metadata standard and help them to deploy their geospatial metadata. On the other hand, NGDC can provide the efficient and effective query of geospatial metadata for data users in order to make it easy to get the existing geospatial data products of appropriate precision, which are updated and upgraded easily [6].

The Integrated Geospatial Metadata

One of the tasks of NGDC is to develop a set of practical geospatial metadata standard. This metadata standard is the foundation of NGDC geospatial metadata management system. In theory, geospatial metadata standard is divided into two levels. The first level is the directory information, which is mainly the macroscopic description of the data set. It is suitable for National Geospatial Data Center or global management and query spatial information. The second level usually includes the detailed or comprehensive description of the geospatial data set when they are deployed by geospatial data producers.

However, it does not need such detailed metadata fields for most data users to get geospatial information through NGDC web site. In other words, data users are usually asked to become a registered member of NGDC, or they can not login NGDC official website. After login, the user may input some query conditions (such as the scale of the digital map, coverage, etc.) in the web page of geospatial metadata query on NGDC web site. The geospatial metadata management system will filter out some candidates of geospatial data products according to the specified conditions. Then the data users will contact with data producers by telephone or visit them personally in order to consult related issues (such as technical parameters, view sample images, and so on) because the query result includes the contact information of data producers.

After analyzing the query conditions entered by the data user in the query page, we can extract a portion of geospatial metadata in order to form "the Integrated Geospatial Metadata"(hereinafter referred to as "IGM"), which includes the basic information of the data set, data publishers, geographical coverage, elevation range four aspects. The details of IGM are shown in Table 1.

Implementation of NGDC Geospatial Metadata Management System

As mentioned earlier, the main task of NGDC is to collect geospatial metadata provided by data producers, and to extract the relevant metadata information into the formation of IGM saved in NGDC database. At the same time, the official website of NGDC provides a hyperlink to geospatial metadata query page in order to provide geospatial metadata query services.

The NGDC geospatial metadata management system includes three components which are the user management function module, the function module of geospatial metadata maintenance, the function module of geospatial metadata query. The system functional block diagram is shown in Fig. 1.

The functions of user management module include user registration, user login, and user information maintenance and other functions. The types of NGDC user include administrators, data producers, data users, guests. Administrators can modify or delete the information for all users. Data users can query metadata after registering as a NGDC user. Guest can only browse NGDC site. The functions related to metadata maintenance and query will be available to someone only after he is a registered user in NGDC official web site.

The functions of geospatial metadata maintenance module consist primarily of functions of adding, modifying and deleting geospatial metadata. Geospatial metadata referred to here mainly refers to IGM, which is mainly used to support the geospatial metadata query function provided by NGDC through Internet. The

Table 1 The details of the integrated geospatial metadata

No.	Metadata item	Data item description	Remarks
1	Name	Name of data set or products	Basic information of data set
2	Version	Version of data set or products	
3	Scale	Scale of data set or products	
4	Abstract	Brief summary of the contents of the data set	
5	Objective	A brief description of the purpose to establish the data set	
6	Subject	The thematic name of data set, such as elevation, cadastral, geology, hydrology, vegetation, marsh, transportation, soil, etc.	
7	Key words	The feature words which can summarize the particular aspect of data sets	
8	Name of deployer	The name of individual or organization that are directly responsible for the contents of the data set	Information of data deployers
9	Province	The province, municipalities, district or Special Administrative Region of data deployer	
10	City	The city name of data deployer	
11	Post address	The detailed post address of data deployer	
12	Postal code	The postal code of data deployer	
13	Contact telephone	The telephone number of data deployer	
14	Fax	The fax phone number of data deployer	
15	E-mail	The E-mail address of data deployer	
16	Web site	The web site address of data deployer if exists	
17	Contact	The name of the contact person	
18	Longitude minimum	The minimum value of longitude for geospatial coverage	Geographical coverage of data set
19	Longitude maximum	The minimum value of longitude for geospatial coverage	
20	Latitude minimum	The minimum value of latitude for geospatial coverage	
21	Latitude maximum	The minimum value of latitude for geospatial coverage	
22	Minimum elevation	Minimum elevation value of data set (unit: m)	Elevation range
23	Maximum elevation	Maximum elevation value of data set (unit: m)	

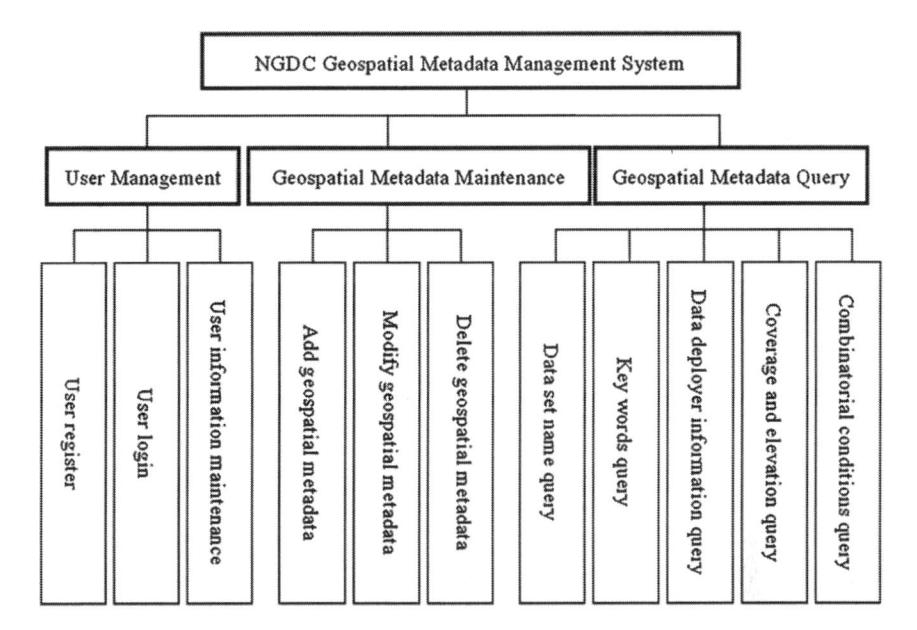

Fig. 1 Functional block diagram of NGDC geospatial metadata management system

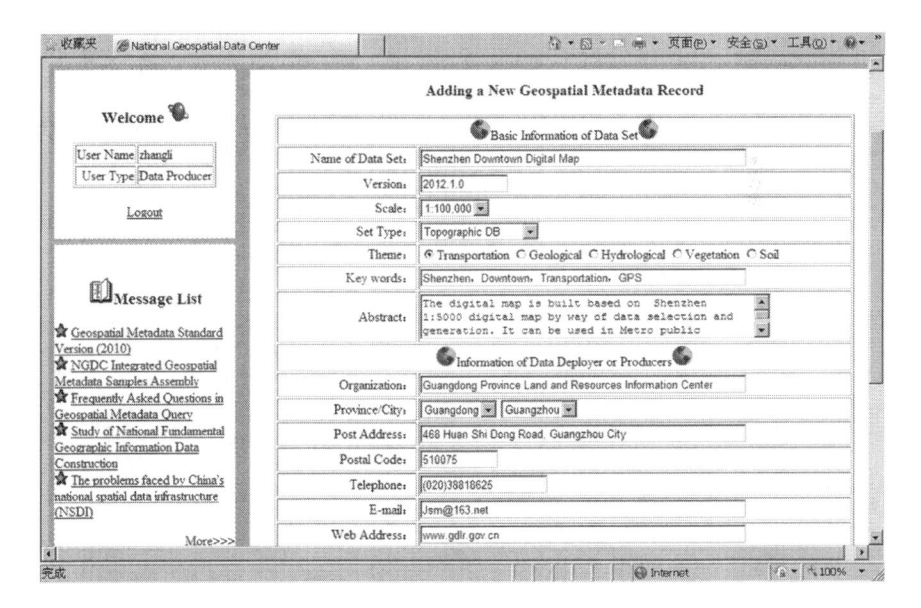

Fig. 2 Web page of adding a new geospatial metadata record

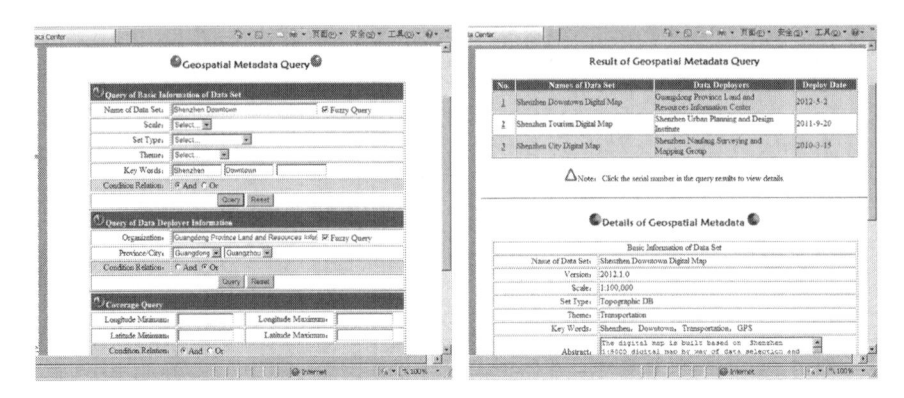

Fig. 3 Web pages of geospatial metadata query and results

detailed geospatial metadata information can be maintained by data producers in accordance with the relevant metadata national standards. The adding Geospatial Metadata records page is shown in Fig. 2.

The functions of geospatial metadata query module include the queries of data set name, keywords, data producer information, coverage and elevation. Of course, combinatorial conditions query should be also supported. It is shown in Fig. 3.

Conclusion

In summary, as part of the Digital City Geospatial Framework the main task of NGDC is to help data users to search the data sets which meet their application needs from lots of data products. The distributed information technology based on Internet provides a strong guarantee for this task. The geospatial metadata management system is developed based on "the Integrated Geospatial Metadata" which is extracted from normal geospatial metadata. As a reference, local NGDC websites can expand the corresponding functions in this basic website according to their operation mode. For example, in order to allow data users get a direct feeling about the contents of the specified data set, local NGDC websites can provide local screenshots of digital products.

References

1. Liang J, Zhang Z, Cheng Y (2012) Research on framework data update system on public services platform. Bull Surveying Mapp 2012(12):79–83
2. Chen Z (2012) Research on urban planning information platform on geospatial framework. Bull Surveying Mapp 3:92–94

3. Xu Y, Jianbang H, WU P, Deng Y, Cao Y, MA L (2010) Research on geographic information sharing environment. Bull Surveying Mapp 6:20–22
4. Wang X, Chen H, Liu L, Li W (2012) Research and discussion on digital region geo-spatial framework construction-taking Hainan international tourism island digital geo-spatial framework for example. Bull Surveying Mapp 6:28–30
5. Li W (2011) On innovation of establishment of digital city geospatial framework. Bull Surveying Mapp 9:1–5
6. Feng X, Gu X, Fan W (2008) Discussion on the key problems of service oriented informationized mapping. Beijing Surveying Mapp 3:1–4

Apply Genetic Algorithm to Cloud Motion Wind

Jiang Han, Ling Li, Chengcheng Yang, Hui Tong, Longji Zeng and Tao Yang

Abstract Cloud Motion Wind (CMW) is a very important issue in the meteorology. In this paper, we firstly apply Genetic Algorithm (GA) to the CMW searching to reduce the computational complexity. We propose a novel CMW method, namely GA-CMW. Compared with the traditional Exhaustive CMW (E-CMW) algorithm, GA-CMW can obtain almost the same performance while with only 11 % of the computational complexity required. Generally speaking, the proposed GA-CMW method can obtain the wind vector picture in shorter time, which makes a lot of sense to the resource saving in the practical application.

Keywords Image matching · Cloud motion wind · Genetic algorithm · Cross-correlation coefficient · Computational complexity

Introduction

Nowadays, with the development of the economies, disaster warning and disaster control are becoming more and more important [1–3]. Hence, how to make the accurate and rapid forecast draws a lot of attentions to the governments.

J. Han (✉) · C. Yang
Key Laboratory of Universal Wireless Communications, Ministry of Education, Beijing University of Posts and Telecommunications, Beijing 100876, China
e-mail: bjtu@bupt.edu.cn

L. Li
College of Mathematic and Information, China West Normal University, Sichuan, China

H. Tong
School of Science, Beijing University of Posts and Telecommunications, Beijing, China

L. Zeng · T. Yang
School of Electronic and Information Engineering, Beijing Jiaotong University, Beijing, China

Y.-M. Huang et al. (eds.), *Advanced Technologies, Embedded and Multimedia for Human-centric Computing*, Lecture Notes in Electrical Engineering 260, DOI: 10.1007/978-94-007-7262-5_88, © Springer Science+Business Media Dordrecht 2014

Cloud Motion Wind (CMW) is a widely used method in meteorology, which aims to judge the speed and direction of the wind from the displacement of anchor cloud [1]. CMW has already been widely used in practical weather forecast and typhoon warning systems. In the traditional Exhaustive CMW (E-CMW) method, the meteorological satellite obtains the earth grey-scale maps on two different moments (commonly 30 min intervals). Then with image matching inside a specific searching window, the displacement of anchor cloud can be calculated and the speed and direction of wind is obtained.

Weather forecast and disaster warning have quite a high request to the instantaneity and computing speed. Algorithm with high speed can provide more authentic and valuable results. However, in the E-CMW algorithm, anchor cloud has to search the whole space with exhaustive method, which costs a lot of resource and time. Also, the matching algorithms are usually with high complexity. Numerous times of loops are required with the Sequence Similarity Detection Algorithm (SSDA algorithm) in [4] and Cross-correlation Coefficient (CC) method in [5]. Hence, searching for a method which can decrease the computational complexity of the traditional E-CMW algorithm is quite beneficial in both meteorology forecast and resource saving.

Genetic Algorithm (GA) is a classical intelligent algorithm to obtain the optimal or nearly optimal value, which has been widely used in mathematics [6], artificial intellegence [7, 8] and telecommunication engineering [9–11]. It is shown in [11] that with GA, a large percentage of computational complexity can be reduced. In this paper, we firstly apply GA to CMW searching and provide the details of the proposed GA-CMW algorithm. Simulations of both GA converging curve and the wind vector figure in a local area show that with little performance lost, only 11 % of the complexity is needed with the proposed GA-CMW.

The paper is organized as follows. Section Traditional E-CMW Algorithm describes the traditional E-CMW algorithm. Section GA based CMW Searching provides the details of the proposed GA-CMW algorithm, including the initialization of GA. Numerical analysis and simulation results on the performance and the computational complexity are shown in section Analysis and Numerical Simulation Results. Finally, conclusion is drawn in section Conclusion.

Traditional E-CMW Algorithm

The Principle of E-CMW

(1) *Model of E-CMW*: E-CWM is a method in meteorology which describes the parameters of wind vector, including its speed and direction. As the movement or displacement of a cloud depends on the wind around it, the parameters of wind vectors can be achieved by tracing the anchor cloud. As shown in Fig. 1, cloud M on time T is the anchor cloud, which is surrounded by small window

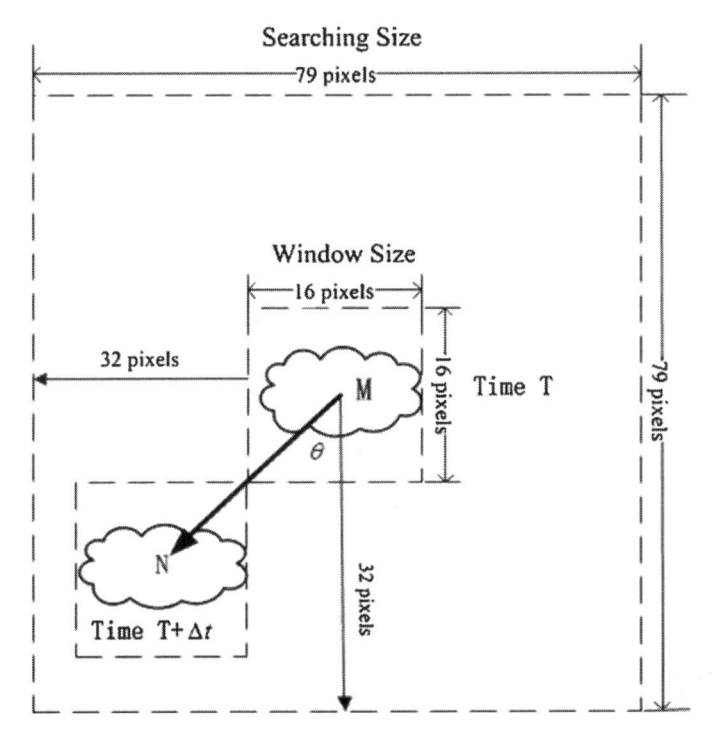

Fig. 1 The model of CMW matching process

with 16×16 pixels. After Δt interval, the anchor cloud moves to the new position N, by calculating the angle θ and distance MN of the movement, parameters of the wind vector are obtained.

Let's assume the coordinate of the anchor cloud on time T is $M(x_M, y_M)$, while its coordinate on time $T + \Delta t$ is $N(x_M, y_M)$. As shown in Fig. 1, considering MN is just a short distance on the sphere surface, vector $\overrightarrow{MN}$ is approximately equivalent to the arc $\overset{\frown}{MN}$. Then we can obtain the angle and speed of the movement as following:

$$\vec{v} = \frac{\sqrt{(y_N - y_M)^2 + (x_N - x_M)^2}}{\Delta t} \tag{1}$$

$$\cos\theta = \frac{|(y_N - y_M)|}{\sqrt{(y_N - y_M)^2 + (x_N - x_M)^2}} \tag{2}$$

(2) *Window Size and Searching Size*: The choice of window size and searching size can directly influence the matching performance and the searching complexity. In the E-CMW method, proper window size should cover most parts of the anchor cloud and the proper searching size should cover the refreshed position on $T + \Delta t$ moment.

With traditional E-CMW searching method, the searching times needed are:

$$\gamma_{CMW} = (S_{size} - W_{size} + 1)^2 \qquad (3)$$

where S_{size} denotes the searching size and W_{size} denotes the window size.

In this paper, we apply the window size with 16 pixels and searching size with 79 pixels. Hence, with formula (3), 64×64 times of searching are totally required with the traditional E-CMW method. Further more, we will get numerous anchor clouds on the whole satellite cloud picture. Hence, the computational complexity will increase tremendously with the increasement of the predicted region.

Matching Algorithm

Above all, we have illustrated the principle of the traditional E-CMW algorithm. In this subsection, the commonly used matching algorithm CC is shown in (4).

$$CC(i,j,a,b) = \frac{\sum_{a=1}^{A} \sum_{b=1}^{B} ((S^{i,j}(a,b) - \bar{S}^{i,j}) \times (T(a,b) - \bar{T}))}{\sqrt{\sum_{a=1}^{A} \sum_{b=1}^{B} (S^{i,j}(a,b) - \bar{S}^{i,j})^2 \times \sum_{a=1}^{A} \sum_{b=1}^{B} (T(a,b) - \bar{T})^2}} \qquad (4)$$

where $T(a,b)$ denotes the original template, $S^{(i,j)}(a,b)$ denotes the matching template. i and j in (4) denote the coordinate of the matching template on $T + \Delta t$ moment, then we have $i \leq 64$, $j \leq 64$ in the whole searching space. A and B denote the template size, which are all equal to 16 in the paper. $\bar{T}$ and $\bar{S}^{i,j}$ denote the average value of the template T and S.

GA Based CMW Searching

Basic GA Based CMW Searching

In this subsection, details of the proposed GA-CMW are illustrated and some of the GA's initialization parameters are introduced.

(1) *Coding Scheme*: In the E-CMW searching, searching range is a window with $S_{size} \times S_{size}$ pixels. With exhaustive method in this paper, γ_{CMW} times of matching, which has been defined in (3), are needed with exhaustive method.

As CMW is a discrete searching problem, it is quite proper to apply the Gray Coding Scheme. Gray Coding has a quite good characteristic in suppressing the error propagation. With S_{size} equals to 79, W_{size} equals to 16, 64×64 times of searching are required. Hence, we can encode the coordinate of each template in the searching space with 8 bits.

(2) *Fitness Evaluation*: Each individual of the generation in GA has a fitness to indicate its weight, and the cost function varies from different problems. In the proposed GA-CMW method, the cost function is used to indicate the matching degree between the original template and the matching template, which can still be the CC matching formula in (4).

Individual with bigger value has the better fitness in GA-CMW matching.

(3) *Selection*: In the natural evolution, individuals die or live with natural selection, which has been indicated in details in the well-known Darwin's theory of evolution. In the GA algorithm, it also has the selection process to choose out those individuals with better fitness. With N_G nodes each of the generation, N_S nodes will survive through the selection, which as follows:

$$N_G = N_S \times P_S \tag{5}$$

where P_S is the selection probability in the algorithm. After the selection process, those $N_S - N_G$ nodes with the bottom fitness will be weeded out to release the space for new bloods.

(4) *Crossover*: The process crossover is used to produce the new individuals from those who have survived from the selection. By exchanging the genes randomly between two individuals, new individual is generated from its parents. P_C is used to indicate the probability of crossover in the N_S nodes and is often set to a high value.

(5) *Mutation*: Mutation is another process to produce the new bloods. Mutation also happens in the nature with a random change to the individual's chromosome which can thus generate unpredicted effect on the final results. In some situations, mutation can work to help the current generation jump out the local optimal areas. P_M is used to indicate the probability of mutation in the N_S nodes and is often close to 0.

(6) *Decision Making*: With the continuous iteration, new individuals keep on replacing those members with worse fitness in the previous generation. The iteration ends with the predefined iteration times or the optimization of objective function is achieved. Then node with the best fitness in previous generation is output to be the final decision.

Analysis and Numerical Simulation Results

In this section, we compare the performance and computational complexity between the traditional E-CMW method and GA-CMW method.

Performance Evaluation

Figure 2 shows the output matching value versus iteration time with different population sizes. The optimal matching value boundry with E-CMW method in Fig. 2 is 0.8085, and we assume the output value above 0.80 as converged. It is shown in Fig. 2 that with population size of 80, the iteration time to converged is 6. For population size of 40, 13 times of iteration are required. While with 20 individuals per generation, it takes 48 times to converged. In the simulation results shown in Fig. 2, if the iteration takes long enough time, the output matching value will anyway converged to the optimal matching value boundry, which means should be no performance lost between GA-CMW searching and the traditional E-CMW algorithm.

Figure 3 shows the traditional E-CMW searching simulation result. Figure 4 shows the simulation result of the proposed GA-CMW algorithm with predefined iteration times.

For E-CMW results shown in Fig. 3, The matching results are in a quite good order. For GA-CMW results shown in Fig. 4, it is shown that GA-CMW results may have several errors because of the suboptimal output of the GA searching. The yellow ellipses marked in Fig. 4 show the bad points of the GA-CMW results. But considering the overall trend of the wind, it is almost the same between Figs. 3 and 4. Hence, the performance of GA-CMW should anyway be accepted.

Fig. 2 Iteration time versus output matching value with different population size

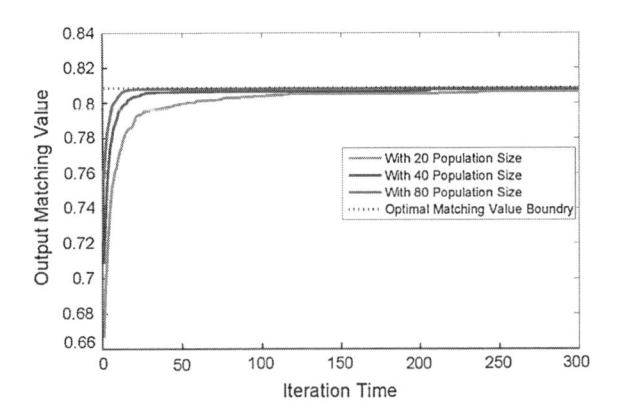

Fig. 3 Matching result with traditional E-CMW algorithm

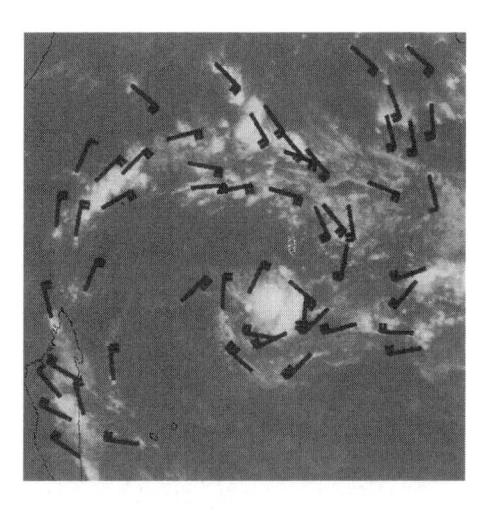

Fig. 4 Matching result with GA-CMW algorithm

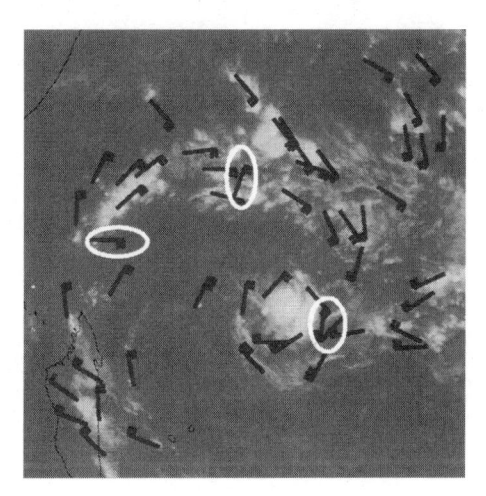

Complexity Evaluation

It is shown in Fig. 2 that with specific iteration time, the proposed GA-CMW has almost no performance lost compared with the traditional E-CMW method. In this subsection, we evaluate the computational complexity of the proposed algorithm. As the computational complexity cost by the selection, crossover, and mutation process is quite small compared with the fitness calculation. We use the running times of the matching function shown in (4) to indicate the computational complexity. Formula (3) and (6) indicate the complexity of traditional E-CMW and GA-CMW.

$$\gamma_{GA-CMW} = N_G \times Iter\_time \tag{6}$$

Fig. 5 Complexity
comparison between
GA-CMW and E-CMW

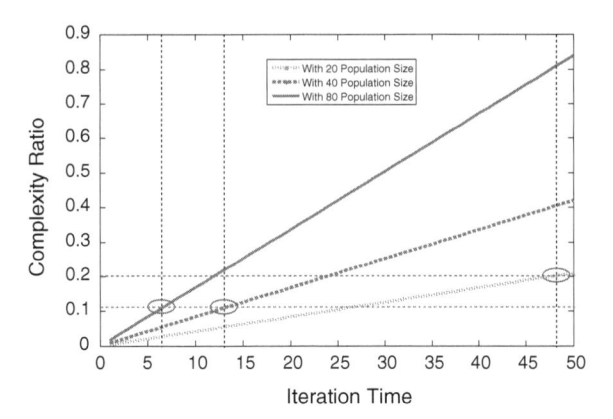

We define the complexity ratio as following:

$$\beta = \frac{\gamma_{GA-CMW}}{\gamma_{E-CMW}} \tag{7}$$

Figure 5 shows the complexity ratio between GA-CMW and E-CMW. It is shown that with the iteration time whose output value firstly beyond 0.80 as the converged time, only 11 % of the complexity is needed with GA-CMW. Even for only 20 individuals each generation, 16 % of the complexity is needed.

For the weather forecast, the satellite data is changing all the time and the CMW result needs to be refreshed quite often. Thus the decrease of the calculating time of CMW does making a lot of sense. By applying GA to the CMW searching, very small percentage of the computational resource will be taken compared with the traditional E-CMW. GA-CMW can save the computational resource and take quite less time to obtain the final CMW result.

Conclusion

In this paper, we firstly apply the GA algorithm to the CMW searching, which is a very important issue in the meteorology. We propose GA-CMW method to reduce the searching complexity in traditional E-CMW method. Simulation results show that with only 11 % of the computational complexity needed, there is almost no performance lost between GA-CMW and E-CMW. Generally speaking, the proposed GA-CMW method can obtain the wind vector picture with shorter time and quite acceptable performance, which makes a lot of sense to the resource saving and disaster control in the practical application.

Acknowledgments This paper is supported by the National Science and Technology Major Project of the Ministry of Science and Technology of China under Grand 2012ZX03001039-002, which is kindly acknowledged.

References

1. Purdom JF (1996) W. Detailed cloud motions from satellite imagery taken at thirty second one and three minute intervals. In: Proceeding to the 3rd international wind workshop in Ascona, Switzerland, 10–12 June 1996, pp 137–146
2. Wang ZH, Browning KA, Kelly GA (1997) Verification of the tracking technique used in an experimental cloud motion wind inferring system. JCMM Report. University of Reading, 1997
3. Wang Z, Zhou J (2000) A preliminary study of Fourier series analysis for cloud tracking with GOES high temporal resolution images. Acta Meteoro Sin 14(1):82–94
4. Leese JA, Novak CS, Clark BB (1972) An automated technique for obtaining cloud motion from geosynchronous satellite data using cross correlation. J Appl Meteor 10(1):118–132
5. Jianmin X, Qisong Z (1996) Calculation of cloud motion wind with GMS-5 images in China. In: Proceedings to the 3rd International Wind Workshop in Ascona Switzerland, pp 45–52, 10–12 June 1996
6. Revello TE, McCartney R (2002) Generating war game strategies using a genetic algorithm. In: Proceeding Congress Evolutionary Computation, 2002, vol. 2, pp 1086–1091
7. Campbell MS, Hoane AJ, Hus FH (2002) Deep blue. Artif Intell 134(1–2):57–83
8. Shibata T, Fukuda T, Tanie K (1997) Chapter 108: Synthesis of fuzzy, artificial intelligence, neural networks, and genetic algorithm for hierarchical intelligent control. CRC Press, Boca Raton, pp 1364–1368
9. Binelo MO, de Almeida ALF, Cavalcanti FRP (2011) MIMO array capacity optimization using a genetic algorithm. IEEE Trans Veh Technol 60(6):2471–2481
10. Mangoud MAA (2009) Optimization of channel capacity for indoor MIMO systems using genetic algorithm. Prog Electromagn Res C 7:137–150
11. Bashir S, Khan AA, Naeem M, Shah SI (2007) An application of GA for symbol detection in MIMO communication systems. In: Third International Conference on Natural Computation, ICNC 2007, Aug

A Topic Evolution Model Based on Microblog Network

Qingling Zhou, Genying Wang and Haiqiang Chen

Abstract Fission mode of information transmission in the Microblog network presents a challenge to the existing public opinion diffusion model. In order to depict the mechanism of the public opinion transmission in the Microblog network, this paper proposes a topic evolution model based on the complex networks and infectious disease dynamics theory. The model takes the directed scale-free network as the carrier, whi ch synthetically considers the characteristics of topic propagation and its influence factors. In addition, the research performs simulation analysis, of which result mirrors the topic evolution process from common topic into hot topic, and verifies its validity.

Keywords Microblog network · Topic evolution · Hot topic

Introduction

Microblog, namely micro blog, is a form of blog allowing users to update brief texts (usually less than 200 words) and can be published openly, which mainly provides posting, forwarding, comments, attention, etc. Fission mode of information transmission in the Microblog network presents a challenge to the existing

Q. Zhou · G. Wang
School of Electronic and Information Engineerring, Beijing Jiaotong University, Beijing 100044, China
e-mail: 11120232@bjtu.edu.cn

Q. Zhou · G. Wang (✉)
Key Laboratory of Communication and Information Systems, Beijing Municipal Commission of Education, Beijing Jiaotong University, Beijing 100044, China
e-mail: gywang@bjtu.edu.cn

H. Chen
China Information Technology Security Evaluation Center, Beijing 100085, China
e-mail: chenhq@itsec.gov.cn

Y.-M. Huang et al. (eds.), *Advanced Technologies, Embedded and Multimedia for Human-centric Computing*, Lecture Notes in Electrical Engineering 260, DOI: 10.1007/978-94-007-7262-5_89, © Springer Science+Business Media Dordrecht 2014

public opinion diffusion model. And Microblog walks into every aspect of people's life in form of "fragmentation" information, so that it gradually becomes the important position in public opinion. So, studying topic dissemination, modeling its evolution process and correspondingly making the quantitative analysis, have an important guiding significance for discovering and controlling the vital public sentiment.

Presently, most related works study information dissemination only from one perspective on interpersonal network, such as, information dissemination in the network topology [1], individual interaction rules [2], degree assortativity characteristic of collaborative network [3, 4], etc. But, actually, information dissemination and formation are a typical evolution process in the complex system, which needs synthetically consider individual interaction and network structure so as to describe topic evolution more accurately in the Microblog network. In this paper, we preliminarily establish the Microblog evolution simulation model based on topic spreading rules and network topology's influence on the spreading behavior. Above all, the result reflects the real situations, and verifies its validity.

Model

Evolution Mechanism

In the Microblog network, once one topic is posted, promptly, it will be discovered by bloggers and they will share, spread, or comment on it with corresponding probability. Meanwhile, if the user's friends found this topic, they would also determine whether to transmit or forward it depending on personal preferences. In this way, the topic is forwarded and propagated along with friend relationship, finally, it is judged whether the common topic evolves into a hot topic by the amount of forwarding and comments.

Based on topic transmission characteristics and whether it's a hot topic, we put the topics into two categories: common topic and hot topic. Common topic is a general topic, presently, on which attention is low, but it may become a hot topic over time. Hot topic is a topic that's paid attention quite high within a certain period of time, and also it's recovering into a common state at a certain speed. Now, we make the following sets:

(1) Common topic evolves into hot topic with the probability $\lambda 1$;
(2) Hot topic renews into a common topic with the speed v.

Topic Evolution Model Based on Microblog Network

For common topic, namely c (t); hot topic, namely h (t). Therefore, having the following topic evolution modes:

$$(1)\, \mathrm{c}\,(t)\, \rightarrow\, \mathrm{h}\,(t);$$
$$(2)\, \mathrm{h}\,(t)\, \rightarrow\, \mathrm{c}\,(t). \tag{1}$$

For the above formula, c (t) evolves into h (t) with the probability $\lambda 1$. Meanwhile, h (t) renews into a common topic with the speed v over time. Then, we can obtain our model, namely common-hot-common (CHC) model, which is similar to the susceptible-infected-susceptible (SIS) model [5].

Microblog network is a directed scale-free network. Based on users' attention directions, the users' node degree can be divided into two categories, namely in-degree and out-degree [6]. As the amount of node in-degree is larger, users have the greater influence on topic evolution. Meanwhile, as the amount of node out-degree is larger, the information source is richer. This paper studies the propagation properties of user posting topics, so we only consider the situations of users' node in-degree.

Suppose one user with in-degree k posts a topic, which is in common state at time t. During the period of $[t,\, t + \Delta t]$, p_{cc} denotes the probability of the topic remaining the same, and p_{ch} represents the probability of the topic evolution from common topic into hot topic, and $p_{ch} = 1 - p_{cc}$. Where,

$$p_{cc} = (1 - \Delta t \lambda 1)^m, \tag{2}$$

Where, $m = m\,(t)$ represents the topic forwarding amount. In addition, m is a random variable with the binomial distribution, as follow:

$$S(m,\, t) = \binom{k}{m} [\alpha \beta q(t)]^m [1 - \alpha \beta q(t)]^{k-m}, \tag{3}$$

Where, $q(t)$ denotes the probability user's friends discover the topic at time t, α represents the users' interest factor [7] in the topic, and β is the sensitive topic factor. $Q(t)$ is a function of time, obviously, of which process yields Poisson distribution, as below shown:

$$q(T = t) = \lambda^t e^{-\lambda}/t!, \tag{4}$$

So, we can get user's average transition probability p'_{cc} with in-degree k in $[t,\, t + \Delta t]$, as follow:

$$p'cc = \sum_{m=0}^{k} \binom{k}{m} (1 - \Delta t \lambda 1)^m [\alpha \beta q(t)]^m [1 - \alpha \beta q(t)]^{k-m}$$
$$= [1 - \Delta t\, \lambda 1 \alpha \beta q(t)]^k \tag{5}$$

This, yields the solution of (4) to (5),

$$p'_{cc} = [1 - \Delta t \lambda 1 \alpha \beta (\lambda^t e^{-\lambda}/t!)]^k, \tag{6}$$

Assuming $N(t)$, $C(t)$, $H(t)$ denotes the amount of the total topic, common topic, hot topic, respectively, at time t. Then, we can obtain:

$$C(t) + H(t) = N(t), \tag{7}$$

So, we can get the changes in the number of common topic in $[t, t + \Delta t]$. As below:

$$
\begin{aligned}
C(t + \Delta t) &= C(t) - C(t)(1 - p'_{cc}) + H(t)v\Delta t \\
&= C(t) - C(t)[1 - (1 - \Delta t\lambda 1\alpha\beta(\lambda^t e^{-\lambda}/t!))^k] + H(t)v\Delta t,
\end{aligned} \tag{8}
$$

Similarly, we can get the changes in hot topic's amount in $[t, t + \Delta t]$. That is:

$$
\begin{aligned}
H(t + \Delta t) &= H(t) + C(t)(1 - p'_{cc}) - H(t)v\Delta t \\
&= H(t) + C(t)[1 - (1 - \Delta t\lambda 1\alpha\beta(\lambda^t e^{-\lambda}/t!))^k] - H(t)v\Delta t,
\end{aligned} \tag{9}
$$

Yielding the solution of (7) to (8), then we can get:

$$
\begin{aligned}
\frac{C(t + \Delta t) - C(t)}{N(t)\Delta t} &= -\frac{C(t)}{N(t)\Delta t}\left[1 - (1 - \Delta t\lambda 1\alpha\beta(\lambda^t e^{-\lambda}/t!))^k\right] \\
&+ \frac{H(t)}{N(t)\Delta t}v\Delta t
\end{aligned} \tag{10}
$$

When Δt converges to zero, we carry on the Taylor expansion for (9) at the right side. Then,

$$\frac{d\rho_c(t)}{dt} = -k\lambda 1\alpha\beta\rho_c(t)(\lambda^t e^{-\lambda}/t!) + \rho_h(t)v, \tag{11}$$

Where, $\rho_c(t)$ represents the density of common topic in the Microblog network at time t.

Similarly, from (9), we can obtain:

$$\frac{d\rho_h(t)}{dt} = k\lambda 1\alpha\beta\rho_c(t)(\lambda^t e^{-\lambda}/t!) - \rho_h(t)v, \tag{12}$$

Here, $\rho_h(t)$ denotes the density of hot topic in the Microblog network at time t.

From Eqs. (11), (12), we can get the simultaneous equations of the dynamics of topic evolution, which is used to describe the changes of topic density with time. Meantime, topic evolution process is affected by users' node in-degree, sensitive topic factor and so on.

The Simulation and Results

Data Source

We use automatic search procedure to obtain a directed subnet of Sina Microblog, namely $G = (V, E)$, where, node $v \in V$ denotes one Microblog ID, directed edge (u, v) ∈ E represents user u is v's fan, namely u → v. Obtained subnet has the following features: (1) total nodes n = 59986; (2) average in-degree din = 312; (3) average out-degree dout = 208. As Fig. 1, of which shows the accumulated distribution of in-degree.

From Fig. 1, we know the distribution of users' node in-degree exhibits a power-law behavior in the Microblog network. Most of users' node in-degree is low, namely grass-root node, but only a few users' nodes have a larger in-degree, namely famous bloggers in the Microblog network. So the model constructs the microblog network structure consistently with the characteristics of real network. Based on this, we construct a topic evolution model, and perform the quantitative analysis.

Topic Density Distribution

1. At initial time, we assume that $\rho_c(0) = 1$, $\rho_h(0) = 0$. Then, we give the model parameters, as follows: $\lambda 1 = 1/20$, $\alpha = 1/10$, $\beta = 1/4$, v = 0.009, $\lambda = 6$. And the average in-degree din = 312, we can get the changes of $\rho_c(t)$, $\rho_h(t)$ with time, respectively. As the following Fig. 2:

From Fig. 2, during one week, we can see that $\rho_c(t)$, $\rho_h(t)$ is changing with time t. Once one user posts a topic, the amount of topic's forwarding or comments increases almost linearly during [0, 12 h]. Finally, common topic may evolve into a hot topic. And during this period, $\rho_h(t)$ grows with the approximate linearity, ultimately, $\rho_h(t)$ comes to its peak at some time point. Then, attention on this topic tends to go down. About one week later, generally, hot topic recovers back to the common situation again. At the same time, during this period, $\rho_h(t)$ is decreasing continually, and finally it tends to zero. This result is consistent to the characteristics of public opinion evolution we observed (As what $\rho_c(t)$-real, $\rho_h(t)$-real show in Fig. 2, which we get by the means of statistical average), when I worked

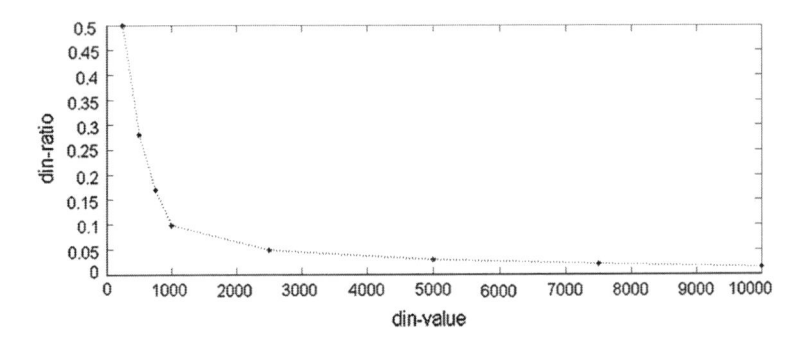

Fig. 1 Din-value, din-ratio represents the value of in-degree and its corresponding ratio, respectively

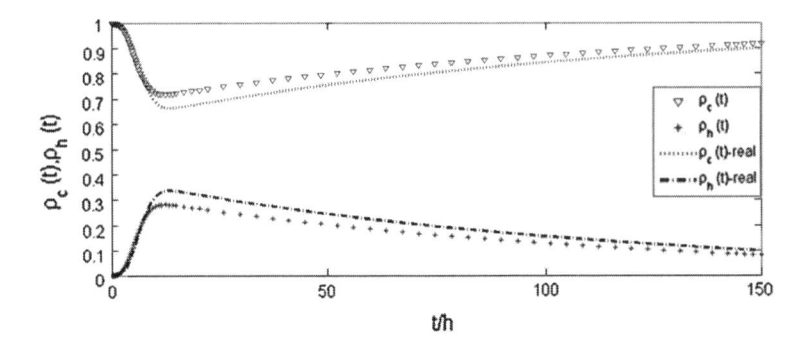

Fig. 2 $\rho_c(t)$, $\rho_h(t)$ changes along with time t, respectively

as an intern in a confidential company for monitoring the Microblog public opinion.

2. Under the condition (1), we discuss the influence on $\rho_h(t)$ for different β. Here, β takes 1/2, 1/8, 1/20, 1/50, respectively. Then we can get Fig. 3. As below:

By Fig. 3, with the increase of sensitive topic factor, topics are easily evolving from common topic into hot topic in a very short time, and then, this topic can be still discussed by everyone in a subsequent week, of which heat plays out slowly. As we know, the simulation results are accord with the real situations.

3. Under the condition (1), we discuss the influence on $\rho_h(t)$ for different k. Here, k takes 50, 300, 2000, 6000, respectively. Then we can obtain Fig. 4. As follow:

From Fig. 4, with the increase of the users' node in-degree, if one user posts a topic, this topic easily evolves from common topic into a hot topic in an extremely short time. What's more, if the users' node in-degree is large enough, the common topic will definitely turn into a hot topic, so we even can ignore the other factors. The results reflect great influence of famous bloggers on topic evolution, playing

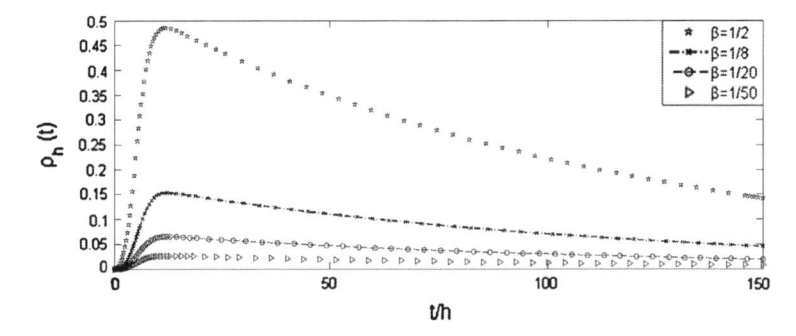

Fig. 3 When β takes different values, $\rho_h(t)$ changes with time t

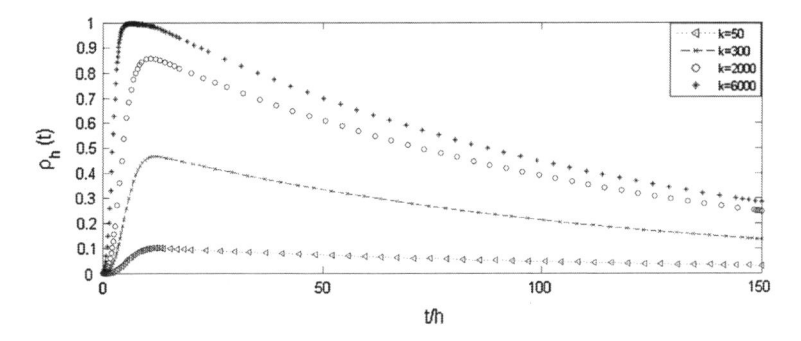

Fig. 4 When k takes different values, $\rho_h(t)$ changes with time t

the roles of opinion leaders [8]. And also it can create conditions for "water army" spreading rumors. Even one topic is not a real event, after several paid posters with large in-degree jacking up, it may also completely be hyped into a real hot one.

Conclusion

This paper studies the characteristics of topic propagation in the microblog network, and proposes a topic evolution model, namely CHC model. The model combines complex network with infectious disease dynamics theory, further to establish the differential equations, which can better describe the topic evolution characteristics. The simulation results show that: all the topic evolution procedure is transforming from common topic into hot topic, and then the hot topic slowly recovers to the common state. In addition, the more sensitive the topic, the more easily common topic evolving into hot topic in extremely short time, and the topic heat keeps high for quite so long time. The larger the user's node in-degree, the more easily common topic evolving into hot topic to play the roles of opinion leaders. Meantime, it can create conditions for spreading rumors in the microblog network. Even one topic is not a real event, after several "water army" with large in-degree jacking up, it also may completely be hyped into a real hot topic. This study will help us understand the topic propagation characteristics deeply, and be meaningful to discover and conduct the crucial public sentiment.

Acknowledgments This work has been supported by the National Natural Science Foundation of China under Grant 61172072, 61271308, the Beijing Natural Science Foundation under Grant 4112045, the Research Fund for the Doctoral Program of Higher Education of China under Grant W11C100030, the Beijing Science and Technology Program under Grant Z121100000312024.

References

1. Jaewon Y, Leskovec J (2010) Modeling information diffusion in implicit networks. IEEE 10th international conference on data mining. Berlin, Germany, IEEE, 2010, pp 599–608
2. Haddadi H, Benevenuto F, Gummadi KP (2010) Measuring user influence in twitter: the million follower fallacy. In: Proceedings of international AAAI conference on weblogs and social. Washington, DC, ICWSM, 2010, pp 97–105
3. Hu H, Wang X (2009) Evolution of a large online social network. Phys Lett A 373:1105–1110
4. Alojairi A, Safayeni F (2012) The dynamics of inter-node coordination in social networks: a theoretical perspective and empirical evidence. Int J Project Manage 30:15–26
5. Zhou J, Liu Z (2008) Epidemic spreading in complex networks. Front Phys China 3(3):331–348
6. Liu Z, Lai Y (2002) Connectivity distribution and attack tolerance of general networks with both preferential and random attachments. Phys Lett A 303:337–344
7. Yan Q, Yi L (2012) Human dynamic model co-driven by interest and social identity in the Microblog community. Phys A 391:1540–1545
8. Iñiguez G et al (2011) Modelling opinion formation driven communities in social networks. Comput Phys Commun 182:1866–1869

A Platform for Massive Railway Information Data Storage

Xu Shan, Genying Wang and Lin Liu

Abstract With the development of national large-scale railway construction, massive railway information data emerge rapidly, and then how to store and manage these data effectively becomes very significant. This paper puts forward a method based on distributed computing technology to store and manage massive railway information data, builds massive railway information data storage platform by using the Linux cluster technology. This system consists of three levels including data access layer, data management layer, application interface layer, enjoying safety and reliability, low operation cost, fast processing speed, easy expansibility characteristics, which shall satisfy the massive railway information data storage requirement.

Keywords Massive railway information data storage · Hadoop distributed technology · Cluster system

X. Shan · G. Wang
School of Electronic and Information Engineering, Beijing Jiaotong University, Beijing 100044, China
e-mail: 11120075@bjtu.edu.cn

X. Shan · G. Wang (✉)
Key Laboratory of Communication and Information Systems, Beijing Municipal Commission of Education, Beijing Jiaotong University, Beijing 100044, China
e-mail: gywang@bjtu.edu.cn

L. Liu
China Information Technology Security Evaluation Center, Beijing, China
e-mail: liul@itsec.gov.cn

Y.-M. Huang et al. (eds.), *Advanced Technologies, Embedded and Multimedia for Human-centric Computing*, Lecture Notes in Electrical Engineering 260, DOI: 10.1007/978-94-007-7262-5_90, © Springer Science+Business Media Dordrecht 2014

Introduction

The Medium and Long-Term Railway Network Plan has established the goal to finish railway construction in 2020, when the high standard and large-scale railway construction will appear in full swing and will generate huge amounts of railway information data. These data, being vast and complex, diverse, heterogeneous, dynamic, relate to various aspects such as railway geographic information, railway construction, railway operation and maintenance, and railway dispatching. However, the current situation is lack of the unified collection and storage criterion or standard, leading to the data island phenomenon. How to store and manage massive railway information data and how to make more efficient use of these data, become one of the key even the bottleneck projects in railway department, and that's what this paper is about.

Traditional methods to deal with massive data mostly use distributed high performance computing and grid computing technology [1], which consume expensive computing resources, need tedious programming to realize effective segmentation of massive data and reasonable distribution of computing tasks. Fortunately, the new development of Hadoop distributed technology can solve these problems better [2].

Basing on Linux cluster technology and using the Hadoop distributed technology, this platform effectively processes the massive amounts of railway information data and stores them in the distributed database, which designs and implements an easily extended and effective massive data storage management system. In the section Platform Architecture of this article, the author gives an introduction to the massive data storage platform architecture. The section The Key Technologies of the Platform Development focuses on the key technologies of system implementation needed. The section The Platform Test and Result Analysis illustrates the implementation and performance characteristics of the system. Finally, the section Conclusion presents the conclusions.

Platform Architecture

Platform Overall Framework

According to the actual needs, we employ the MVC three-tier framework for system design. The platform is divided into data resource layer, business logic layer and presentation layer. Each layer has a clear division of responsibilities, high cohesion and low coupling, making the structure more clear and easier to maintain. The platform overall Framework is shown in Fig. 1.

Data resource layer, composed of several storage nodes, storing and managing massive railway information data, is the foundation of the whole platform.

Fig. 1 Platform overall
framework

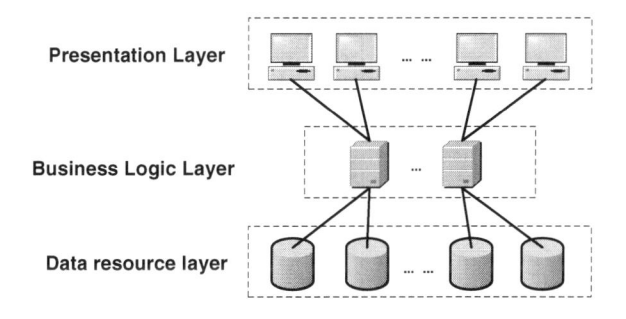

Presentation Layer

Business Logic Layer

Data resource layer

Business logic layer, provides the parallel loading storage of the massive railway information data and the management and support services to ensure the normal operation of the system, is the core of the whole platform.

Presentation layer, provide users with user-friendly interface, convenient for users to query the railway information data and extend system, which meets users directly.

Network Topology of the Platform

The platform adopts the distributed, hierarchical structure in the design, which stores the resources in multiple nodes in different parts of the cluster server. The main service node schedules and manages the distributed server cluster nodes in an unified way. With the data volume increasing and the complex application requirements changing, this platform can be easily extended, and the existing relational database can also be integrated into the platform [3]. And through the de-isomerization process, the platform and the existing relational database can jointly provide storage services for users, which serves the users transparently massive railway information data by storage and management functions (Fig. 2).

Overall Functional Design of the Platform

According to the system function, the system can be divided into three layers, including data access layer, data management layer and application interface layer, as is shown in Fig. 3.

Data Access Layer, with the support of the upper data management, connects the storage nodes distributed in different parts through the local area network (LAN) and the Internet to form a distributed cluster system, which provides shielding for different sources of various databases, and provides database access by service function, only to provide a convenient management and deployment chance.

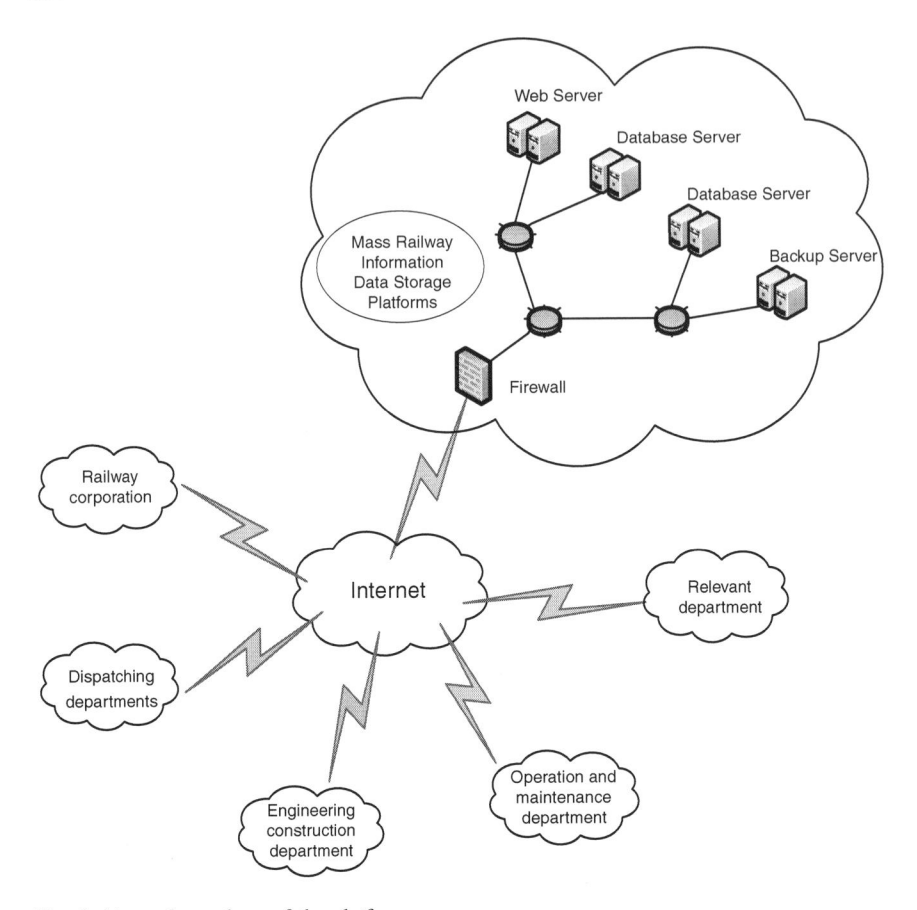

Fig. 2 Network topology of the platform

Data Management Layer, after the huge amounts of data parallel processing, stores the processed data in the distributed database of this system, while it provides management support service to guarantee the system run normally. It is formed of six function modules, including system management module, log management module, parallel load module, load balancing module, storage module, parallel query module, backup recovery module. System management module is used for software management, including running state monitoring, remote system deployment and autonomous running and maintenance, etc. Log management module is used for software operation log management, including the system trajectories, key events and state records, etc. Load balancing serves load balancing and fault tolerance of management of storage node. Parallel loading storage module provides parallel loading and storage for huge amounts of data. Parallel query module provides parallel query, user-defined function such as transaction processing. Backup recovery module provides system stored data backup management, backup storage, backup restore, etc. [3].

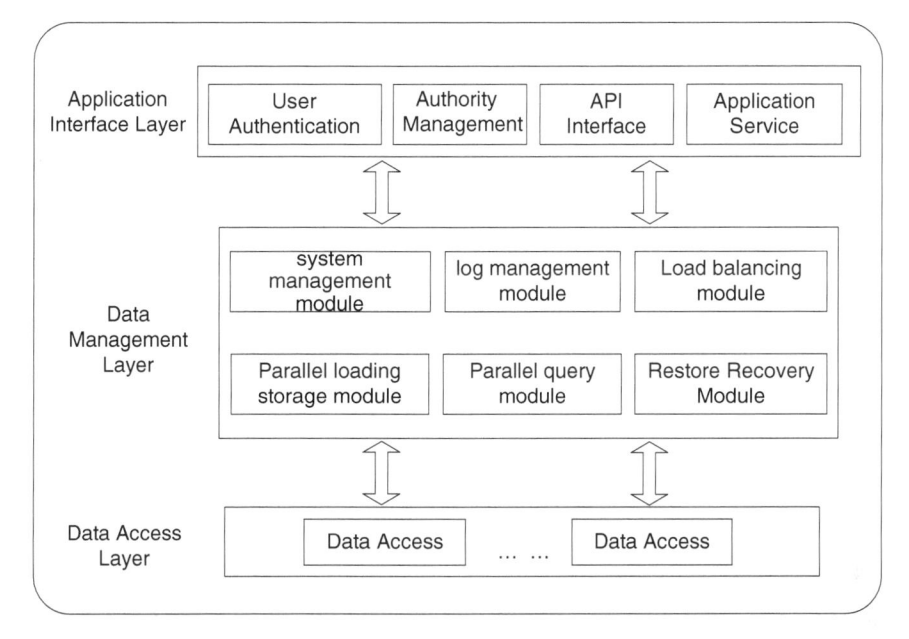

Fig. 3 Overall functional design of the platform

Application Interface Layer, according to the actual business types, provides different application service interface, and provides various application service on the basis of user authentication to meet the needs of different users.

The Key Technologies of the Platform Development

According to the front introduction of Hadoop and functional design of the platform, the most important part is the data management layer. When to manage it, parallel loading storage module of the data become the core of the entire platform. The Hadoop distributed technology provides models and methods of data storage and data processing for the platform. We use the Hadoop distributed file system to store a huge number of source data, and use MapReduce distributed computing model to deal with these data, then use HBase distributed database to store the processed data, in order to realize the storage and management of massive railway information data. Hadoop storage architecture is shown in the Fig. 4 [4].

Fig. 4 Hadoop storage
system architecture diagram

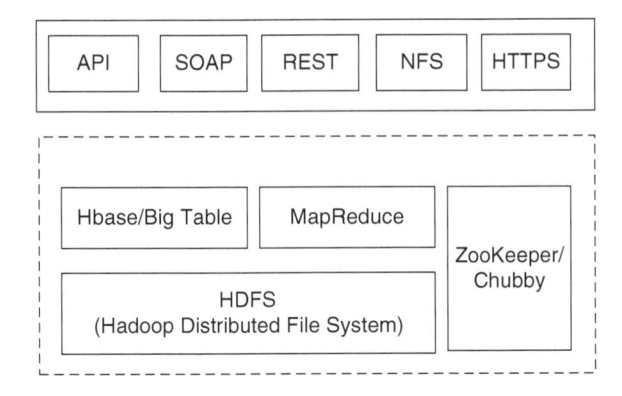

The Hadoop Distributed File System

HDFS is the storage basis of distributed computing, which has high fault tolerance and can be deployed on cheap hardware equipment, to store massive data [4].

HDFS uses a structural model of master–slave (Master/Slave). A HDFS cluster consists of a NameNode and several DataNodes. NameNode plays as the primary server, which manages the file system namespace and the client operation access to the file. DataNode manages stored data. From internal point, the file is divided into a plurality of data blocks, and the plurality of data blocks are stored in a set of DataNode. The NameNode execute file system namespace operations, such as opening, closing, renaming file or directory. It is also responsible for mapping data block to the specific DataNode. DataNode is responsible for handling the file system client's file read and write requests, and carries on the data block create, delete, and copy job under the unified dispatching of the NameNode. The HDFS architecture is shown in Fig. 5.

HDFS is of high throughput of data reading and writing, provides a basis to massive storage for railway information data, stored as an unprocessed set of source data in the Hadoop distributed file system. In our platform, we use HDFS to store these large amounts of source data.

The MapReduce Distributed Computing Model

MapReduce is a summary of task decomposition and results. Map is to break the task down into multiple tasks, and Reduce is to sum the results of the breakdown multitasking together to get the final result. Calculation process can be concluded as the Map (in_key in_value)— > list (inter_key inter_value) and Reduce (inter_key, list (inter_value))— > list (out_value). In this platform, we will firstly read vast railway information data from the HDFS and divide them into M pieces to operate in parallel Map, and secondly form the state intermediate pair <k, value

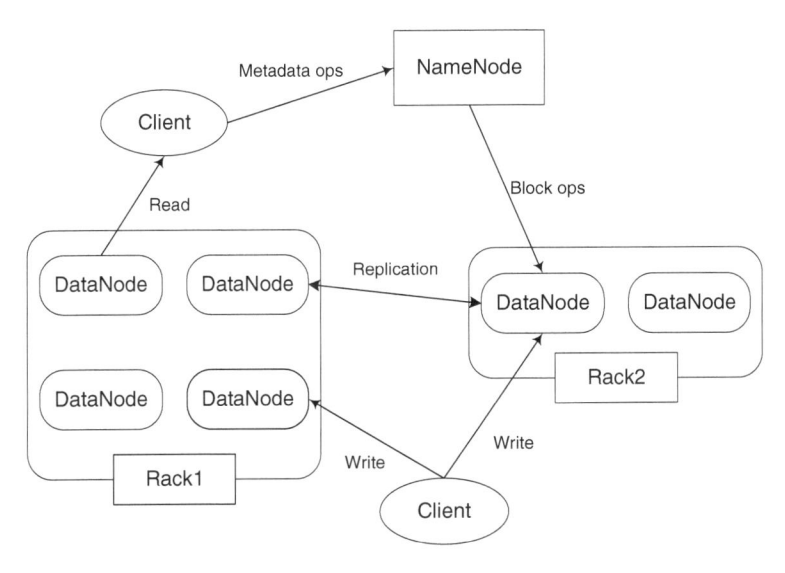

Fig. 5 HDFS architecture diagram

Fig. 6 MapReduce
computing model

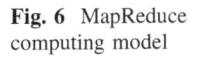

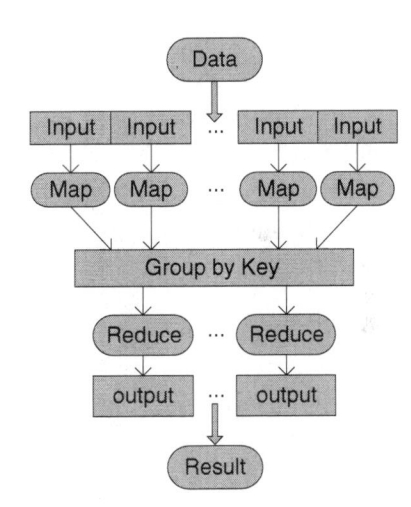

> , and then operate in Group on k value, just form new <k list (value) > tuple, and then break the tuples into R segments to operate in parallel Reduce, and finally, the processed results shall be stored in the distributed database. Calculation model is shown in (Fig. 6).

The implementation of the MapReduce computing model in this platform is composed of JobTracker running on the primary node singly and TaskTracker running on nodes of each cluster [5]. The primary node is responsible for all the tasks scheduling, which constitutes the whole work, and these tasks are distributed in different nodes. The master node monitors their execution, and reruns the failing tasks; sub-node is responsible for the tasks assigned by the master node. When a Job is submitted, the JobTracker receives operation and configuration information, it will distribute the configuration information to the sub-nodes, and schedule the task and monitor TaskTracker's execution. Our platform use this way to achieve the massive railway information data processing.

HBase Distributed Database

HBase is of high-reliability, high performance, oriented column, scalable distributed storage system [6]. The data line includes three basic types, Row Key, Timestamp and Column Family. Each line includes a sortable line keywords that uniquely identifies the data row in a table. An optional time stamp, each data operation has an associated timestamp. One or more column clusters, each column clusters can consist of any number of columns, they can have data or not. Vast amounts of railway information data after the MapReduce computation can use the value of k as a line keyword for distributed storage, implements massive data storage and management functions. Railway information data storage sample as shown in Table 1.

Line keywords represent railway geographic, construction, operation and maintenance, and dispatching information. Timestamp means the time it cost to operate the data. As is shown in the table, at time t3 Jinan to Qingdao direction at 73 km happens Roadbed damages, and in the moment of t6 shows the dispatching information that the G88 train starts from Shanghai to Beijing at 15:00.

The Platform Test and Result Analysis

Platform Performance Test

When testing the system, the data files are divided into different order of magnitude to get rule numeration, and time consuming of the single machine and of Hadoop cluster shall get comparative analysis. Test results is shown in (Fig. 7).

Table 1 Railway information data storage sample

Row Key	Timestamp	Column family		
		Location	Value	
Geographic Construction	t1 t3	Jinan	Jinan-Qingdao 73 km roadbed damage	
Maintenance Dispatching	t2 t4 t5	Beijing Shanghai	Shanghai-Beijing G18 15:00	

Fig. 7 Clustering
performance test results

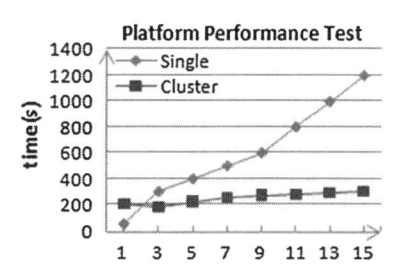

We can see from the diagram that when the system deals with 1 GB data, the elapsed time the cluster takes is about 4 time of single machine, which is because the distributed architecture of cluster costs some time when the system initialization and intermediate files generated and passed. When the data quantity is small, the Hadoop cluster cannot play out the advantages of distributed computing. As the amount of input file data, Hadoop cluster advantages of distributed parallel computing plays out gradually. When the amount of entering data increases to 15 GB from 5 GB, single machine processing time increases significantly, while processing time of cluster system increases in a tiny amplitude. When data volumes get close to 20 GB, cluster system takes about a quarter time of single one. Data test shows that with the amount of data increasing, the cluster saves more time than single machine, which embodies the advantage of the Hadoop cluster on the large amount of data processing speed.

The Result Analysis

Through performance tests, this platform not only can efficiently store and manage massive data, but also has the following features:

(1) High safety and reliability. System will save file in different server in the form of multiple copies to ensure the security and integrity of data.
(2) Data processing speed is fast. System makes documents distribution to different local compute nodes to process the data, which reduces data transfer amount and improves the speed of data processing through the MapReduce model.
(3) Operation cost is low. Using distributed computing architecture, server performance requirements are lower, just to reduce the cost.
(4) God extensibility. System adopts parallel expansion method, which can extend cluster scale and storage capacity at any time according to need.

Conclusion

This paper bases on the Linux cluster technology, uses Hadoop distributed technology, employs the HDFS distributed file system, Map/Reduce distributed computing model and HBase distributed database technology to deal with huge amounts of data, and designs and develops the vast railway information data storage platform based on Hadoop. By doing a lot of ordinary test on cheap computers, it meets the requirements of railway information efficient storage and management of massive data. This platform has the characteristics of high safety and reliability, fast data processing speed, low running cost, good scalability, which will provide certain reference value to railway department for data storage.

Acknowledgments This research is supported by National Natural Science Foundation of China under Grant 61071076, the National High-tech Research And Development Plans (863Program) under Grant 2011AA010104-2, the Beijing Municipal Natural Science Foundation under Grant 4132057.

References

1. Dean J, Ghemawat S (2008) MapReduce: simplified data processing on large clusters. Commun ACM 51(01):107–113
2. Apache Hadoop, http://hadoop.apache.org/core/
3. Papadimitriou S, Sun J (2008) DisCo: distributed co-clustering with Map-Reduce. IEEE ICDM'08, 2008, pp 512–521
4. Yang HC, Dasdan A, Hsiao RL, Parker DS (2007) Mapreduce-merge: simplified relational data processing on large clusters. SCMD'07, 2007, pp 1029–1040
5. Li Y (2010) Research on parallelization of clustering algorithm based on MapReduce. Zhongshan University 2010, pp 30–33
6. Xue-song D, Jing Z, Qiang G (2010) A massive data management system based on the hadoop. Microcomput Inf 26(05-I):202–204

Hybrid Data Fusion Method Using Bayesian Estimation and Fuzzy Cluster Analysis for WSN

Huilei Fu, Yun Liu, Zhenjiang Zhang and Shenghua Dai

Abstract Data fusion is the process of combining data from multiple sensors in order to minimize the amount of data and get an accurate estimation of the true value. The uncertainties in data fusion are mainly caused by two aspects, device noise and spurious measurement. This paper proposes a new fusion method considering these two aspects. This method consists of two steps. First, using fuzzy cluster analysis, the spurious data can be detected and separated from fusion automatically. Second, using Bayesian estimation, the fusion result is got. The superiorities of this method are the accuracy of the fusion result and the adaptability for occasions.

Keywords Data fusion · Fuzzy cluster analysis · Bayesian estimation · Spurious data

H. Fu · Y. Liu (✉) · Z. Zhang · S. Dai
School of Electronic and Information Engineering, Beijing Jiaotong University,
Beijing 100044, China
e-mail: liuyun@bjtu.edu.cn

H. Fu
e-mail: 12120065@bjtu.edu.cn

Z. Zhang
e-mail: zhjzhang1@bjtu.edu.cn

S. Dai
e-mail: shdai@bjtu.edu.cn

H. Fu · Y. Liu · Z. Zhang
Key Laboratory of Communication and Information Systems, Beijing Municipal
Commission of Education, Beijing Jiaotong University, Beijing 100044, China

Y.-M. Huang et al. (eds.), *Advanced Technologies, Embedded and Multimedia
for Human-centric Computing*, Lecture Notes in Electrical Engineering 260,
DOI: 10.1007/978-94-007-7262-5_91, © Springer Science+Business Media Dordrecht 2014

Introduction

Sensor networks usually have a large number of sensor nodes to observe the interest parameter in the environment. This often results in data redundancy and repetition. As a consequence of that, the energy of the network is wasted. It is important to minimize the amount of data transmission so that the lifetime can be extended and the bandwidth can be saved. Data fusion is the process of combining data from multiple sensors in order to minimize the amount of data [1]. Many algorithms about data fusion have been proposed in literatures and books include arithmetic mean, Kalman filter, Bayesian estimation, Entropy theory, Dempster-Shafer theory, fuzzy logic and neural network [2–8]. As the energy of the network is limited, the data fusion algorithm for WSN is required to be easy. The objective of multisensor data fusion is to obtain an accurate, consistent and meaningful information. This information cannot be achieved by any single sensor in the network because of the uncertainties in the network. These uncertainties are mainly caused by two aspects, device noise and spurious measurement. Device noise here includes device inaccuracy and the noise in the environment. Spurious measurement is caused by sensor failure or even security attack. If any of the aspects is not considered during data fusion, it might lead to an inaccurate or erroneous result. Hence, a good data fusion algorithm should consider both of the aspects.

Considering the device noise and the spurious measurement, this paper proposes a new method for data fusion. The two aspects have different performance on the observation. The observation of each sensor is a combination of the true value of the interest parameter and the device noise. The aspect of device noise is reflected in the variance of the measured data. The quantification of the device noise can be obviously got by the variance of the measured data. The aspect of spurious measurement is mainly reflected in the mean value of the measured data. The quantification of the spurious measurement can be got by comparing the value of the observation between sensors. This method uses fuzzy cluster analysis to deal with the spurious data and then uses the Bayesian Estimation or random weighting method in [7] to deal with the device noise.

Related Work

A random weighting method for multisensory data fusion is proposed in [7]. This method achieves the fusion result in least mean square error. However, this method does not consider the probability that the sensor will provide spurious measurement and it will fail under that condition. Dealing with the spurious measurement, there are several methods reported in the literature. A Bayesian method is proposed in [8], it is effective in sensor validation and identification of inconsistent data. However, the spurious data is not eliminated totally and still has some impacts on the fusion result. Also, a constant value b_k is unsolved, different value will lead to different results. Fuzzy cluster analysis is another one of the most effective methods to recognize the inconsistency. A comparison of all the method in fuzzy cluster analysis is made in [9]. This paper finds out the best method in fuzzy cluster analysis. The conclusion of [9] is used in this paper. And the methods in [7, 8] are used as comparisons during simulation.

Hybrid Data Fusion Method

General Distribution Under Gaussian White Noise

The observation of each sensor is a combination of the true value of the interest parameter and the noise. Assume that x is the true value of the interest parameter, w is the value of device noise, y is the measured value. Then the equation of the measured data can be written as:

$$y = x + w \tag{1}$$

Assume that the device noise of each sensor is Gaussian white noise. σ_w is the standard deviation of Gaussian white noise and the noise has a mean value 0.

The distribution of the measured data is also Gaussian, and it can be written as

$$P(Y = y | X = x) = \frac{1}{\sqrt{2\pi}\sigma} e^{-\frac{(y-x)^2}{2\sigma^2}} \tag{2}$$

σ is the standard deviation of the measured data and it is equal to the standard deviation of Gaussian white noise. x is the true value of the interest parameter and it is also the mean value of the observation

$$Var[y] = E(y - E[y])^2 = E(y - x)^2 = E(w)^2 = \sigma_w^2 = \sigma^2 \tag{3}$$

Spurious Measurement Elimination

In fuzzy math, the fuzzy cluster analysis can give a quantitative determination of sample relationships using mathematical methods. According to incomplete statistics, there are 13 methods in fuzzy cluster analysis now. Such as Hamming distance, Euclidean distance, Chebyshev distance, absolute reciprocal... In the previous research [9], the author has given an analysis and comparison of 13 methods. It puts forward three principles to compare the 13 methods. The result shows that only the absolute reciprocal meet the requirement of the three principles. The formula of the absolute reciprocal is

$$r_{ij} = \frac{c}{\sum_{k=1}^{m} |y_{ik} - y_{jk}|} \quad i \neq j \tag{4}$$

r_{ij} is the similarity between sample y_i and y_j. c is a constant value. k is the round of measurement.

$$r_i = \sum_{j \neq i} r_{ij} \tag{5}$$

r_i is the overall similarity between sensor i and other sensors, or the support degree of sensor i from other sensors.

From the formula we know that if a measurement has a clear difference from other sensors, r_{ij} will be decreased. And if the sensor continues to provide spurious data, as the number of sensors and the amount of measurement increase, the decrement will become larger. This means if the sensor continues to provide spurious data, the support degree will decreases significantly. But if the sensor only provides a few spurious data, the support degree will still decrease but not very obvious and then this sensor will not be eliminated. So, if a sensor measurement has a clear difference from the other sensors and the other sensors' measurements are in agreement, the support degree of the sensor is low. Here, the support degree consists of k rounds measurements, because the measurement of one round has too much randomness, and the effect of fuzzy cluster analysis is not good.

For the purpose of spurious measurement elimination, a threshold k is needed to separate the spurious measurement.

In order to eliminate the effect of the constant c and give an appropriate threshold for all range of value, r_i needs to be normalized.

$$v_i = \frac{r_i}{\sum_i r_i} \tag{6}$$

Hence, if $v_i < k$, that means the support degree of sensor i from other sensors is low, the measurement of sensor i is considered as spurious measurement. The measurement should be eliminated. If $v_i > k$ the measurement of sensor i is considered in data fusion.

Threshold Selection

In Gaussian distribution, the probability that the point fall in $[x - \sigma, x + \sigma]$ is 0.6826. And the probability that the point fall in $[x - 3\sigma, x + 3\sigma]$ is 0.9974. Hence, we consider $x, x + \sigma, x + 3\sigma$ for setting the threshold k.

$$k = v_{x+3\sigma} = \frac{\frac{c}{3\sigma} + \frac{c}{2\sigma}}{\left(\frac{c}{3\sigma} + \frac{c}{2\sigma}\right) + \left(\frac{c}{\sigma} + \frac{c}{2\sigma}\right) + \left(\frac{c}{3\sigma} + \frac{c}{\sigma}\right)} = 0.2273 \tag{7}$$

Bayesian Estimation for Data Fusion

According to the Bayesian formula in Statistics,

$$P(x|y) = \frac{P(x)P(y|x)}{\sum P(x)P(y|x)} = \frac{P(x)P(y|x)}{P(y)} \tag{8}$$

x is the true value of the interest parameter, y is the measured value. Since the denominator depends only on the measurement (the summation is carried out over all possible values of state), an intuitive estimation can be made by maximizing this posterior distribution, by maximizing the numerator. This is called maximum a posteriori (or MAP) estimate [8], and the fusion formula is given by

$$x_{MAP} = (\sigma')^2 \left(\frac{z_1}{(\sigma_1)^2} + \frac{z_2}{(\sigma_2)^2} + \ldots + \frac{z_n}{(\sigma_n)^2} \right) \tag{9}$$

$$(\sigma')^2 = \left[(\sigma_1)^{-2} + (\sigma_2)^{-2} + \ldots + (\sigma_n)^{-2} \right]^{-1} \tag{10}$$

From above, we know that the standard deviation of the fused distribution is smaller than any of the individual sensor measurement distribution. Hence, the fused data is more accurate with less uncertainty.

Process of Hybrid Data Fusion Method

Step one: Get sensor measurement samples.
Step two: For each sensor i, calculate v_i. If $v_i < k$, sensor i is eliminated from data fusion. If $v_i > k$, sensor i is considered in data fusion.
Step three: For each sensor i considered in data fusion, calculate the σ_i.
Step Four: Use MAP estimate or random weighting in [7] to get fusion result.

Simulation Results

A simulation is made to show the performance of the proposed method. A comparison is made to show the improvement of the method. Four methods are compared together. They are arithmetic mean, random weighting proposed in [7], method proposed in [8] and the method proposed in this paper.

In order to simplify, assume there are 3 sensors in the sensor network. In simulation, the variance of the measurement is assumed. It is feasible because, in particular, the measurement variance can be calculated from the sample. The curves in the figures are not the standard Gaussian distribution because the samples are randomly selected. And according to the samples, the probability density is recalculated. This is in order to match the reality. Assume that each sample contains 100 points. In all figures, x label represents the value of the parameter, y label represents the probability of the fused data and the measured data. So the curves mean the probabilities of the measured and fused value at that point around the true value. Let x be the true value (Figs 1, 2, 3).

From the above simulations, it is obviously that the proposed new method can give an accurate result. In situation one, there is no spurious measurement. It is shown in the figure that the new method can give a result which is almost the same with the random weighting. In situation two and three, there is one sensor that provides spurious measurement. Although the ranges of the value are different, the new method can distinguish and eliminate the spurious measurement automatically. The new method can give a result which is more accurate than the random weighting and arithmetic mean. The random weighting method gives more weight to the sensor which has a smaller variance. This method only applies to the situation that none of the sensor provides spurious measurement. And when the variance of the spurious measurement is the smallest, it performs even worse than the normal arithmetic mean. Also the result is better than the method proposed in [8] and it does not need to set the constant value. Hence, the new method can be applied to all the situations.

However, there are still some weaknesses of the proposed method. If two of the three sensors provide spurious measurement, this method will fail. For extension, if the amount of the failed sensor is larger than the normal, this method will lead to inaccurate estimation. It is also the weakness of the fuzzy cluster analysis.

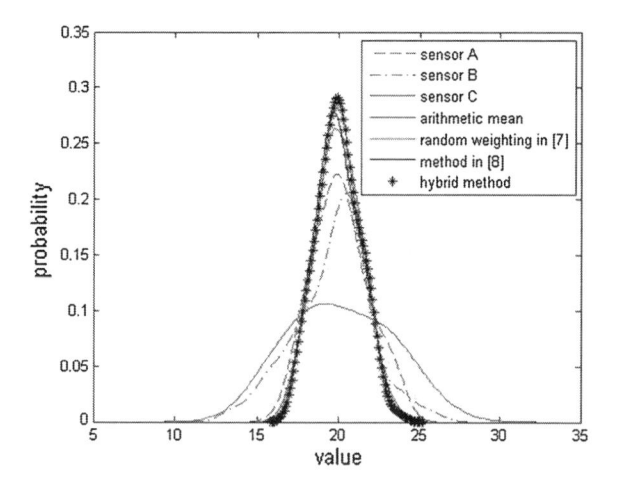

Fig. 1 Situation one.
Sensor A: $\sigma_A = 2, x = 20$,
Sensor B: $\sigma_B = 2.5, x = 20$,
Sensor C: $\sigma_C = 3, x = 20$

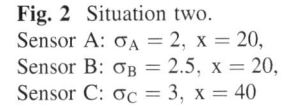

Fig. 2 Situation two.
Sensor A: $\sigma_A = 2$, $x = 20$,
Sensor B: $\sigma_B = 2.5$, $x = 20$,
Sensor C: $\sigma_C = 3$, $x = 40$

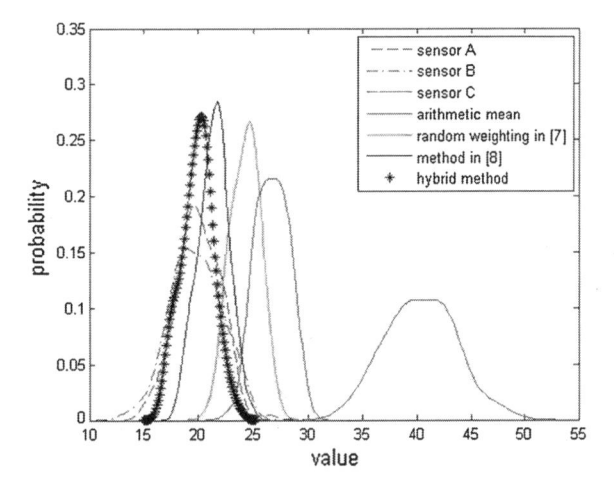

Fig. 3 Situation three.
Sensor A:
$\sigma_A = 0.02$, $x = 0.1$,
Sensor B:
$\sigma_B = 0.025$, $x = 0.1$,
Sensor C:
$\sigma_C = 0.03$, $x = 0.3$

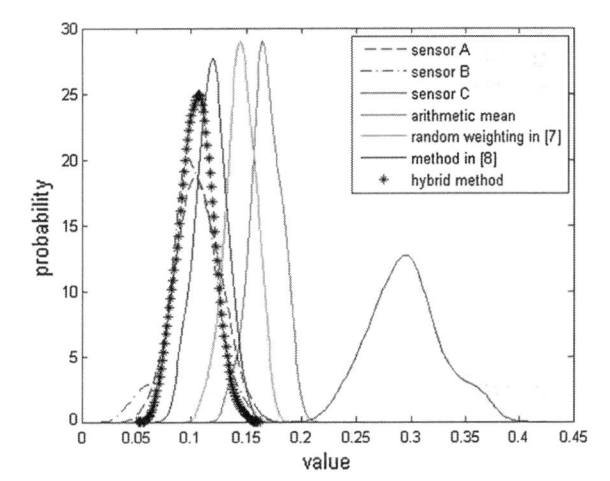

Conclusion

This paper gives a new method for multisensory data fusion. This method can distinguish and eliminate the spurious measurement automatically, regardless of the range of the value. And whether there is spurious measurement, the method can give an accurate result.

Acknowledgments This research is supported by National Natural Science Foundation of China under Grant 61071076, the National High-tech Research And Development Plans (863 Program) under Grant 2011AA010104-2, the Beijing Municipal Natural Science Foundation under Grant 4132057.

References

1. Ozdemir S, Xiao Y (2009) Secure data aggregation in wireless sensor networks: a comprehensive overview. Comput Netw 53:2022–2037
2. Bahador K, Khamis A, Karry FO, Razavi SN (2013) Multisensor data fusion: a review of the state-of-the-art. Inf Fusion
3. Welch G, Bishop G (1995) An Introduction to the Kalman Filter. Department of Computer Science, University of North Carolina, North Carolina
4. Yen J (1990) Generalizing the Dempster–Shafer theory to fuzzy sets. IEEE Trans SMC 20:559–570
5. Maskell S (2008) A Bayesian approach to fusing uncertain, imprecise and conflicting information. Inf Fusion 9:259–277
6. Zhu H, Basir O (2006) A novel fuzzy evidential reasoning paradigm for data fusion with applications in image processing. Soft Comput J—A Fusion of Foundations, Methodologies and Applications, 2006
7. Gao S, Zhong Y, Li W (2011) Random weighting method for multisensor data fusion. IEEE Sens J 11:1955–1961
8. Kumar M, Grag DP, Zachery RA (2007) A method for judicious fusion of inconsistent multiple sensor data. IEEE Sens J 7:723–733
9. Xinzhou W, Haichi S (2003) Construction of Fuzzy Similar Matrix. J Jishou Univ (Nat Sci edn)

Enhancements of Authenticated Differentiated Pre-distribution Key Methodology Based on GPSR

Lin Sun and Zhen-Jiang Zhang

Abstract Wireless sensor network consists of plentiful energy and computing power-constrained tiny sensor nodes. On the basis of the protection of security, the main goal of our study is to maintain the network life time at a maximum with the appropriate routing protocol. In this article, we will propose a new routing algorithm on the basis of the original routing algorithm. Considering the security of the wireless sensor network, the residual energy of nodes as well as issues such as physical distance, we try to extend the maximum lifetime of a wireless sensor network with all the combination of these factors for routing. This energy-balance routing algorithm takes into account the number of pre-distributed shared keys between the sending node and receiving node, the residual energy of the receiving node and the physical distance between two nodes in order to protect the security at the same time, to maintain the network life time at a maximum.

Keywords WSN · Energy saving · Life time · RKP

Introduction

With the rapid development of wireless sensor network, it has too many crucial applications in many aspects, i.e. military applications, environmental applications, health applications, home applications and numerous commercial applications [1].

L. Sun · Z.-J. Zhang
School of Electronic and Information Engineering, Beijing Jiaotong University, Beijing 100044, China
e-mail: 11120222@bjtu.edu.cn

L. Sun · Z.-J. Zhang (✉)
Key Laboratory of Communication and Information Systems, Beijing Municipal Commission of Education, Beijing Jiaotong University, Beijing 100044, China
e-mail: zhjzhang1@bjtu.edu.cn

Y.-M. Huang et al. (eds.), *Advanced Technologies, Embedded and Multimedia for Human-centric Computing*, Lecture Notes in Electrical Engineering 260, DOI: 10.1007/978-94-007-7262-5_92, © Springer Science+Business Media Dordrecht 2014

Compared with traditional networks, wireless sensor network has many advantages, but also because of its features of battery-powered, large number of nodes, mobility, and low processing capacity defect, so that there are a lot of restrictions on its applications. This requires complete the maximum amount of data transfer with minimal energy cost under the resource-constrained condition such as computing power and storage capacity.

At the same time, wireless sensor nodes are often deployed in public, untrusted or even hostile environments, which often bring a large quantity of security problems. The security problems will be more important in wireless sensor network than in other networks. In all safety problems, the security of data transmission is particularly significant. The main target in the design of the security agreement is to provide a safe and reliable security data communication between sensor nodes. The first key agreement protocol was introduced by L. Eschenauer and V. D. Gligor called random key pre-distribution (RKP) which has been very popular during these years.

Random key pre-distribution (RKP) is the one of the key distribution that is developed for the sake of safe communication of WSN. The main characteristic of RKP [2] is that each sensor is pre-distributed with k distinct keys randomly chosen from a key pool with the size of K keys before deployment. Then two neighboring sensors attempt to establish a pair-wise key with the pre-distributed keys, for secure communications between themselves [3].

Our contributions: In this paper, we show some improvements in their routing protocol and the differentiated key pre-distribution algorithm. We demonstrate that the next hop will be chosen not only considering the residual energy of the receiving node but also the physical distance between the two nodes. Moreover, we propose a simple modification to the original protocol to protect the security while taking the problem of network lifetime into account.

Paper Organization: The remainder of the paper is organized as follows: section Background and Related Work focuses on the background of RKP Scheme and routing protocol GPSR and section Review of the Methodology reviews the differentiated key pre-distribution methodology. In the next section the weaknesses of the former methodology are discussed while the our revised protocol are proposed in section Our Revised Version of the Protocol. Final section concludes the paper.

Background and Related Work

Something About RKP

In this section, we present a basic introduction of the RKP algorithm., including an extension of the concrete steps and a variety of RKP algorithm.

Basic RKP Scheme.

RKP is one of the key distribution that is developed for the sake of safe communication of WSN. RKP specific algorithm can be divided into two steps.

The first step is the key pre-distribution and the second is the pairkey establishment. In the first step the process of the key pre-distribution, each node randomly selects the same number of k keys from key pool with the size of K keys as its pre-distribution keys. After the first step, we carried out the second step pairkey establishment. Nodes communicate with its neighbors within their communication range in order to find the specific neighbor which has the same pre-distributed keys with them, then nodes, with their neighbors, establish pairkeys which can be used to encrypt/decrypt communications between neighboring sensors in each hop.

Variants of RKP Scheme.

In [4], it demonstrates a probabilistic unbalanced distribution of keys throughout the network, which distributes different number of keys to nodes according to the function of the specific each node. In the area of network security, the cluster head nodes are responsible for key distribution, updating the keys, the expulsion of nodes and seeking common features such as recovery. Taking into account that the cluster head node is in the important role of wireless sensor network, the cluster head node needs to ensure that the information is not procure, through encryption and key management.

In [5], considering in homogeneous sensor networks, where all sensor nodes have identical capabilities in terms of with limited computation, communication, energy supply and storage capability, nodes are divided into the H nodes and L nodes. H sensors are more powerful nodes and have more computation, communication, energy supply and storage capability, while L nodes have limited computing power, communication capabilities, energy supply and storage capability.

In [3], three random key pre-allocation algorithm, the q-composite random key pre-distribution scheme, the multi-path key reinforcement scheme and the random-pairwise keys scheme are presented, proposing the idea of multi-key path to increase the flexibility of the link between the two points, which can better able to protect the security of data transmission under attack.

In [6], each node with a simple calculation ability based on a symmetric key system only has a small number of pre-shared keys regardless of the size of the entire sensor network. While the multipath key reinforcement scheme in [7] uses multiple physically disjoint paths between two sensor nodes and decides the number of pre-distributed keys according to the specific security level.

Something About GPSR

This article will present some improvements to the original routing protocol GPSR. In this well known location centric routing protocol, the node uses greedy algorithm to make the choice of the next hop from its neighbour nodes. If the distance from all the neighbors to the destination are further than the current node, then there will be the existence of a "hole". The node will follow the right-hand rule to forward along the "hole" in the border until it finds a distance closer to the destination node, and then forward packets using greedy algorithm.

Energy Saving in WSN

In the design of network routing protocols, the node energy consumption and the network energy balance problems will be considered to avoid frequent use of a path or a few nodes. The energy of these nodes will soon run out [8], which can not only result in the incomplete coverage of the network to obtain monitoring data, but also seriously affecting the entire wireless sensor network security.

We will do a brief introduction to a few formulas in this article concerning the energy-efficient wireless sensor networks. In [9], based on the preliminary report on this work in [10] and comparing with [11], the methodology gives a weight value to each of its neighbor based on the initial energy level of node energy and the current energy level, used to indicate the node energy consumption of the neighbor how much, so that in the routing. Then in [12], node, in the different pre-distribution key methodology, are divided into c classes, so the difference is to abandon the node graded idea in this paper.

Review of the Methodology

In this section, we depict the intuition and technical details of the differentiated key pre-distribution methodology and the GPSR routing protocol.

The core idea of the original text methodology is to pre-distribute different number of keys to different nodes. The quantity of N nodes is divided into c classes, and each class of each node is assigned with k_i ($1 \leq i \leq c$) keys. In the key pre-distribution phase, each node of the first class randomly selected k1 keys from the key pool with the size of K. Then the first class k nodes are divided into two parts: A part consists of $n_1 - (k_i - \lfloor k_i/n_1 \rfloor \cdot n_1)$ nodes, well B part possess $(k_i - \lfloor k_i/n_1 \rfloor \cdot n_1)$ nodes. In order to increase the probability of the key share between nodes, other class of nodes apart from class 1 nodes, follow the semi-random key pre-distribution process. The process is as follows: each node in class i ($1 \leq i \leq c$) randomly chooses $\lfloor k_i/n_1 \rfloor$ keys from each of the A part nodes, and at the same time chooses $\lceil k_i/n_1 \rceil$ keys from each of the B part ones. If some of the chosen keys are the same, repeat the above steps until all k_i keys are distinct. We define $\lceil x \rceil$ as the smallest integer no less than x, while $\lfloor x \rfloor$ is the largest integer no more x.

After the key pre-distribution phase, the node begins to communicate with its neighbors within its communication range γ and gets the ID of its neighbors as well as their pre-distributed keys. Node i establishes a one-hop or two-hop secure communication link by sending request to its secure neighbors which have the same pre-distributed key with node i. We assume that there are p two-hop links between node i and node j, and define s_l ($1 \leq l \leq p$) as the proxy of each link, and k(i, j) as the number of the shared keys between node i and node j. Then the protection keys can be obtained as follows,

$$\text{key}(i,j) = k(i,j) + \sum_{l=1}^{p} \min(k(i,s_l), k(s_l,j))$$

Routing

The original text presents some improvements to GPSR in order to better protect the security of point to point communication. We give weight values to secure neighbors which are closer to the sink than the node itself. $U(i)$ is defined as the set of the secure neighbors which are closer to the sink than the node itself and key(i, j) as the number of the protection keys between node i and node j. So the weight value can be obtained as,

$$\omega = \frac{key(i,j)^{\alpha}}{\sum_{m \in U(i)} key(i,m)^{\alpha}}$$

ω_j is the probability that node j is chosen as the proxy of node i. When α is zero, all nodes have the same probability of being selected independent of the resilient of the link. While when α is positive, the bigger the resilient of the link, the higher the probability that the node will be selected. Only the highest resilient link can be chosen when α trends toward infinity. Mean while the selection of α also has something with the attention of the system to the network security and the lifetime.

Weaknesses of the Differentiated Key Pre-distribution Methodology

In the original text, the weight value only pays attention to the number of the protection keys on the link, as a result the energy consuming will be much higher than the links with lower resilient and energy of nodes on the high resilient link drain out rapidly. The disappearing of the high resilient links brings a more negative impact on the security of the whole network.

Moreover the α in the formula takes the balance of the network lifetime and the security into consider. When α takes to a relatively high value, it means the link with high resilient is given more priority and the wireless sensor network considers more about the security, as a result the energy of the nodes on these links drain out rapidly, which, to a very great extent, contraries to our starting point for trying to improve network security.

Our Revised Version of the Protocol

Data transmission in the wireless sensor network needs the help of the organic collaboration between nodes. While sensor nodes powered by batteries, have limited computing, communications, energy and other resources. Considering that, an important problem at the same time ensuring the security of data transmission we also needs to solve is how to save energy, in order to prolong the life cycle of wireless sensor network.

In this paper, we will preprocessing the data collected by the nodes before the data transmission, and define a new weight value. The new weight value W_{ij} not only takes the protection of security on the basis of saving energy into account, but also able to balance the residual energy. Moreover the W_{ij} adjusts the routing protocol according to the residual energy of the node itself, which makes a contribution to entend the lifetime of the node. The weight value W_{ij} can be obtained as,

$$W_{ij} = \theta^{\alpha} \omega^{\beta}$$

In this formula α, β are two parameters that control the relative weight of security and network lifetime. When α is zero, the weight value considers more about the number of the protection keys on the link which means the security of the link. On the contrary, when β is zero, the residual energy of the receiving node and the distance between the transmitting node and the receiving node will be considered more, which pay more attention to the problem of energy. When α tends to infinity, the situation is almost the same when β is zero. While when α, β are positive, the weight value makes a combination of the security and the network lifetime. An appropriate α, β can be used to make a good trade-off in the energy and life and security, on the basis of the protection of security, and at the same time as far as possible to extend the life of the nodes to reduce energy consumption.

This weight value in this paper uses parameter ω as the considering of the security and the ω is defined in the original text.

θ is introduced in this algorithm as the consider of the residual energy of the receiving node and distance of the two nodes when choosing the next hop routing in order to make a better trade off between the security and the energy of the node. θ can be defined as follows,

$$\theta = \frac{\eta(i,j)^{-\alpha_2}}{d(i,j)^{\beta_2}}$$

In order to protect the energy balance in the data transmission, this algorithm is dedicated to the residual energy of the receiving node to calculate the weight values, and find the node with highest weight value to transfer data. Improvements on the basis of this algorithm to get rid of the assumptions on the grade if nodes, but only consider the physical distance between the nodes and the residual energy of the receiving node, have been made. Suppose that node j is the receiving node of node i, and $\eta(i, j)$ is the energy consumption of the secure neighbor of node i, $d(i, j)$ is the distance between node i and node j. In the formula, α and β represent the rest energy factor and the distance factor. α_2, β_2 are the parameters make a perfect trade off between the residual energy of the receiving node and the distance between the two nodes. The system will only consider the physical distance between two nodes on the condition that when α_2 is zero and in the other case when β_2 is zero, the residual energy will be the only consideration. Moreover when α_2 is positive, the node with more residual energy will get a bigger weigh value than the one with the closer distance, and at the same time will be chosen preferentially when routing. When α_2 tends to infinity, only the node with the most residual energy can be chosen. The value of parameter β_2 is roughly the same with α_2. During the routing, node i will take the residual energy of the receiving node as well as the distance between two nodes into consideration because of the the the existence of α_2 and β_2. On the basis of the original algorithm, taking into account the residual energy issues as well as physical distance, we can ensure the nodes with less residual energy while on the high resilient link bear less communication traffic, but the surplus energy nodes on the relatively lower resilient link take part in the data communication tasks to make a better balance of energy. In this way, it may extend the network life time.

The weight coefficient $\eta(i,j)$ of all the nodes j in communication radius of i is defined as the following,

$$\eta(i,j) = \frac{(I - e_j)^{-1}}{\sum_{n \in U_{(i)}} (I - e_j)^{-1}}$$

In this formula, $\eta(i, j)$ is a weight value about the condition of the energy consumption of the secure neighbor of node i. I can be defined as the initial energy level and with the passage of time, the energy consumed gradually. Then e_j is the current energy level of the receiving node j. To simplify the problem, we set a proper time T as the time period and after time T, the node would broadcast to its neighbors their energy levels, so that its neighbors will use the value of its residual energy during routing as a reference. At last choosing $U_{(i)}$ as the set of the secure neighbor of node i. Above all, the node with lower residual energy is less likely to be chosen as the next hop than the node with the higher residual energy level. $\eta(i, j)$ can be a crucial reference as the measure of the secure neighbor of node I when routing.

Description of the Algorithm:

Step 0. Initial the network and then divide the nodes into c classes.

Step 1. Pre-distribute the keys to each node and then establish the pairwise keys between secure neighbors.

Step 2. Collect the information from its secure neighbors including the residual energy level and the distance between them.

Step 3. Get the weight value according to the formula 6 to choose the perfect next hop.

Step 4. Circulate step 2 until reach the destination sink node.

This algorithm is dedicated to finding a routing algorithm, not only in the protection of security on the basis of energy conservation, but also to make a perfect energy balance. More importantly, the algorithm in this paper is to adjust the routing protocol according to the residual energy level of the node itself, so we can extend the lifetime of the node to the maximum.

Conclusion

In this paper we show the weakness of the differentiated key pre-distribution protocol which only takes the resilience of links into consideration. First it is overly dependent on the link resilience to choose the next hop during routing, as a result the energy of nodes on the high resilient links will drain out rapidly. Then the security of the whole network will seriously affected. In general, the weaknesses of protocols in the original text are arisen from the only consideration of resilience of link during routing.

We also provide a revised version of this protocol which adds two crucial parameters into the routing weight value which are the residual energy level and the distance between the two nodes. The studied protocol, on the basis of ensuring the security of the network, makes a wonderful energy balance in the whole network.

So designing a perfect secure routing protocol has been taken into account as our future work.

Acknowledgments This research is supported by National Natural Science Foundation of China under Grant 61071076, the National High-tech Research and Development Plans (863 Program) under Grant 2011AA010104-2, the Beijing Municipal Natural Science Foundation under Grant 4132057.

References

1. Du W et al (2003) A pairwise key pre-distribution scheme for wireless sensor networks. In: Proceedings 10th ACM conference computer and communication security, Oct 2003, pp 42–51
2. Eschenauer L, Gligor VD (2002) A key-management scheme for distributed sensor networks. In: Proceedings 9th ACM conference computer and communication security. Security, Nov. 2002, pp 41–47
3. Chan H, Perrig A, Song D (2003) Random key predistribution schemes for sensor networks. Department of Electrical and Computer Engineering, Paper 20
4. Patrick Traynor, Heesook Choi, Guohong Cao, Sencun Zhu and Tom La Porta (2006) Establishing pair-wise keys in heterogeneous sensor networks. In: INFOCOM 2006. Proceedings 25th IEEE international conference on computer communications
5. Poornima AS, Amberker BB (2008) Tree-based key management scheme for heterogeneous sensor networks. In: Networks, 2008. ICON 2008. 16th IEEE international conference on 2008
6. Zhu S, Xu S, Setia S, Jajodia S (2003) Establishing pairwise keys for secure communication in ad hoc networks: a probabilistic approach. In: The 11th IEEE international conference on network protocols, IEEE, 2003
7. Chan H, Perrig A, Song D (2003) Random key predistribution schemes for sensor networks. In: Proceedings of the IEEE security and privacy symposium 2003, May 2003
8. Wu CX, Liu Y (2012) WSN on-demand multipath routing protocol based on energy-aware. In: Comput Eng 38(9)
9. Gu W, Dutta N, Chellappan S, Bai X (2011) Providing end-to-end secure communications in wireless sensor networks. IEEE Trans Netw Serv Manag 8(3):205–218

10. Camilo T, Carreto C, Silva J, Boavida F (2006) An energy-efficient ant base routing algorithm for wireless sensor networks. In: ANTS 2006—Fifth international workshop on ant colony optimization and swarm intelligence, 4150, pp 49–59
11. Okdem S, Karaboga D (2006) Routing in wireless sensor networks using ant colony optimization. In: First NASA/ESA conference on adaptive hardware and systems—AHS, pp 401–404
12. Xue J, Qi X, Wang C (2011) An energy-balance routing algorithm based on node classification for wireless sensor networks. J Comput Inf Syst 7:2277–2284

Research on Kernel Function of Support Vector Machine

Lijuan Liu, Bo Shen and Xing Wang

Abstract Support Vector Machine is a kind of algorithm used for classifying linear and nonlinear data, which not only has a solid theoretical foundation, but is more accurate than other sorting algorithms in many areas of applications, especially in dealing with high-dimensional data. It is not necessary for us to get the specific mapping function in solving quadratic optimization problem of SVM, and the only thing we need to do is to use kernel function to replace the complicated calculation of the dot product of the data set, reducing the number of dimension calculation. This paper introduces the theoretical basis of support vector machine, summarizes the research status and analyses the research direction and development prospects of kernel function.

Keywords Support vector machine · High-dimension data · Kernel function · Quadratic optimization

Introduction

Support vector machine (SVM) was introduced into the field of machine learning and its related area in 1992 [1], having received widespread attention of researchers in later time and has made great progress in many fields. It uses a

L. Liu · B. Shen
School of Electronic and Information Engineering, Beijing Jiaotong University, Beijing 100044, China
e-mail: 11120126@bjtu.edu.cn

L. Liu · B. Shen (✉)
Key Laboratory of Communication and Information Systems, Beijing Municipal Commission of Education, Beijing Jiaotong University, Beijing 100044, China
e-mail: bshen@bjtu.edu.cn

X. Wang
China Information Technology Security Evaluation Center, Bejing, China
e-mail: wangx@itsec.gov.cn

Y.-M. Huang et al. (eds.), *Advanced Technologies, Embedded and Multimedia for Human-centric Computing*, Lecture Notes in Electrical Engineering 260, DOI: 10.1007/978-94-007-7262-5_93, © Springer Science+Business Media Dordrecht 2014

nonlinear mapping to map original training data into high-dimensional data space in order to find the optimal classification hyper plane separating those data belonging to different categories. Support vector machine is based on SLT (statistical learning theory) [2, 3] VC dimension theory and structural risk minimization principle. Compared with traditional neural networks, support vector machine gains great enhancement in generalization ability and overcomes some problems existing in feed-forward neural networks, such as local minimum and the curse of dimensionality [4]. The introduction of kernel function greatly simplifies the complexity of dot product operation in support vector machine for nonlinear data classification, and it can be used to distinguish and enlarge the useful features, and support vector machine based on kernel function is playing a powerful role in the field of data mining.

Support Vector Machine

If the training data set is linear separable, the given data set is $D: (X_1, y_1)$, $(X_2, y_2), \ldots, (X_{|D|}, y_{|D|})$, among which X_i is training data with class label y_i. The scope of each y_i is $+1$ or -1, namely $y_i \in \{+1, -1\}$. In dealing with classification problem, the optimal classification hyper plane can be denoted as follows:

$$W \cdot X + b = 0. \tag{1}$$

W is a weight vector, that is to say, $W = \{w_1, w_2, \ldots, w_n\}$, where w_i is the weight of X_i; n is the number of attributes; parameter b is a scalar, and is often referred to as the bias. The formula $W \cdot X$ stands for the dot product of W and X. From geometric point of view, the entire input space can be divided by hyper plane into two parts: one part is positive value data set; another part is the negative one. Hyper plane is a line in two-dimensional space, a surface in three-dimensional space. The biggest edge distance between two types of training data set is $\frac{2}{\|W\|}$. Support vector machine discovers the optimum classification hyper plane by means of support vectors and the edges between them [5] and gets the maximum edge distance of two classes of data sets at the same time.

Research on Kernel Function

For linear separable data, support vector machine can directly classify the data set into two categories in the input space; for those nonlinear separable data, SVM has to map the original input data X with nonlinear mapping ($\Phi : X \to F$) into another high-dimensional space where we can solve the maximum interval of classification, and this new high-dimensional space is the feature space. A dot product operation can be directly substituted by kernel function in feature space, and we needn't know the concrete eigenvector and mapping function, which is also known as kernel trick. Frequently-used kernel functions include the following ones:

$$\text{Linear kernel function: } K(X_i, X_j) = X_i \cdot X_j. \tag{2}$$

$$\text{Polynomial kernel function: } K(X_i, X_j) = (X_i \cdot X_j + 1)^h. \tag{3}$$

$$\text{Gaussian radial basis function kernel function: } K(X_i, X_j) = \ell^{-\|X_i - X_j\|^2 / 2\sigma^2}. \tag{4}$$

$$\text{Sigmoid kernel function : } K(X_i, X_j) = \tanh(\kappa X_i \cdot X_j - \delta). \tag{5}$$

The performance of support vector machine mainly depends on model selection, including the selection of the kernel function type and the kernel parameter selection [6]. In the study of kernel function selection, the kernel alignment is a good method [7]. Kernel alignment method is based on a hypothesis: the kernel matrix of a good kernel function should be as similar as possible to the calibration matrix. Under normal circumstances, the first consideration of choosing kernel function is the Gaussian radial basis function kernel function (RBF), which is because RBF has fewer parameters to select, and what's more, for some parameters, RBF has similar performance to the Sigmoid kernel function.

The development and application of support vector machine have been greatly promoted since the introduction of kernel function, and its application area has extended from hand-written numeral recognition, reference time series prediction test and other traditional application area to new areas such as information image processing [8], industrial process control, etc. The following content in this paper will center on the discussion of kernel function in support vector machine and put forward its future research directions.

Kernel Clustering

Clustering analysis divides data objects into different subsets and the data objects in the same subset are similar to each other, while those located in different subsets have different properties. Kernel clustering combines kernel function and clustering together, which is based on the characteristics of clustering [9]. In the first stepwise, kernel clustering clusters the training data and test data, and then constructs the kernel function on the basis of clustering results. Chapell et al. [10] proposed an overall framework of constructing kernel clustering, using different conversion functions to change the eigenvalue decomposed by kernel matrix. Jason Weston et al. came up with the bagged clustering kernel [10] to overcome the time complexity problem existing in Chapell's clustering kernel. Although the bagged clustering kernel shortened the kernel clustering time, there still is much room for improvement in terms of classification accuracy.

Fuzzy C-means (FCM) clustering algorithm [11] introduces fuzzy set theory to the process of clustering. Fuzzy kernel clustering algorithm firstly maps the data of input space into high-dimensional feature space to enlarge pattern differences between classes, and then carry through fuzzy clustering in the feature space [12, 13]. Fuzzy kernel clustering algorithm is able to highlight the differences of different sample characteristics, increasing clustering accuracy and speed [14]. It has always been an important research problem to expand SVM classifier to multi-class classification [15, 16], so Zhao et al. [17] applied fuzzy kernel clustering to multi-class classification method, solving the serious problem of fuzzy overlapping, but didn't have a large increase in classification speed.

The objective function of the fuzzy kernel clustering algorithm is as follows:

$$J_n(U, \ v) = \sum_{i=1}^{c} \sum_{j=1}^{m} u_{ij}^n \left\| \Phi(x_j) - \Phi(v_i) \right\|^2. \tag{6}$$

In the above formula, parameter c is the clustering number initially set; v_i is the initialized clustering center; u_{ij} is the membership function of sample j belonging to i category; $U = \{u_{ij}\}, v = \{v_1, v_2, \ldots, v_c\}$, parameter n is weighted index, and $n > 1$. The criterion of fuzzy clustering algorithm is for the minimum value of the above objective function until each membership value stabilized.

Kernel clustering support vector machine has been successfully utilized in biomedical, text classification [18] and many other application fields, undoubtedly, it will involve a wider range of application fields in later time.

Super-Kernel Function

Support vector machine for data classification often brings about two dilemmas using one single kernel function: one is unable to complete effective nonlinear mapping; the other is over-fitting or under-fitting [19]. The extrapolation ability of Gaussian radial basis function kernel function weakens along with the increase of σ parameter [20], so it has strong locality. Polynomial kernel function regulates different mapping dimensions through adjusting h parameter, and computation grows with h parameter, thus having strong global property and poor locality.

In order to adapt to the increase on data set and high efficiency requirements, many algorithms have been developed, such as large data set training [21] algorithm, super kernel learning [22] algorithm, fast convergence algorithm [23], etc. Combining several kernel functions of different categories with a polynomial composition can give full play to the excellent characteristics of different kernel functions in dealing with data classification, and overcomes the shortage of single kernel function while maintaining translational invariance and rotation invariance, which provides a fresh effective way to studying the construction of kernel function in support vector machine. A simple form of super-kernel function is like the following formula:

$$K(X_i, X_j) = \beta_1 \ell^{-\left\|X_i - X_j\right\|^2 / 2\sigma^2} + \beta_2 (X_i \cdot X_j + 1)^h. \tag{7}$$

It is the linear combination of Gaussian radial basis function kernel function and h polynomial kernel function. Super-kernel function parameters were regulated in the form of parameter vector, that is to say, super-kernel function adjusts all parameters at the same time, and those several more parameters just add to the length of parameter vector but not affect the determination time of parameters.

Kernel Parameter Selection

Selection of kernel parameter has a direct impact on the performance of the SVM classifier. A commonly used parameter selection method is based on the Generalization Error estimates [24]. The Generalization Error estimates predicate and forecast the generalization capability of classification decision criteria by means of training data set [25]. Vapnik et al. [26] estimate generalization ability by span of support vectors on the basis of Generalization Error estimates, which has an advantage of higher accuracy but has more complex computation. A method of utilizing data set to evaluate the optimal kernel parameter method [27] is put forward on the basis of Vapnik's method, determining the optimal kernel parameter choice from a geometric point of view.

Qi et al. [28] proposed a kernel parameter selection method to solve LOO (leave-one-out) upper bound minimum point based on genetic algorithm, where we can choose the reproduction operator and combine genetic algorithm with steepest descent method, improving the accuracy of forecast but without leading to local optimal solution. Chen et al. [29, 30] adopted different generalization ability estimates as the fitness function of genetic algorithm, and it not only reduced the computation time to choose parameters but also the dependence on the initial value.

Parameter selection method based on kernel matrix similarity measure starts from research on kernel matrix in order to search for the optimal kernel parameter and learning model, and this method improved the calculation speed of SVM. Liu Xiangdong et al. [31] eventually found the optimal kernel parameters and the kernel matrix by means of experimenting on UCI standard data set and FERET standard faces library. Parameter selection method based on kernel matrix similarity measure can serve as a feasible method to choose the optimal SVM model, and it also has certain reference value for choosing other kernel parameters.

Conclusion

The study of kernel function of SVM is an important data mining research, so choosing the appropriate kernel function and its parameters can give full play to the performance of SVM and even has remarkable significance in promoting the popularization and application of data mining. This paper does research on the kernel function of SVM and does some summary comments on the kernel clustering, super kernel function and the selection of kernel parameters. Judging from the current study, the author believes that the study of kernel function in the following areas is to be further developed:

1. Finishing data mapping efficiently and reliably in the environment of big data. "Big data" has features of giant, high growth rate and diversification, and it needs new processing mode to excavate useful information from the massive data and get an insight in them. In this case, the conventional transformation of kernel function will face with new bottleneck in processing speed and processing quality. On the basis of existing research, it remains further study to extend the scale of the expansion of kernel function processing data and select the appropriate kernel parameters and further improve the quality and speed of processing data.

2. Giving full play to the advantages of different kernel functions in super-kernel functions. On the perspective of present research on the selection of kernel parameters, Gaussian radial basis function (RBF) and its parameter selection have been more detailed studied in the field by virtue of its favorable advantages in computer vision. How to expand the application area of polynomial kernel function and Sigmoid kernel function, especially how to give full play to the advantage of each kernel function in the super-kernel function still needs deeper exploration. It is worth studying that selecting and optimizing super-kernel function parameters and applying the concept of constructing super-kernel function to support vector machines.

3. Selecting appropriate kernel function of support vector machines for specific applications. The scope of data mining processing data is developing from structured data to the direction of semi-structured and unstructured data. As one part of the data mining classification algorithms, the application fields of SVM continue to expand, thus it appears particularly important to select the appropriate kernel functions of specific domain. The choice of kernel function is closely related to the data field [32], in the meantime, the performance of kernel function depends largely upon the selection of parameters. Future studies are required to determine the kernel function and its parameters according to different application areas of the support vector machine for the sake of reducing the consuming of storage space and computing time of computers.

Big data processing is the future research tendency, and with the arrival of the cloud era, big data is attracting more and more attention. In future work, the author will mainly focus on the study of achieving efficient and accurate classification with support vector machine in the environment of big data.

Acknowledgments This work has been supported by the National Natural Science Foundation of China under Grant 61172072, 61271308, and Beijing Natural Science Foundation under Grant 4112045, and the Research Fund for the Doctoral Program of Higher Education of China under Grant W11C100030, the Beijing Science and Technology Program under Grant Z121100000312024.

References

1. Boser B, Guyon I, Vapnik V (1992) A training algorithm for optimal margin classifiers [C]. In: Proceedings of the 5th annual ACM conference on computational learning theory, Pittsburgh, pp 144–152
2. Vapnik VN (2000) The nature of statistical learning theory [M]. Translated by Zhang Xuegong (trans: Zhang X). Tsinghua University Press, Beijing
3. Vapnik VN (2004) Statistical learning theory [M]. Translated by Xu Jianhua, Zhang Xuegong (trans: Xu J, Zhang X). Publishing House of Electronics Industry, Beijing
4. Vapnik V (1995) The nature of statistical learning theory [M]. Springer, New York
5. Ju C, Guo F (2010) A distributed data mining model based on support vector machines DSVM [J]. 30(10):1855–1863
6. Zhu S, Zhang R (2008) Research for selection of kernel function used in support vector machine [J]. Sci Technol Eng 8(16):4513–4517
7. Cristianini, N, Taylor Shawe J, Kandola J et al. (2002) On kernel target alignment. In: Proceedings neural information processing systems. MIT Press, Cambridge, pp 367–373
8. Yang Z (2008) Kernel-based support vector machines [J]. Comput Eng Appl 44(33):1–6
9. Li T, Wang X (2013) A semi-supervised support vector machine classification method based on cluster kernel [J]. Appl Res Comput 30(1):42–45
10. Tison C, Nicolas JM, Tupin F et al. (2004) A new statistical model for Markovian classification of urban areas in high-resolution SAR images [J]. IEEE Trans Geosci Remote Sens 42(10):2046–2057
11. Bezdek JC (1981) Pattern recognition with fuzzy objective function algorithms. Plenum Press, New York
12. Wu Z, Gao X, Xie W (2004) A study of a new fuzzy clustering algorithm based on the kernel method. Journal of Xi 'an university of electronic science and technology magazine, 31(4):533–537
13. Zhang N, Zhang Y (2010) Support vector machine ensemble model based on KFCM and its application [J]. J Comput Appl 30(1):175–177
14. Cao W, Zhao Y, Gao S (2010) Multi-class support vector machine based on fuzzy kernel clustering [J]. CIESC Journal 61(2):420–424
15. Angulo C, Parra X (2003) K-SVCR Andreu Catala. A support vector machine for multi-class classification [J]. Neurocomputing 55(9):55–77
16. Platt JC, Cristianini N, Shawe-Taylor J (2000) Large margin DAGs for multiclass classification [J]. Adv Neural Inf Proc Syst 12(3):547–553
17. Zhao H, Rong L (2006) SVM multi-class classification based on fuzzy kernel clustering [J]. Syst Eng Electron 28(5):770–774
18. Yang Z (2008) Research progress of the kernel function support vector machine [J]. Sci Technol Inf 19:209–210
19. Jia L, Liao S (2008) Support vector machines with hyper-kernel functions [J]. Comput Sci 35(12):148–150
20. Guo L, Sun S, Duan X (2008) Research for support vector machine and kernel function [J]. Sci Technol Eng 8(2):487–489
21. Collobert R, Bengio S (2001) SVM torch: support vector machines for large-scale regression problems. J Mach Learn Res 1:143–160
22. Cheng SO, Smola AJ, Williamson RC (2005) Learning the kernel with hyper-kernel. J Mach Learn Res 6:1043–1071
23. Platt J, Burges CJC (1998) Fast training of support vector machines sequential minimal optimization. In: Sholkpof B, Smola AJ (eds) MIT Press, Cambridge
24. Tao W (2003) Kernels' properties, tricks and its application on obstacle detection [J]. National University of Defense Technology, Changsha
25. Y Fu, D Ren (2010) Kernel function and its parameters selection of support vector machines [J]. Sci Technol Innov Herald 9:6–7

26. Chapelle O, Vapnik V, Bousquet O et al (2002) Choosing multiple parameters for support vector machines [J]. Mach Learn 46(1):131–159

27. Men C, Wang W (2006) Kernel parameter selection method based on estimation of convex [J]. Comput Eng Des 27(11):1961–1963

28. Qi Z, Tian Y, Xu Z (2005) Kernel-parameter selection problem in support vector machine [J]. Control Eng China 12(44):379–381

29. Chen PW, Wang JY, Lee HM (2004) Mode selection of SVNs using GA approach [C]. Proceedings of 2004 IEEE international joint conference on neural networks. IEEE Press, Piscataway, pp 2035–2040

30. Zheng CH et al. (2004) Automatic parameters selection for SVM based on GA [C]. Proceedings of the 5th World congress on intelligent control and automation, IEEE Press, Piscataway, pp 1869–1872

31. Liu X, Luo B, Qian Z (2005) Optimal model selection for support vector machines [J]. J Comput Res Dev 42(4):576–581

32. Wang T, Chen J (2012) Survey of research on kernel selection [J]. Comput Eng Des 33(3):1181–1186

Improved Multi-dimensional Top-k Query Processing Based on Data Prediction in Wireless Sensor Networks

Zhen-Jiang Zhang, Jun-Ren Jie and Yun Liu

Abstract Since the scale of wireless sensor networks is expanding and one single node can sense a variety of data, selecting the data of interest to users from a tremendous data stream has become an important topic. With further development in the field of WSN query, extensive research is being conducted to solve different kinds of query issues. Skyline is a typical query for multi-criteria decision making, and many applications have been developed for it. Studies of multi-dimensional top-k query processing have proven it to be more efficient than traditional centralized scheme. In some cases, variations of observed conditions, such as temperature and humidity, are related to time. Thus, we used a data- prediction method to establish the bi-boundary filter rule, which helps filter the data that may be dropped by the final result set. The bi-boundary filter rules determine whether the received or generated data will be transmitted. We analyzed the simulation results and concluded that the bi-boundary filter rules can be more energy-efficient in situations in which temporal correlation exists.

Keywords WSNs · Top-k · Data filter

Z.-J. Zhang · J.-R. Jie (✉) · Y. Liu
Department of Electronic and Information Engineering, Key Laboratory of Communication and Information Systems, Beijing Jiaotong University, Beijing, China
e-mail: 10120101@bjtu.edu.cn

Z.-J. Zhang
e-mail: zhjzhang1@bjtu.edu.cn

Y. Liu
e-mail: liuyun@bjtu.edu.cn

Y.-M. Huang et al. (eds.), *Advanced Technologies, Embedded and Multimedia for Human-centric Computing*, Lecture Notes in Electrical Engineering 260, DOI: 10.1007/978-94-007-7262-5_94, © Springer Science+Business Media Dordrecht 2014

Introduction

The preliminary ideas of promoting the Internet for the use of the general population without restrictions will come to fruition, and this gradual process has validated the concept of ubiquitous computing that was first proposed by Mark Weiser [9]. In the future, people can access the Internet, acquire information, and conduct official and private business seamlessly through the processors that currently are and that will be available in almost any location. The wireless sensor network (WSN) is an expansion of the ubiquitous computing concept, and it consists of many sensors that allow us to meet the requirements of exploring, utilizing, and managing the physical world. The database theory associated with WSNs, proposed by Berkeley in 2002, heightens the prospects extensive applications of WSNs and brings the research related to WSNs one step closer to the vision of ubiquitous computing [3].

The database theory of WSNs attempts to transplant the mature database technology of the Internet to the data stream environment of WSNs. The concept is that users propose queries to the WSNs in the form of structured query language, after which the gateway server analyzes the requests and sends the queries to the sink node. The sink node broadcasts the queries to wireless sensor networks and begins the aggregation of the data. Finally, the results are generated, and the server classifies them for corresponding users.

Top-k and skyline are two popular optimal data query approaches that have common features, and both return the representative or special results to users. In many situations, we need to combine the two querying technologies to satisfy the user's preferences, [8] describe an algorithm to achieve top-k skyline query, [4] gave a novel types of skyline queries called the scored k-dominant skyline query to solve the problem in high dimensions. A framework based on the filter rules with dominant graph has been improved to solve the problem associated with multi-dimensional top-k query processing problem [1, 2]. Actually, this problem is more of an issue for the skyline approach than the top-k approach. The simulation results showed that the schemes are more effective than the traditional schemes. Considering that WSNs are task-based networks, they may be useful and effective for applications in which the observed object varies regularly or remains stable. Normally our workplace or the production workshop in a factory requires a relatively stable environment; otherwise, anomalous situation could occur and would have to be dealt with in some way. For example, the temperature of our lab for one day approximates a sine curve. In this case, we can use the predicate method to estimate future values.

The use of time series for making predictions has been researched extensively in many fields, including financial analysis, stock trend forecasting, and environmental monitoring. Reference [5] addressed the modeling of complex environmental data. Reference [6] was focused on one task associated with financial time-series analysis, i.e., forecasting future stock prices based on historical data. Reference [7] proposed prediction rules based on a hybrid model.

All of these prior research efforts inspired us to pursue the research reported in this paper.

In this paper, we used historical data to conduct predictions by time–space relativity and to optimize data retrieval and processing on the basis of [1] with keeping dominate graph locally, and attempt to reduce more energy consumption. We conducted a multi-dimensional top-k query under regular variation environmental conditions. The contributions of this paper were related to the proposals summarized below:

(1) We propose a data-dominated prediction probability to describe the possibility that the future data dominate history data.
(2) We propose a bi-boundary filter to help limit the transmission of data that match the result condition momentarily but that may not be accepted by the final result set.
(3) We simulate these bi-boundary filter algorithms and compare them to the original skyline algorithms. The simulation results will be analyzed thoroughly.

The rest of the paper is organized as follows. Second section provides background information and defines the multi-layer skyline problem briefly. Section Design of Algorithms is devoted to the multi-layer skyline query processing algorithms with bi-boundary filter rules. The simulations are discussed in fourth section, and our conclusions are presented in fifth section.

Background Information

The k-Layers Extraction Algorithm

Given two data points $p = (p.x_1, p.x_2...p.x_n)$ and $q = (q.x_1, q.x_2...q.x_n)$, if p has no dimensions worse than q but has at least one dimension that is better than q, then we say p dominates q, which is written as $p \succ q$ [11]. For convenience, in this paper, we used the term "better" to mean "a greater value", and this condition will not affect the correction of our algorithm or its results.

In the following graph, $\{P_1, P_2, P_3, P_4\}$ is a subset that contains all of the data that cannot be dominated by other data, which is the traditional skyline scheme. The skyline scheme identifies the data that are of special concern to us. But in WSNs, we may need more information that is provided by just one skyline dataset. Then, we have the following definition:

Given a data set S and a data point $p \in S$, if p is dominated by k other data points in S, p belongs to the $(k + 1)$ layer [1]. The $(k + 1)$-layer of S is the subset that contains all of the data that are dominated by the other k data in S. (Fig. 1).

In the following picture, the 1-layer skyline is $\{P_1, P_2, P_3, P_4\}$, the 2-layer skyline is $\{P_7\}$, the 3-layer skyline is $\{P_5, P_6, P_9\}$ and the 4-layer skyline is $\{P_8\}$.

We give the k-layers extraction algorithm as Algorithm 1 which will help us get the sets of top-k skyline result.

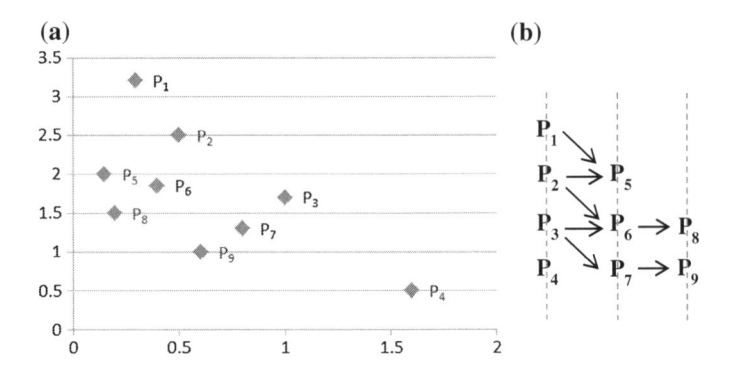

Fig. 1 An example of skyline

Algorithm 1 *k*-layers Extraction Algorithm

Require: Input:a dataset S and k
    Output: k skylines of dataset S : $\{L_1 \ldots L_i \ldots L_k\}$
  1: All data points in S are put into $Q_{candidate}$.
  2: set iteration i= 1; $L_1, L_{2\ldots}L_k = \emptyset$;
  3: while $Q_{candidate} \neq \emptyset$ and i ≤ k do
  4:   for evey data d in $Q_{candidate}$ do
  5:    if d has less than k predecessors in DG then

  6:     $L_i = L_i \cup \{d\}$;

  7:     $Q_{candidate} = Q_{candidate} / \{d\}$;
  8:    end if
  9:   i = i+1;
10: end for
11: end while
12: set $Q_{candidate} = \emptyset$;
13: return $\{L_1 \ldots L_i \ldots L_k\}$

The results that were retrieved were in the form of strata instead of collections because it was important for us to generate the details associated with the bi-boundary.

The Problem

Users send a query to the WSN and want the k-layer skyline data during some future time T. The server must know: (1) the dimensions of the user's interest; (2) the k; and (3) the query time, T. The server allows the sink node to broadcast this query to every sensor belongs to the WSNs.

The nodes generate data when they are awake, and the data di will be packed in the structure, $(di.x_1, di.x_2...di.x_m, di.Id, di.t, di.P)$. In this structure, $di.x_1$, $di.x_2...di.x_m$ means the data in each of the dimensions, di.Id is the unique number for every node, di.t is the timestamp of these data, and di.E is the dominate expectation of the next sample, which we will introduce in detail in the next section. The data are always effective during the query time. T; this is unlike continuous query, in which the lifetime of the data is limited.

Assume that the data set D contains all the data generated during the query time, T, by the entire WSN, but the the user needs only the subset D' that contains the top-k skyline layer, $\{L_1...L_i...L_k\}$. So, our target is to reduce the communication of data that are not included in the results.

The problem description is simplified and omits the situations of time sliding and query continuing. This is done just to stress the new filter rules, and the new rules can be transplanted directly to the original frame.

Design of Algorithms

Our improved algorithms stress the design of the bi-boundary filter based on the data prediction rules. We assumed that the routing tree already has been constructed and that all the sensor nodes that participated in query processing are awake.

Prediction Probability Algorithms

We use ε to designate prediction coefficients in the range of 0–1 to show the strength of the relationship between historical data and future predictions. This is a simple statistic or empirical value. This is a simple statistical or empirical value. We can test a node in a specific place for an m sampling period. If the result has n inflection points, we have $\varepsilon = (m - n)/m$, usually ε is affected by sensor quality and environmental conditions. For example, in a manufacturing shop, we can divide the environment into different areas, such as the production line, offices, storage, and laboratory. Environmental factors vary between these locations, but they are relatively stable within a given location.

In the one dimension situation, a data point d1 is generated by sensor Si during the first sampling period. In the next sampling period we assume that d2 have the half probability better than d1. We use DOMi (n) to describe the dominate state of dn to dn − 1, if dn dominates dn − 1, then DOMi (n) = 1; if dn is dominated by dn − 1, then DOMi (n) = −1, else DOMi (n) = 0.

$$P = 1 * \frac{1}{2} + \left(\frac{1}{2}\right)\left[\frac{\varepsilon}{2}DOM\,(n)\right] + \left(\frac{1}{2^2}\right)\left[\frac{\varepsilon^2}{2}DOM\,(n-1)\right]$$
$$+ \left(\frac{1}{2^3}\right)\left[\frac{\varepsilon^3}{2}DOM\,(n-2)\right]. \tag{1}$$

After the merger:

$$P = \frac{1}{2} + \frac{1}{2}\sum_{i=1}^{n} \left(\frac{\varepsilon}{2}\right)^{i} DOM(n-i+1) \tag{2}$$

It is apparent that the probability of dominate status in the next time increases when the sensor Si keeps generating a series of data that dominate the old data. If ε is close to 1, that means the relationship of historical and future values is very strong. If ε is close to 1, the extreme value, so the probability will also approach 1 that predicts the new data will always dominate the old data. Note that the probability is not actually the accurate dominated probability in the real world; rather, it is a predicted probability that we designed to serve as a rough estimate.

In the 2-d situation, if sensor S_i gets two data points, d_1 and d_2, and d_2 dominates d_1, d_2 is not worse than d_1 in any two of the dimensions and gets at least one dimension better than d_1. We used x and y to note the two dimensions temporarily, so the relationship constants are εx and εy. DOMx (n) denotes the dominant status between dn and dn $- 1$ in dimension x, as was defined before, so as the DOMy (n). We have the probability that the dn $+ 1$ can dominate dn as follows:

$$P_x = \frac{1}{2} + \frac{\varepsilon_x}{2^2}(DOM_x(n)) + \frac{\varepsilon_x^2}{2^3}(DOM_x(n-1)) + \frac{\varepsilon_x^3}{2^4}(DOM_x(n-2)). \tag{3}$$

Similarly, we have the probability in the y dimension as:

$$P_y = \frac{1}{2} + \frac{\varepsilon_y}{2^2}(DOM_y(n)) + \frac{\varepsilon_y^2}{2^3}(DOM_y(n-1)) + \frac{\varepsilon_y^3}{2^4}(DOM_y(n-2)). \tag{4}$$

So, it is obvious that the probability that d_{n+1} dominates d_n is PxPy, as shown below:

$$P_x P_y = \frac{1}{4} + \frac{1}{2}\sum_{i=1}^{n}\left(\frac{\varepsilon_x}{2}\right)^{i}DOM_x(n-i+1) + \frac{1}{2}\sum_{i=1}^{n}\left(\frac{\varepsilon_y}{2}\right)^{i}DOM_y(n-i+1)$$
$$+ \left[\sum_{i=1}^{n}\left(\frac{\varepsilon_x}{2}\right)^{i}DOM_x(n-i+1)\right]\left[\sum_{i=1}^{n}\left(\frac{\varepsilon_y}{2}\right)^{i}DOM_y(n-i+1)\right] \tag{5}$$

The formula may be somewhat complicated when conducting multi-dimensional computations. So, some simplifying modifications are discussed below.

Still using the 2-dimensional case as an example, when d_n dominates d_{n-1}, we have DOM (n) $= 1$. We want to use DOM (n) instead of DOM_x (n) and DOM_y (n) for simplicity. We note $P_x P_y{}'$ as the simplified calculation. $P_x P_y{}' \le P_x P_y$ because

DOM (n) = 1 is an unnecessary and sufficient condition for DOM_x (n) = 1 or DOM_y (n) = 1. Then we have P_xP_y' as shown below:

$$P_xP_y' = \frac{1}{4} + \frac{1}{2}\sum\nolimits_{i=1}^{n}\left[\left(\frac{\varepsilon_x}{2}\right)^i + \left(\frac{\varepsilon_y}{2}\right)^i\right]DOM(n-i+1)$$
$$+ \left[\sum\nolimits_{i=1}^{n}\left(\frac{\varepsilon_x}{2}\right)^i\right]\left[\sum\nolimits_{i=1}^{n}\left(\frac{\varepsilon_y}{2}\right)^i\right]DOM(n-i+1)^2 \qquad (6)$$

In addition, if we think that ε_x is equal to ε_y, the formula can be simplified further. If we use ε instead of ε_x and ε_y, the formula becomes: $\varepsilon_x + \varepsilon_y \geq 2\sqrt{\varepsilon_x\varepsilon_y}$. We set $\varepsilon = 2\sqrt{\varepsilon_x\varepsilon_y}$, so:

$$P_xP_y'' = \frac{1}{4} + \sum\nolimits_{i=1}^{n}\left(\frac{\varepsilon}{2}\right)^i DOM(n-i+1) + \left[\frac{\varepsilon^2\left(1-\left(\frac{\varepsilon}{2}\right)^n\right)^2}{(2-\varepsilon)^2}\right]DOM(n-i+1)^2$$
$$(7)$$

In three dimensions, the polynomial will be a little more complex, but when the time relationship constant ε is small, terms with high time power can be ignored. In addition, uncertainties increase when the number of dimensions increases. The predicted probability will be very small when there are more than four dimensions, and the algorithm will be inefficient. So, another problem to be solved is making the use of these concepts more efficient and effective.

Algorithm 2 One Dimension Local Dominate Prediction

Require: Input:a dataset of point i $S_i = \{d_{i1}, d_{i2} ... d_{in}\}$
        prediction coefficients $\boldsymbol{\varepsilon}$ of this dimension x
        precison **pre**
        Output: $P_i(n+1)$;
1: All data points in S_i are put into S_i'
2: set iteration j= 1; $P_i(n+1) = \frac{1}{2}$
3: **while** $S_i' \neq \emptyset$ and j $\leq$ **pre do**
4:   **if** $d_{(n+1-j)}.d_x > d_{(n+1-j)}.d_x$ **then**
5:     $P_i(n+1) = P_i(n+1) + \frac{\varepsilon^j}{2^{j+1}}$;
6:   **end if**
7:   **if** $d_{(n+1-j)}.d_x = d_{(n+1-j)}.d_x$ **then**
8:     $P_i(n+1) = P_i(n+1)$;
9:   **end if**
10:   **if** $d_{(n+1-j)}.d_x < d_{(n+1-j)}.d_x$ **then**
11:     $P_i(n+1) = P_i(n+1) - \frac{\varepsilon^j}{2^{j+1}}$;
12:   **end if**
13:   j++;
14: **end while**
15: **return** $P_{ix}(n+1)$

Bi-boundary Filter Algorithms

We used two boundaries to filter the local data. One boundary is the real boundary, which is calculated by all the data points sink node has already received so far. The other boundary is the predicted boundary, which is calculated by the exception of recently received data points at sink node.

Use bi-boundary filter will delay the data not filter by the real boundary but have a strong possibility not to be adopted by the final result set. Following is the detailed description of the bi-boundary filter calculation.

(1) *current information table kept at sink node*

We regard the sink node as unlimited with respect to energy and as providing sufficient storage. In addition to the essential data-record table, we keep a global dominant information table at the sink node. This table has four labels:

- node ID i;
- Exception about layer increase of node i's data last received at sink: this value is calculated by algorithm 3 when predicting filter generation.
- Correction factor η: this fixed value helps us acquire a predicted value that is closer to the actual value. It is described in more detail later.
- The count for invalid data: used to directionally update the local filter; detailed information is not provided here, but it is available in Ref. [1] (Table 1).

(2) *correction factor η*

The predicted dominant possibility is a conjecture, not the real possibility, w, as we have emphasized before. When we use this possibility to generate the predicted filter, we could perhaps encounter the situation in which the data's layer is different with our prediction, so we use η to fix it.

$$\eta_i = \frac{\Delta Layer}{E_i} \tag{8}$$

This algorithm is runing at sink node, which always has a strong energy supply and limitless memory. So this algorithm can be used efficently and quickly at sink node with its dominant graph. The generated bi-boundary consist of the fixed prediction line and the actual line. The sink node only has to send these packages to the designated nodes.

Table 1 Current information table

ID	E	η	Count
1	0.71	1.63	0
2	0.21	0.29	3
3	0.56	0.92	2
...	...	...	...

Here, we have to set a redundance value because the estimations are always decimals. We use rounded data instead of the approximated values.

(3) Generation of the Bi-boundary Filter

As we know, if a data point cannot be included in the result at any timestamp, this data point will never be accepted by the sink node. So the data that are filtered by the true boundary have no chance to be sent during the query time. Algorithm 3 is used to count the predicted boundary. The sink node will count $\{L_k, L_k'\}$ for every time period and respond to the nodes' requests for filter updates.

Algorithm 3 Predicted k-layer skyline Boundary

Require: Input:a dataset S_{sink} at (n-1), S_{sink}' at n
 Time correlation ε of dimension $x_1 x_2 ...$
 η and k
 Output: predicte k-layer $\{L_1', L_2' ... L_k'\}$
 1: $S_{sink} = S_{sink} + S_{sink}'$
 2: update DG in sink
 3: caculate $\{L_1...L_i....L_k\}$ with Algorithm1
 4: $\{L_1', L_2' ... L_k'\}=\{L_1...L_i....L_k\}$
 5: **for** every data $d_i \in L_j$ (j=1,2...k) **do**
 6: **if** j = 1**then**
 7: $E = E_i$
 8: **else** use DG count E_{max};E= E_{max}
 9: **if** $[\eta E] \geqslant 1$**then**
 10: $L_1 = L_1 / \{d_i\}$
 11: **if** $[\eta E] \leqslant k$ **then**
 12: $L_{[\eta E]+1} = L_{[\eta E]+1} \cup \{d_i\}$
 13: **end if**
 14: **end if**
 15: **end for**
 16: **return** $\{L_1', L_2' ... L_k'\}$

Simulation

We used the nodes in our laboratory to collect the temperature and humidity data from 20 different places, including the working space, the server cabinet, a meeting room, and outspace. These environments were relatively stable, but the data set was not enough for simulation. So, we generated an extensive amount of simulated data with existing data, and the maximum size of the data dimensionality was five. We classified the data to three degrees, i.e., (1) range A: $\varepsilon \sim [0.8 - 1]$, stable; (2) range B: $\varepsilon \sim [0.5 - 0.8]$, medium; (3) Range C: $\varepsilon \sim (0 - 0.5)$ frequent change. We used three methods to conduct this simulation. The first was the traditional, centralized, exact method in which all data were sent to the sink node [10]. The second method was the original scheme proposed in [1]. The third method was the improved scheme proposed in this paper.

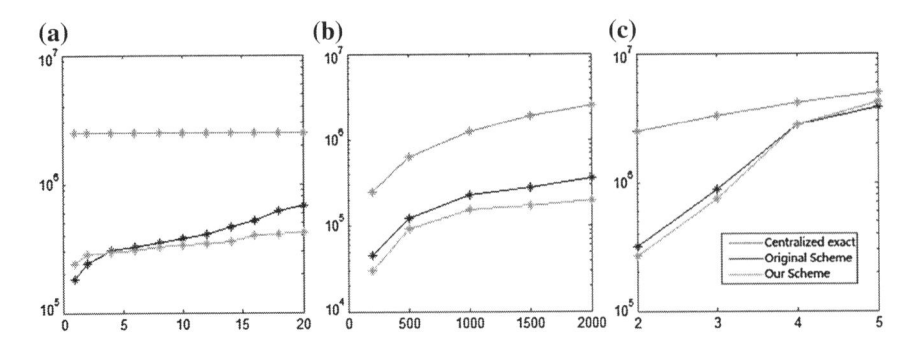

Fig. 2 The simulation about the 3 algorithms. **a** N = 2,000, D = 2, ε ~ A; **b** D = 2, ε ~ A k = 10; **c** N = 2,000, k = 10, ε ~ A)

We assumed that each packet contained a single, double-precision, floating-point number (8 bytes). Accordingly, a 2-dimensional data point takes three packets to transmit, i.e., two for di.x1, di.x2, one for di.E, and one for the time-stamp di.t and ID di.id. The query time was set to one hour. We conducted the simulation in MATLAB 2009.

All of the data used in Fig. 2 came from rank A, i.e., ε ~ [0.8,1]. In Fig. 2a, the size of the nodes is 2,000, and there are two dimensionalities. Our improved scheme cost more energy when k is less than 3, because, when k is too small, the two boundaries are not much different and the packets have one more byte than the original scheme. When k increases, the effect of the bi-boundary filter is apparent. The efficiency of our scheme increases as the value of k increases. In Fig. 2b, there are two dimensionalities, and the value of k is 10. Our scheme is approximately 20 % better than the original scheme. In Fig. 2c, the size of the nodes is 2,000, and the value of k is 10. It is apparent that the meaning of the prediction decreases as the number of dimensions increases. When the number of dimensions is more than four, the communications exceed those of the traditional scheme and the original

Fig. 3 The simulation on different η

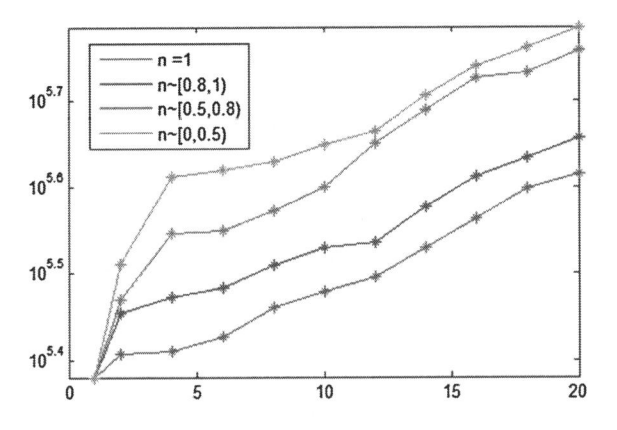

scheme. This occurred because the results of the predicted exception algorithm become invalid as the uncertainty increases.

The simulation results in Fig. 3 show that the relationship between historical and predicted data is very strong when ε becomes closer to 1, so our bi-boundary will be more effective. That means that our scheme is better suited than the other schemes for a relatively stable monitoring environment, and it can save energy cost by reducing communications.

Conclusion

In the research reported in this paper, we applied the original scheme proposed in [1] to a stable environment. We designed a dominant prediction algorithm of node data and used it to improve the filter rules. According to the simulation results, our scheme can save energy costs by reducing communications when there is a large quantity of data, the time relativity of the data is strong, and the dimensions do not exceed four.

Since time and funds are limited at the present time, we expect to conduct additional research in this area in the future focused on (1) making the prediction more accurate and (2) taking into account other WSN data that also are related to the distribution area.

References

1. Jiang H, Cheng J, Wang D, Wang C, Tan G (2011) Continuous multi-dimensional top-k query processing in sensor networks. In: Proceedings of IEEE INFOCOM
2. Zou L, Chen L (2008) Dominant graph: an efficient indexing structure to answer top-k queries. In: Proceedings of IEEE ICDE
3. Madden S, Franklin MJ (2002) Fjording the stream: an architecture for queries over streaming sensor data. In: Proceedings of IEEE ICDE
4. Kim YS, Jung HR, Sung MK, Chung YD (2011) On processing scored k-dominant skyline queries. In: Proceedings of IEEE conference
5. Peralta J, Gutierrez G, Sanchis A (2009) Shuffle design to improve time series forecasting accuracy. In: IEEE congress on digital object identifier, pp 741–748
6. Huang Y, Wang H, McClean S (2009) Neighborhood counting for financial time series forecasting. Evolutionary computation. CEC '09. IEEE Congress, pp 815–821. doi: 10.1109/CEC.2009.4983029
7. Wang JJ, Wang JZ, Zhang ZG, Guo SP (2012) Stock index forecasting based on a hybrid model. Omega Int J Manage Sci 40:758–766
8. Sun YQ, Li Q, Chen ZY (2009) The top-k skyline query in pervasive computing environments. Pervasive Computing (JCPC), 335–338
9. Weiser M (2002) The computer for the 21st century. Pervasive Computing, IEEE
10. Madden SR, Franklin MJ, Hellerstein JM, Hong W (2005) TinyDB: an acquisitional query processing system for sensor networks. ACM Trans Database Syst 30(1):122–173
11. Wu M, Xu J, Tang X, Lee W-C (2007) Top-k monitoring in wireless sensor networks. IEEE Trans Knowl Data Eng 19(7):962–976

An Improved LMAP^{++} Protocol Combined with Low-Cost and Privacy Protection

Fei Zeng, Haibing Mu and Xiaojun Wen

Abstract With the fast development of the Radio Frequency Identification (RFID) technology, the cost and privacy of RFID systems have become an obstacle that prevents it from being deployed in a larger scale. In this paper, we introduce the attacks which the LMAP^{++} protocol suffers from briefly. We modify the LMAP^{++} protocol by importing hash function on the parameters A and B as well as setting the lowest bit of ID as "1". Analysis shows that the protocol can improve the security for the traceable and desynchronized problems with reasonable overhead.

Keywords RFID · LMAP^{++} · Privacy · Traceability · Desynchronization

Introduction

The Internet of things is called the third revolution in the information industry after computers and the Internet. And RFID is the most important technology of the Internet of things. It provides instant identification functions for objects. Despite of a good prospect, the cost and privacy problems constrain the development of the Internet of things. Hence, a perfect combination between low cost and high privacy protection has been the target of the academic circles.

F. Zeng · H. Mu (✉)
Beijing Key Laboratory of Communication and Information Systems, School of Electronic and Information Engineering, Beijing Jiaotong University, Beijing 100044, China
e-mail: hbmu@bjtu.edu.cn

F. Zeng
e-mail: 12125056@bjtu.edu.cn

X. Wen
School of Computer Engineering, Shenzhen Polytechnic, Shenzhen 518055, China
e-mail: szwxjun@sina.com

Y.-M. Huang et al. (eds.), *Advanced Technologies, Embedded and Multimedia for Human-centric Computing*, Lecture Notes in Electrical Engineering 260, DOI: 10.1007/978-94-007-7262-5_95, © Springer Science+Business Media Dordrecht 2014

In this paper we proposed a protocol which is improved from LMAP^{++}. The hash function is used to confuse the parameters and the exchanged information which will prevent attackers from getting useful information. It is low-cost and can avoid from traceability attack and desynchronizaion attack.

The rest of the paper is organized as follows: Section Related Work is the related work about RFID protocols which focus on the effective combination between cost and privacy. Section LAMP^{++}, Traceability Attack and Desynchronization Attack is the description of traceability attack and desynchronizaion attack against LAMP^{++} protocol. Section The Improved Protocol is the detailed process of my protocol. We analyze the protocol we propose in Section Analysis.

Related Work

There are large quantities of related studies which focus on the balance of cost and privacy of the RFID systems.

Peris et al. [9] introduced a protocol which was lightweight. But it had privacy protection problems. Because of the plaintext transmission of the tag, it can't guarantee the users' privacy, such as the location.

The protocol produced by Juels et al. [10] was traceable because users that used their blocker tags would need to acquire an opt-out mechanism to protect their privacy.

Mubarak et al. [11] introduced a new idea that combined security, trust and privacy to protect the users' privacy. They applied Trusted Platform Module (TPM) to provide trusted computing. Through the interaction between the reader, tag and the TPM module, the protocol had higher security.

In [1], Peris et al. put forward a Lightweight Mutual Authentication Protocol named LMAP. Besides, they proposed an extension of this protocol—LMAP^{+}. These protocols were indeed lightweight and used only simple bitwise operations. However, it was discovered before long that the claimed security was not achieved. Later, following the LMAP designing strategy, Li [2] proposed a new lightweight protocol. Li [2] also called the proposed scheme LMAP^{+}. However, to avoid confusion with the extension of LMAP proposed by Peris et al. in [1], we call Li's scheme LMAP^{++} protocol in the rest of this paper. The LMAP^{++} protocol can be seen as a modified version of SLMAP protocol [3] which has been analyzed in [4, 5].

In [6], Masoumeh et al. used the model for traceability proposed by Jules and Weis in [7] to successfully attack LMAP^{++} and presented a desynchronization attack which has the success probability of 1/16 on each run of the protocol.

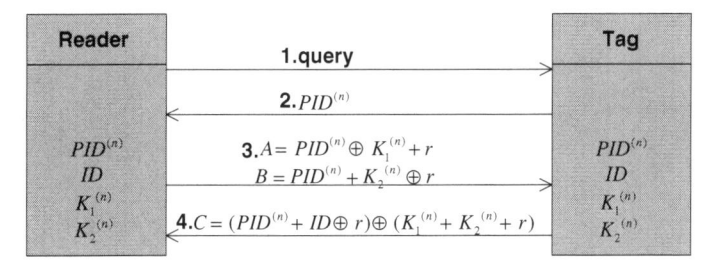

Fig. 1 The process in a single run of LMAP++ protocol

LAMP++, Traceability Attack and Desynchronization Attack

LMAP++

LMAP++ is a lightweight mutual authentication protocol. It assumed the channel between the database and reader is secure. The process of the protocol is described in Fig. 1.

Notation:

$PID^{(n)}$	The pseudonym of the tag at the nth run
ID	The tag's identifier
$K_1^{(n)}$, $K_2^{(n)}$	Two secret keys at the nth run
$\oplus$	XOR operation
+	Addition mode 2^m
r	A random number
‖	Concatenation operator

The process that LMAP++ protocol runs is described as below:

1. The reader sends a query message to the tag.
2. The tag responses with a $PID^{(n)}$.

   The reader looks for $PID^{(n)}$ from the table. If there is a matching option, calculate A, B as follows:
   $$A = PID^{(n)} \oplus K_1^{(n)} + r$$
   $$B = PID^{(n)} + K_2^{(n)} \oplus r$$
   Next, passes $A\|B$ to the tag.

3. The tag extracts r from A and B:
   $$r_1 = A - \left(PID^{(n)} \oplus K_1^{(n)}\right)$$

$r_2 = \left(B - PID^{(n)}\right) \oplus K_2^{(n)}$

If $r_1 = r_2$, the tag authenticates the reader. Else, the protocol will be terminated.

The tag computes $C = \left(PID^{(n)} + ID \oplus r\right) \oplus \left(K_1^{(n)} + K_2^{(n)} + r\right)$ and passes

C to the reader. And the tag updates $PID^{(n)}$, $K_1^{(n)}$, $K_2^{(n)}$ as follows:

$PID^{(n+1)} = \left(PID^{(n)} + K_1^{(n)}\right) \oplus r + (ID + K_2^{(n)}) \oplus r$

$K_1^{(n+1)} = K_1^{(n)} \oplus r + \left(PID^{(n+1)} + K_2^{(n)} + ID\right)$

$K_2^{(n+1)} = K_2^{(n)} \oplus r + \left(PID^{(n+1)} + K_1^{(n)} + ID\right)$

4. The reader calculates $C' = \left(PID^{(n)} + ID \oplus r\right) \oplus \left(K_1^{(n)} + K_2^{(n)} + r\right)$

If $C = C'$, the reader authenticates the tag. At last, the reader updates $PID^{(n)}$, $K_1^{(n)}$, $K_2^{(n)}$ as above.

Traceability Attack and Desynchronization Attack

In [6], Masoumeh et al. used the traceability model proposed by Jules and Weis in [8]. The attacking process can be summarized as this. They assumed that there were two tags which the lowest bits were respectively "0" and "1". If they could distinguish the tags after the attacker eavesdropped a whole run of the protocol, they declared LMAP++ had suffered from traceability attack.

In addition, they analyzed that under the probability of 1/16, attackers could desynchronize the updating of the keys through tampering with the data which the reader transferred to the tag.

The Improved Protocol

The improved protocol proposed is as below:

1. Set the lowest bit of ID as "1".
2. The reader sends a query message to the tag.
3. The tag responses with a $PID^{(n)}$.

   The reader looks for $PID^{(n)}$ from the table. If there is a matching option, calculate A, B using PID, K_1 and K_2. And then, hash A‖B and pass it with r to the tag. Else, the protocol will be terminated.

4. In the tag, use PID, K_1, K_2, r to compute A_1, B_1. If $H(A‖B) = H(A1‖B1)$, the tag authenticates the reader.

   In the tag, compute $C = \left(PID^{(n)} + ID \oplus r\right) \oplus \left(K_1^{(n)} + K_2^{(n)} + r\right)$ and

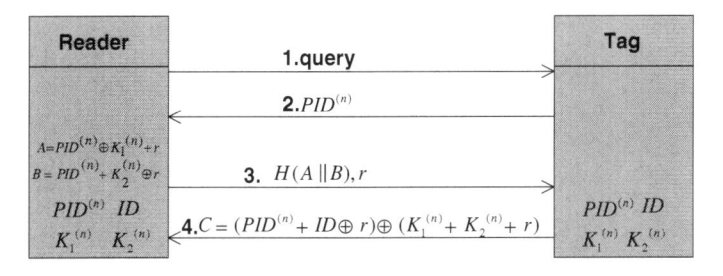

Fig. 2 The process in a single run of the improved protocol

passes C to the reader. And the tag updates $PID^{(n)}$, $K_1^{(n)}$, $K_2^{(n)}$ as follows:

$$PID^{(n+1)} = (PID^{(n)} + K_1(n)) \oplus r + \left(ID + K_2^{(n)}\right) \oplus r$$

$$K_1^{(n+1)} = K_1^{(n)} \oplus r + \left(PID^{(n+1)} + K_2^{(n)} + ID\right)$$

$$K_2^{(n+1)} = K_2^{(n)} \oplus r + \left(PID^{(n+1)} + K_1^{(n)} + ID\right)$$

5. The reader calculates $C' = \left(PID^{(n)} + ID \oplus r\right) \oplus \left(K_1^{(n)} + K_2^{(n)} + r\right)$

If $C = C'$, the reader authenticates the tag. Then, updating $PID^{(n)}$, $K_1^{(n)}$, $K_2^{(n)}$ as above.

The process of the improved protocol is shown in Fig. 2.

Analysis

On one hand, the desynchronization attack has a premise that $(PID)_0 = (K_1)_0 = (K_2)_0 = (ID)_0 = 0$ while our protocol sets $(ID)_0 = 1$ at first. Hence, our protocol is immune to this attack.

On the other hand, our protocol can resist from the traceability attack because the reader passes H (A‖B) to the tag instead of A‖B in my protocol. Under this circumstance, although the attacker can intercept A‖B, he has no way to get a knowledge of the lowest bit of A or B. Furthermore, we transmit the random number r with H (A‖B) at the same time. R will be generated from the random number generator every time the protocol successfully runs. Though attackers eavesdrop the channel between readers and tags, there is no approach to guess the effective information. Besides, the secret keys and the pseudonym stored in the reader and the tag will be updated synchronously, too. Therefore, attackers aren't able to distinguish the two tags which the lowest bit is separately "0" and "1".

Compared with LMAP^{++}, my protocol adds hash operation modules to enhance security. But hash operation modules are not so complicated that we can apply it to the design of the RFID system. So, our protocol still costs little. Moreover, we solve the traceable problems that LMAP^{++} is suffering from.

As mentioned above, my protocol not only remains the low-cost advantages but also possess a more secure performance.

Conclusion

In this paper we propose an improved protocol that is based on LMAP^{++}. To fix the vulnerability, we import hash operations and encrypt the exchanging information. It fulfills the aim that finishing a good RFID protocol is a balance of low-cost and privacy preserving. However, this is not enough to put our protocol into large-scale applications. The next step of our work will to be design a better RFID protocol in the future.

Acknowledgments This work is supported by National Natural Science Foundation of China under Grant 61201159, and Fundamental Research Funds for the Central Universities under Grant 2012JBM016. The authors also gratefully acknowledge the helpful comments and suggestions of the reviewers, which have improved the presentation.

References

1. Peris-Lopez P, Hernandez-Castro JC, Estevez-Tapiador JM, Ribagorda A (2006) LMAP: a real lightweight mutual authentication protocol for low-cost RFID tags. In: Proceedings of RFIDSec06 workshop on RFID security, Graz, Austria, pp 12–14
2. Li T (2008) Employing lightweight primitives on low-cost RFID tags for authentication. In: Proceedings of vehicular technology conference fall, pp 1–5
3. Li T, Wang G (2007) SLMAP—a secure ultra-lightweight RFID mutual authentication protocol. In: Proceedings of Chinacrypt, vol 07, pp 19–22
4. Hernandez-Castro JC, Estevez-Tapiador JM, Peris-Lopez P, Clark JA, Talbi E-G (2009) Metaheuristic traceability attack against SLAMP, an RFID lightweight authentication protocol. In: Proceedings of 23rd IEEE international symposium on parallel and distributed processing (23rd IPDPS'09), workshop on nature inspired distributed computing, pp 1–5
5. Hernandez-Castro JC, Estevez-Tapiador JM, Peris-Lopez P, Clark JA, Talbi E-G Metaheuristic traceability attack against SLAMP, an RFID lightweight authentication protocol. Int J Found Comput Sci
6. Safkhani M, Bagheri N, Naderi M (2011) Security analysis of LMAP^{++}, an RFID authentication protocol. In: Proceedings of 6th international conference on internet technology and secured transactions, 11–44 Dec 2011, Abu Dhabi, United Arab Emirates
7. Juels A, Weis SA (2007) Defining strong privacy for RFID. In: Proceedings of IEEE computer society, PerCom workshops, pp 342–347
8. Juels A, Weis SA (2007) Defining strong privacy for RFID. In: Proceedings of IEEE PerCom' 07, pp 342–347
9. Peris-Lopez P, Lee LT, Li T (2008) Providing stronger authentication at a low-cost to RFID tags operating under the EPCglobal framework. In: Proceedings of IEEE/IFIP international

symposium on trust, security and privacy for pervasive applications—TSP'08, Shanghai, China, pp 159–166

10. Juels A, Rivest R, Szydlo M (2003) The blocker tag: selective blocking of RFID tags for consumer privacy. In: Proceedings of conference on computer and communications security—ACM CCS, USA, pp 103–111

11. Mubarak MF, Manan J, Yahya S (2011) A critical review on RFID system towards security, trust, and privacy (STP). In: Proceedings of 2011 IEEE 7th international colloquium on signal processing and its applications

Empirical Analysis of User Life Span in Microblog

WeiGuo Yuan and Yun Liu

Abstract The aim of this work is to study two kinds of user life spans and their connection to the distribution of followers, friends and statuses in microblog. Both the user activity spans and user age approximately follow a two-part exponential distribution. Moreover, the users' average number of followers and statuses increases linearly with the active span, but the average number of friends does not change significantly during the life span. We plot the distribution of users, followers, friends, and statuses as a cumulative sum to obtain a strict power-law form, which indicates an allometric growth phenomenon. These new findings show that the users' production capacity is consistent and with self-similar growth ability in different user activity spans. We argue that the scale effect of user count development is the reason for the allometric growth phenomenon in microblog.

Keywords Life spans · Power-law · Allometric growth · Microblog

Introduction

Online social networks (OSN) have changed how we communicate. Research on OSNs is very popular and highly interdisciplinary. Some researchers focus on the network topological structure [1, 2]. Most OSNs are scale-free, which means that

W. Yuan · Y. Liu (✉)
Key Laboratory of Communication and Information Systems, Beijing Municipal Commission of Education, Beijing Jiaotong University, Beijing, China
e-mail: liuyun@bjtu.edu.cn

W. Yuan
e-mail: ywg@cnnic.cn

W. Yuan
Computer Network Information Center, Chinese Academy of Sciences, Beijing, China

Y.-M. Huang et al. (eds.), *Advanced Technologies, Embedded and Multimedia for Human-centric Computing*, Lecture Notes in Electrical Engineering 260, DOI: 10.1007/978-94-007-7262-5_96, © Springer Science+Business Media Dordrecht 2014

the node degrees follow a power-law distribution. OSNs provide useful information for the study of user behavior and human dynamic. Recently, there have been an increasing number of studies in this active field [3, 4]. In particular, empirical analyses of user behavior in a variety of OSNs have received extensive attention [5, 6].

Microblog, a new kind of OSN applications in the age of Web 2.0, has attracted much attention from researchers studying user behavior characteristics. Kwak et al. [7] studied the correlation between the number of users' friends, followers, and statuses. Yan et al. [8] found that the time interval of two consecutive publishing statuses followed a power-law distribution. These studies are important to understand the structure and evolution of OSNs.

Allometric growth is widespread in many natural and social phenomena [9]. For example, growth in urban populations and areas, volume, and energy consumption present a power-law relationship in the growth process. Recently, Bettencourt et al. [10] found an allometric growth relationship between consumption and population size, indicating that the emergence of a city makes people get together in order to consume less energy and create more output. Similarly, Wu et al. [11] argued that the total human online activity grows faster than the active population in OSNs, indicating an allometric growth phenomena.

Sina Weibo, the largest microblog website in China, attracted over 500 million registered users as of December 2012. Our paper focuses on the relations between user life span and user characteristics in Sina Weibo. This paper offers rich empirical materials to improve understanding of user behavior and human dynamic features in OSNs.

The paper is organized as follows. Dataset and Measurement Methodology describes the data sets and methods. User Life Span Data Analysis presents two kinds of user life span statistics: user activity span and user age. User Profile Features and Growth with Activity Spans analyzes the distribution of some user features with the same user activity spans and examines the allometric growth phenomenon with different user activity spans. Finally, we provide conclusions with a brief look at future works in Conclusions.

Dataset and Measurement Methodology

In this paper, we collected a dataset based on the Sina Open API interface to enhance accuracy and efficiency. By using the snowball crawling strategy, we selected the author as a starting point and then collected his detailed information as well as the profile information of his friends and followers. We obtained 881,146 user profiles from February 9, 2012 to February 26, 2012. The basic attributes of users include user ID, registration time, and the number of followers, friends, and statuses. At the same time, we collected users' latest status information including the publishing time.

In order to study the user life span distribution, we draw from complex networks theory and statistical theory. To facilitate discussion, we introduce the complementary cumulative distribution function (CCDF). The power-law distribution CCDF can be defined as follows:

$$F_x(x) = P(X > x) \sim x^{-\alpha} \tag{1}$$

User Life Span Data Analysis

In this section, we analyze two kinds of users life span statistics collected in our dataset:

- Activity span $T_{Activity}$, which is the time between the creation of a registered user t_{reg} and the latest time of the user published statues t_{status}, namely

$$T_{Activity} = t_{status} - t_{reg}. \tag{2}$$

- Age T_{Age}, which is the time between the creation of a registered user t_{reg} and when it is collected $t_{collect}$, namely,

$$T_{Age} = t_{collect} - t_{reg}. \tag{3}$$

User activity span means how long the user has been active, while user age means how long the user has existed since the user registration. As Sina Weibo was launched in August 2009, and the earliest user registration time collected is August 14, 2009. The dataset was collected in February 2012; the user latest status time collected was on February 26, 2012. The longest user activity span and user age is about 30 months. More detailed information on the user time of the dataset is shown in Table 1.

Table 1 User time satistics results

User time type	Min (in days)	Max (in days)
$t_{collect}$ (user collected time)	09-Feb-2012	26-Feb-2012
t_{reg} (user registration time)	14-Aug-2009	26-Feb-2012
t_{status} (user lastest status time)	27-Sep-2011	26-Feb-2012
$T_{Activity}$ (user activity span)	0	914
T_{Age} (user age)	0	926

User Activity Span Distribution

The user activity span, one of the most important statistics, describes the length of time that the user account has been active. As shown in Fig. 1, the CCDF distribution of user activity spans (in months) follows an exponential distribution, with more than 70 % of users who have been active less than 15 months. The distribution can be divided into two sections: the first part with activity spans less than 15 months, which obeys a linear exponential distribution ($R^2 = 0.98$), and the second part with activity spans of more than 15 months, which follows a parabola curve exponential distribution ($R^2 = 0.99$) in log-scale. Ribeiro et al. [12] found a similar result with Myspace, which is one of the largest OSNs in the world.

User Age Distribution

User age describes the growth of the number of users. Figure 2 shows the distribution of user ages (in months). Similar to the activity spans, the distribution of user ages is approximately an exponential distribution. Less than 20 % of the total number of accounts are older than 18 months. They were created during Sina Weibo's early years and experienced quick exponential growth. The remaining 80 % of accounts are newer than 18 months and have been on a slower exponential growth. Huberman [13] believes that the WWW sites' exponential growth can be explained by the number of pages on its website to produce power-law distributions. Thus, these two modes of growth in the user age may be the reason behind the double power-law distribution [14] of the followers count in Sina Weibo accounts.

Fig. 1 The CCDF distribution of user activity spans ($T_{Activity}$)

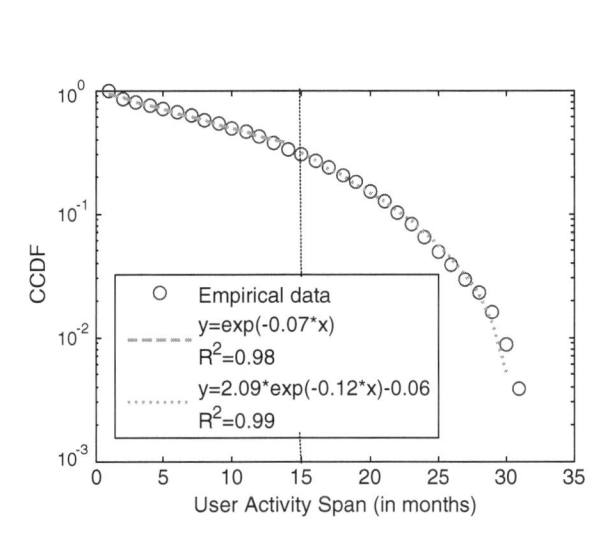

Fig. 2 The CCDF
distribution of user age (T_{Age})

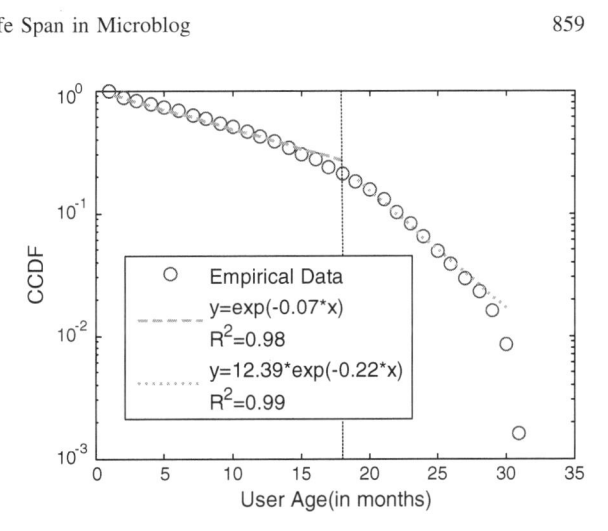

User Profile Features and Growth with Activity Spans

User Profile Features Distribution with Activity Spans

A user profile on Sina Weibo displays the user name, the number of followers and friends, and the number of statuses. In this section, we focus on the impact of user activity spans on the profile features distribution, which includes followers count, friends count, and statuses count. While the user profile features distribution of OSNs has been extensively studied in the literature [7, 14], we show some statistical properties that have been excluded from previous works.

In Fig. 3, we plot the average counts of three user profile features with the same activity spans. The number of statuses and friends increases exponentially in the initial stage (1–2 months), indicating that the new user published statuses with high interest. Thus, a new user account is more likely to attract other users' interests. Over time, the average number of followers and statuses changes to increase linearly and the growth slows down, indicating that the users' status interests and the ability of attract fans declines. The average number of user friends does not change over time.

User Allometric Growth with Activity Spans

In this section, we focus the allometric growth characteristics in the microblog. The number of statuses represents the production capacity for the information; the number of friends represents the user's ability to establish a relationship; and the number of followers represents the user's ability to attract other users' attention. We sorted users with different activity spans (in days) and gradually added the user

Fig. 3 Distribution of average user profile features with activity spans

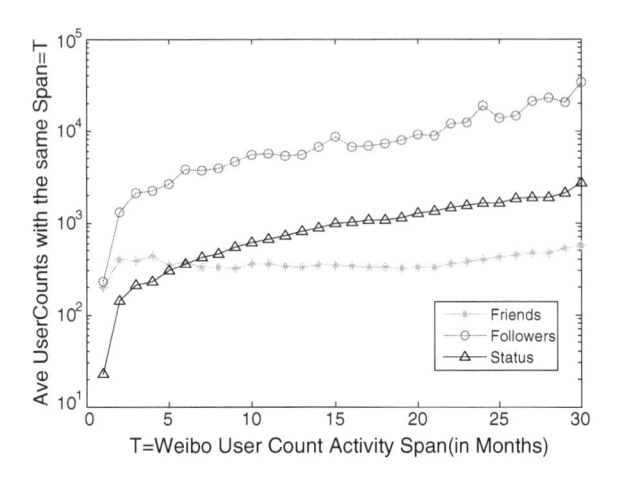

count in a larger active span. Then, we calculated the sum of the production capacity of the added users. Figure 4 shows the distribution of cumulative sum of user count and three user features count, which represent users' information production capacity. All of these user feature counts increase gradually with the user count and obtain a strict power-law form. The goodness-of-fit R^2 is greater than 0.99, and only the power-law exponent is different.

These relationships can be condensed into a unified expression as follows:

$$P(t) \propto N(t)^{\gamma} \tag{4}$$

where $P(t)$ means a certain cumulative production capacity of the network at different times of the users', $N(t)$ is the cumulative number of users of the network in different periods, and γ is the regression coefficients of the power-law. The

Fig. 4 Accumulation distribution of user count and user profile features count with same activity spans

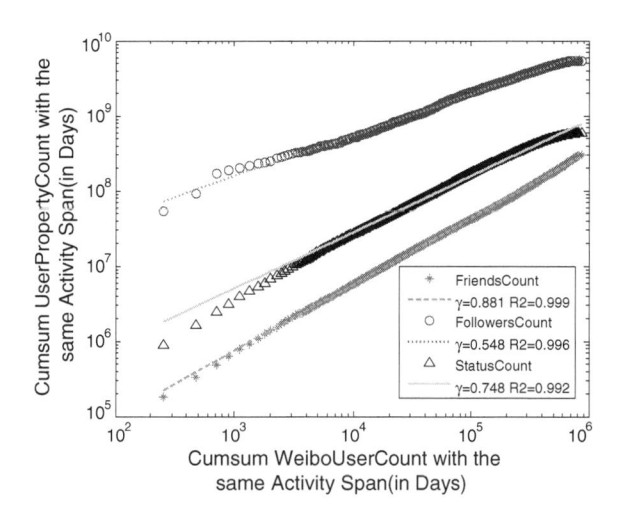

users' production capacity is very consistent in different active spans; this includes creating the relationship, acquiring others' attention, and producing information. The whole production growth is a non-stop and self-replicating process, and users in different stages show self-similar growth characteristics. The size-dependent distribution reflects the scale-free feature and allometric growth characteristics in different user activity spans.

When $\gamma < 1$, this means negative allometric growth. In Fig. 4, we find the exponent value of the power-law as $\gamma_{followers} < \gamma_{status} < \gamma_{friends} < 1$, which means that the user's average production capabilities, including information production, relationship creation, and interest attraction, are increasing more than the rate of the user count growth. The user's ability to attract interest grows the fastest, followed by the user's ability to create information. Finally, the user's ability to create friend relations grows relatively slowest. We argue that this phenomenon is consistent with the scale-effect of user count development. Namely, when an increasing number of users join a microblog, the user productivity improves quickly.

Conclusions

In this work, we studied two kinds of user life span in Sina Weibo and its connection to the distribution of the number of followers, friends, and statuses. User activity span and user age could be well approximated by two sections of exponential distribution. These finding sheds light on the double power-law distribution of followers, friends, and statuses. We also found an allometric growth phenomenon in the cumulative sum of the user count with the user's status, friends, and followers, all of which obtain a strict power-law form. These new findings show that the users' production capacity is consistent and that users show self-similar growth ability in different activity spans. We argue that the user's average production capabilities grow faster than the rate of user accounts. This can be explained by the scale-effect of user count development. In future works, we will introduce an evolution model based on user behavior and interest to explain the allometric growth characteristics described in this paper.

Acknowledgments This work has been supported by the National Natural Science Foundation of China under Grant 61172072, 61271308, the Beijing Natural Science Foundation under Grant 4112045, the Research Fund for the Doctoral Program of Higher Education of China under Grant W11C100030, the Beijing Science and Technology Program under Grant Z121100000312024.

References

1. Mislove A, Marcon M, Gummadi KP, Druschel P, Bhattacharjee B (2007) Measurement and analysis of online social networks. In: Proceedings of the 7th ACM SIGCOMM conference on Internet measurement, ACM, San Diego, California, USA, pp 29–42
2. Ahn Y-Y, Han S, Kwak H, Moon S, Jeong H (2007) Analysis of topological characteristics of huge online social networking services. In: Proceedings of the 16th international conference on world wide web, ACM, Banff, Alberta, Canada, pp. 835–844
3. Jiang J, Wilson C, Wang X, Huang P, Sha W, Dai Y, Zhao BY (2010) Understanding latent interactions in online social networks. In: Proceedings of the 10th annual conference on Internet measurement, ACM, Melbourne, Australia, pp 369–382
4. Benevenuto F, Rodrigues T, Cha M, Almeida V (2012) Characterizing user navigation and interactions in online social networks. Inf Sci 195:1–24
5. Kumar R, Novak J, Tomkins A (2006) Structure and evolution of online social networks. In: Proceedings of the 12th ACM SIGKDD international conference on knowledge discovery and data mining, ACM, Philadelphia, PA, USA, pp 611–617
6. Grabowski A (2009) Human behavior in online social systems. Eur Phys J B: Condens Matter Complex Syst 69:605–611
7. Kwak H, Lee C, Park H, Moon S (2010) What is twitter, a social network or a news media? In: Proceedings of the 19th international conference on world wide web, ACM, Raleigh, North Carolina, USA, pp 591–600
8. Yan Q, Yi L, Wu L (2012) Human dynamic model co-driven by interest and social identity in the MicroBlog community. Phys A 391:1540–1545
9. Gayon J (2000) History of the concept of allometry. Am Zool 40:748–758
10. Bettencourt LM, Lobo J, Helbing D, Kühnert C, West GB (2007) Growth, innovation, scaling, and the pace of life in cities. Proc Natl Acad Sci 104:7301–7306
11. Wu L, Zhang J (2011) Accelerating growth and size-dependent distribution of human online activities. Phys Rev E 84:026113
12. Ribeiro B, Gauvin W, Liu B, Towsley D (2010) On MySpace account spans and double Pareto-like distribution of friends. In: Proceedings of INFOCOM IEEE conference on computer communications workshops, pp 1–6
13. Huberman BA, Adamic LA (1999) Internet: growth dynamics of the world-wide web. Nature 401:131
14. Fan P, Li P, Jiang Z, Li W, Wang H (2011) Measurement and analysis of topology and information propagation on Sina-Microblog. In: Proceedings of IEEE international conference on intelligence and security informatics (ISI), Beijing, pp 396–401

Study and Implement of Topology Analysis Based on Hyper-Nodes in GIS

Li Zhang

Abstract In order to take full advantage of the topology analysis mechanism provided by the Geographical information system (GIS) platform software, the graphical topology method is advised to express the topology in pipe network GIS. This paper puts forward the concept of hyper-node, which has two graphic objects which each belongs to different graphics worlds. It can help implement the topology analysis across different graphics worlds. Finally, the power supply range analysis of electrical equipment (such as transformers in substation) in electric power GIS is implemented, which will validate the feasibility of the topology analysis method based on hyper-nodes.

Keywords Geographical information system (GIS) · Topology analysis · Hyper-node · Internal world

Introduction

There are many Geographical Information System (GIS) constituted by pipe network facilities such as power lines, water pipes, gas pipe and so on. The network topology analysis is one of the most important functions in pipe network GIS. GIS platform software (such as GE Smallworld, ArcInfo) provides some objects or controls with powerful topology analysis capability [1]. But it does not provide the network analysis across different graphics world, such as the topology analysis in electric power GIS from switch inside substation to feeder-lines outside which do not exist in the same graphics system [2].

L. Zhang (✉)
School of Computer Engineering, Shenzhen Polytechnic, Shenzhen,
518055 Guangdong, China
e-mail: zhangli@szpt.edu.cn

Y.-M. Huang et al. (eds.), *Advanced Technologies, Embedded and Multimedia for Human-centric Computing*, Lecture Notes in Electrical Engineering 260, DOI: 10.1007/978-94-007-7262-5_97, © Springer Science+Business Media Dordrecht 2014

Graphical Representation and Topology in GIS

In the pipe network GIS, the main window usually displays the geographical layout map whose main task is to reflect the geographical location and connection relationship of the pipeline network in the real world. There are many methods to express the topology in GIS [3]. One of them is the graphical topology method. This method combines the graphic expression with the topological relations, which requires the graphical position of pipeline facilities must be precise. The lines or equipments with the connection relationship should be really connected together in GIS. (i.e. the two geometries must overlap or cover on the coordinate position), otherwise it means that they do not connect topologically with each other [4].

In the graphical topology method, the maintenance of the topology is completely contained (hidden) in the maintenance of graphics data. So topology analysis can completely depend on the mechanisms of topology analysis provided by the GIS platform software. As a result, GIS application development workload becomes smaller.

Creating a Hyper-Node

In order to connect the feeder lines outside and the equipments inside substations topologically, we need to create such a "hyper-node" object, which has two geometries existing in two different graphics systems (called "Graphics World" in GE Smallworld). For example, *hypernode* class defined in Magik (the programming language of GE Smallworld) code is as follows:

```
def_slotted_exemplar( :hypernode, {}, { :rwo_record } )
$
_method hypernode.new_between_worlds( a_view,coord1,
world1, coord2, world2)
    _dynamic !current_dsview! << a_view
    _dynamic !current_world!
    drec << _self.target_collection.new_detached_record()
    new << _self.target_collection.insert( drec )
    !current_world! << world1
    new.make_geometry( _self.hypernode_info[:pin1], coord1
)
    !current_world! << world2
    new.make_geometry( _self.hypernode_info[:pin2], coord2
)
    >> new
_endmethod
$
```

The following Magik code is used to create a class named as *elec_hypernode* in the electric power GIS. It is a subclass of *hypernode* class. A shared constant named as *hypernode_info* should be defined. According to the code, the table name in Smallworld database is *elec_hypernode* and two point geometry field name are *pin1* and *pin2*. Another shared constant is *associated_grs* including some graphics worlds which the geometries of *elec_hypernode* may belong to. In other words, an *elec_hypernode* object has two point geometries belonging to different graphics worlds. One is *pin1* which belongs to the geographic layout map, the other is *pin2* which belongs to the substation internal world.

```
def_slotted_exemplar (:elec_hypernode,{},:hypernode)
$
elec_hypernode.define_shared_constant(:hypernode_info,
    property_list.new_with(
        :name, :elec_hypernodes,      # table name
        :pin1, :pin1,                 # 2 point geometries
        :pin2, :pin2
        ), _false)
$
elec_hypernode.define_shared_constant(:associated_grs,{
        :gis,               # geographic layout map
        :subinternal,       # substation internal diagram
        }, _false)
$
```

Topology Analysis Based on Byper-Nodes

Here is an example to explore how to achieve the topology analysis based on hyper-nodes. In electric power GIS, the power range analysis of the substation equipments (e.g. a transformer inside substation) is a topological track, whose goal is to get all objects whose power is supplied by the specified station equipment.

In reality, the overhead lines or cables outside are connected to the switch inside substations to complete the power transmission. When you create a hyper-node, the geometry *pin2* should be connected to a switch inside substation and the *pin1* should be connected with the end point of an overhead line or cable.

It is more complex to express the topological relations of substation equipments, such as buses, transformers, breakers and switches. In substation internal world in GIS, the bus object is a linear object (*chain*). The other objects are *two_pin_device* objects, which have two point geometries named as *point1* and *point2*. The shape of *two_pin_device* objects are drawn based on the coordinates of the two points as a reference by the appropriate Magik code. So the inside geometry (i.e. *pin2*) of the hyper-node must be connected with the *point1* or *point2* of a switch inside the substation when a hyper-node is created.

In order to implement the power supply range analysis of substation equipments such as transformers, it is necessary to define a class named as *analyse_sub_device_tool* inherited from the *network_follower_menu* class. It is an operating dialog in which users can execute topology analysis. As a subclass of *network_follower_menu*, the *analyse_sub_device_tool* object will have a property named as follower, which is actually a *network_follower_manager* object. So you can easily to analyze the topology by calling the methods defined in *network_follower_manager*, such as *trace_out()* and *shortest_path ()* method.

```
def_slotted_exemplar( :analyse_sub_device_tool,{
    { :actual_start_point,   _unset},
    { :selected_sub_device, _unset},
    { :trace_results, _unset,     :readable},

    ......

    },:network_follower_menu )
$
```

Several properties are added when *analyse_sub_device_tool* class is defined. The *selected_sub_device* property is used to save the currently selected station equipment and the *trace_results* property will include all objects involved in tracking. The *actual_start_point* property is used to save the point geometry of the selected station equipment which is the actual start point of the track for the lower-voltage side. In addition, it includes a custom method named as *analyse_sub_device ()*, whose main task is to call the custom method *trace_down_from_subdevice()* defined in the *network_follower_manager* class.

```
_method analyse_sub_device_tool.analyse_sub_device()

    ......

    .actual_start_point<<.follower.trace_down_from_subdevi
ce(.selected_sub_device,_true,_true)

    ......

_endmethod
$
```

The idea of defining custom method *trace_down_from_subdevice()* can be descibed as follows: firstly the voltage of the selected equipment (e.g. a breaker) inside substation is saved to a variable named as *current_voltage*. Then the stop condition *stop_predicates* is set as *feeder_switch* with the status as *OFF* and breakers with the status as *OFF*. The parameter *stop_nodes* is set as the geometry *point2* of the selected station equipment. The *trace_out()* method defined in *network_follower_manager* object is called to execute topological track in one direction. The *trace_in_lower_voltage_direction?()* method will be called to determine whether it is the track for the lower-voltage side or not. If not true, the *trace_out()* method will be called again to execute topological track in the other direction while the parameter *stop_nodes* is set as the geometry *point1* of the selected station equipment. As a result, all objects whose power is supplied by the selected station equipment are saved in the property named as *trace_results* of *analyse_sub_device_tool* class. The Magik code of *trace_down_from_subdevice ()* method is as follows:

```
_method
network_follower_manager.trace_down_from_subdevice(a_subd
evice,_optional draw?)
    draw?<<draw?.default(_false)
    through_hypernode?<<through_hypernode?.default(_true)
    a_stop_predicates_group<<hash_table.new_with(:breaker,
{predicate.new(:on_off_status,:eq,"off")},
            :feeder_switch,{predicate.eq( :normal_status,
"on" )})
    .draw_trace? << _false
    current_voltage << a_subdevice.voltage.as_number()
    a_start_point<<a_subdevice.point1
    _self.trace_out(a_subdevice.point1.node,
        :via_node_rwo?,_true,
        :stop_predicates,a_stop_predicates_group,
        :stop_nodes,{a_subdevice.point2.node})
    _if
_self.trace_in_lower_voltage_direction?(current_voltage)
_is _false
    _then
        a_start_point<<a_subdevice.point2
        _self.trace_out(a_subdevice.point2.node,
            :via_node_rwo?,_true,
            :stop_predicates,a_stop_predicates_group,
            :stop_nodes,{a_subdevice.point1.node})
    _endif
    _return a_start_point
_endmethod
$
```

In *trace_in_lower_voltage_direction?()* Method, it checks the direction of the topology tracking by the following conditions: (1) If there is not any one breaker in the tracking result set, this is the track for the lower-voltage side. (2) If there exist one (or more) breaker in the tracking result set, and the highest voltage of the breakers is equal to the voltage of the check point, it is still the track for the lower-voltage side. Otherwise it is the track for the high-voltage side.

Conclusion

Because GIS platform software itself does not provide the topology analysis across different graphics worlds [5], hyper-node objects with two geometries are created to connect topologically two objects inside and outside station house in GIS. In this way, topology analysis can reflect the topology relationship in reality, such as the connection between the feeder line outside and the switch inside the substation. The hyper-nodes make it easy to topology tracking according the connection rules of pipe network. At the same time the topology analysis mechanism in GIS will play a greater role in the applications of topology relationship.

References

1. Lin F, Guo B, Qian W (2011) Common electric power grid GIS platform oriented architecture for electric power grid GIS applications. Autom Electr Power Syst 35(24):63–67
2. Liu W, Zhang J, Liu N, Zhang L (2008) Design of a distribution network intelligent planning platform based on COTS. Automation of Electr Power Syst 32(12):48–51
3. Zhang K (2008) Construction of distribution network GIS based on SmallWorld. Comput Era (1):38–41
4. Qun Y, Wei L, Guonian L (2003) Topology analysis of distribution management system based on geographical information system. Autom Electr Power Syst 27(18):80–82
5. Wu Q, Chen T, Su Y (2005) Building of spatial topological relationship based on mapobjects. Comput Simul 22(1):73–75

A New Lightweight RFID Grouping Proof Protocol

Ping Huang, Haibing Mu and Changlun Zhang

Abstract Radio Frequency Identification (RFID) is a key technology for the Internet of things. With the spread of RFID system, the security and privacy related to RFID plays a vital role in applications. At present, EPC C1G2 tags have been widely spread in the commodity retail, pharmaceutical sector, medicine distribution and many other fields. This kind of tags is passive, which means that they have limited power and memory space. Under these conditions, we present a new lightweight protocol for the RFID system with two tags and one reader. In the protocol, we develop a new function to realize the exchange between bits of a vector, which uses fewer resources in computing. In addition pieces of analysis show that the protocol reaches the capabilities of the grouping proofs and is resistant to a variety of attacks.

Keywords RFID · Lightweight · Grouping proof protocol · Privacy · Security

Introduction

Radio Frequency Identification (RFID) is a key technology of the Internet of things. An RFID system is consist of three parts, which are RFID tag, RFID reader and a backend server. RFID tags are used for preserving information of things and

P. Huang · H. Mu (✉)
Beijing Key Laboratory of Communication and Information Systems, School of Electronic and Information Engineering, Beijing Jiaotong University, Beijing 100044, China
e-mail: hbmu@bjtu.edu.cn

P. Huang
e-mail: 12120086@bjtu.edu.cn

C. Zhang
Science School, Beijing University of Civil Engineering and Architecture,
Beijing 100044, China
e-mail: zclun@bucea.edu.cn

Y.-M. Huang et al. (eds.), *Advanced Technologies, Embedded and Multimedia for Human-centric Computing*, Lecture Notes in Electrical Engineering 260, DOI: 10.1007/978-94-007-7262-5_98, © Springer Science+Business Media Dordrecht 2014

can be sorted into three types: active tags, semi-passive tags and passive tags. Every active tag is powered by a battery, which support the tags maintain their memories and transmitting signals. Its computing power is the strongest of all. Semi-passive tags also have a battery within them but it is just used for maintain the information in the chip. Passive tags, the most widespread kind of all, without any devices to offer power, need non-volatile memory. Moreover, because of its limited ability and transmission distance, we should design special protocols for this kind of tags. This paper is based on the passive tags, considering the power limited problem.

At present, RFID is widely applied in scientific research, business, medical and many other industries. Presenting the medicine distribution of the inpatients for instance, accuracy and correctness are very important for the hospitalized patients. Only reaching the required standards can the hospitals ensure the security of patients. In other words, if privacy and security couldn't be promised during the process of distribution, a malicious adversary will prevent or interfere the corresponding medicine's being distributed to the right patients by forging information or any other means.

Aiming at the above questions, our paper presents a new lightweight RFID grouping proof for two tags. Using our protocol in medicine distribution, privacy and security can be guaranteed. There is a shared secret key between each pair of a tag and a reader. Our protocol is efficient and useful for three accounts. Firstly, considering that the EPC C1G2 tags have limit power and low computation capacities, we define a new function Con to replace the use of Hash function, which can also reach confusion and diffusion. Secondly, taking privacy and security into account, there are shared secret keys between the tag and reader. The shared information will not be transmitted in the channel during the protocol run. The last one, as for two tags in each system, information computed by the two tags depends on each other. Synthesizing the above three aspects, our lightweight protocol is efficient and secure.

The rest of this paper is organized as follows. In section Related Works, the related works are reviewed. In section Model, a proper adversary model is presented. In section A Novel Protocol, the protocol is described in details. In section Analysis, we analyze the cost and security of our protocol. In section Conclusion, we give a conclusion.

Related Works

The first approach to protect RFID users' privacy is the implementation of "Kill" [1]. That means the Reader send a command of kill the tag and the tag will be inactive forever. In this way, privacy can be well protected from adversary's scanning the tag illegally. But the tag will never work anymore. Afterwards, according to the thought of cryptography presented by Shannon, amount of researchers come up with many RFID protocols for privacy and security of RFID system. Such as the Hash-Lock protocol [2, 3]. Weis S.A. et al. [2] also described a new research direction in RFID for researchers with cryptography. An randomized Hash-Lock protocol [4] was put forward, and the authors said that low-cost Radio Frequency Identification (RFID) systems would become pervasive in our daily lives when affixed to everyday consumer items as "smart labels". It predicts low-cost is important for RFID system's being applied in our daily life. The hash-chain protocol [5] can ensure forward-secure privacy, because hash-function has one-way property. Hash-based enhancement of location privacy for RFID devices using varying identifiers [6] and so on [2–6] are based on Hash that cost more than some protocols using bitwise operations and didn't consider about the limited resources in EPC C1G2 tags. Yun Tian et al. [7] created a new lightweight RFID authentication protocol with a new definition of Permutation. Kai Fan et al. [8] presented a lightweight RFID authentication Scheme to achieve efficiency. Erik-Oliver Blass et al. [9] showed a novel framework for lightweight RFID authentication protocol without using of Hash or any other complex functions. Erik-Oliver Blass et al. [10] provided a provably privacy-preserving protocol that applied for passive tags. [7–10] can decrease cost of the protocol. Roberto Di Pietro and Refik Molva [11] described two schemes that can be used independently to enhance the security of the RFID protocols. However, researches on RFID grouping authentication protocols have not been mature. The concept of "Yoking proof" was put forward by Jules [12] firstly. Soon afterwards, Saito and Sakurai [13] improved the former one with the use of timestamp and produced their protocol named "grouping proof". Their approach can solve some problems and make up some of the vulnerabilities. Above most of the existing protocols for RFID group authentication, Pedro Peris-Lopez et al. [14] pointed the flaws in some of the existing grouping proofs and summarize a guideline for the design of this kind of protocol. Piramuthu [15] raised a new improved group proof model, but it can be suffered from attack on the communication between the tags and reader and the attack by disassembling the tags. Most of the approaches have limits, either high cost or cannot be used for more than one tags' system.

Model

This paper is based on the EPC C1G2 tags, and can be used for medicine distribution for inpatients. So in this section, some proper assumptions are presented about the system and adversary as follows.

System Model

1. A pallet presents a hospitalized patient. In addition, Tag_A means the medicine to be distributed and Tag_B is the corresponding pallet in our protocol.

2. Every tag stores an ID and a secret key within its memory, while the reader has all the IDs and keys for the protocol.
3. A tag can operate three kinds computation, rotation Rot(*,*), exclusive or XOR($\oplus$) and function Con.
4. Through our protocol, a proof P_{AB} is finished and submitted to the backend verifier to that the two tags participate in the protocol simultaneously.

Adversary Model

1. A adversary can eavesdrop and intercept what reaches him. Furthermore, he can also send counterfeit information to any reader or tags.
2. As ISO14443 required, the distance between reader and tag is no more than 4 inches. Thus we assume that the adversary cannot stand between the reader and tags and he is out of the tags' communication range. Under the condition, an adversary can eavesdrop the information sent by the reader but can't eavesdrop or intercept the messages sent by the tags.

A Novel Protocol

For passive RFID tags, limited power is a vital problem for protocol designers. In order to reduce the amount of calculation and power needed, we define a novel function Con to replace the Hash used in most of RFID authentication protocols. At the same time, we just use other two kinds of bitwise operation, bitwise XOR($\oplus$) and Rot(*,*). There are two variables in Con, for example X, Y. We rearrange the vector X according to Y.

Definition Suppose there are two l-bit vectors X, Y, where
$X = x_1x_2...x_l$, where $x_i \in \{0, 1\}$ and $i = 1, 2, ..., l$;
$Y = y_1y_2...y_l$, where $y_j \in \{0, 1\}$ and $j = 1, 2, ..., l$;

Moreover, the Hamming weight of Y is m, that means $\sum_{j=1}^{l} y_j = m$ and $y_{k_1} = y_{k_2} = ... = y_{k_m}$ and $y_{k_{m+1}} = y_{k_{m+2}} = ... = y_{k_l}$ (k_i is discontinuous, i.e. y_{k_i} just presents one of all the y_j, that equals to 0 or 1.).

Thus, $Con = x_{k_1}x_{k_3}...x_{k_{m+1}}x_{k_{m+2}}...x_{k_l}...x_{k_4}x_{k_2}$.

Example (Figure 1)   X = 01101011,   Y = 11010110,   where   m = 5,   and Con(X,Y) = 10011110.

There are three kinds of entities. They are RFID tags (Tag_A, Tag_B), RFID reader and backend server. We suppose the channel between the reader and the server is secure. Under the ISO14443, we assume that the adversary cannot carry out any attack on the system between the reader and the tag on the physical. That

Fig. 1 An example of function con

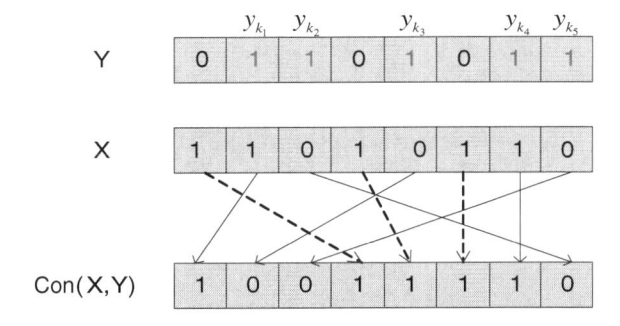

also means the distance between malicious adversary and tag is longer than that between the tag and its reader. So we can adjust the tag's transmission power to keep the adversary out of the signal range.

In our protocol, every tag keeps a shared secret key k_i and its identity ID_i. Every server stores all the keys and identities.

The protocol runs as Fig. 2 shows:

1. Reader transmits request and random number (r_1, r_2) as challenges to the two tags, Tag_A and Tag_B.
2. After receiving the request and r_1, Tag_A generates a random number r_A and computes $A = Con(rot(r_A, r_1) \oplus ID_A, r_A)$ with r_A, r_1, ID_A, then he sends A and r_A as a reply to the Reader.
3. Upon receiving the A and random number r_A, the reader computes $V_{R_A} = Con(r_A, ID_A)$ and transmits V_{R_A} to Tag_B.
4. When receiving the V_{R_A}, Tag_B generates a random number r_B and computes a $B = Con(rot(V_{R_A}, r_B) \oplus ID_B, r_B \oplus r_2)$ with the challenge number r_2, V_{R_A}, the new random number r_B and its identity ID_B, after computing the B, it computes $V_B = Con(rot(B, r_2), k_2)$ with its secret key k_2, B and r_2. Then it transmit B, r_B and V_B to the Reader.
5. After receiving the three numbers, the reader computes $V_{R_B} = Con(r_B, ID_B)$ with r_B and ID_B and transmits V_{R_B} to the Tag_A.

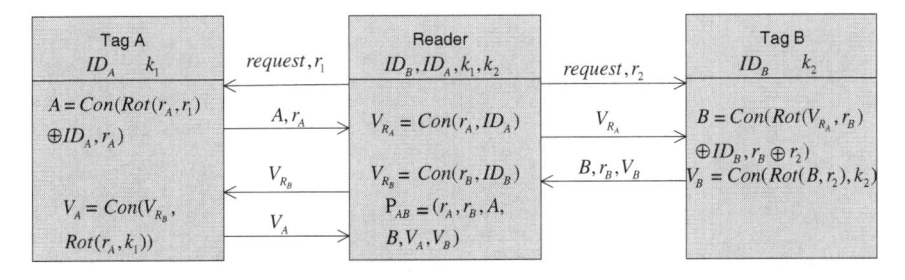

Fig. 2 A new lightweight RFID protocol for two tags

6. Then Tag_A computes $V_A = Con(V_{R_B}, rot(r_A, k_1))$ with V_{R_B}, the random number generated by Tag_A and its secret keyk_1. After this, it transmits V_A to the Reader.
7. The Reader collects r_A, r_B, A, B, V_A, V_B as a proof P_{AB}, then submits the P_{AB} to the backend server.
8. When the server receives P_{AB}, he computes a local P'_{AB} using the shared information in his database and computes the local value with what he received before. If they are equal, the protocol is successful, otherwise, the communication cannot continue anymore.

Analysis

In this section, we will evaluate our protocol on the following eight aspects. They are cost of the protocol, attack on the Tags, attack on the Reader, attack on the communication between the tags and the Reader, attack on user's privacy, tracking the tags, attack against the key and the last is disassembling the tags.

1. Overhead of the protocol.

In our protocol, we use three kinds of bitwise operations, bitwise XOR ($\oplus$), Rotation Rot (A, B) and a new designed function Con(X, Y). Compared with the existing protocols, most of them contain the hash to protect the privacy and security. But for passive tags, it is not appropriate.

2. Attack on the tags.

An adversary pretends to be a legal Reader and attack the tags with its range. From the protocol, we can learn that the legal reader computes V_{R_A} and V_{R_B} using the tags' IDs, but the adversary can't get the ID_i or r_i transmitted by the tags because the signal sent by the tags cannot reach the adversary. So the adversary cannot compute the V_{R_i} (i = A or B). Thus, our protocol can fight against this attack.

3. Attack on the Reader.

When an adversary tries to pretend a Tag and cheat the Reader, he can't success. Because the entities shared secret keys and they are not transmitted during the protocol.

4. Attack on the communication between tags and reader.

According to the ISO14443, we can assume that the adversary stays farther from the Tags than the Reader that he can't capture or intercept the information transmitted from the tags to the Reader.

5. Attack against the user's privacy.

What we can get from the description of the protocol is that there is no private information transmitted in the channel.

6. Tracking the tags.

Because the random numbers r_1, r_2, r_A, r_B are refreshed when a new protocol begins, the adversary can't use the previous information to carry out a new attack.

7. Attack against the keys.

The secret keys are stored in the backend server and the memory of each tag. When the protocol runs, they do not appear separately. No matter what kind of ability does the adversary own, he still can't get the keys.

8. Disassembling the tags.

Although the tags are not tamper-resistant, we can implant a kill-circuit. When the adversary try to get the tags physical structure by taking apart the tag, the circuit will output a signal to set up the kill mechanism and make the tag invalid forever.

9. Separating the tags.

In grouping proof protocols, resistance against separating the tags is one of the most important aims. Our protocol provides dependence on each other that can protect the tags from separating. If one tag is separated from the other, the proof P_{AB} submitted to the server in the end won't be verified successfully and the protocol will fail.

Conclusion

In this paper, we presented a lightweight RFID grouping proofs protocol that can be used for inpatient. The reader sends two different challenge numbers to the tags and the protocol begins. During the protocol runs, each tag computes verification messages using the information sent by the other one last time. At last, the reader collects enough information and submits it to the backend server to finish the protocol. Compared with other existing protocols for such RFID system that owns one reader and multiple tags, our approach achieves higher security and privacy requirement and improves efficiency. There are only three kinds of bitwise operation within the novel protocol. In the future, we will explore new schemes to solve some other practical problems and try to improve the RFID system.

Acknowledgments This work is supported by National Natural Science Foundation of China under Grant 61201159 and Fundamental Research Funds for the Central Universities under Grant 2012JBM016. The authors also gratefully acknowledge the helpful comments and suggestions of the reviewers, which have improved the presentation.

References

1. Juels A (2006) RFID security and privacy: a research survey. IEEE J Sel Areas Commun 24(2):381–394
2. Sarma SE, Weis SA, Engels DW (2003) RFID systems and privacy implications. In: Kaliski BS, Koc CK, Paar C (eds) Proceedings of the 4th international workshop on cryptographic hardware and embedded systems (CHES 2002). Lectures notes in computer science 2523. Springer, Berlin, pp 454–469
3. Sarma SE, Weis SA, Engels DW (2003) Radio-frequency identification: Secure risks and challenges. RSA Lab Cryptobytes 6(1):2–9
4. Weis SA, Sarma SE, Rivest RL, Engels DW (2004) Security and privacy aspects of low-cost radio frequency identification systems. In: Hutter D, Muller G, Stephan W, Ullmann M (eds) Proceedings of the 1st international conference on security in pervasive computing. Lectures notes in computer science 2802, Springer, Berlin, pp 201–212
5. Ohkubo M, Suzuki K, Kinoshita S (2004) Hash-chain based forward-secure privacy protection scheme for low-cost RFID. In: Proceedings of the 2004 symposium on crytography and information security (SCIS 2004), Sendai, pp 719–724
6. Henrici D, Muller P (2004) Hash-based enhancement of location privacy for radio-frequency identification devices using varying identifiers. In: Proceedings of the 2nd IEEE annual conference on pervasive computing and communications workshops (PERCOMW'04), Washington, DC, USA, pp 149–153
7. Tian Y, Chen G, Li J (2012) A new ultralightweight RFID authentication protocol with permutation. IEEE Commun Lett 16(5):702–705
8. Fan K, Li J, Li H, Liang X, Shen XS, Yang Y (2012) ESLRAS: a lightweight RFID authentication scheme with high efficiency and strong security for internet of things. In: Proceedings of 4th international conference on intelligent networking and collaborative systems, pp 323–328
9. Blass E-O, Kurmus A, Molva R, Noubir G, Shikfa A (2011) The Ff-family of protocols for RFID-privacy and authentication. IEEE Trans Dependable Secure Comput 8(3):466–480
10. Blass E-O, Elkhiyaoui K, Molva R (2012) PPS: Privacy-preserving statistics using RFID tags, world of wireless. In: Proceedings of 2012 IEEE international symposium mobile and multimedia networks (WoWMoM), pp 1–6
11. Di Pietro R, Molva R (2007) Information confinement, privacy, and security in RFID systems. In: Proceedings of the 12th European symposium on research in computer security, pp 187–202
12. Juels A (2004) Yoking proofs" for RFID tags. In: Proceedings of the first international workshop on pervasive computing and communication security, IEEE Press
13. Saito J, Sakurai K (2005) Grouping proof for RFID tags. In: Proceedings of the 19th international conference on advanced information networking and applications (AINA'05), pp 621–624
14. Peris-Lopez P, Orfila A, Hernandez-Castro JC, Van der Lubbe JC (2011) Flaws on RFID grouping-proofs. Guidelines for future sound protocols. J Netw Comput Appl 34(3):833–845
15. Piramuthu S (2007) Protocols for RFID tag/reader authentication. Decis Support Syst 43:897–914

Iris Recognition

Hwei Jen Lin, Yue Sheng Li, Yuan Sheng Wang and Shih Min Wei

Abstract An iris recognition system uses the iris to distinguish the identity of a person using the rich iris texture feature. To effectively remove noise and precisely segment the stable iris region is a crucial stage prior to recognition. Most noises on iris images are caused by occlusion of eyelids or eyelashes in certain areas. In this paper, we propose an iris recognition system which precisely locates and segments iris regions. We extract the iris feature from a relatively reliable portion of the iris region using a DoG filter. Experimental results show that the proposed iris recognition system has satisfactory results in terms of time efficiency and recognition rate.

Keywords Biometric recognition · Iris recognition · Iris segmentation · Iris normalization · Feature extraction · Difference of Gaussian (DoG)

H. J. Lin (✉) · Y. S. Li · S. M. Wei
Department of Computer Science and Information Engineering, Tamkang University, Taipei, Taiwan, Republic of China
e-mail: 086204@mail.tku.edu.tw

Y. S. Li
e-mail: 699410519@s99.tku.edu.tw

S. M. Wei
e-mail: wesley78527@gmail.com

Y. S. Wang
Office of Physical Education, Tamkang University, Taipei, Taiwan, Republic of China
e-mail: 119391@mail.tku.edu.tw

Y.-M. Huang et al. (eds.), *Advanced Technologies, Embedded and Multimedia for Human-centric Computing*, Lecture Notes in Electrical Engineering 260, DOI: 10.1007/978-94-007-7262-5_99, © Springer Science+Business Media Dordrecht 2014

A LDoS Detection Method Based on Packet Arrival Time

Kun Ding, Lin Liu and Yun Liu

Abstract Low-rate Denial of Service (LDoS) attack is a new type of Denial of Service attack, which is difficult for the router and victim site to detect because the attack packets are as many as the valid packets. In consideration of the fact that the features in time domain between attack traffic and valid traffic are different, we present a method to identify such kind of attack and try to trace the potential location of the attacker. We also carry out a simulation to illustrate the usability of this method.

Keywords Denial of service · Low-rate · TCP · Retransmission timeout · Arrival time

Introduction

As a serious challenge of the Internet security, Denial of Service (DoS) attacks abuse resources in the networks. Recently, the evolution of such type of attack, especially the Low-rate Denial of Service (LDoS) attack [1, 2], has eluded lots of

K. Ding · Y. Liu (✉)
School of Electronic and Information Engineering, Beijing Jiaotong University,
Beijing 100044, China
e-mail: liuyun@bjtu.edu.cn

K. Ding
e-mail: 12120057@bjtu.edu.cn

K. Ding · Y. Liu
Key Laboratory of Communication and Information Systems, Beijing Municipal
Commission of Education, Beijing Jiaotong University, Beijing 100044, China

L. Liu
China Information Technology Security Evaluation Center, Beijing, China
e-mail: liul@itsec.gov.cn

Y.-M. Huang et al. (eds.), *Advanced Technologies, Embedded and Multimedia for Human-centric Computing*, Lecture Notes in Electrical Engineering 260, DOI: 10.1007/978-94-007-7262-5_100, © Springer Science+Business Media Dordrecht 2014

detection methods. The common flood-based DoS attacks crash the victim site by sending high-rate invalid packets continuously which produce a large number of packets and the number is quite larger than the valid packets, so that the routers can easily detect the attack by recording the traffic counters [3, 4]. While the LDoS attacks try to repeatedly provoke the network enter the retransmission timeout phase of TCP congestion control mechanism by sending high-rate packets periodically. The burst transmission instead of continuous transmission ensures that the total number of the attack packets during the detection period of the routers is as large as the number of valid packets, so that it is difficult for the routers to identify the LDoS attacks.

LDoS Attacks Model

Generously, LDoS attacks perform at a particular rate periodically and the burst length is fixed. According to these features, we use a "square wave" model [5]. As shown in Fig. 1, we denote the duration of attack bursts as l, the rate as R, and the period as T. Consider the number of invalid packets which are transmitted to the victim site during each attack is $\mathbf{X}$. Thus, we can easily get the corresponding equation as follows:

$$\mathbf{X} = l \cdot \mathbf{R} \tag{1}$$

In most cases, the attackers send packets in a pulse pattern. While in some way, the rate pattern can be sinusoidal wave or cosine wave [6].

Characteristics Analysis in Time Domain

Since the attack packets during the router detection period are almost as many as the valid packets, even less than the valid packets, it is difficult for the router to detect the attack behavior. Although the packets from the normal subscriber are

Fig. 1 LDoS attack model

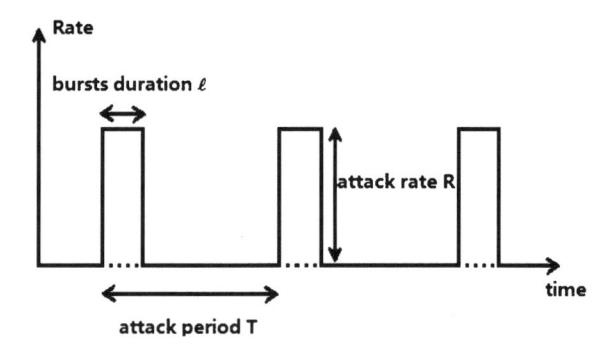

also burst data, there are some different characteristics between the valid packets and the attack packets. In time domain, the most difference is obviously the arrival time. The duration of the attack burst is about several hundred milliseconds. The arrival time of the attack packets is intensive during each attack, while the valid packets' arrival time is relatively dispersed.

Figure 2 depicts the arrival time of valid packets to the victim site in a simulation model. The distribution of the arrival time is random. And the simulation time is from 0 to 5.5 s. Each red line in the figure represents a packet. When there is no attack, the throughput of the network is quite high. Figure 3 depicts the arrival time of the attack packets. The most prominent feature in time domain is the periodicity and the aggregation.

Influenced by the attack burst, the throughput of the network will be degraded. The arrival time of all packets is shown in Fig. 4. Due to the TCP congestion control mechanism, the receive window is reduced.

Detection Method

According to the above LDoS attack model, we present a detection method which is based on packet arrival time. This method is deployed in the uplink routers of the victim site or the victim site itself. Because the attack burst is high-rate traffic, the arrival time is relatively intensive. During the attack, the number of packets which enter the router per unit of time is larger than the usual.

When the traffic enters the queue of the router or passes through the router, log information will be recorded. The log information contains the source link, the destination, the arrival time, the packet type, etc. In our method, only the packets which reached the participating routers with the destination of the victim site are considered.

Fig. 2 The arrival time of valid packets

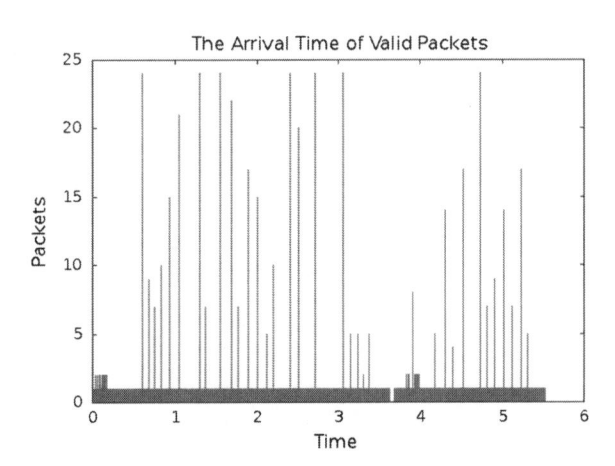

Fig. 3 The arrival time of
attack packets

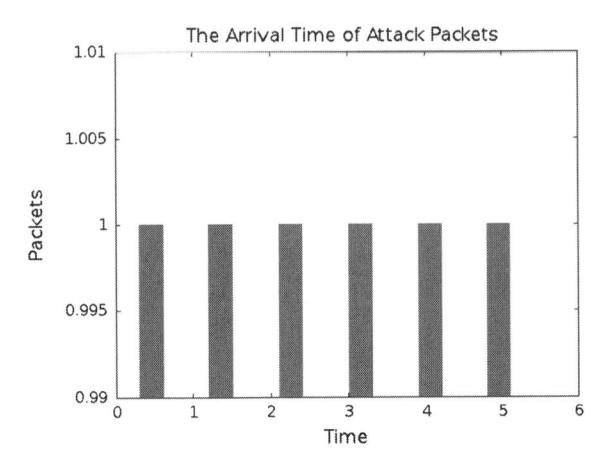

Fig. 4 The arrival time of all
packets

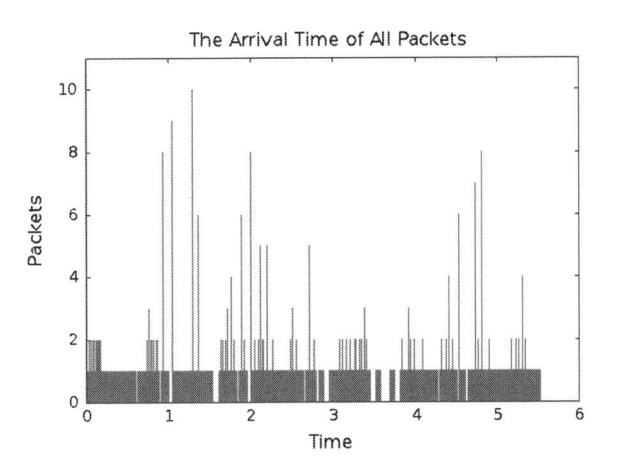

When the network enters the retransmission timeout phase of TCP congestion
control mechanic, the victim site sends control packets to all of the participating
routers. The routers implement the algorithm as followed:

Definite P_i as the ith packet that reached the router, and let t_i be the time instant
of the ith packet that reached the router. Thus, the time interval of the two adjacent
packets can be depicted by Δ_i as followed:

$$\Delta_i = t_i - t_{i-1} \tag{2}$$

Definite parameter r_{ci} as the consequent ratio of Δ_i, r_{ai} as the antecedent ratio of
Δ_i:

$$r_{ci} = \frac{\Delta_{i+1}}{\Delta_i} \tag{3}$$

$$\mathbf{r_{ai}} = \frac{\Delta_j}{\Delta_{j+1}} \tag{4}$$

According to the characteristic of the attack packets mentioned in the previous part, $\mathbf{r_{ci}}$ will be greater at the instant when the current attack burst is over. Correspondingly, $\mathbf{r_{ai}}$ will be another numerical peak which represents the beginning of an attack burst.

Let δ_c be the threshold of $\mathbf{r_{ci}}$ and δ_a be the threshold of $\mathbf{r_{ai}}$. Analyze the log information and figure out $\Delta_1, \Delta_2, \Delta_3, \ldots, \Delta_i, \ldots$ and then figure out $\mathbf{r_{c1}}, \mathbf{r_{c2}}, \mathbf{r_{c3}}, \ldots, \mathbf{r_{ci}}, \ldots$ and $\mathbf{r_{a1}}, \mathbf{r_{a2}}, \mathbf{r_{a3}}, \ldots, \mathbf{r_{ai}}, \ldots$ Compare $\mathbf{r_{c1}}, \mathbf{r_{c2}}, \mathbf{r_{c3}}, \ldots, \mathbf{r_{ci}}, \ldots$ with δ_c, the one which is greater than δ_c would be the potential attack feature. If $\mathbf{r_{ck}}$ (k = 1, 2, 3...) is greater than δ_c, put the corresponding instant t_k of $\mathbf{P_k}$ in the "potential beginning attack" set $\mathbf{C}[m]$:

$$\mathbf{C}[m] = t_k\ m = 1, 2, 3\ldots \tag{5}$$

m represents the mth element of set $\mathbf{C}$. Theoretically, $\mathbf{P_k}$ represents the last packet of the current attack burst and t_k is the instant that the current attack burst is over.

Compare $\mathbf{r_{a1}}, \mathbf{r_{a2}}, \mathbf{r_{a3}}, \ldots, \mathbf{r_{ai}}, \ldots$ with δ_a and if $\mathbf{r_{ak}}$ (k = 1, 2, 3...) is greater than δ_a, put the corresponding instant t_k of $\mathbf{P_k}$ in the "potential ending attack" set $\mathbf{A}[n]$:

$$\mathbf{A}[n] = t_k\ n = 1, 2, 3\ldots \tag{6}$$

n represents the nth element of set $\mathbf{A}$. $\mathbf{P_k}$ may be the first packet of the current attack burst.

Analyze all the log information and get set $\mathbf{C}$ and set $\mathbf{A}$, and count the num of elements in these two sets. Definite threshold η, if

$$\min\{\text{ element num of } \mathbf{C},\ \text{element num of } \mathbf{A}\} > \eta$$

the alarm would be started which means the potential LDoS attack is detected.

Then, the router will analyze the log information and find out the source domain of attack packet. Finally, the router will send reply to the victim site to inform the victim site that a potential LDoS attack is on. In the reply, the attack source domain information will be included. After all of the participating routers sending replies, the victim site will get the attack route.

Simulation

Let us define the network model. As shown in Fig. 5, Router 1 is the uplink router of the victim site; Client 1 is in the local domain of Router 1; Client 2 and Attacker are in the different two domains of Router 2. Router 1 can be regarded as a gateway of the victim site. The duration of the simulation is set to 5.5 s. The arrival time of the packets from the valid clients is shown in Fig. 2. The arrival time of the packets from the attacker is shown in Fig. 3.

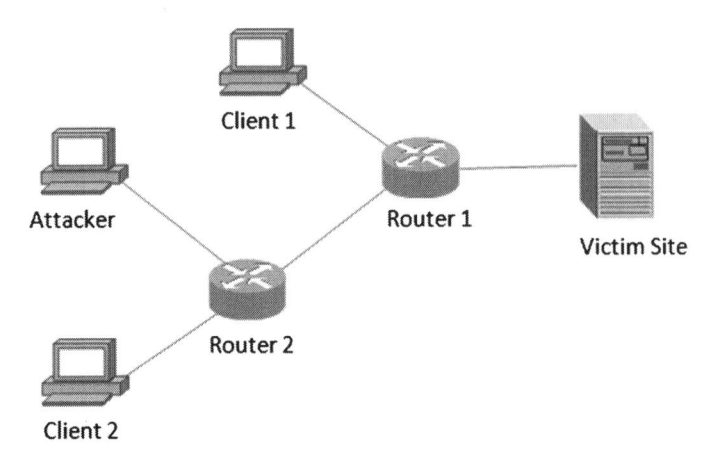

Fig. 5 The network topology of the simulation

Fig. 6 The consequent ratio

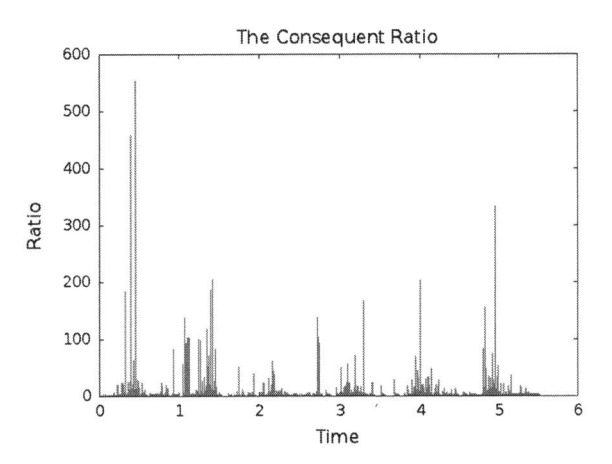

Client 1 and Client 2 send burst packets to the victim site randomly at the rate of 2 Mbps.

When there is no LDoS attack, no packet loses in the network. While the attack bursts periodically result in packets lost and invoke the network enters the retransmission timeout phase of TCP congestion control mechanism.

The period of the attack is set to 0.3 s, and according to Eq. (1), parameters are shown as followed:

$$T = 0.9\,\text{s}$$
$$R = 3\,\text{Mbps}$$
$$l = 0.3\,\text{s}$$
$$X = 900\,\text{Kb}$$

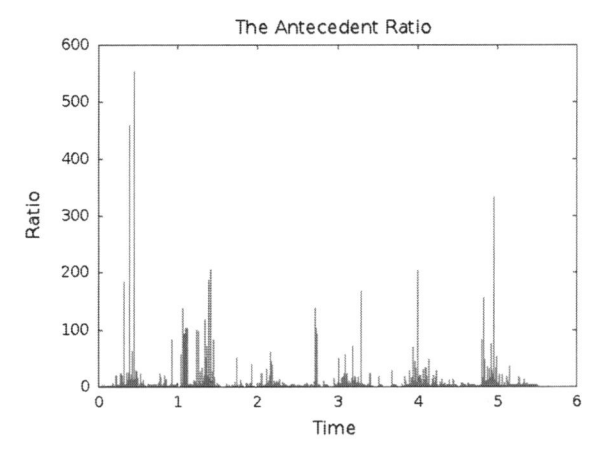

Fig. 7 The antecedent ratio

Table 1 Set A and set C

A[1]	0.328992	C[1]	0.399715
A[2]	0.399712	C[2]	0.433000
A[3]	0.453792	C[3]	0.453795
A[4]	1.394333	C[4]	1.343667
A[5]	1.423656	C[5]	1.394344
A[6]	3.295667	C[6]	1.423667
A[7]	3.998320	C[7]	3.095683
A[8]	4.824912	C[8]	3.295683
A[9]	4.949000	C[9]	4.115824

By analyzing the log information of Router 1, the consequent ratio and the antecedent ratio can be shown in Figs. 6 and 7. δ_c and δ_a are adjustable parameters. Different routers can set different value. In this model, they are both set to 150 in Router 1. And the adjustable parameter η is set to 6. According to Eqs. (5) and (6), set **C** and set **A** can be shown in Table 1.

$$\min\{ \text{ element num of C, element num of A}\} = 9 > \eta$$

In this model, Router 2 does not detect the attack behavior. After Router 1 sends reply to the victim site. The victim site will get the message that the attacker is in the local domain of Router 1.

Conclusion

The LDoS attacks, whose number of attack packets is as small as possible, are quite different from the well-known flood-based DoS attacks. Methods which are based on statistics of traffic do not work in such attacks because the number of attack packets is too small to be detected by the routers.

We have analyzed one of the characteristics in time domain between the attack traffic and the valid traffic. By take advantage of the difference of the arrival time, we present a method to detect the potential LDoS attack behavior and try to trace the location of the attacker. Simulations show the usability of this method.

This method also provides a reference to rebuild the attack model by evaluating parameters of the attack model. And we will try to improve this method in the efficiency and reliability in the future.

Acknowledgments This work has been supported by the National Natural Science Foundation of China under Grant 61172072, 61271308, the Beijing Natural Science Foundation under Grant 4112045, the Research Fund for the Doctoral Program of Higher Education of China under Grant W11C100030, the Beijing Science and Technology Program under Grant Z121100000312024. The authors are also grateful for the comments and suggestions of the reviewers.

References

1. Yang G, Gerla M, Sanadidi MY (2004) Defense against low-rate TCP-targeted denial-of-service attacks. In: Proceedings of ninth international symposium on computers and communications ISCC 2004, vol 1. IEEE, pp 345–350
2. Kuzmanovic A, Knightly EW (2006) Low-rate TCP-targeted denial of service attacks and counter strategies. IEEE/ACM Trans Networking 14(4):683–696
3. Wong TY, Law KT, Lui JCS et al (2006) An efficient distributed algorithm to identify and traceback ddos traffic. Comput J 49(4):418–442
4. Chang RKC (2002) Defending against flooding-based distributed denial-of-service attacks: a tutorial. Commun Mag IEEE 40(10):42–51
5. Kuzmanovic A, Knightly EW (2003) Low-rate TCP-targeted denial of service attacks: the shrew vs. the mice and elephants. In: Proceedings of the 2003 conference on applications, technologies, architectures, and protocols for computer communications. ACM, pp 75–86
6. Sun H, Lui J, Yau DKY (2006) Distributed mechanism in detecting and defending against the low-rate TCP attack. Comput Netw 50(13):2312–2330

Simplifying Data Migration from Relational Database Management System to Google App Engine Datastore

Yao-Chung Chang, Ruay-Shiung Chang and Yudy Chen

Abstract Cloud computing has been widely introduced because of its ability to increase resource utilization. Furthermore, cloud computing offers resources as services that taken part as one of the next generation computing technologies. Before delivery application to cloud computing environment, the first step is the migration of the data. Migrating data from relational database management system (RDBMS) to Google App Engine is time-consuming problem. Hence, Google App Engine (GAE) Datastore provides NoSQL data storage with configuration file that contains table schema and CSV or XML file. This study presents the method for simplifying data migration from RDBMS to GAE including blob data migration. The proposed method leverages AppCfg to provide convenience way for data migration. As a result, user has eliminated at least 75 % task effort for data migration.

Keywords Cloud computing · Data migration · Google App Engine · Task queue

Y.-C. Chang (✉)
Department of Computer Science and Information Engineering,
National Taitung Unviersity, Taitung, Taiwan, Republic of China
e-mail: ycc@nttu.edu.tw

R.-S. Chang · Y. Chen
Department of Computer Science and Information Engineering,
National Dong Hwa University, Hualien, Taiwan, Republic of China
e-mail: rschang@mail.ndhu.edu.tw

Y. Chen
e-mail: m9921081@ems.ndhu.edu.tw

Y.-M. Huang et al. (eds.), *Advanced Technologies, Embedded and Multimedia for Human-centric Computing*, Lecture Notes in Electrical Engineering 260, DOI: 10.1007/978-94-007-7262-5_101, © Springer Science+Business Media Dordrecht 2014

Introduction

Since cloud computing was introduced, it has been becoming more and more popular every year. Easy to use (setup, maintenance, reliability and scalability) and low cost (pay per use) are the reasons for companies to move their systems into cloud [1–3]. Along with cloud computing growth, data also become larger and more complex every day. The need of database that can handle growth of data is unavoidable. The challenges are not only data growth, data processing and analyzing but also very hard when data become bigger and bigger.

Google App Engine (GAE) is a cloud computing provider that offers platform as a service for user to host their web applications in Google's infrastructures [4]. Moreover, GAE allows developer to write and run an application above Google's infrastructure which is used by Google [5]. GAE supports distributed data storage that provides NoSql schemaless object which is built on the top of Bigtable called Datastore [6]. Datastore is used for storing common data types like integers, floats, strings, dates and binary data; and for storing binary large objects such as image, audio or video, GAE provides Blobstore.

Since data structures between RDBMS and Datastore are not the same, data migration from RDBMS to Datastore seems to be troublesome. Therefore, GAE provides one tool named "AppCfg" that can be used for uploading and downloading either data or application. Before use AppCfg for uploading data, two files should be prepared for every table. One is configuration file that contains information about table, data type of each column, etc. The other is CSV or XML file that contains the data to migrate. The objectives of this work are: (1) Minimizing error-prone while migrating data from RDBMS to GAE. (2) Reducing the time for data migration. (3) Currently, AppCfg doesn't support files bulk uploading, it is also important to provide files bulk uploading from other server into GAE blobstore.

The rest of this study is organized as follows: An overview of migrating data to GAE is described in section Related Studies. Section System Framework provides the system framework. The implementation of proposed method and subsequent result are sent out and analyzed in section Implementation. Finally, section Conclusions concludes the paper.

Related Studies

Data migration is known as one time process for migrating formatted data whereas both structure between databases are different on a conceptual and/or technical level [7]. Generally, it has some reasons for data migration [8]. Study in [9] presented that more than 75 % of their respondents encounter the problems during data migration. Some companies tackle those issues from affecting business operations by doing data migration during off-hours or weekends. However, this method increase the migration costs as a result of staff overtime, and this can negatively impact IT staff morale. Therefore, the good methodology for data migration is needed.

The relational database is a database that stores data into tables [10]. In relational database, one database can have multiple tables and if needed, the data between tables are related each other. The example of this database is MySQL. RDBMS fits for structured data, while now the data is transforming into structured (data come from RDBMS), semi structured (XML data) and unstructured data (files). Hand in hand with RDBMS, NoSQL came when the data complexity was becoming overhead.

Google App Engine is a platform as a service cloud provider. As a cloud provider, Google offers convenience services for users. Without software and hardware to buy or maintain, user can use GAE without limitation of application scaling. Rich API and simple to use is one of the advantages from GAE [11]. GAE performs load balancing and cache management automatically by allocating resources depends on traffic and resource use patterns. Every applications runs in a sandbox to prevent malicious operations. It also can improve CPU and memory utilization for multiple applications on the same physical machine [12].

Google Datastore is derived from Google Bigtable and it provides SQL-like query language called Google Query Language (GQL). The data are stored on High Replication Datastore (HRD) and are replicated across multiple data centers using Paxos algorithm and most queries are eventually consistent [13]. GAE has improvement from ordinary NoSQL database by providing transaction and roll back. Every transaction is committed if all of query on the transaction are success, otherwise it will be rolled back. The differences between traditional databases and GAE Datastore are shown in Table 1.

Since the data stored in database has different format, reliable tools for data migration is needed [14]. Some previous studies about data migration have been proposed. A middleware is created in Ref. [14] for data migration between RDBMS. It helps to reduce the time of migration, which is important when dealing with huge volumes of information. The methodology for migration database tables and their data between various types of RDBMS also has been proposed in Ref. [15]. It provides decision support during the process of data migration. Another study about data migration between document oriented and RDBMS has been proposed in Ref. [16]. It provides a model for the flexible specification of transformation rules and transforming the data schema from RDBMS into object oriented model.

Table 1 Differences between traditional databases and GAE datastore

	GAE datastore	Proposed system
Automatic scaling	Yes	No
Flexible queries	Yes	Yes
Strict data integrity	Customized on application	Yes

Moving data and applications from one cloud provider to another is possible when user has better choices [17]. When the migration between cloud providers occurs, it is not easy to move the data. The study in Ref. [17] proposed cloud broker to process the data migration between different cloud providers.

Because the Google AppCfg and AppRocket can not be used for migrating blob data, therefore the proposed approach in this study provides a convenience way to uploading data into Datastore and supporting blob data migration. This work provides simplicity and convenience way for data migration. The graphical user interface is introduced for user to input database's and Datastore's information. Also, user can upload their blob data during migration process. The configurations files and data files are automatically generated by the systems and user is welcomed to edit the configurations as needed. Not only uploading the automatic generated files, the proposed system can upload the data that has been prepared by user. Finally, the logs of data migration processes will exhibit the processes is succeed or failed.

System Framework

In this study, graphical user interface (GUI) of AppCfg is proposed for data migration from RDBMS to GAE Datastore. This method adds one layer above AppCfg that interacts with AppCfg. Some modifications of AppCfg are needed to eliminate interaction of user's authentication when migration occurs.

The proposed system architecture is separated into two parts (shown in Fig. 1), one is server side, and the other is client side. Server is the application runs on the Google App Engine. Client is the application runs on the client-side that interacts with end user.

- Server's Side

For processing blob data migration, Blob Downloading Manager is created on the server. Three additional columns will be added. The first column is "blobUrl". Due to the name of column for blob URLs is different from every user, this column holds URLs of blob data that is retrieved from original column from database that contains blob URLs. The second column is "newBlobUrl". This column is empty and will be filled with blobstore's key when the data is uploaded into blobstore. The third column is "fetchedBlob". This column also empty and will be filled "Y" when the data is uploaded into blobstore.

- Client's Side

  - Graphical User Interface (GUI) is the interface that interacts with users. It provides the input from users about RDBMS's information, GAE's information and blob's information. It also shows the result of each process.
  - Data Uploading Agent is an agent that controls each process on the client-side. It provides the GUI, queries the information and data from RDBMS,

Fig. 1 Proposed system architecture

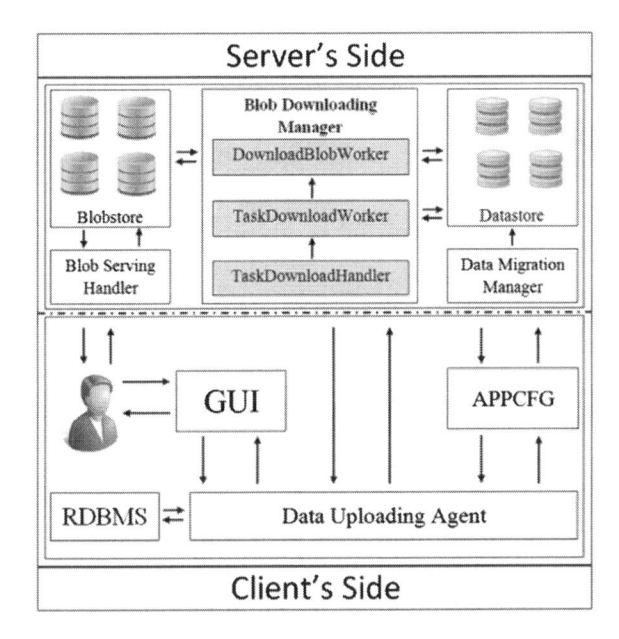

calls the TaskDownloadHandler on GAE to download the blob data and passes the message from and to AppCfg.

The correlation between TaskDownloadHandler, TaskDownloadWorker and DownloadBlobWorker is shown in Fig. 2. First, variables such as names of table and column are declared. After that, data uploading agent queries the data schema from RDBMS and gets the table's name and columns' names. For each row, the agent compares the variable tableName and the value on field table's name. If the variable tableName is empty, it means this is the new table and needs to set variable writeToFile with the variable declarations which contains the declaration needed for AppCfg's configuration file. If "tableName" is equal with the value of the field table's name, it means the current row is still belonging with same table. Otherwise, the table is new or no table anymore. Finally, it is the time for write the configuration file to local disk. Then, the system will show the configuration files to user. User can change or provide the information about blob data migration. The system needs column's name that contains blob data and/or server's path. If the server's path is blank, the proper blob path including server address should be provided as blob's path.

Due to the table's name from one user and another user is different, and GAE uses database model from the application on the TaskDownloadWorker, the database model's name from default name with the user table's name should be changed. For every row, DownloadBlobWorker will check if there is any blob has been downloaded or not. If has, the system will use the same blob's key as a "newBlobUrl". Otherwise, the system will download the blob data.

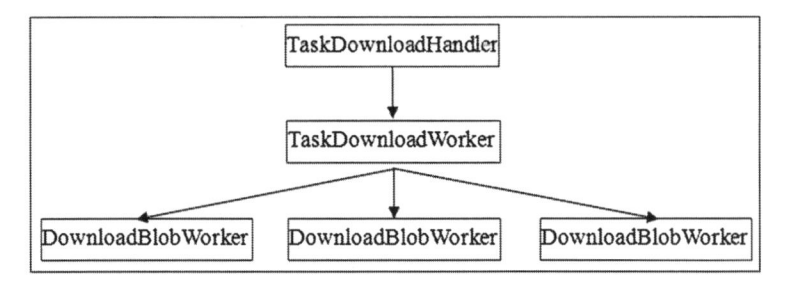

Fig. 2 Correlation between handler and workers

Implementation

- Data Types Translation

Data Types Translation is the process of creating configuration file. The system will run the query from RDBMS to get three columns. The columns needed for every table are "TABLE_NAME", "COLUMN_NAME" and "DATA_TYPE". TABLE_NAME is used for kind's value on GAE. COLUMN_NAME is used for property value and DATA_TYPE is used for "import_transform" on GAE.

- Lahman's Baseball Data Migration

Non blob data means that data types are not binary large objects data, e.g., text and number. On the other hand, blob data means data types are binary large objects data, e.g., audio, video and image. Non blob data migration is the process for migrating non blob data from RDBMS to GAE. The data come from Lahman's Baseball Database. It contains the history of batting and pitching statistics back to 1871, plus fielding, standings, team's statistics, managerial records, post-season data, and more [18]. In this implementation, some tables from Lahman's database are used.

- ImageNet's Database Migration

For blob data uploading, user needs to provide the information about column that holds the blob's path and or server's address. The system will add three additional columns on the configuration file and data file. In this implementation, the proposed system uses the source from ImageNet. ImageNet is an image database organized according to the WordNet hierarchy, in which each node of the hierarchy is depicted by hundreds and thousands of images [19].

Figure 3 shows the user interface of data migration from RDBMS to GAE. Users need to provide RDBMS information such as RDBMS's address, username, password, and database. On the right side, user needs to provide GAE's information including application ID, email and password. On the bottom, user can see the logs of the processes. User can also choose "Load Data From Local Disk" if they have prepared data on the same directory with the application.fig.

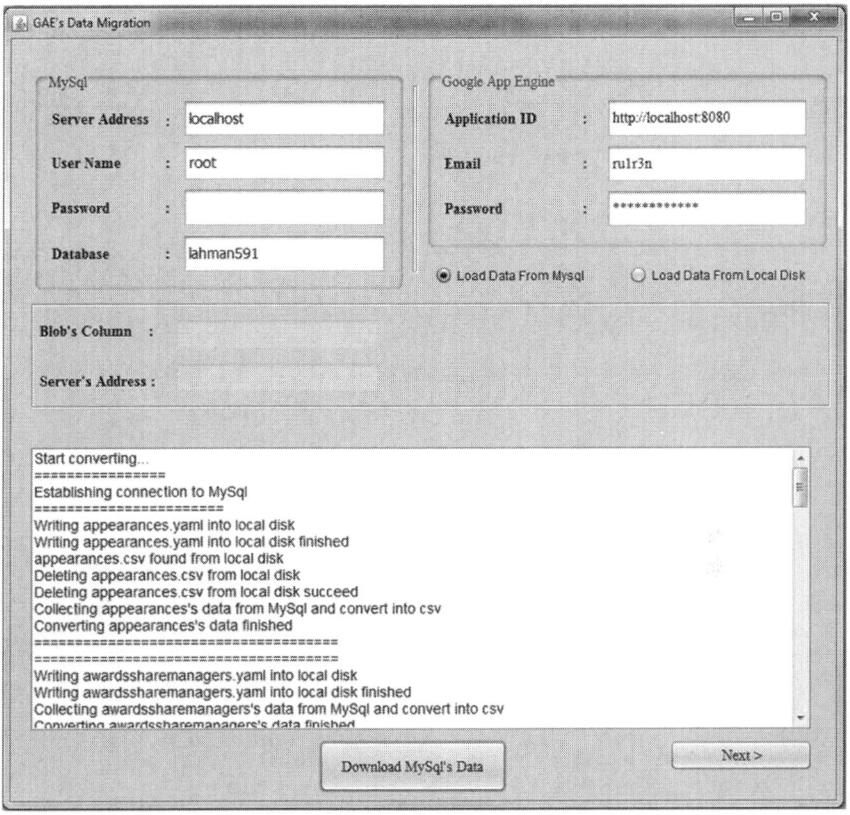

Fig. 3 GUI for data migration from RDBMS to GAE

- Jobs Efficiency

The jobs demanded for migration are shown in Table 2 including "Creating configuration file", "Preparing and migrating the data", "Migrating blob data" and "Debugging". By using the proposed method for migration of the data, user has eliminated at least 75 % task effort. User provides the information about RDBMS and GAE and the rest of migration process will be done by proposed system. User only needs to debug if the migration process failed to run.

Table 2 Jobs needed comparison

	Manual	Our system
Creating configuration file	Yes	No
Preparing & migrating the data	Yes	No
Migrating blob data	Yes	No
Debugging	Partially yes	Partially yes

Conclusions

Simplify data migration from traditional database to Google App Engine Datastore is presented in this study. The blob data migration using task queue are also implemented in the proposed system. User only needs to interact to one system for data migration without dealing with many complicated systems. Moreover, the proposed system prepares the needed files for data migration. User is allowed to modify the prepared data, even the data types dictionary can be edited manually. Convenience way for data migration has been proposed in this work. As a result, user has eliminated at least 75 % task effort for data migration.

Acknowledgments The authors would like to thank the National Science Council of the Republic of China, Taiwan for financially/partially supporting this research under Contract No. NSC101-2221-E-143-005-, NSC101-2221-E-259-003- and NSC101-2221-E-259-005-MY2.

References

1. Elsenpeter R, Velte AT, Velte TJ (2010) Cloud computing—a practical approach. McGraw-Hill, USA, p 23
2. Ingthorsson O (2012) 5 Reasons cloud computing is key to business success. http://www.datacenterknowledge.com/archives/2012/06/25/5-reasons-cloud-computing-is-key-to-businesss-success/. Retrieved 5 Nov 2012
3. Paul F (2012) 8 Reasons why cloud computing is even better for small businesses. http://readwrite.com/2012/04/06/8-reasons-why-cloud-computing. Retrieved 5 Nov 2012
4. Google (2012) What is Google App Engine? https://developers.google.com/appengine/docs/whatisgoogleappengine. Retrieved 29 Dec 2012
5. Severance C (2009) Using Google App Engine. O'Reilly Media, USA, p 14
6. Burrows M, Chandra T, Chang F, Dean J, Fikes A, Ghemawat S, Gruber RE, Hsieh WC, Wallach DA (2008) Bigtable: a distributed storage system for structured data. J ACM Trans Comput Syst 26(2), article no. 4, June 2008
7. Matthes F, Schulz C, Haller K (2011) Testing & quality assurance in data migration projects. In: 2011 27th IEEE international conference on software maintenance (ICSM), pp 438–447, Sept 2011
8. Kang S, Reddy ALN (2008) User-centric data migration in networked storage systems. In: IEEE international symposium on parallel and distributed processing, IPDPS 2008, pp 1–12, June 2008
9. IBM (2007) Best practices for data migration: methodologies for planning, designing, migrating and validating data migration. https://www-935.ibm.com/services/us/gts/pdf/softek-best-practices-data-migration.pdf. Retrieved 29 Jan 2013
10. Wikipedia (2013) Relational database. http://en.wikipedia.org/wiki/Relational_database. Retrieved 9 Jan 2012
11. Google (2012) Google App Engine—the platform for your next great idea
12. Prodan R, Sperk M, Osterman S (2012) Evaluating high-performance computing on Google App Engine. Software IEEE 29(2):52–58
13. Google (2012) Datastore overview. https://developers.google.com/appengine/docs/python/datastore/overview. Retrieved 15 Dec 2012

14. Elamparithi M (2010) Database migration tool (DMT)—accomplishments & future directions. In: 2010 International conference on communication and computational intelligence (INCOCCI), Dec 2010, pp 481–485

15. Walek B, Klimes C (2012) A methodology for data migration between different database management systems. Int J Comput Inf Eng 6:85–90

16. Walek B, Klimes C (2012) Data migration between document-oriented and relational databases. World Acad Sci, Eng Technol (69):894–898, Sept 2012

17. Shirazi MN, Kuan HC, Dolatabadi H () Design patterns to enable data portability between clouds' databases. In: 2012 12th International conference on computational science and its applications (ICCSA), June 2012, pp 117–120

18. Lahman S (2012) Download Lahman's baseball database. http://www.seanlahman.com/baseball-archive/statistics/. Retrieved 30 Dec 2012

19. Imagenet (n.d) ImageNet. http://www.image-net.org. Retrieved 1 Jan 2012

20. Couchbase (2012) NoSQL database technology

21. Google (2012) Blobstore python API overview. https://developers.google.com/appengine/docs/python/blobstore/overview. Retrieved 15 Dec 2012

22. Dancis K (2009) AppRocket 2.0.0. http://kaspa.rs/. Retrieved 16 Dec 2012

A Degree-Based Method to Solve Cold-Start Problem in Network-Based Recommendation

Yong Liu, Fan Jia and Wei Cao

Abstract Recommender systems have become increasingly essential in fields where mass personalization is highly valued. In this paper, we propose a model based on the analysis of the similarity between the new item and the object that the users have selected to solve cold-start problem in network-based recommendation. In order to improve the accuracy of the model, we take the degree of the items that have been collected by the user into consideration. The experiments with *MovieLens* data set indicate substantial improvements of this model in overcoming the cold-start problem in network-based recommendation.

Keywords Recommender systems · Network-based filtering · Similarity · Item degree · Cold-start

Introduction

With the rapid development of the computer network and other mass media, the amount of information that we can get is growing exponentially [1]. We are facing so much information that we have to spend a lot of time and effort to obtain the

Y. Liu · F. Jia (✉)
School of Electronic and Information Engineering, Beijing Jiaotong University,
Beijing 100044, China
e-mail: fjia@bjtu.edu.cn

Y. Liu
e-mail: 12120117@bjtu.edu.cn

Y. Liu · F. Jia
Key Laboratory of Communication and Information Systems, Beijing Municipal
Commission of Education, Beijing Jiaotong University, Beijing 100044, China

W. Cao
China Information Technology Security Evaluation Center, Beijing, China
e-mail: caow@itsec.gov.cn

Y.-M. Huang et al. (eds.), *Advanced Technologies, Embedded and Multimedia for Human-centric Computing*, Lecture Notes in Electrical Engineering 260, DOI: 10.1007/978-94-007-7262-5_102, © Springer Science+Business Media Dordrecht 2014

information that are most appropriate for us. Recommendation system is regarded as the most promising way to solve information overload problem. Recommendation systems recommend items of interest to users based on users' own explicit and implicit preferences, the preferences of other users, the attributes of users, and the attributes of items [2]. For example, a book recommender might integrate explicit ratings data (e.g., Tom rates *Introduction to Algorithms* a 3 out of 5), implicit data (e.g., Tom purchased *Algorithms in a Nutshell*), user demographic information (e.g., Tom is male), and book content information (e.g., Introduction to Algorithms is marketed as a computer-related book) to make recommendations to specific users. As recommendation system has enormous significance to the development of economy and society, a wide variety of recommender algorithms have been proposed, such as collaborative filtering algorithm [3, 4], content-based filtering algorithm [5], spectral analysis, principle component analysis, network-based algorithm [6–9], and so on.

In recent years, the network-based recommender algorithm, which was proposed by Tao Zhou in Ref. [4], has been extensively investigated over the past several years and it tends to be one of the most successful technologies for recommendation system. Some physical dynamics, including heat conduction process and mass diffusion, have been used in network-based personal recommendation. The recommender algorithm that adopted these physical approaches have been demonstrated to be both highly efficient and of low computational complexity. However, just as the collaborative filtering algorithm, such pure network-based recommendation also has the problem of cold-start. The newly added items that haven't been collected by any user are not suggested to the users.

Fortunately, content information of the newly added items can help to bridge the gap between existing and new items by inferring similarities among them. Therefore the network-based recommendation can introduce into the concept of similarities to solve the cold-start problem. In order to improve the accuracy of recommending the new items, we introduce the degree of the items that the user have collected to adjust the similarities of the user and the new items. Moreover, we introduce a free parameter β to regulate the contributions of objects degree to user-new items similarities. We carry out a benchmark experiment on *MovieLens* data set and compare it against the conventional methods to demonstrate the effectiveness. The numerical results indicate that decreasing the influence of popular objects can further improve the algorithmic accuracy.

The rest of this paper is organized as follows. In section Background, we introduce the process of network-based recommender algorithm and explain why it has cold-start problem. In section The Model, we present our model to resolve the cold-start problem. The fourth section is numerical results where we display the experiment results and explain how well our model performs. Finally, we make a summary of our work.

Fig. 1 Network-based
recommender model

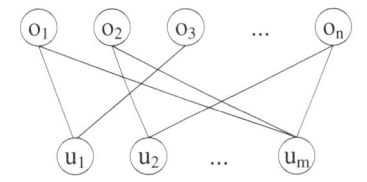

Background

In this section we describe the process of network-based recommender system and introduce the cold-start problem in it. The network-based recommendation was proposed by Tao Zhou in 2007 [4] and its model is shown in Fig. 1. Denote the object-set as $O = \{o_1, o_2, \ldots, o_n\}$ and user-set as $U = \{u_1, u_2, \ldots, u_m\}$. We give up with a binary variable a_{ij} to representation the relation of u_j with o_i. If u_j has already collected $o_i, a_{ij} = 1$ and $a_{ij} = 0$ otherwise. For example, u_1 is connected to o_1 and o_3 in Fig. 1. It represents that u_1 have collected o_1 and o_3, so $a_{11} = 1$ and $a_{31} = 1$. In this way, the relation of n items and m users can be described by an $n \times m$ adjacent matrix $\{a_{ij}\}$. Then, for a given user u_j the initial resource located on each item is a_{ij}. That's to say, if the object o_i has been collected by u_j, its initial resource is unit, otherwise it is zero. The initial resource can be understood as giving a unit recommending capacity to each collected object. According to the weighted resource allocation process discussed in Ref [4], the final resource of every object can be obtained easily. For any user u_j, all the uncollected items are sorted in descending order of the final resource, and those items with highest value of final resource are recommended. The new items haven't been collected by any user, so their final resource is zero and they wouldn't be suggested forever. This problem is the so-called cold-start.

The Model

In this section we describe a model to address the cold-start problem. The model utilizes content information of items and the record of what the user have collected In order to keep consistency with the second section, we denote the object set as $O = \{o_1, o_2, \ldots, o_n\}$ and user set as $U = \{u_1, u_2, \ldots, u_m\}$. A recommendation system can be fully described by an adjacent matrix $A = \{a_{ij}\} \in R^{n,m}$. If o_i is collected by $u_j, a_{ij} = 1$ and $a_{ij} = 0$ otherwise. As we all know, every object has its own features and they can be divided into different categories. For example, a movie may be tagged with "Action" and "Comedy". We can acquire the main features of the movie from these keywords. So if we can obtain the interest of the users, we can suggest specific items that are consistent with user hobbies. The user hobbies can be obtained from the objects that the user has collected. For instance, if most of the movies that a user has seen belong to romance, recommender system

Fig. 2 The proposed model

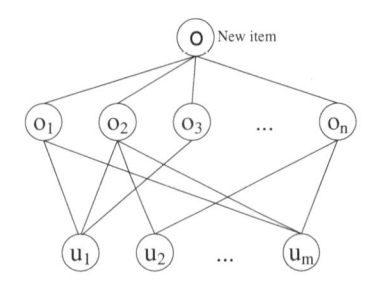

should suggest romantic movies to the user. Based on this method, we can calculate the similarities of new items and items that have been collected first and then get the similarities of new items and users. For a new item, all the users are sorted in the descending order according to their similarities with the new item, and the new item is recommended to those users, which are ranked in the front of all the users. This model can be described as Fig. 2.

Data related to the style of items are gathered and a "style-item matrix" is built, as shown in Fig. 3. The matrix consists of binary data. For example, object o_1 belongs to style 1, style 2 and object o_2 belongs to style 1, style Q. The "style-item matrix" can be used to calculate the similarities of two different items. Here we use the cosine theorem to determine the similarity of the two items. We use F to represent the "style-item matrix" and its column vector can be represented as f_i, which indicates the features of o_i. The elements of f_i are expressed as f_{ji}, which indicates whether o_i belongs to style j or not. Defined in the same way, the feature vector of the new item is written as fn and its elements are expressed as fn_j.

By using the cosine expression, the similarities between the new object and o_i can be written as:

$$s_{in} = \frac{1}{\sqrt{|f_i||fn|}} \sum_{j=1}^{Q} f_{ji} fn_j. \tag{1}$$

In this equation, $|f_i|$ and $|fn|$ is the modulus of f_i and fn respectively and they can be expressed as:

Fig. 3 Style-item matrix

	O_1	O_2	...	
Style 1	1	1		
Style 2	1			1
⋮				
StyLe Q		1		

$$|f_i| = \sum_{j=1}^{Q} f_{ji}, \tag{2}$$

$$|fn| = \sum_{j=1}^{Q} fn_j \tag{3}$$

The predicted score, to what extent the new object will be suggested to u_j, is given as:

$$v_{jn} = \frac{\sum_{i=1}^{n} s_{in} a_{ij}}{k(u_j)}. \tag{4}$$

In Eq. (4), $k(u_j)$ represents the degree of u_j and it means how many items u_j have collected. Equation (4) can be regarded as an average of the similarities between the new item and the previous items that the users have collected.

A common problem of Eq. (1) is that they have not taken into account the influence of an item's degree, so that items, which a user has collected previously, with different degrees have the same contribution to the similarity of the new item and the user. If user u_l have selected item o_i and o_j, that is to say, they are all in line with the user's interest. Provided that item o_i is very popular (the degree of o_i is very large) and item o_j is relatively less popular (the degree of o_j is relatively small), s_{in} (the similarity between o_i and the new item) is a very ordinary similarity and it does not mean the new item and u_l are very similar. Therefore, its contributions should be small. On the other hand, s_{jn} (the similarity between o_j and the new item) is a peculiar similarity and it means the new item and u_l are very similar. Therefore its contributions should be large. In other words, it is not meaningful if the new item is similar to a very popular item that the user has collected, while if the new item is similar to an unpopular item that the user has collected, the new item is much more likely to correspond with user's interest and it should be suggested to the user with the higher probability. Accordingly, the contribution of s_{in} to the similarity v_{ln} (if u_l has collected o_i) should be negatively correlated with its degree $k(o_i)$. We suppose the s_{in}'s contribution to v_{ln} being inversely proportional to $k^{\alpha}(o_i)$ where α is a freely tunable parameter. Therefore, the proposed similarity equation should be written as:

$$v_{jn} = \frac{\sum_{i=1}^{n} s_{in} a_{ij} k^{\alpha}(o_i)}{k(u_j)}. \tag{5}$$

From now on, for a given new item, we use Eq. (5) to calculate its similarities with every user and we can get m numbers (they are $v_{1n}, v_{2n}, v_{3n}, \ldots, v_{mn}$) in the end. Then the m users should be sorted in the descending order according to the value of $v_{1n}, v_{2n}, v_{3n}, \ldots, v_{mn}$ and the new item should be suggested to the top users.

Numerical Results

In this paper, we use the standard MovieLens [10] data set to evaluate the accuracy of the proposed model. There are 1,682 movies and 943 users in this data set. In fact, MovieLens is a rating system, where each user votes movies in five discrete ratings 1–5. Hence we suppose that, the movie is collected by a user if the rating given by the user is at least 3 [11]. The data set contains 100,000 ratings and 82.52 % of the ratings are at least 3, so it contains 82,520 user-item pairs. In order to test the proposed model, we randomly select 20 numbers from 1 to 1682 as the movie ids. Then we select the corresponding records according to the 20 movie ids as test set and the remaining data serve as training set. The records in test set are seen as the new items that haven't been collected by any user. The training set is treated as known information, while no information in the probe set is allowed to be used for prediction. A recommendation algorithm generally can be evaluated with ranking score, recall and precision. As our main goal in this paper is to solve the cold-start problem and ranking score can reflect the effectiveness of the algorithm in solving this problem best, we use ranking score to evaluate the given algorithm.

Here new movies can be represented as $NewO_i$ where $i = 1, 2, \ldots, 20$. For a given $NewO_i$, we can obtain an ordered list, which contains all the users. If the relation $NewO_i - u_j$ is in the test set, we measure the position of u_j in the ordered list. For instance, if u_j is the 20th from the top, the position of u_j can be denoted by $r_{ij} = 20/943 = 0.02$. The mean value of the r_{ij}, which is calculated over all the records in test set, can be used to evaluate the effectiveness of the algorithm. The smaller the mean value of r_{ij}, the more accurate the algorithm is.

The result is shown in Fig. 4 where horizontal axis is the freely tunable parameter and longitudinal axis is the mean value of r_{ij}.The red solid line represents the situation where we don't take the item degree into account and the black

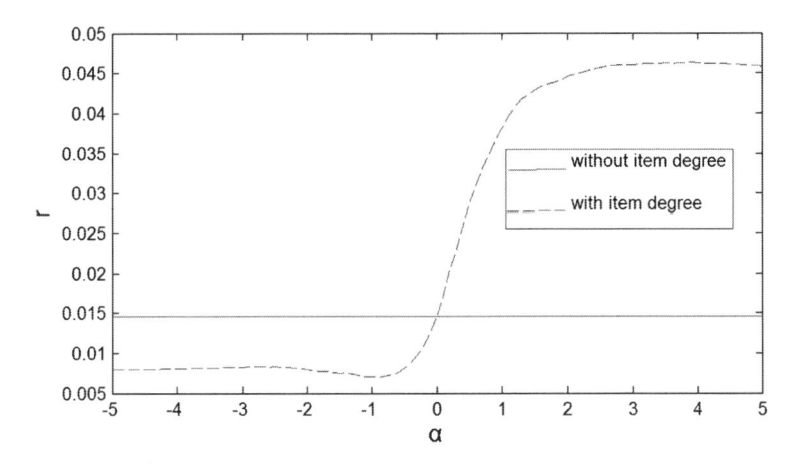

Fig. 4 The simulation results

dotted line reflects the change of r when α varies from -5 to 5. When $\alpha = 0$, the value of the solid curve is the same as the value of the dotted one. Compared with $\alpha = 0$, a positive α strengthens the influence of large-degree items, while a negative α weakens the influence of large-degree objects. We can see that, the algorithm has a better performance when $\alpha < 0$ and the curve has a clear minimum around $\alpha = -1$. When $\alpha = 0$, it means that we don't take item degree into account and $r = 0.0146$ in this case. When $\alpha = -1$, $r = 0.007$ and it is reduced by $52.05\ \%$. It is indeed a greater improvement in solving cold-start problem.

Conclusion

In this paper, we propose a model to solve cold-start problem in network-based recommendation. The item degree is taken into account when we calculate the similarities between new items and the users. Furthermore, we introduce a free parameter α to regulate the influence of item degree. Numerical results indicate that decreasing the influence of large-degree items further improves the recommendation accuracy: When $\alpha = -1$, the r can be reduced by $52.05\ \%$ compared with the case where item degree isn't taken into account.

Acknowledgments This work has been supported by the National Natural Science Foundation of China under Grant 61172072, 61271308, the Fundamental Research Funds for the Central Universities under Grant 2013JBM006, the Beijing Natural Science Foundation under Grant 4112045, the Research Fund for the Doctoral Program of Higher Education of China under Grant W11C100030, the Beijing Science and Technology Program under Grant Z121100000312024. The authors are also grateful for the comments and suggestions of the reviewers.

References

1. Broder A, Kumar R, Moghoul F, Raghavan P, Rajagopalan S, Stata R, Tomkins A, Wiener J (2000) Comput Netw 33:309
2. Schein AI, Popescul A, Ungar LH, Pennock DM (2002) In: SIGIR '02 Proceedings of the 25th annual international ACM SIGIR conference on research and development in information retrieval, pp 253–260
3. Adomavicius G, Tuzhilin A (2005) IEEE Trans Knowl Data Eng 17:734
4. Zhou T, Ren J, Medo M, Zhang Y-C (2007) Bipartite network projection and personal recommendation. Phys Rev E 76:046115
5. Herlocker JL, Konstan JA, Terveen LG, Riedl JT (2004) ACM Trans Inf Syst 22:5
6. Ren J, Zhou T, Zhang Y-C (2008) Europhys Lett 82:58007
7. Zhou T, Jiang L-L, Su R-Q, Zhang Y-C (2008) Europhys Lett 81:58004
8. Zhang Y-C, Medo M, Ren J, Zhou T, Li T, Yang F (2007) Europhys Lett 80:68003
9. Liu C, Zhou W-X (2012) Heterogeneity in initial resource configurations improves a network-based hybrid recommendation algorithm. Physica A, pp 5704–5711
10. The MovieLens data can be downloaded from the website of GroupLens Research. http://www.grouplens.org

11. Blattner M, Zhang Y-C, Maslov S (2007) Phys A 373:753
12. Said A, Jain BJ, Albayrak S (2012) Analyzing weighting schemes in collaborative filtering: cold start, post cold start and power users. In: SAC '12 Proceedings of the 27th annual ACM symposium on applied computing, pp 2035–2040

Security Flaws of Off-Line Micro Payment Scheme with Dual Signatures

Shin-Jia Hwang

Abstract Wuu et al. proposed their off-line micro-payment scheme with dual signatures to provide customers' anonymity. However, some security flaw is pointed out. To remove this flaw, the channel between bank and trusted party and the channel between the bank and customers should be authenticated and secure.

Keywords Blind signatures · Payword chains · Micro payment

Introduction

Micro payment provides electronic payment mechanisms for small value transactions over networks. Due to the small value of each transaction, the computation and communication costs of a micro payment should be low. Among the proposed micro payment schemes [1–15], the payword chain [11] is the famous technique to reduce the computation cost of each transaction.

Wuu et al. [14] proposed the off-line micro payment scheme with dual signatures to provide anonymity for customers. To provide customers' anonymity, a trusted authority and a trusted issuer, are involved. The trusted authority authenticates and authorizes the pseudo public keys for each customer. Since the authorized pseudo public keys are validated only by the issuer, the bank needs the help of the issuer to validate the customer's pseudo public key. The issuer also generates some anonymous payword chain and obtains the bank's authorization by the blind signature scheme [2] to break the link between the payword chain and the customer. Since the bank cannot authenticate the customer and knows the blind signature on payword chains, a security flaw occurs.

S.-J. Hwang (✉)
Department of Computer Science and Information Engineering, TamKang University, Tamsui, New Taipei City 251, Taiwan, Republic of China
e-mail: sjhwang@mail.tku.edu.tw

Y.-M. Huang et al. (eds.), *Advanced Technologies, Embedded and Multimedia for Human-centric Computing*, Lecture Notes in Electrical Engineering 260, DOI: 10.1007/978-94-007-7262-5_103, © Springer Science+Business Media Dordrecht 2014

Table 1 Notation description

Notation	Description
N	An amount of coins withdrawn by a consumer
Life	Coin expiration date
T	Timestamp
H()	Public hash function
$H^n(\bullet)$	$H^n(\bullet) = H(H^{n-1}(\bullet))$ and $H^1(\bullet) = H(\bullet)$
PK_X, SK_X	Long-term public and secret key pair of the participate X
PPK_X, PSK_X	Pseudo public and secret key pair chosen by some customer
<Data>K	The ciphertext of data encrypted/decrypted with key K in public cryptosystems
Sig_X(Data)	A digital signature generated by participate X and Sig_X(Data) = Data||<H(Data)>SK_X

Section Review of Off-Line Micro Payment Scheme with Dual Signatures. Section Security Flaw describes the security flaws of the off-line micro payment scheme. The final section is our Conclusion.

Review of Off-Line Micro Payment Scheme with Dual Signatures

The off-line micro payment scheme with dual signatures includes five kinds of participants: Consumer C, merchant M, bank B, issuer I, and trusted authority TA. The scheme contains four phases: Register, withdrawal, payment, and deposit phases. Table 1 defines the notations.

Register Phase

A customer C has to register at TA to obtain an encrypted payment certificated. C generates his/her pseudo public secret key pair (PPK, PSK), and submits TA his/her real identity C and PPK. Then TA gives C <PCert> PK_C, where PCert = <Sig_{SKTA}(C, PPK)> PK_I.

Withdrawal Phase

The withdrawal protocol is described below.

Step W1: C sends B the message (C, PCert, Sig_{SKPSK}(N, T)).
Step W2: B sends I the message (B, Life, PCert, Sig_{SKPSK}(N, T)).

Step W3: I recovers C and PPK by decrypting PCert and validates Sig_{SKTA}(C, PPK). If Sig_{SKTA}(C, PPK) is valid, then I randomly chooses two primes K_P, K_{SI} and computes $PK = K_P \times K_{SI}$. Finally, send B (H(CCert) $\times$ <r> PK_B, PK), where CCert = (B, <K_P> SK_I, PPK, Life) and r is the blind factor.

Step W4: B chooses two primes p and q such that gcd(PK, ϕ(n)) = 1, and computes the product n = pq and $K_{SC} = PK^{-1}$ mod ϕ(n), where ϕ(n) = (p − 1)(q − 1). Then send I the message (<H(CCert)> $SK_B \times r$, n).

Step W5: I prepares a payword chain Coin = ((c_0, c_0'), (c_1, c_1'), ..., (c_N, c_N')) by randomly choosing a integer c_N and computing $c_{i-1} = H(c_i)$ for i = N, N − 1, ..., 1 and $c_i' = <c_i> K_{SI}$ for i = 0, 1, ..., N. Remove the blind factor r from <H(CCert)> $SK_B \times r$ to obtain <H(CCert)> SK_B. Then send B <Coin, Sig_{SKB}(CCert)> PPK, where Sig_{SKB}(CCert) = CCert||<H(CCert)> SK_B.

Step W6: B sends <Coin, Sig_{SKB}(CCert)> PPK and Sig_{SKB}(K_{SC}) to C.

Finally, customers obtains the key K_{SC}, and Coin and Sig_{SKB}(CCert).

Payment Phase

Assume that C spends j coins ((c_0, c_0'), (c_1, c_1'), ..., (c_j, c_j')), where $c_i' = <c_i'> K_{SI} = (c_i)^{KSI}$ mod n for i = 0, 1, 2, ..., j. The customer wants to spend another k coins on the merchant M with the transaction timestamp TT.

Step P1: C sends M the transaction <Order, Sig_{PSK}(PI)> PK_M, where Order is the order information and the payment information PI = (M, TT, Sig_{SKB} (CCert), $<c_j' > K_{SC} = (c_j')^{Ksc}$ mod n, c_{j+k}).

Step P2: M obtain Order and Sig_{PSK}(PI) by decrypting <Order, Sig_{PSK}(PI)> PK_M. Then M verifies Sig_{SKB}(CCert) to check whether or not PPK is authorized by B. M checks the whether or not PI is authorized by the owner of PPK. M also obtains the authorized public key K_P from CCert and checks whether or not $\ll c_j' > K_{SC} > K_P = H^k(c_{j+k})$. If all the verifications hold, M accepts transaction; otherwise rejects the transaction. Finally M acknowledges C the final result.

Deposit Phase

M deposits coins by sending Sig_{SKM}(M, TT, Sig_{PSK}(PI)) to B. B not only checks whether or not $\ll c_j' > K_{SC} > K_P = H^k(c_{j+k})$ but also detects the occurrence of double spending. If double spending occurs, B asks TA to find out the owner of PPK.

Security Flaw

In the withdrawal phase, B and I do not authenticate one another, so a malicious customer can cheat B. Suppose that some malicious customer C performs the withdrawal protocol once, so C has CCert = (B, $<K_P>$ SK_I, PPK, Life) and the public key K_P. The cheating attack is described below.

Step W1: C sends B the message (C', PCert, $Sig_{PSK}(N, T')$), where C' is another customer real identity and T' is another timestamp.

Step W2: B sends I the message (B, Life, PCert, $Sig_{PSK}(N, T')$) but C intercepts (B, Life, PCert, $Sig_{PSK}(N, T')$).

Step W3: C constructs another pseudo public and secret key pair (PPK', PSK'). C also chooses another prime K_{SI}' and computes $PK' = K_P \times K_{SI}'$. C sends B (H(CCert') $\times$ $<r>$ PK_B, PK'), where CCert' = (B, $<K_P>$ SK_I, PPK', Life') and r' is the blind factor.

Step W4: B chooses two primes p' and q' such that gcd(PK, $\phi(n')$) = 1, and computes the product n' and $K_{Sc}' = PK'^{-1}$ mod $\phi(n')$, where $\phi(n') = (p' - 1)(q' - 1)$. Then send I the message ($<H(CCert') > SK_B \times r'$, n') that is intercepted by C.

Step W5: C prepares Coin' = $((c_0, c_0'), (c_1, c_1'), ..., (c_N, c_N'))$ by randomly choosing a integer c_N and computing $c_{i-1} = H(c_i)$ for i = N, N − 1, ..., 1 and $c_i' = <c_i > K_{SI}'$ for i = 0, 1, 2, ..., N. Remove the blind factor r' from $<H(CCert')> SK_B \times r'$ to obtain $<H(CCert')> SK_B$. Then C sends $<Coin', Sig_{SKB}(CCert')>$ PPK' to B, where
$Sig_{SKB}(CCert') =$
CCert'‖ $<H(CCert')> SK_B$.

Step W6: B sends $<Coin, Sig_{SKB}(CCert')>$ PPK' and $Sig_{SKB}(K'_{SC})$ to C.

Since B cannot authenticated the received messages, the malicious customer C entraps the bank and an innocent customer C'. To overcome the cheating flaw, the communication link between B and I should be authenticated. B relies on I to bind the pseudo public key and the real customer identity with the encrypted payment certificate PCert, the communication link between B and C should be also secure; otherwise, $Sig_{SKB}(K_{SC})$ leases the important secret key K_{SC}.

Conclusion

In Wuu et al.'s off-line micro payment scheme with dual signatures, the customer easily impersonates the issuer to entrap the bank and some innocent customer. To remove this security flaw, the channel between the bank and issuer must be secure and authenticated in the withdrawal protocol.

References

1. Bellare M, Garay J, Hauser R, Herzberg A, Krawczyk H, Steiner M, Tsudik G, Waidner M (1995) iKP—A family of secure electronic payment protocols. In: Proceeding of 1st USENIX workshop on electronic commerce, pp 89–106
2. Chaum D (1983) Blind signatures for untraceable payments. In: Advances in cryptography—Proceeding of Crypto'82. Springer, New York, pp 199–203
3. Chaum D, Fiat A, Naor M (1988) Untraceable electronic cash. In: Advances in crytology—CRYPTO'88, LNCS, vol 403. Springer, New York, pp 21–25
4. Chen L, Kudla C, Paterson KG (2004) Concurrent signatures. In: Advances in cryptology—EUROCRYPT 2004, LNCS, vol 3027. Springer, Berlin, pp 287–305
5. Furche, A, Wrightson G (1996) SubScrip—an efficient protocol for pay-per-view payments on the internet. In: Proceedings of 5th international conference on computer communications and networks (ICCCN'96), Rockville, MD, 16–19 Oct 1996
6. Glassman S, Manasse MS, Abadi M, Gauthier P, Sobalvarro P (1996) The millicent protocol for inexpensive electronic commerce. In: World wide web journal, proceeding of 4th international world wide web conference. O'Reilly, Boston, MA, pp 603–618
7. Herberg A (1998) Micropayment. In: Kou W (ed) Payment technologies for E-commerce. Springer, New York, pp 245–282
8. Huang C-W (2006) A postpaid micropayment scheme with revocable customers' anonymity. Master Thesis, Tamkang University, Taiwan, R.O.C
9. Lin S-Y (2004) Design and cryptanalysis of micropayment schemes. Master Thesis, National Central University, Taiwan, R.O.C
10. Manasee MS (1995) The millicent protocols for electronic commerce. In: Proceeding of 1st USENIX workshop on electronic commerce, pp 117–123
11. Rivest RL, Shamir A (1997) PayWord and MicroMint: two simple micropayment schemes. In: Proceeding of security protocols workshop, LNCS 1189. Springer, New York, pp 69–87
12. Stern J, Vaudenay S (1997) SVP: a flexible micropayment scheme. In: Proceeding of financial cryptography, LNCS, vol 1318. Springer, New York, pp 161–172
13. Tsou J-H (2005) The study of electronic payment scheme. Master Thesis, Tamkang University, Taiwan, R.O.C
14. Wuu L-C, Chen K-Y, Lin C-M (2008) Off-line micro payment scheme with dual signature. J Comput 19(1):23–28
15. Yen S-M (2001) PayFair: a prepaid internet micropayment scheme ensuring customer fairness. Comput Digital Tech, IEE Proc 148(6):207–213

Mobile Reference Based Localization Mechanism in Grid-Based Wireless Sensor Networks

Ying-Hong Wang, Yi-Hsun Lin and Han-Ming Chang

Abstract Wireless sensor networks (WSNs) are based on monitoring or managing the sensing area by using the location information with sensor nodes. These sensor nodes are sometimes random deployed, so they have to be aware to their location before starting their tasks. Most sensor nodes need hardware support or receive packets with location information to estimate their location, and this needs lots of time or costs, and may have a huge error. In this paper we present a localization mechanism in wireless sensor networks (MRN). This mechanism can cooperate with node localization algorithm and mobile reference node moving direction scheme. We use a mobile reference node with GPS to move to the whole environment, and we use RSSI and trilateration to estimate unknown nodes' location. We can obtain more unknown nodes location by mobile reference node moving scheme, and will decreases the energy consumption and average location error.

Keywords Wireless sensor networks (WSN) · Localization · Mobile sensor node · Received signal strength indicator (RSSI)

Introduction

Recent years wireless sensor networks [1] are getting more convenient, and the applications or researches are become skillful. So in this paper we proposed a mechanism in wireless sensor networks localization.

Y.-H. Wang (✉) · Y.-H. Lin · H.-M. Chang
Department of Computer Science and Information Engineering, Tamkang University,
Tamsui, Taiwan, Republic of China
e-mail: inhon@mail.tku.edu.tw

Y.-H. Lin
e-mail: kidsle0830@gmail.com

H.-M. Chang
e-mail: chmcicel74723@gmail.com

Y.-M. Huang et al. (eds.), *Advanced Technologies, Embedded and Multimedia for Human-centric Computing*, Lecture Notes in Electrical Engineering 260, DOI: 10.1007/978-94-007-7262-5_104, © Springer Science+Business Media Dordrecht 2014

In recent researches, we find out localization can be classified as range-based and range-free approaches. Range-free approaches do not assume the availability or validity of distance information and only rely on the connectivity measurements undetermined sensors to a number of seeds [1]. Having lower requirements on hardware, the accuracy and precision of range-free approaches are easily affected by the node densities and network conditions, which are often unacceptable for many WSN applications that demand precise localizations. Range-based approaches calculate node distances based on some measured quantity [2], whereas they usually require extra hardware support; thus, they are expensive in terms of manufacturing cost and energy consumption. And how to reduce the extra cost becomes an important task for us to find out. When the sensor node position is estimate, it would have a significant increase for the data transfer speed or other works need to do.

In this paper we proposed a mechanism using a mobile reference node (MRN) with RSSI [6] and trilateration in wireless sensor networks localization to reduce energy consumption costs and reduce the location error. In section Related Work will introduce some range-based and range-free approaches. In section Localization Mechanism with Mobile Reference Node will introduce the mechanism we proposed. And final are the simulation and conclusion.

Related Work

Many approaches have been proposed to determine sensor node locations, falling into two categories: (a) range-free approaches and (b) range-based approaches.

Range-Free Approaches

Knowing the hardware limitations and energy constraints required by range-based approaches, researchers propose range-free solutions as cost-effective alternatives. Having no distances among nodes, range-free approaches depend on the connectivity measurements from sensor nodes to a number of reference nodes, called seeds. For example, in Approximate Point-in-Triangulation Test (APIT) [2], some sensor nodes have high frequency to transmit signal with GPS or other ways to obtain sensor nodes location called Beacon. As an alternate solution, DV-Hop [3] only makes use of a constant number of seeds. Instead of single-hop broadcasts, seeds flood their locations throughout the network, maintaining a running hop count at each node along the path.

Range-Based Approaches

Range-based approaches assume that sensor nodes can measure the distance and/or the relative directions of neighbor nodes. Various techniques are employed to measure the physical distance. For example, TOA obtains range information through signal propagation times [4], and TDOA estimates the node locations by utilizing the time differences among signals that are received from multiple senders [4]. As an extension of TOA and TDOA, AOA allows nodes to estimate the relative directions between neighbors by setting an antenna array for each node [5]. All those approaches require expensive hardware. For example, TDOA needs at least two different signal generators. AOA needs antenna arrays and multiple ultrasonic receivers. RSSI is utilized to estimate the distance between two nodes with ordinary hardware [6].

Therefore, no interaction is required between nodes, avoiding cumulative errors of coordinate calculations and unnecessary communication overhead. The localization accuracy can also be improved by multiple measurements that are obtained when the mobile nodes are at different positions. With the RSSI values from mobile node to an unknown node in an ideal sense, the distance between other unknown nodes should be calculated according to the log-normal shadowing model in (1), which is widely used in range-based localization approaches [6], i.e.

$$\mathrm{RSSI}(d) = \mathrm{P}_T - \mathrm{P}_L(d_0) - 10\alpha\log_{10}\frac{d}{d_0} + X_\sigma \tag{1}$$

where **PT** is the transmission power, **PL(d0)** is the path loss for a reference distance of **d0**, and η is the path-loss exponent. The random variation in RSSI is expressed as a Gaussian random variable **Xσ = N(0, σ2)**. All powers are in given in decibels relative to 1mW, and all distances are given in meters. η is set between 2 and 5. σ is set between 4 and 10, depending on the specific environment [6].

And last step we use the trilateration [7]. The trilateration is a common mathematical formula, using the distances we estimate, and then calculate the unknown node location.

Localization Mechanism with Mobile Reference Node

In this section presents our proposed mechanism. The mechanism can divide into two phases. First phase is Node localization Algorithm, and in the second phase is Mobile Reference node moving direction scheme, according to the environment size to choose which algorithm we should use. In Fig. 1 is whole mechanism structure.

Network Environment

In this section, introduce our proposed localization mechanism in the environmental setting conditions. As shown in Fig. 2, we random deployment 20 unknown sensor nodes, and give each unknown sensor node its unique node ID, and mark the upper left corner is the initial position of the mobile reference node.

As shown in Fig. 2, after the sensor nodes deployed, the sink will know the length and width of the entire environment, assuming that the width of the environment define as M, and the length of the environment as L, each grid length is define as the mobile reference node transmission radius R, and then we use the $M/R = k$, $L/R = p$ these two formulas to get the values as k, p, and use the results of the values of k to do the following decisions: (1) divided each virtual grid and tag an grid ID on each virtual grid (2) determine the mobile reference node start position (3) help mobile reference node to set a first direction need to turn.

Node Localization Algorithm

In this section we introduce node localization algorithm, it has two phases (a) Mobile Reference node Broadcast Algorithm, (b) Sensor node Localization Algorithm. They are parallel execution in the localization mechanism in the first phase.

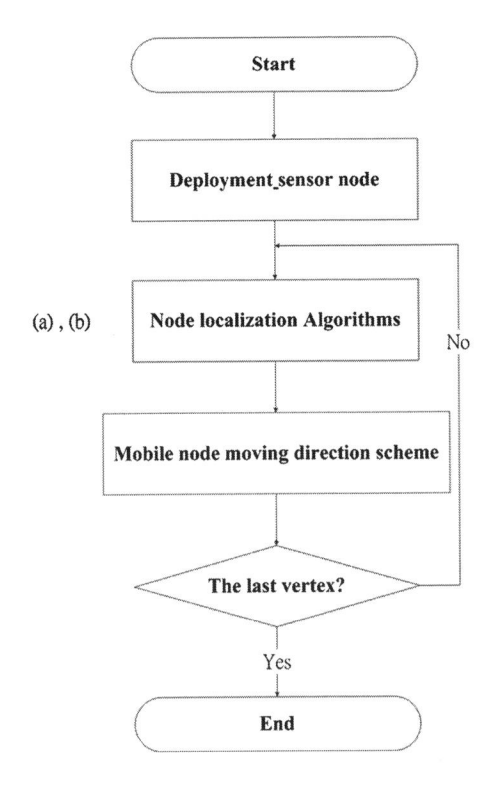

Fig. 1 Grid-based mobile sensor node localization mechanism

Fig. 2 Mobile reference
node and Sink initial position

Sink & Mobile reference
node initial location

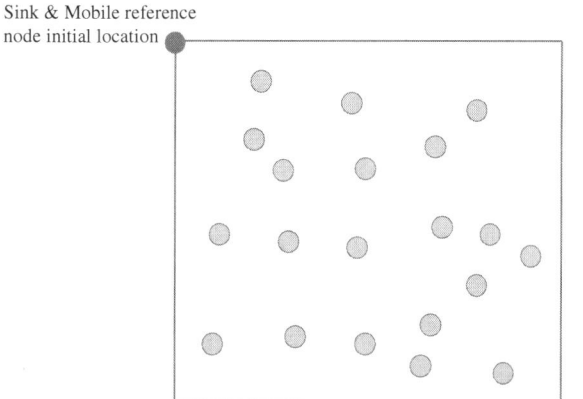

Mobile Reference Node Broadcast Algorithm

First, mobile reference node start broadcast a Wake_up of beacon to wake up the unknown sensor nodes where are in the virtual grid, then broadcast a Initial_start signal, then mobile reference node will move to reach the R/2 position, as the half transmission radius. To connect with unknown sensor nodes, and show the location mobile reference node is. After that mobile reference node will start moving to the end point of a virtual grid's length. Last when mobile reference node move to the end point will broadcast a Middle_stop signal means mobile reference node has finished moving a grid side length, shown in Fig. 3 is a single virtual square broadcast.

In Table 1 is the introduction of start signal, start signal packet contains two fields Start_signal_flag, and Mobile node coordinate, first Start_signal_flag is the one used to decide what kind of signal type. As shown in Fig. 5, first we use the k value to determine if it's an odd number, if so, to determine whether the special region of the grid in the grid area, as shown in Fig. 4, assume that we want to locate unknown sensor nodes in the position of x, first mobile reference node came to (1) location, unknown sensor node can receive the first start signal, then the unknown sensor node will use sensor node localization algorithm to statistics the number of receiving signal.

Sensor Node Localization Algorithm

In this section we introduce sensor node localization algorithm, the main function of this algorithm is for unknown sensor node performance in a virtual grid, through the broadcast receive from mobile reference to calculate the signal strength and signal attenuation formula to calculate the distance. We will make unknown sensor nodes that are waiting for the start signal into the sleep state until the Wake_up beacon accept in the next action. Finally, when an unknown sensor node use three signal values and calculate its node coordinate of the location through trilateration, will wait to receive any end signal to do the action, to transmit a packet, and the

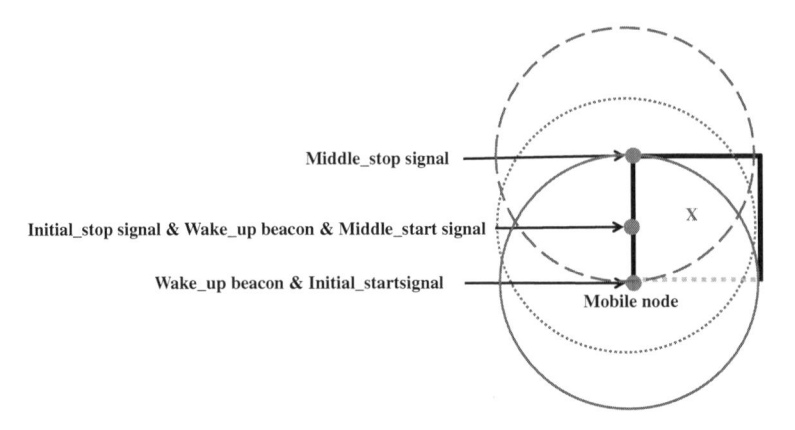

Fig. 3 Single virtual square broadcast

Table 1 Start signal packet

Start signal packet	
Start_signal_flag	Mobile node coordinate

Fig. 4 Exception grid

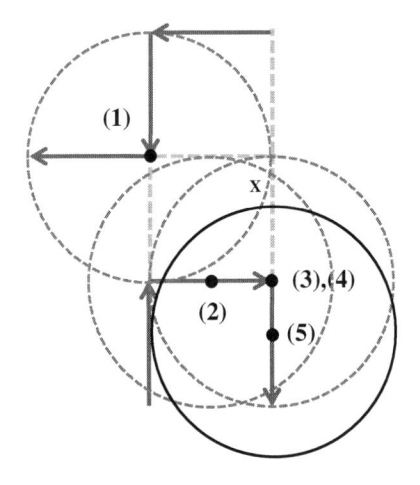

format as shown in Table 2, while waiting for end signal, the unknown sensor node will continue to wait for and won't enter the sleeping mode.

Mobile Reference Node Moving Direction Scheme

After mobile reference node moves an edge, we need to determine the next move and the point of view, therefore, we propose Mobile reference node moving direction scheme, and is divided into two choices: (a) Mobile reference node

Table 2 Unknown sensor node return packet

Sensor_node_position packet	
Sensor node ID	Sensor node coordinate

moving direction for even algorithm (b) Mobile reference node moving direction for odd algorithm. And the flow charts are shown in Fig. 5.

Mobile Reference Node Moving Direction for Even Algorithm

As shown in Fig. 6, when the environment size is 4×4. We find out a regularity for mobile, mobile reference node change direction, when we will continue to make twice the same direction of rotation.

Mobile Reference Node Moving Direction for Odd Algorithm

We use solid lines in Fig. 7 to determine place, when the mobile reference node moves to the sides of the environment, the moving situation is different with the front, just move down linear, and we define a special switch as SPMD, shown in Fig. 8, the real line the circle marked SPMD will set to 1, to control the rotation of the position.

The mobile reference node has move in the bottom of the environment shown in Fig. 9. Left part of the execution flow, is the same we talk in the front when even $\times$ even environment size situation.

Simulation and Analysis

We use NS2 vision 2.29 as the simulator to analyze our proposed localization mechanism. The simulation parameters are similar to a method we compare in [8], and shown in Table 3.

Average Location Error Analysis

In this paper we to hope that can reduce the estimate location error rate of the unknown sensor nodes, and has reached a more complete experimental data through the different size of the environment.

Shown in Fig. 10, in different size of environments and compare the average location error in each environment size, can be found between the MRN and the PI have almost the same location error, because the PI is uninterrupted broadcast to find the strongest signal strength and use the triangle moving path in order to achieve the lowest error rate.

Fig. 5 Mobile reference node moving direction scheme

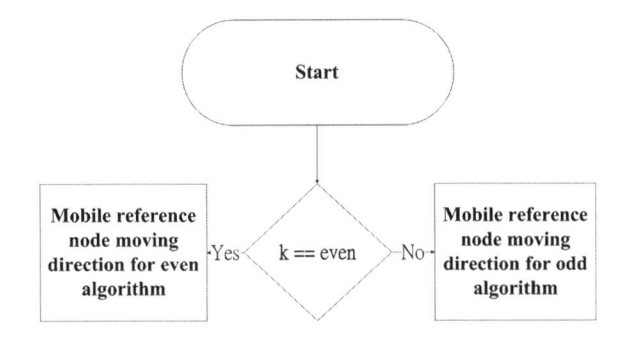

Fig. 6 Environment size (p × k) is 4 × 4

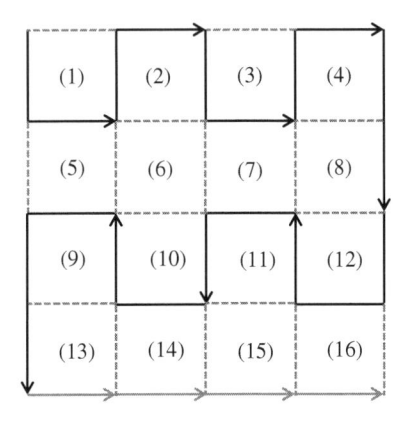

Fig. 7 Environment size (p × k) is 5 × 5

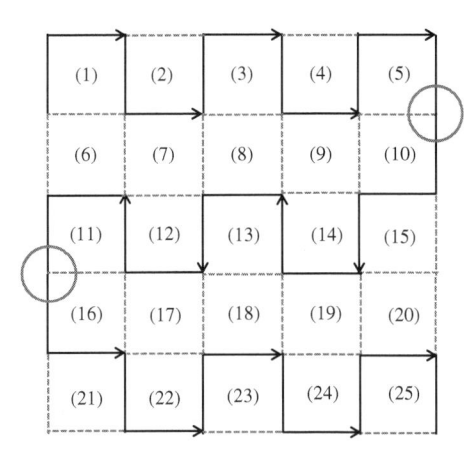

Energy Consumption Analysis

Shown in Fig. 11, in this simulation we only compared with PI and MBBGC, because the location error rate of PI and MBBGC are close with our proposed

Fig. 8 Environment size
$(p \times k)$ is 5×5

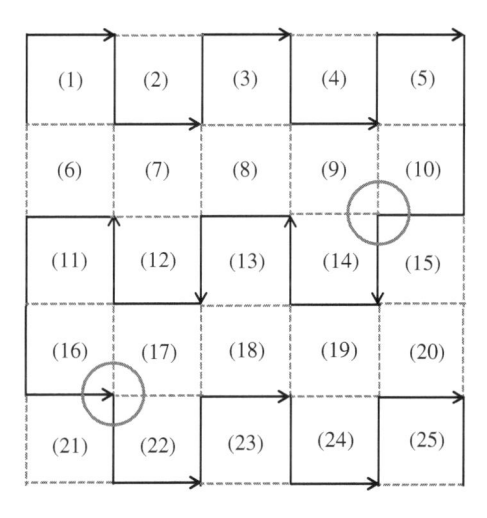

Fig. 9 Environment size
$(p \times k)$ is 4×5

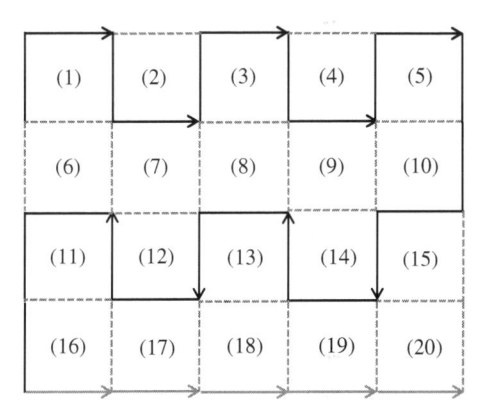

mechanism, the average location error between each other are 0.5–1 m. So in Fig. 11 can see the MRN we proposed has lower power consumption than other methods.

Conclusion and Future Work

In wireless sensor networks, it will increase lots of throughput when we can handle all sensor nodes' location in the whole environment. Nowadays there are a lot of localization technology are restriction by the cost or the natural environment, therefore in some cases made the location error can't be avoided, to reduce the location error need to find other direction to make the breakthrough.

Table 3 Parameter setting

Parameter	Value
Communication range	10 m
Mobile node moving rate	0.1 m/s
Transmission frequency	1 Hz
Sensor node power	3.6 V
Transmit energy consumption	40 mA
Receive energy consumption	35 mA
Sleep energy consumption	0.1 uA
Number of sensor nodes	$80 \sim 130$

Fig. 10 Average location error

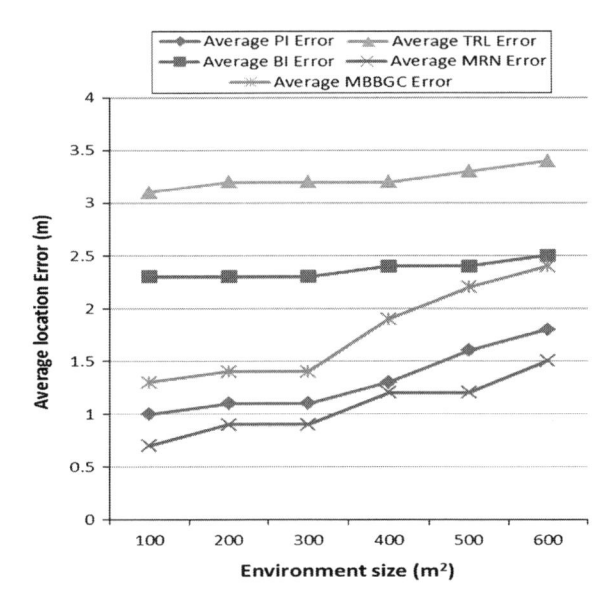

Fig. 11 Mobile node energy consumption

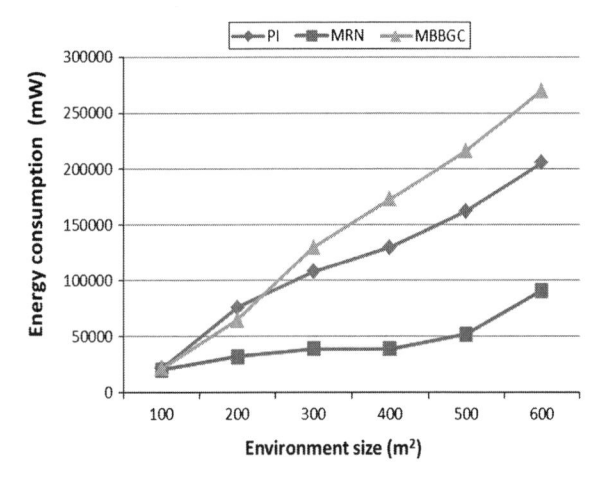

In this paper, we proposed a localization mechanism (MRN) using a mobile reference nod with RSSI method to estimate distances, finally we use Trilateration to ensure the location more correctly. According to the simulation results can find out that the mechanism we proposed can have the same location error between methods we compare. And in energy consumption comparison, have a very significant reduce, whether in the mobile node or unknown nodes.

In the future, we can improve our mechanism in the three-dimensional size of the environment, or Obstacles in the moving path to make sure the moving algorithm can cover whole environment.

References

1. Akyildiz IF, Su W, Sankarasubramaniam Y, Cayirci E (2002) A survey on sensor networks. IEEE Commun Mag 40(8):102–114
2. He T, Huang C, Blum BM, Stankovic JA, Abdelzaher TF (2005) Range-free localization schemes for large scale sensor networks. ACM Trans Embed Comput Syst 4(4):877–906
3. Niculescu D, Nath B (2003) DV-based positioning in ad hoc networks. Kluwer J Telecommun Syst 22(1):267–280
4. Frampton KD (2006) Acoustic self-localization in a distributed sensor network. IEEE Sens J 6(1):166–172
5. Dai F, Wu J (2006) Efficient broadcasting in ad hoc wireless networks using directional antennas. IEEE Trans Parallel Distrib Syst 17(4):335–347
6. Hightower J, Borriello G, Want R (2000) SpotON: an indoor 3D location sensing technology based on RF signal strength, University of Washington, Seattle, Univ. Washington, Tech. Rep. UW CSE 00-02-02, Feb 2000
7. Sun J, Yu J, Zhu L, Wu D, Cao Y (2012) Construction of generalized Ricci flow based virtual coordinates for wireless sensors network. IEEE Sens J 12(6):2109–2112
8. Guo Z, Guo Y, Hong F, Jin Z, He Y, Feng Y, Liu Y (2010) Perpendicular intersection: locating wireless sensors with mobile beacon. IEEE Trans Veh Technol 59(7):3501–3509
9. Lee S, Kim E, Kim C, Kim K (2009) Localization with a mobile beacon based on geometric constraints in wireless sensor networks. IEEE Trans Wireless Commun 8(12):5801–5805

A Delegation-Based Unlinkable Authentication Protocol for Portable Communication Systems with Non-repudiation

Shin-Jia Hwang and Cheng-Han You

Abstract For portable communication systems, the delegation-based authentication protocol provides efficient subsequent login authentication, data confidentiality, User privacy protection, and non-repudiation. However, in all proposed protocols, the non-repudiation of mobile users is based on an impractical assumption that home location registers are trusted. To reduce the HLR's trust assumption and enhance the non-repudiation of the mobile users, our new delegation-based authentication protocol is proposed. Our protocol also removes the exhaustive search problem in the subsequent login authentication to improve the subsequent login authentication performance. Moreover, the user unlinkability in the subsequent login authentication is also provided in our protocol to enhance the user identity privacy.

Keywords Delegation-based authentication · Concurrent signatures · Concurrent signcryption scheme · Portable communication systems

Introduction

Portable communication systems (PCSs) provide roaming services in wireless communication networks. In PCSs, User has to first register in some home location register (HLR) to get its legality. Before roaming, a user must login some visit location register (VLR) and VLR validates the user's legality with the help of the

S.-J. Hwang · C.-H. You (✉)
Department of Computer Science and Information Engineering, Tamkang University,
Tamsui, New Taipei City 251, Taiwan, Republic of China
e-mail: 699420435@s99.tku.edu.tw

S.-J. Hwang
e-mail: sjhwang@mail.tku.edu.tw

Y.-M. Huang et al. (eds.), *Advanced Technologies, Embedded and Multimedia for Human-centric Computing*, Lecture Notes in Electrical Engineering 260, DOI: 10.1007/978-94-007-7262-5_105, © Springer Science+Business Media Dordrecht 2014

HLR. If the user is legal in some HLR, VLR allows this user to use its roaming services.

Global system for mobile communications has some drawbacks [1]: No non-repudiation property, no users' identity privacy, and no mutual authentication between users and VLR. Some protocols try to improve the GSM protocol [2, 3] based on symmetric cryptosystems. But, it is hard to achieve non-repudiation property. Some public key cryptosystems-based protocols are proposed [4, 5] to provide both non-repudiation and mutual authentication.

The delegation-based authentication protocol for PCSs [1] is proposed by exhibiting off-line authentication processes to reduce the communication load between VLR and HLR; and keeping the lower computation load for users. The protocol satisfies the non-repudiation, mutual authentication, data secrecy, and identity privacy, but does not satisfy the non-repudiation property in off-line authentication processes [6]. So the enhanced protocol [6] is proposed. Both [7] and [8] indicated that Lee protocol suffers the linkable problem. To remove the link problem, [7] and [8] state two protocols, respectively. However, Wang et al. protocol still suffers forgery attack [9]. Both two protocols suffer the exhausted search problem in off-line authentication process [10]. To overcome exhaustive search problem, Chen et al. [10] proposed their protocol.

In the proposed protocols [1, 6–8, 10], a registered user obtained the proxy public key and the proxy private key generated by HLR. The proxy public key is used as the user's alias for roaming and the proxy private key is used to generate the proxy signatures to provide non-repudiation. However, HLR also knows the proxy key pair. To avoid misuse of the proxy key pair and provide the user's non-repudiation, those protocols must assume that HLR is trustworthy by all users.

To guarantee the HLR's trust means that the staffs of the HLR also are trusted. However, if some malicious staff in HLR steals the proxy private keys, then the staff can successfully forge the proxy signatures for some legal user and impersonate the legal user for roaming. Then the legal user owning the proxy private key cannot deny the roaming records from the malicious staff. The trust HLR assumption weakens users' non-repudiation. So, to reduce the trusted HLR assumption to semi-trust HLR assumption also enhances users' non-repudiation.

Our Contribution

Our new delegation-based authentication protocol for PCSs is proposed by first designing the embedded concurrent signcryption scheme with anonymity, adopting the concept of the confidential deniable authentication protocol [10]. In the embedded concurrent signcryption scheme, an initial signer and a matching signer can exchange their signatures in a fair and confidential way. So the concurrent signcryption scheme satisfies the following security properties [11, 12]: unlinkability, correctness, fairness, unforgeability, and confidentiality.

The new concurrent signcryption scheme is used to fairly exchange the user's signature on the proxy public key application and the HLR's signature on the proxy public key delegation warrants. Our delegation-based authentication protocol also satisfies the following security properties [1, 6–8, 10]: non-repudiation, mutual authentication, session key security, user identity privacy, user unlinkability, and no exhausted search.

Our new delegation-based authentication protocol is proposed in section Our Delegation-Based Unlinkable Authentication Protocol for PCS with User's Non-repudiation. The security analysis of our protocol is given in section Security Analysis and Proofs. The final section is our conclusions.

Our Delegation-Based Unlinkable Authentication Protocol for PCS with User's Non-repudiation

The parameter l is a security level parameter. The public parameters p and q are two large primes such that $q \mid (p-1)$. The public parameter g is an element in Z_p^* with order q. Two hash functions $h()$ and $H()$ are published, where $h(.)$ maps from $\{0,1\}^*$ to Z_q^* and $H(.)$ maps from $\{0, 1\}^*$ to $\{0, 1\}^l$. Our protocol publishes a symmetric cipher satisfying indistinguishable security against adaptive chosen ciphertext attacks (IND-CCA2). The symmetric encryption function is denoted by $[M]_K$, where M is a message and K is the symmetric secret key. Some public key based signature scheme is also published for all legal users. Notation $A = Sig[M]_x$ denotes the signature generation function and $Verify[A, M]_y$ is the signature verification function, where M is the message, A is the signature on M, x is the signer's private key, and y is the signer's public key. Notation K_{VH} denotes the shared secret key between VLR and HLR. Notations ID_V and ID_H denote the identities of VLR and HLR, respectively. Notation $m_1 \| m_2$ denotes the message m_1 is concatenated with the message m_2. Notation $A \to B$: M denotes that A sends the message M to B. Notations $(x_M, y_M = g^{x_M} \bmod p)$ and $(x_H, y_H = g^{x_H} \bmod p)$ denote the certificated private-public key pair of MS and HLR, respectively. Our protocol contains three phases: Initialization, login authentication, and subsequent login authentication phases.

Initialization Phase

Some mobile user submits its anonymous private-public key pair to the HLR for the first registration by the following registration protocol.

Step 1: User constructs an anonymous private-public key pair (x_K, y_K) by randomly selecting the private key $x_K \in Z_q^*$ and computing the public key $y_K = g^{x_K} \bmod p$.

Step 2: User generates the promise of Schnorr signature σ_M on the registration for (x_K, y_K).

Step 2.1: Select a random number $r_M \in Z_q^*$.

Step 2.2: Compute $V_M = h(g^{r_M} \bmod p, m_M \| ID_M \| y_K)$, the keystone $s_M = r_M + V_M x_M \bmod q$, and $S_M = g^{s_M} \bmod p (= g^{r_M + V_M x_M} \bmod p)$, where m_M is the registration letter for the anonymous private-public key pair (x_K, y_K).

Step 2.3: Store $\sigma_M = (S_M, V_M)$ and s_M in User's local database.

Step 3: User transmits $(\sigma_M, m_M \| ID_M \| y_K)$ to HLR through secure channels.

Step 4: HLR verifies σ_M by checking whether or not $V_M = h(S_M y_M^{-V_M} \bmod p, m_M \| ID_M \| y_K)$. If the equation does not hold, then stop.

Step 5: HLR generates the promise of Schnorr-like signature.

Step 5.1: Select a random number $r_H \in Z_q^*$.

Step 5.2: Compute $S_H = S_M^{X_H} \bmod p$, $V_H = h(g^{r_H} S_H \bmod p, m_H \| ID_H \| y_K)$, $k = (r_H - V_H) x_H^{-1} \bmod q$, where m_H is the authorization warrant for (x_K, y_K).

Step 5.3: Store $(\sigma_H = (S_H, k, V_H), \sigma_M, m_H, m_M, ID_M, y_K)$ in HLR's temporary database by y_K's order

Step 6: HLR transmits (σ_H, m_H) to User through secure channels.

Step 7: User verifies σ_H by checking whether or not $V_H = h(g^{VH} y_H k S_H \bmod p, m_H \| ID_H \| y_K)$. If the equation does not hold, then stop.

Step 8: User sets his/her proxy private key as x_K and his/her proxy public key as y_K.

Step 9: User computes HLR's Schnorr-like signature $\rho_H = (s_H = s_M + k \bmod q, V_H)$. Then HLR easily recovers User's Schnorr signature $\rho_M = (s_M = s_H - k \bmod q, V_M)$ after obtaining ρ_H.

Login Authentication Phase

User contracts VLR to obtain services, and VLR checks the User's legality by the login authentication protocol. Suppose that the current unused proxy public key is y_K.

Step 1: User randomly selects an integer n_1 and computes a one-way hash chain $h^1(n_1), h^2(n_1), h^3(n_1), \ldots, h^{(n+1)}(n_1)(=N_1)$, where $h^{i+1}(n_1) = h(h^i(n_1))$ for $n \geq i \geq 1$ and $h^1(n_1) = h(n_1)$.

Step 2: User constructs a new anonymous private-public key for the next round by selecting a random number $x_{K,new} \in Z_q^*$ and computing $y_{K,new} = g^{x_{K,new}} \bmod p$.

Step 3: User generates promise of Schnorr signature for new key pair $(x_{k,new}, y_{k,new})$.

Step 3.1: Select a random number $r_{M'} \in Z_q^*$.

Step 3.2: Compute $V_M' = h(g^{r_M'} \bmod p, m_M \| ID_M \| y_{K,new})$ the keystone $S_M' = r_M' + V_M' x_M \bmod q$, and $S_M' = g^{S_M'} \bmod p = g^{r_M' + V_M' X_M} \bmod p$, where m_M is the registration letter for $(x_{K,new}, y_{K,new})$. Store $\sigma_M' = (S_M', V_M')$ and s_M' in its database

Step 4: User generates the promise of signcrytext.

Step 4.1: Select a random number $x \in Z_q^*$.

Step 4.2: Compute $Y = g^x \bmod p$, $S_K = H(S_M' \| (y_H)^{S_M'} \bmod p)$, and $MK = H(SK)$.

Step 4.3: Generate a MAC $= H(MK, m_M \| Y \| ID_V \| ID_M \| V_M' \| y_{K,new})$ and encrypt them both by $C_M = [m_M \| Y \| ID_V \| ID_M \| V_M' \| y_{K,new} \| MAC]_{SK}$.

Step 5: User transmits $(\rho_H = (s_H, V_H) = (s_M + k \bmod q, V_H), m_H, ID_H, y_K)$ to VLR.

Step 6: VLR verifies ρ_H by checking whether or not $V_H = h(g^{V_H} y_H{}^{s_H} \bmod p, m_H \| ID_H \| y_K)$. If the equation does not hold, then stop.

Step 7: VLR selects a random number n_2 and transmits User n_2, the period of validity Per, and ID_V.

Step 8: User generates the signature $A = Sig[N_1 \| n_2 \| Per \| ID_V]_{xK}$ then sends $(A, (C_M, S_M'), ID_V, N_1)$ to VLR.

Step 9: VLR validates A on the message $N_1 \| n_2 \| Per \| ID_V$.

Step 9.1: Generate $[A \| Per \| N_1 \| n_2 \| y_K \| (C_M, S_M') \| \rho_H \| Dig]K_{HV}$, if Verify$[A, N_1 \| n_2 \| Per \| ID_V]_{y_K}$ is valid for $Dig = H(A \| Per \| N_1 \| n_2 \| y_K \| (C_M, s_M') \| \rho_H)$.

Step 9.2: Send $([A \| Per \| N_1 \| n_2 \| y_K \| (C_M, S_M') \| \rho_H \| Dig]K_{HV}, ID_H, ID_V)$ to HLR.

Step 10: HLR decrypts $[A \| Per \| N_1 \| n_2 \| y_K \| (C_M, s_M') \| \rho_H \| Dig]K_{HV}$.

Step 10.1: Obtain $(A \| Per \| N_1 \| n_2 \| y_K \| (C_M, S_M') \| \rho_H \| Dig)$ by decrypting $[A \| Per \| N_1 \| n_2 \| y_K \| (C_M, S_M') \| \rho_H \| Dig]_{k_{HV}}$.

Step 10.2: Validate the recovered message by checking whether or not $Dig = H(A \| Per \| N_1 \| n_2 \| y_K \| (C_M, S_M') \| \rho_H \| Dig)$.

Step 11: HLR validates the certificate ρ_H of User's anonymous public key y_k.

Step 11.1: Find $\sigma_H = (S_H, k, V_H)$ and $\sigma_M = (S_M, V_M)$ in HLR's temporary database using y_K as the searching key.

Step 11.2: Recover the User's signature $\rho_M = (s_M, V_M)$ by computing $s_M = s_H - k \bmod q$.

Step 11.3: Verify ρ_M by checking whether or not If $V_M = h(g^{s_M y_M - V_M} \bmod p, m_M \| ID_M \| y_K)$. If ρ_M is valid, HLR believes that User is some legal User.

Step 11.4: Validate the signature A by checking *Verify* $[A, N_1 \| n_2 \| Per \| ID_V]_{y_K}$ to confirm whether the specified User is the one knowing the secret key x_K.

Step 12: HLR validates User's promise of the Schnorr signature on $y_{k,\,new}$.

Step 12.1: Compute $S_H' = s_M'^{x_H} \bmod p$ and $SK = H(S_M | S_H' \bmod p)$.

Step 12.2: Obtain $m_M \| Y \| ID_V \| ID_M \| V_M' \| y_{K,new} \| h(MK, m_M \| Y \| ID_V \| ID_M \| V_M' \| y_{K,new})$ by decrypting C_M with SK.

Step 12.3: Compute $MK = H(SK)$ and check $h(MK, m_M \| Y \| ID\text{-}v \| ID_M | V_M' \| y_{K,new})$ to authenticate message.

Step 12.4: Check ID_V.

Step 12.5: Validate the promise (S_M', V_M') by the equation $V_M' = h(S_M' y_K^{-V_M'} \bmod p, m_M \| ID_M \| y_{K,new})$.

Step 13: HLR generates the response

Step 13.1: Select two random numbers r_H' and R'.

Step 13.2: Compute $V_H' = h(g^{r_H'} S_M' \bmod p, m_H \| ID_H \| y_{K,new})$ and $K' = (r_H' - V_H') x_H^{-1} \bmod q$. So the new promise of the certificate is $\sigma_H' = (S_H', K', V_H')$

Step 13.3: Compute $SK' = H(Y^{X_H} \bmod p)$.

Step 13.4: Choose a nonce n_3, compute $RK_1 = H(N_1 \| n_2 \| n_3 \| y_K)$ and store $L = N_1$.

Step 13.5: Generate and transmit $([[N_1 \| n_3 \| ID_V \| R' \| \sigma_{H'}]_{SK'} \| Y_K \| n_2 \| N_1 \| RK_1]_{k_{HV}}, ID_H, ID_V)$ to VLR.

Step 14: VLR validates HLR's response.

Step 14.1: Obtain $[N_1 \| n_3 \| ID_V \| R' \| \sigma_{H'}]_{SK'} \| Y_K \| n_2 \| N_1 \| RK_1$ by decrypting the ciphertext $([[N_1 \| n_3 \| ID_V \| R' \| \sigma_{H'}]_{SK'} \| Y_K \| n_2 \| N_1 \| RK_1]_{k_{HV}}$.

Step 14.2: Check the freshness of n_2. If n_2 is not fresh, stop.

Step 14.3: Compute $ID_1 = H([N_1 \| n_3 \| ID_V \| R' \| \sigma_{H'}]_{SK'} \| RK_1)$.

Step 14.4: Store $((\rho_H, m_H, ID_H, ID_1), RK_1, A, L = N_1, 1)$ into its local database according to the order of ID_1.

Step 14.5: Transmit $([N_1 \| n_3 \| ID_V \| R' \| \sigma_{H'}]_{SK'}, ID_V$ to User.

Step 15: User decrypts HLR's ciphertext.

Step 15.1: Get $N_1 \| n_3 \| ID_V \| R' \| \sigma_{H'}$ by decrypting $[N_1 \| n_3 \| ID_V \| R' \| \sigma_{H'}]_{SK'}$, where $SK' = H(y_H^x \bmod p)$.

Step 15.2: Check the freshness of N_1. If N_1 is not fresh, then stop.

Step 16: User validates the promise σ_H' on new key $y_{K,new}$ by checking
$$V_H' = h(g^{V_{H'}} y_H^{K'} S_{H'} \bmod p, m_H \| ID_H \| y_{K,new}). \text{If}$$
$$V_H' = h(g^{V_{H'}} y_H^{K'} S_{H'} \bmod p, m_H \| ID_H \| y_{K,new}). \text{ then accept; otherwise,}$$
reject.

Step 17: User sets the proxy private key as $x_{K,new}$ and the proxy public key as $y_{K,new}$ for the next round.

Step 18: User computes HLR's Schnorr-like signature $\rho_H' = (S_H', V_H')$, where
$$S_H' = s_M' + K' \bmod q.$$

Step 19: User stores $(RK_1 = H(N_1\|n_2\|n_3\|y_K),\ (h^1(n_1),\ h^2(n_1),\ h^3(n_1),\ ...,\ h^{(n+1)}(n_1)),\ y_K,\ \text{round number} = 1,\ ID_1 = H([N_1\|n_3\|ID_V\|R'\|\sigma_{H'}]_{SK'}\|RK_1).)$ into User's database.

When $\rho_H' = (S_H', V_H')$ is used, HLR recovers User's Schnorr signature $\rho_M' = (s_M' = S_H' - K' \bmod q, V_M')$ after obtaining ρ_H'.

Subsequent Login Authentication Phase

VLR authenticates User repeatedly without contracting HLR in the subsequent login authentication described below. Suppose that this is ith round subsequent login authentication with the session key RK_i, where $i \leq n$, $RK_i = H(L, RK_{i-1})$, $RK_1 = H(N_1\|n_2\|n_3\|y_K)$, and $L = H^{(n-i+2)}(n_1)$. Therefore, User retrieves the record $(RK_i, (h^1(n_1), h^2(n_1), h^3(n_1), ..., h^{(n+1)}(n_1)), y_K, \text{round number} = (1)$ first.

Step 1: User transmits $(ID_i, [h^{(n-i+1)}(n_1)]_{RK_i})$ to VLR.

Step 2: VLR finds $((\rho_H, m_H, ID_H, ID_i), RK_1, A, L, i)$ by searching its local database according to ID_i and obtain $h^{(n-i+1)}(n_1)$ by decrypting $[h^{(n-i+1)}(n_1)]_{RK_i}$

Step 3: VLR updates $L = h^{(n-i+1)}(n_1)$, $i = i+1$, $RK_{i+1} = h(L, RK_i)$, and $ID_{i+1} = H(ID_i\|RK_{i+1})$ for next round if $h(h^{(n-i+1)}(n_1)) = L$.

Step 4: VLR sends $[ACK]_{RK_i}$ back to User, where $ACK = h(h^{(n-i+1)}(n_1) \| ID_i)$.

Step 5: User obtains ACK by decrypting the ciphertext and validates ACK by checking whether or not $ACK = h(h^{(n-i+1)}(n_1)\|ID_i)$.

If $i = n+1$, then User has to repeat the login authentication by setting $y_K = y_{K,new}$.

Security Analysis and Proofs

Security assumptions of our protocol are stated first. One is the symmetric encryption/decryption scheme is IND-CCA2. One is the DDH assumption that no probabilistic polynomial-time (PPT) algorithm solves DDH problem with non-negligible probability. Our protocol satisfies unlinkability, correctness, fairness, unforgeability, confidentiality, non-repudiation, mutual authentication, session key security, and User identity privacy, and User unlinkability.

Confidentiality

Our protocol satisfies IND-CCA2, if there is no PPT adversary α wins the IND-CCA2 game with the non-negligible probability.

Theorem 1: Assume the underlying symmetric encryption/decryption scheme is IND-CCA2. The proposed protocol provides IND-CCA2 if no PPT algorithm wins the IND-CCA2 game with non-negligible probability based on the DDH assumption in random oracle model.

Unlinkability Between the Promise of Signcrytext and HLR's Ciphertext

Unlinkability among difference services means no one can deduce any two different services belonging to the same User. Two cases cause linkability problem in our protocol. One case is that adversaries may find out the link between ρ_H' and (C_M, s_M') from the same User on two successive roaming. Since (C_M, s_M') is the signcryptext of the promise $\sigma_M' = (s_M', V_M')$, the adversary may find out the link between ρ_H' and σ_M' with the help of the link find between ρ_H' and (C_M, s_M'), if the decryption of C_M is feasible. Therefore the link find between ρ_H' and (C_M, s_M') is at least harder than the link find between ρ_H' and σ_M'. The link find problem between ρ_H' and σ_M' is also the link find problem in Nguyen's scheme that is our underlying scheme. Fortunately, this link find is infeasible for the unlinkability of Nguyen's scheme [12].

The other case is to find out the link between the promise of signcrytext (C_M, s_M') and HLR's ciphertext $[N_1\|n_3\|ID_V\|R'\|\sigma_{H'}]_{SK'}$. To prove this link find is infeasible, the unlinkability game is defined first. Our protocol is unlinkabile between the promise of signcrytext (C_M, s_M') and HLR's ciphertext $[N_1\|n_3\|ID_V\|R'\|\sigma_{H'}]_{SK'}$ if no PPT adversary α wins the unlinkability game with a non-negligible probability.

Unlinkability Game Between the Promise of Signcrytext and HLR's Ciphertext

This game has two participators, Adversary α and Challenger C. C controls all hash oracles whom α is allowed to query. These hash oracles are the same as the hash oracles in the security proof of confidentiality. This game consists of two phases: Setup, and challenging and guessing phases. In the setup phase, all systems parameters, public functions, and the public keys/private keys of all users are constructed. The public parameters and all public keys are sent to the adversary.

In the challenging and guessing phase, the adversary first chooses two anonymous public keys that may belong to two different users to the challenger. The challenges have two types. Type 0 is that the challenge consists of two promises of signcryptext using two anonymous public keys and one HLR's ciphertext for one anonymous public key. The adversary guesses which the promise matches the HLR's ciphertext. Type 1 is that the challenge consists of two HLR's ciphertexts using two anonymous public keys and one promise of signcryptext for one anonymous public key. The adversary guesses that the HLR's ciphertext matches the promise of signcryptext.

Theorem 2: Our protocol is unlinkable between the promise of signcrytext and HLR's ciphertext if no PPT algorithm wins the above unlinkability game with non-negligible probability based on the DDH assumption in random oracle model.

Fairness

The fairness in our protocol means that User (HLR) can obtain the HLR's (User's) signature but HLR (User) cannot. There are two cases violating the fairness.

Case 1: Given the promises $\sigma_M = (S_M, V_M)$ satisfying $V_M = H(S_M y_M^{-V_M} \bmod p, m_M \| ID_M \| y_K)$, HLR generates alone the corresponding $\rho_M = (s_M, V_M)$ satisfying $V_M = H(g^{s_M} y_M^{-V_M} \bmod p, m_M \| ID_M \| y_K)$..

Case 2: Given two promises $\sigma_M = (S_M, V_M)$ and $\sigma_H = (S_H, k, V_H)$ satisfying $V_M = H(g^{s_M} y_M^{-V_M} \bmod p, m_M \| ID_M \| y_K)$. and $V_H = H(g^{V_H} y_H^{k} S_H \bmod p, m_H \| ID_H \| y_K)$, respectively, User obtains $\rho_H = (s_H, V_H)$ satisfying $V_H = H(g^{V_H} y_H^{s_H} \bmod p, m_H \| ID_H \| y_K)$ but HLR cannot obtain $\rho_M = (s_M, V_M)$ satisfying $V_M = H(g^{s_M} y_M^{-V_M} \bmod p, m_M \| ID_M \| y_K)$.

Theorem 3: Our protocol is fair since neither Case 1 nor Case 2 occurs based on the discrete logarithm and DDH assumption.

Unforgeability

The unforgeability means that both User's Schnorr signatures and HLR's Schnorr-like signatures are unforgeability. Fortunately, User's Schnorr signatures are existentially unforgeable against chosen-message attacks [13, 14]. HLR's Schnorr-like signatures are unforgeability [12].

Our protocol also satisfied incorrectness, on-reputation, session key security, the mutual authentication, users' identity privacy, and users' unlinkability.

Conclusion

By utilizing our concurrent signcryption, HLR cannot fore users' signatures during roaming. Consequently, the users' non-repudiation is enhanced. Moreover, the performance of the subsequent login authentication is improved by removing the exhaustive search problem when VLR has to find the alias of an anonymous User. Since the User adopts different aliases to login VLR in subsequent login authentication each time, our protocol supposes the subsequent login User unlinkability to protect the User identity privacy. The concurrent signcryption scheme also supposes the fairness property to fairly protect User's and HLR's benefits.

References

1. Lee W-B, Yeh C-K (2005) A new delegation-based authentication protocol for use in portable communication systems. IEEE Trans Wireless Commun 4(1):57–64
2. Al-Tawill K, Akrami A, Youssef H (1999) A new authentication protocol for GSM networks. In: Proceedings of 23rd annual IEEE conference on local computer networks, pp 21–30
3. Lee C-H, Hwang M-S, Yang W-P (1999) Enhanced privacy and authentication for the global system for mobile communications. Wireless Netw 5(4):231–243
4. Beller MJ, Chang L-F, Yacobi Y (1993) Privacy and authentication on a portable communication system. IEEE J Sel Areas Commun 11(6):821–829
5. Lo C-C, Chen Y-J (1999) Secure communication mechanisms for GSM networks. IEEE Trans Consum Electron 45(4):1074–1080
6. Lee T-F, Chang S-H, Hwang T, Chong SK (2009) Enhanced delegation-based authentication protocol for PCSs. IEEE Trans Wireless Commun 8(5):2166–2171
7. Youn T-Y, Lim J (2011) Improved delegation-based authentication protocol for secure roaming service with unlinkability. IEEE Commun Lett 14(9):791–793
8. Wang R-C, Juang W-S, Lei CL (2011) A privacy and delegation-enhanced user authentication protocol for portable communication systems. Int J Ad Hoc Ubiquitous Comput 6(3):183–190
9. Hwang S.-J, You C.-H (2011) Weakness of Wang et al.'s privacy and delegation enhanced user authentication protocol for PCSs," CSCIST 2011 and iCube 2011, Taipei, 2011
10. Chen H.-B, Lai Y.-H, Chen K.-W, Lee W.-B (2011) Enhanced delegation based authentication protocol for secure roaming service with synchronization. J Electronic Sci Technol 9(4), pp 345–351
11. Hwang S-J, Sung Y-H (2011) Confidential deniable authentication using promised signcryption. J Syst Softw 84:1652–1659
12. Nguyen K (2005) Asymmetric concurrent signatures. In: Proceedings of information and communications security conference (ICICS 2005), LNCS 3783, Springer, New York, pp 181–193
13. Schnorr C (1991) Efficient signature generation by smart cards. J Cryptology 14(3), pp 161–174
14. Pointcheval D, Stern J (2000) Security arguments for digital signatures and blind signatures. J Cryptography 13(3), pp 361–396

A Lightweight Mutual Authentication Protocol for RFID

Changlun Zhang and Haibing Mu

Abstract In order to protect the tag and its communication, the authentication between the tag and the reader as well as its backend database is necessary. The paper proposed a mutual authentication protocol by introducing pointer and check number pool which make a simple random number to hide the ID information and disturb the static answers in challenge-response of the protocol. The check number pair selected from the pool in each authentication turn is also pointed out by a random number which controls the shifting of the pointer in pool. The analysis shows that the protocol can resist the common attacks in RFID communication with low computation overhead.

Keywords IoT · RFID · Check number pool · Pointer · Security

Introduction

The Internet of Things (IoT) is another milestone after the Internet in the development of information technology. It combines the virtual Internet with human and objects in real world to achieve the interrelation of human with human, human with object and objects with objects. The terminal nodes in IoT are extended to a large number of sensors, Radio Frequency Identification (RFID) labels and intelligent equipment besides computers and human. The recipient and analyzer of

C. Zhang
Science School, Beijing University of Civil Engineering and Architecture,
Beijing 100044, China
e-mail: zclun@bucea.edu.cn

H. Mu (✉)
Beijing Key Laboratory of Communication and Information Systems, School of Electronic
and Information Engineering, Beijing Jiaotong University, Beijing 100044, China
e-mail: hbmu@bjtu.edu.cn

Y.-M. Huang et al. (eds.), *Advanced Technologies, Embedded and Multimedia for Human-centric Computing*, Lecture Notes in Electrical Engineering 260,
DOI: 10.1007/978-94-007-7262-5_106, © Springer Science+Business Media Dordrecht 2014

the information in the network may be the device itself rather than the thinking man. The device has limited ability to identify the information source and need new authentication protocols the support the secure communication in IoT.

RFID is a main kind of IoT terminals and provides non-contact data transmission with a certain distance which makes an attack window for intruder. We must implement mutual authentication between tag and reader before data transmission to protect the communication. Due to the limited resources of the RFID label, an ideal authentication protocol should be secure, efficient and lightweight. The simplest form of authentication protocols in the RFID system works in this way: the reader sends inquiry request information to scan the target label, and label responds ID items to help identify their own identity. Although this form of authentication protocol is simple and fast, it is vulnerable to forgery, eavesdropping, tracking, impersonation and other attacks. An RFID authentication protocol should be of confidentiality and forward security to prevent tracking, forging and cloning with good certifying efficiency.

Related Work

In recent years, more and more research focuses on the IoT and the RFID with its security. A variety of RFID security authentication protocols have been proposed. Hash-Lock protocol [1, 2] by Sarma et al. proposed a security protocol, which basically solved the privacy protection and access control, and can be implemented on low-cost tags for the tag needs only hash operation. The data of the answer in each turn of the challenge remains unchanged which made the tag is vulnerable to position tracking attack. Furthermore, the tag with an ID sent in plain text through an insecure channel will be vulnerable to replay and forgery attack. Random Hash-Lock protocol [3] is a modified form for Hash-Lock protocol, the use of random numbers prevents location tracking according to the particular output. But the ID with plain text makes it lack of forward security and an attacker can forge the tag once it got the value of the tag's ID. The protocol also need a large quantity of data communication and computation to get $H(IDj\|R)$ for each tag. The large amount of calculation, the efficiency is not high. Hash chain protocol [4] is also based on shared secret inquiry—response protocol with indistinguishability and forward security. The protocol implements one-way authentication for tag and is vulnerable to replay and forgery attack. In addition, each authentication requires a larger amount of computation and comparison, and is not suitable for a large number of tags in the system. Distributed request-response authentication protocol [5] is a mutual authentication protocol against eavesdropping, replay, tracking and other common attacks with better security. However the main drawback of the protocol is the large number of complex function computation according to the number of the readers and the tags.

Recent studies [6–10] introduce pseudo-random number generator and cyclic redundancy code in EPC to improve efficiency and reduce the overhead at the same time. Tian et al. [11] proposed an ultra-lightweight RFID authentication protocol with permutation. The tags only operate bitwise XOR, left rotation and permutation, a new defined rule that can make diffusion and confusion. In this way, the tags avoid using unbalanced OR and AND operations to improve security. Pietro and Molva [12] made two contributions within their new identification and authentication protocol. Firstly, when the servers can be compromised, they devised a technique that makes RFID identification sever-dependent that can protect the tags from malicious servers. Secondly, a probabilistic tag identification scheme was proposed only using bitwise operations that can speed up the authentication process. Inspired by [12], Blass et al. [13]. put forward a lightweight F_f family of privacy-preserving authentication protocols for RFID system. The protocol offers strong authentication. Although it is based on a algebraic framework, it can resist algebraic attacks. Different from lightweight authentication protocols for RFID system, AX Liu and LRA Bailey [14] put forward a privacy and authentication protocol (PAP). In this paper, they distinguished privacy from authentication. The protocol was placed in a concrete environment, store. It requires only an extremely small amount of computations and can protect privacy from malicious attacks. Molnar and Wagner [15] suggested novel architectures to solve privacy issues related to RFID in libraries. They gave a general scheme to build private authentication using work logarithmic in the number of tags. In addition, they also presented a simple scheme that provided security against a passive eavesdropper using XOR, without any pseudo-random functions. Lee et al. [16]. presented a RFID authentication protocol using both XOR and hash chains to realize low-cost and high security. The scheme can resist to traceability and cloning.

But because of the new characteristics of RFID based IoT, these protocols still cannot meet special environmental requirements in the application. In this paper, a new protocol based on the index is proposed to reduce the computation overhead in tag and database as well as provide necessary security for RFID. The paper is organized as follows: section The Authentication Protocol Based on Check Number Pool describes the details of the protocol, and section Analysis of the Protocol analyzes security and performance of the protocol. At last the conclusion is provided in section Conclusion.

The Authentication Protocol Based on Check Number Pool

The protocol combines small pseudo-random verification number and the random number generator to achieve nearly random number function with low computation overhead.

Initialization and Parameters

There are several parameters to be generated during the system initialization phase, such as encryption and decryption keys K_i and the corresponding check number pools. There are records in each pool with three factors in each record which are the serial number and the random number R_A and R_B. Each RFID tag is written with a key and its corresponding check number pool as well as its ID and EPC information. At the same time, the ID of the tag is also synchronized to the backend database, so that each tag ID is combined with the corresponding key and check number pool. There is a pointer in backend database to keep the synchronization of the pool.

The initialization and the parameters are shown in Fig. 1.

Detail of the Authentication

The authentication process between the RFID tag and the reader supported by backend database can be described as following communication interaction.

1. The reader generates a random number R and sends it together with the communication request to the target tag.

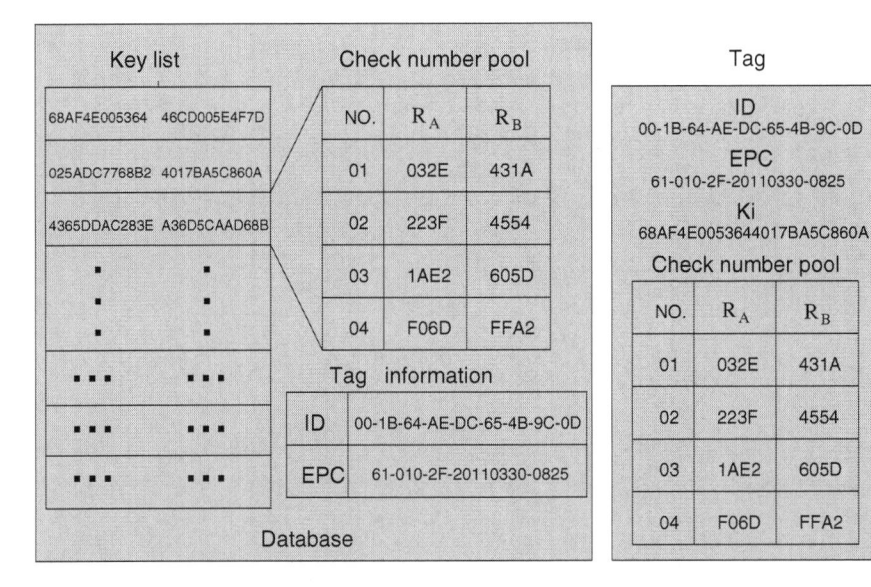

Fig. 1 Initialization and the parameters

2. The tag receives R and takes the check number R_A according to the current pointer. It then calculates $P = E_{ki}(R \oplus R_A \oplus ID)$ and sends the P and R_A to the reader.
3. Reader receives the data and sends P, R, R_A to the back-end database.
4. The database searches the check number pool and gets the subset according to R_A. It then computes $P' = E_{ki}(R \oplus R_A \oplus ID')$ of each record using R, R_A with the corresponding key K_i', and compares results with received P respectively. If there is not any matching record, the tag will be considered as illegal and communication ends. If there is any result equal to P, the back-end database will get the ID of the tag and prove it a legal one.

After authentication, the database will send to the reader another random number R_B in this pair of check number. In order to update the exchange information during authentication turn, the generator in database will generate a random number r to shift the pointer value. The new pointer and the previous one are stored in database until next turn. Database also sends this value to the reader in order to synchronize the pointer in the tag.

5. The reader receives the information and then sends R_B and r to tag.
6. Tag compares the received R_B with the stored one in the pool according to the current pointer. If the two numbers are of the same, the reader is considered as legal one. And the pointer is updated according to r. Check number pools are synchronized completely. Otherwise, if this communication is failed, the tag will use the old pointer in next authentication. This may be also successful because the previous one is still stored in database and the authentication achieved.

Details of one turn of the authentication are shown in Fig. 2.

Analysis of the Protocol

For any newly proposed security protocol, analysis of the protocol on whether it can protect the entities in RFID system from existing security threats or not is necessary.

1. Eavesdropping

Eliminating eavesdropping in most wireless networks thoroughly is impossible. Authentication protocols just limit the amount of information transmitted in unsecure physical channel only. By encrypting authentication messages, communication protocol can lower the value of information eavesdropped by malicious adversary. In this way, effectiveness of such an attack can be reduced.

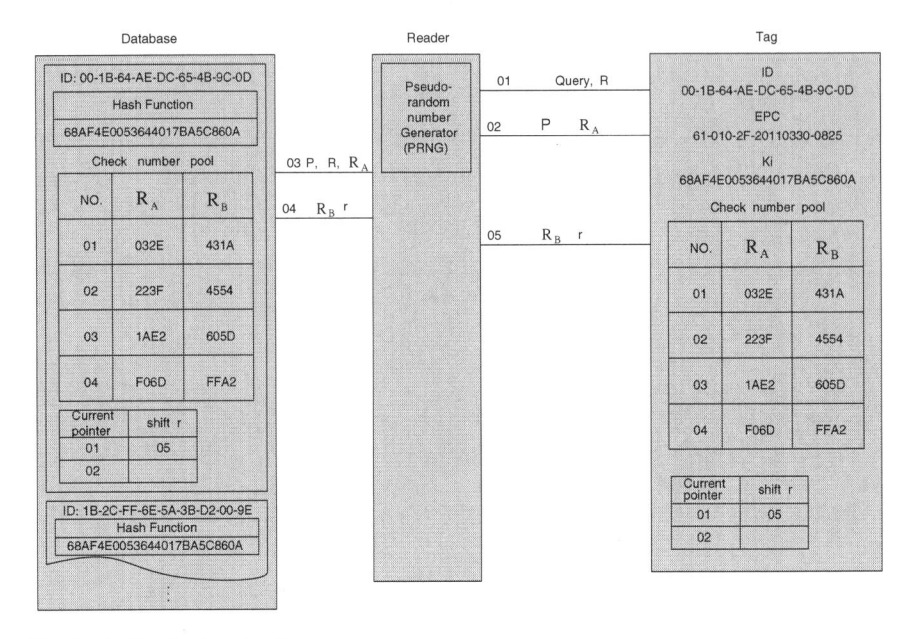

Fig. 2 Authentication detail

2. Replay attack

Our protocol can resist replay attack effectively. The main weakness of random Hash-Lock protocol is lead by the authentication message from reader to tag, which is the static ID of a tag. Our protocol did not send ID and any other stead information during authentication instead of random numbers.

If an adversary can capture a large number of verified sequence pairs sent during every authentication process, when a repeated verified sequence is captured, the adversary will be clear about the corresponding response. Aiming at the situation above, an adversary has to monitoring a mass of verified sequence pairs. However, it is impractical for most of the adversaries.

3. Dos attack

There is no need to compute P for every ID in this protocol, because there is a process of retrieving matching values before computing. There are only simple computations of retrieving instead of a large number of computations. Computing on P won't start until the matching values are obtained. Even if there are amount of challenge information, it can be confirmed whether the data is true or fault by simply filtering, because the adversary doesn't know an accurate RA of next hop. Although Dos attack is possible, the probability of its success has been lower comparing with original protocol.

4. Tracking

This protocol can be effective against identity tracking. Tag ID information cannot be obtained by attacker because it is not transmitted in plain text or other steady forms. The responses to the challenge are no longer the same ID information, but randomized check number thus avoids being tracked.

5. De-synchronization attack

The database will store the previous pointer to shift the check number pairs for one more turn in order to avoid the de-synchronization for the failure of the last information sent to tag.

Although it is impossible to design a security protocol to solve all the problems, different protocol will be suitable to different environment.

Conclusion

In this paper, an authentication protocol for RFID security in IoT focus on improving the original random Hash-Lock protocol security issues, especially against replay attacks. It introduces check number pool with shift pointer which makes a simple random number pair, effectively improve the randomized Hash-Lock protocol against replay, tracking, DoS and other attacks. At the same time the protocol improves the performance in computing greatly by matching check number firstly before other operation. The further work of the protocol is to get more simulation data to compare and analyze its performance in more detail.

Acknowledgments This work is supported by National Natural Science Foundation of China under Grant 61201159, the Beijing Municipal Organization Department of talents training-funded project (2010D005017000008), Beijing Institute of Architectural Engineering School research fund (Z10053) and Jilin University Key Laboratory of Symbolic Computation and Knowledge Engineering of Ministry of Education research fund (93 K-17-2012-02).

References

1. Sarma SE, Wreis SA, Engels DW (2003) Radio frequency identification: Secure risks and challenges. RSA Laboratories Cryptobytes 6(1):2–9
2. Sarma SE, Wreis SA, Engels DW (2003) RFID systems and security and privacy implications. In: Proceedings of the 4th international workshop on cryptographic hardware and embedded systems, pp 454–469
3. Juels A, Rivest RL, Szydlo M (2003) The blocker tag: selective blocking of RFID tags for consumer privacy. In: Proceedings of the 10th ACM conference on computer and communications security, Washington DC, USA, pp 103–111
4. Ohkubo M, Suzuki K, Kinoshita S (2004) Hash-chain based forward-secure privacy protection scheme for low-cost RFID. In: Proceedings of the 2004 symposium on cryptography and information security (scis 2004), Sendai, pp 719–724

5. Rhee K, Kwak J, Kim S (2005) Challenge-response based RFID authentication protocol for distributed database environment. In: Proceedings of the 2nd international conference on security in pervasive computing (SPC 2005). Lectures Notes in Computer Science 3450. Springer, Berlin, pp 70–84
6. Duc DN, Park J, Lee H, et al (2006) Enhancing security of EPC global gen 2 RFID tag against traceability and cloning. In: Symposium on cryptography and information security-SCIS 2006, Hiroshima, Japan
7. Chien H, Chen C (2007) Mutual authentication protocol for RFID conforming to EPC class 1 generation 2 standards. Comput Stand Interfaces 29(2):254–259
8. Yuan S, Dai H, Lai S (2008) Hash based RFID authentication protocol. Comput Eng 12:51
9. Yang L, Chen Z (2010) A mutual authentication protocol for low-cost RFID
10. Juels A (2007) RFID security and privacy: a research survey. IEEE J Sel Areas in Commun 24:381
11. Tian Y, Chen G, Li J (2012) A new ultralightweight RFID authentication protocol with permutation. Commun Lett IEEE 16(5):702–705
12. Di Pietro R, Molva R. Information confinement, privacy, and security in RFID systems. In: Computer security–ESORICS 2007. Springer, Berlin, pp 187–202
13. Blass EO, Kurmus A, Molva R et al (2011) The F_f-family of protocols for RFID-privacy and authentication. IEEE Trans Dependable Secure Comput 8(3):466–480
14. Liu AX, Bailey LRA (2009) PAP: a privacy and authentication protocol for passive RFID tags. Comput Commun 32(7):1194–1199
15. Molnar D, Wagner D (2004) Privacy and security in library RFID: issues, practices, and architectures. In: Proceedings of the 11th ACM conference on computer and communications security, pp 210–219
16. Lee S, Asano T, Kim K (2006) RFID mutual authentication scheme based on synchronized secret information. In: Symposium on cryptography and information security

An Approach for Detecting Flooding Attack Based on Integrated Entropy Measurement in E-Mail Server

Hsing-Chung Chen, Shian-Shyong Tseng, Chuan-Hsien Mao, Chao-Ching Lee and Rendabel Churniawan

Abstract The aim of this study is to protect an electronic mail (email) server system based on an integrated Entropy calculation via detecting flooding attacks. Lots of approaches have been proposed by many researchers to detect packets accessing email whether are belonging to the normal or abnormal packets. Entropy is an approach of the mathematical theory of Communication; it can be used to measure the uncertainty or randomness in a random variable. A normal email server usually supports the four protocols consists of Simple Mail Transfer Protocol (SMTP), Post Office Protocol version 3 (POP3), Internet Message Access Protocol version 4 (IMAP4), and HTTPS being used by remote web-based email. However, in Internet, there are many flooding attacks will try to paralyze email server system. Therefore, we propose a new approach for detecting flooding attack based on Integrated Entropy Measurement in email server. Our approach can reduce the misjudge rate compared to conventional approaches.

Keywords Entropy · Flooding attack · Email server

H.-C. Chen (✉) · S.-S. Tseng · C.-H. Mao · C.-C. Lee · R. Churniawan
Department of Computer Science and Information Engineering, Asia University, Taichung 41354, Taiwan, Republic of China
e-mail: shin8409@ms6.hinet.net; cdma2000@asia.edu.tw

S.-S. Tseng
e-mail: sstseng@asia.edu.tw

C.-H. Mao
e-mail: 101267005@live.asia.edu.tw

C.-C. Lee
e-mail: johnson10723@gmail.com

R. Churniawan
e-mail: rendabel@gmail.com

Y.-M. Huang et al. (eds.), *Advanced Technologies, Embedded and Multimedia for Human-centric Computing*, Lecture Notes in Electrical Engineering 260, DOI: 10.1007/978-94-007-7262-5_107, © Springer Science+Business Media Dordrecht 2014

Introduction

In recent year, the rapid development of technologies helps people to communicate any information and sharing information via Internet. Email has become one of necessary communication services for Internet users. The using of Electronic Mail (email) is a method of exchanging digital messages from one person to one or more recipients, via connecting internet or computer network. There are many kind purposes of using email services, from private purposes to business purposes. Email Service Provider (ESP) is an organization which provides email server to send, receive and store emails for personal and or organization necessity. Some ESP who may provide the services to general public to personal email are Gmail, Yahoo! Mail, Hotmail and many others.

Each email server is able to support many kind of protocol. In 1982, the early stage of email development, the Simple Mail Transfer Protocol (SMTP, for short) which is formulated in RFC (Request for Comments) 821 [1, 2]. SMTP is a protocol for a mail sender communicates with a mail receiver. On certain types of smaller nodes in the Internet it is often impractical to maintain a message transport system [3]. For example, a workstation may not have sufficient resources (cycles, disk space) in order to permit a SMTP server [RFC821] [3]. To solve this problem, The Post Office Protocol—Version 3 (POP3, for short) is intended to permit a workstation to dynamically access a mail drop on a server host in a useful fashion. Usually, this means that the POP3 protocol is used to allow a workstation to retrieve mail that the server is holding for it. POP3 is an application layer Internet standard protocol used by local email clients to retrieve email from remote server over a TCP/IP connection. The other protocol which is also supported by email server is Internet Message Access Protocol (IMAP, for short) [4]. In 2003, M. Crispin [5] proposed the latest version IMAP version 4rev1 (IMAP4, for short) which is formulated in RFC 2060. It has also been publish to the Internet Engineering Task Force (IETF). The MAP4 allows a client to access and manipulate electronic mail messages on a server [5]. IMAP4 permits manipulation of remote message folders, called "mailboxes", in a way that is functionally equivalent to local mailboxes. IMAP4 also provides the capability for an offline client to re-synchronize with the server. There is the other approach for supporting email services called webmail (or web-based email) [6, 7]. It is any email client implemented as a web application accessed via a web browser. For example, when accessing webmail at https://webmail.asia.edu.tw, you will be redirected to a SSL [6] secured address and your connection will be encrypted. The Secure Sockets Layer (SSL) Protocol Version 3.0 was proposed by P. Karlton [6], in 2011, which is formulated in RFC 6101[6]. It is a security protocol that provides communications privacy over the Internet via HTTPS (Hypertext Transfer Protocol Secure). The protocol allows client/server applications to communicate in a way that is designed to prevent eavesdropping, tampering, or message forgery [6].

However, a simple email server has a lot of users; it is an important attacked target in Internet. The methods of attacks include SMTP Flooding Attack, spam attacks and the malicious attachment etc. in email [8–12]. The various flooding attacks will increase the loading of the server. In this paper, we propose an approach for detecting flooding attack based on integrated entropy measurement in email server. Then, we use the entropy operations [13–15] to analyse the received packets, in order to estimate normal packets and abnormal packets from SMTP together with POP3, IMAP4, HTTP messages flows, and then evaluate its corresponding risk information. Therefore, the risk information will be used to describe the status of the serving server. According to the value of this status, the server will determine whether it is suffered by flooding attacks.

The remainder of this paper is organized as follows. "Related Work" describes the SMTP, POP3, IMAP4 and entropy operation related work. In the "Email Server Prevention Against Flooding Attack Based on Entropy", we propose a new approach for detecting flooding attack based on integrated entropy measurement in email server, and describe how to calculate the evaluate value of risk information of email server. Finally, we draw conclusions in "Conclusions".

Related Work

In our proposed approach, we use the entropy measurement to detect the behaviour of the SMTPFA. Therefore, in "Related Work", we will describe the normal message flows of SMTP standard [1, 2], and the entropy operations [13–15].

SMTP

First, SMTP had been defined in the RFC 821[1, 2, 16]. It is an independent subsystem in special communication system. In this communication system, it only needs a reliable channel to transmit the related sequence message flows. SMTP has an important simple delivering email protocol which it can forward an email between two different networks. The architecture of SMTP is shown in Fig. 1. In the SMTP architecture [16], it consists of a Sender, a sender-SMTP, a receiver-SMTP and a Receiver. When a Sender (user or file server) will connect to another receiver, it will send a request message of Establishes Connection to the sender-SMTP. Then, the sender-SMTP will establish a two-way transmission channel in order to connect the Receiver. The receiver-SMTP will be as a destination point or a relay point. Thus, the sender-SMTP will send the related SMTP commands to the receiver-SMTP. Finally, the receiver-SMTP will follow these commands to send back a SMTP response message to sender-SMTP. According to the above steps, if the command-respond pair has been completed during one normal time-period, it means that a round of SMTP session has been completed. The established SMTP message flows are divided into seven stages [16] as below: Establishes Connection, HELO, MAIL FROM, RCPT TO, DATA, DATA TRANSFER, and QUIT. The SMTP message flows are shown in Fig. 2 [16].

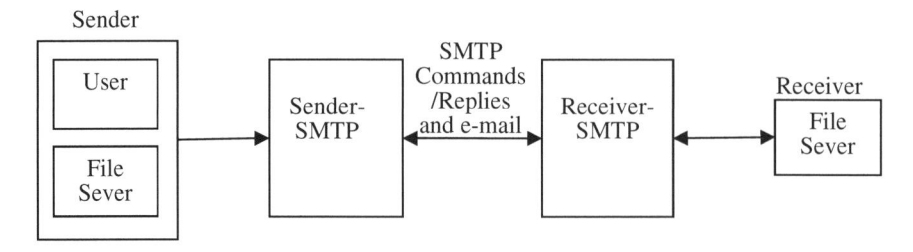

Fig. 1 The SMTP architecture [1, 2, 16]

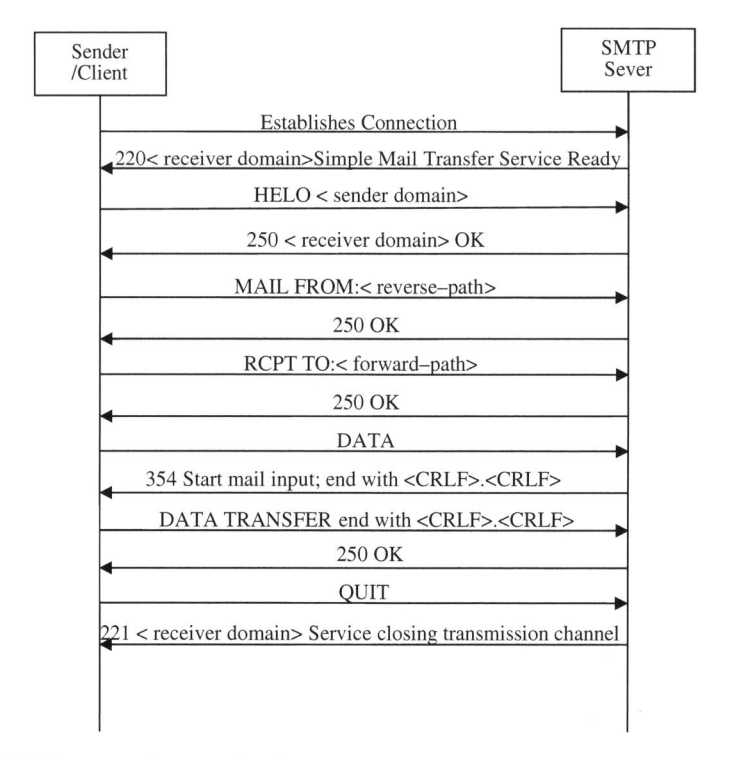

Fig. 2 SMTP message flows [1, 2, 16]

POP 3

Post Office protocol (POP) [3] is an application layer internet standard protocol used by local email clients to retrieve email from a remoter server over TCP/IP connection. The POP 3 messages Flow is shown as Fig. 3 [3]. The POP3 flow of Transaction state is shown as Fig. 4.

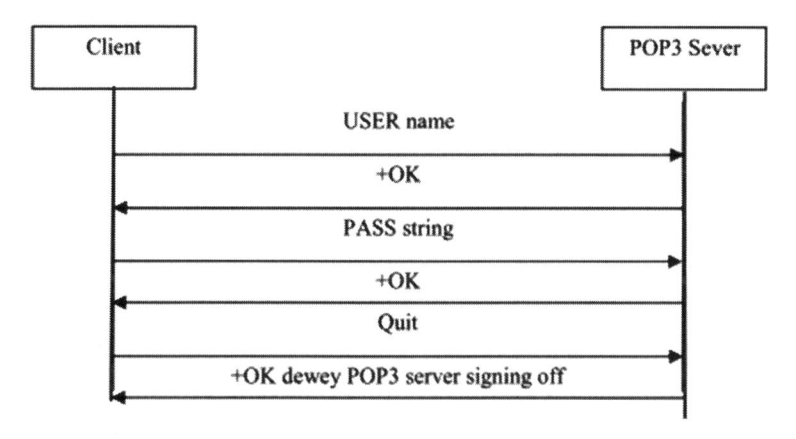

Fig. 3 POP3 flow Authorization state [3]

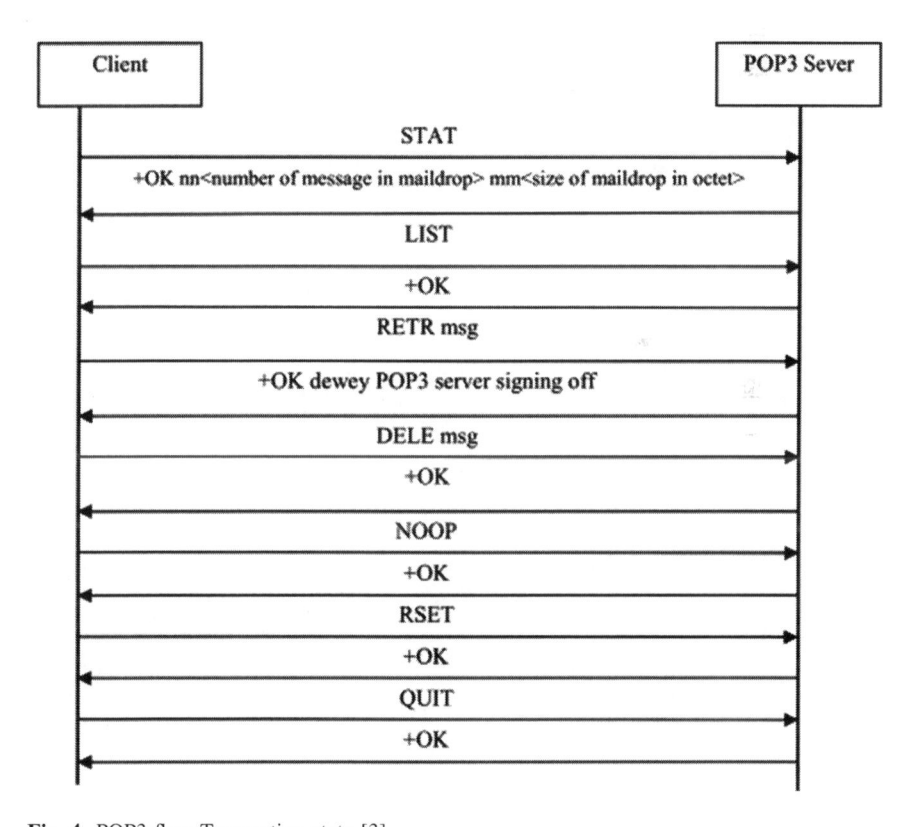

Fig. 4 POP3 flow Transaction state [3]

Internet Message Access Protocol

Internet Message Access Protocol (IMAP) is one of two general protocols for receiving electronic mail. The Internet Message Access Protocol version 4rev1 (IMAP4rev1) [5] allow a client to access and manipulate electronic mail message in the server. IMAP4rev1 permits manipulation of mailboxes in a way that is functionally equivalent to local folders.

a. IMAP Message Flow

IMAP version 4 has been defined in RFC 3501 [5]. This protocol is an inter-action of client/server; consist of a client command, server data and completion result response. The main architecture of IMAP is shown as Fig. 5.
where the information means as below.

1. Connection without pre-authentication (OK greeting)
2. Pre-authenticated connection (PREAUTH greeting)
3. Rejected connection (BYE greeting)
4. Successful LOGIN or AUTHENTICATE command
5. Successful SELECT or EXAMINE command
6. CLOSE command, or failed SELECT or EXAMINE command
7. LOGOUT command, server shutdown, or connection closed

Fig. 5 IMAP architecture [5]

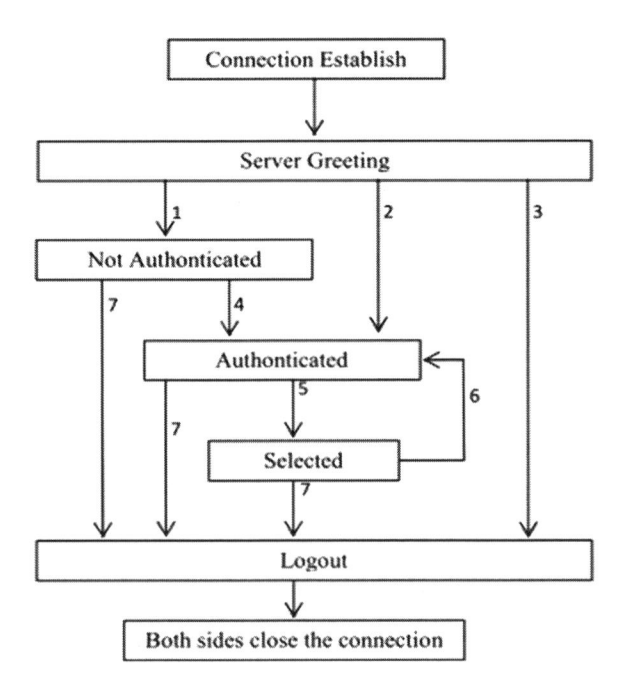

b. IMAP Client Commands—Any State

In the any state client command, there are three commands: CAPABILITY, NOOP, and LOGOUT. These command is always be used in NOT Authenticated state, Authenticated state, and selected state. In the communication flows of client server for any state command are shown in Fig. 6.

c. IMAP Client Commands—Not Authenticated State

In the not authenticated state, the AUTHENTICATED or LOGIN command establishes authentication and enters the authenticated state. The AUTHENTI-CATE command provides a general mechanism for a variety of authentication techniques, privacy protection, and integrity checking [5].

The STARTTLS command is an alternate form of establishing session privacy protection and integrity checking, but doesn't establish authentication or enter the authenticated state.

Server implementations may allow access to certain mailboxes without establishing authentication. This can be done by means of the ANONYMOUS authenticator. An older convention is a LOGIN command using use rid "anonymous"; in this case, a password is required although the server may choose to accept any password. The restrictions placed on anonymous user are implementation-dependent [5].

Once authenticated, it is not possible to re-enter not authenticated state. In addition the universal command (any state command) is valid in the NOT Authenticated state. For description of not authenticated state client server communication, is shown in the Fig. 7 as below.

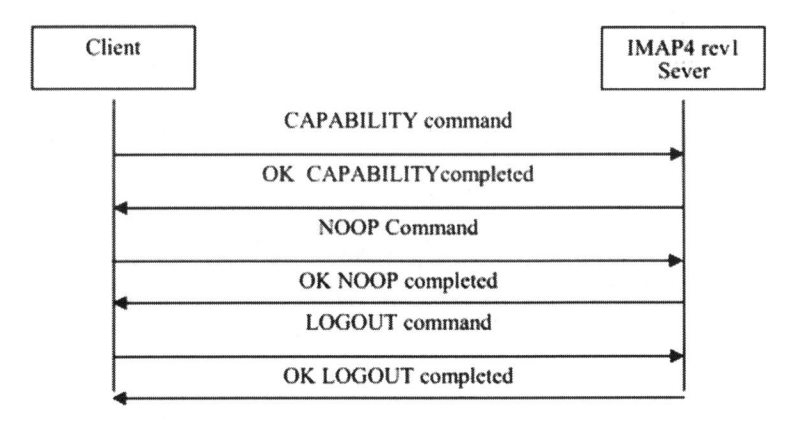

Fig. 6 IMAP flow any state [5]

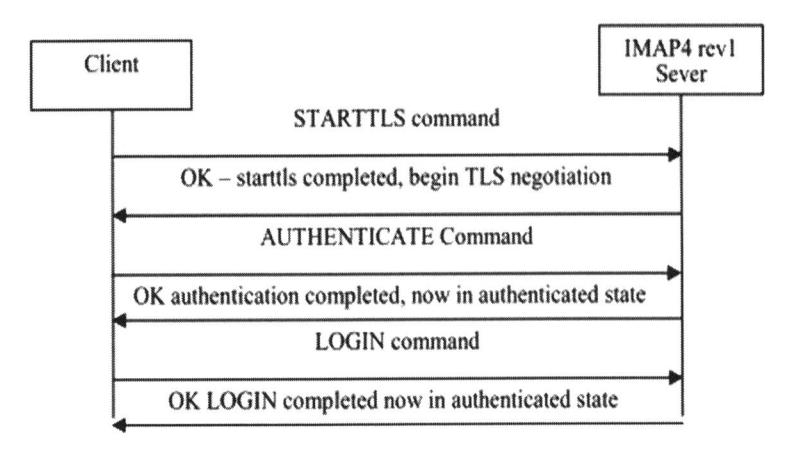

Fig. 7 IMAP flow not authenticated state [5]

d. IMAP Client Commands—Authenticated State

In the authenticated state, commands that manipulate mailboxes are permitted. Of these commands, the SELECT and EXAMINE commands will select a mailbox for access and enter the selected state.

In authenticated state, the universal commands (CAPABILITY, NOOP, and LOGOUT), are also valid to be used. And for the commands in authenticated state are: SELECT, EXAMINE, CREATE, DELETE, RENAME, SUBSCRIBE, UN-SUBSCRIBE, LIST, LSUB, STATUS, and APPEND. For a description of client server communication in the authenticated state for IMAP, is shown in the Fig. 8.

HTTPS

HTTPS is a communications protocol for secure communication over Internet. Technically, it is not a protocol in and of itself; rather, it is the result of simply layering the Hypertext Transfer Protocol (HTTP) on top of the SSL/TLS (Secure Sockets Layer/Transport Layer Security) protocol [6], thus adding the security capabilities of SSL/TLS to standard HTTP communications. In its popular deployment on the internet, HTTPS provides authentication of the web site and associated web server that one is communicating with, which protects against man-in-the-middle attacks [7].

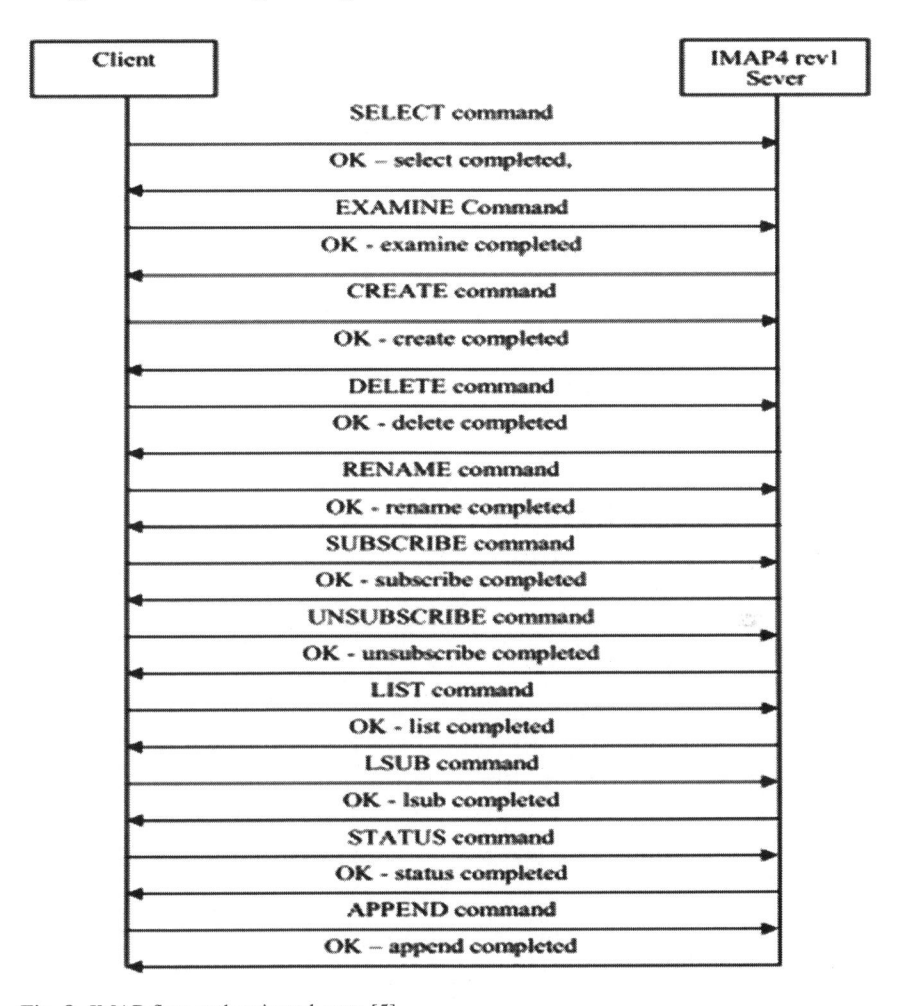

Fig. 8 IMAP flow authenticated state [5]

Entropy Operation

In the Information Theory, Entropy is an approach used to measure the uncertainty or randomness in random variable [13–16]. Entropy measurement approach is proposed by Shannon [13] and Weaver [15]. In the entropy operation, a random entropy value $X \in \{x_1, x_2, x_3, \ldots, x_n\}$, the entropy calculation formula [13–16] as below:

$$H(X) = - \sum_{i=1}^{n} P(x_i) \log P(x_i) \tag{1}$$

where $P(x_i) = \frac{m_i}{m}$, $m = \sum_{i=1}^{n} m_i$, m_i is the observation frequency or numbers of the x_i from X. It can represent [13–16] as:

$$H(X) = -\sum_{i=1}^{n} \left(\frac{m_i}{m}\right) \log P\left(\frac{m_i}{m}\right) \tag{2}$$

From the example above, we know that the coin is thrown according to the positive and negative probability to determine the entropy. The entropy value is inversely proportional to the probability value. With this feature, the value of results we calculate has dependability [16].

Email server Prevention Against Flooding Attack Based on Entropy

In approach of detecting flooding attack in email server, each protocol message flows are divided into two one-group pairing between server and client. Then, we use the entropy operations to calculate the entropy values of normal packets and abnormal packets for each protocol, individually. The formula for normal packet and abnormal packet are listed as below.

$$Entropy : H(X_{K_{T_w},n}) = -\left(\sum_{i=1}^{t} \frac{P_{(K_{T_w},n)}}{P_{K_{T_w}}} \log_2 \frac{P_{(K_{T_w},n)}}{P_{K_{T_w}}}\right), \quad T_w = 1,2,\ldots,t; \tag{3}$$

$$Entropy : H(X_{K_{T_w},a}) = -\left(\sum_{i=1}^{t} \frac{P_{(K_{T_w},a)}}{P_{K_{T_w}}} \log_2 \frac{P_{(K_{T_w},a)}}{P_{K_{T_w}}}\right), \quad T_w = 1,2,\ldots,t; \tag{4}$$

where

$X_{K_{T_w},n} \in \left\{x_{K_{T_w},n1}, x_{K_{T_w},n2}, x_{K_{T_w},n3}, \ldots, x_{K_{T_w},nt}\right\}$ is a random entropy value of normal packet during sampling time duration T_w, and $X_{K_{T_w},a} \in \left\{x_{K_{T_w},a1}, x_{K_{T_w},a2}, x_{K_{T_w},a3}, \ldots, x_{K_{T_w},at}\right\}$ is a random entropy value of abnormal packet during sampling time duration T_w;
$H(X_{K_{T_w},n})$ is an Entropy value set for normal packet; $H(X_{K_{T_w},a})$ is an Entropy value set for abnormal packet;
$K \in \{SMTP, IMAP4, POP3, HTTPS\}$;
$P_{(K_{T_w},n)} + P_{(K_{T_w},a)} = 1$, where $P_{(K_{T_w},n)}$ means the probability of normal packets of K during a sampling time duration T_w; $P_{(K_{T_w},a)}$ means the probability of abnormal packets of K during a sampling time duration T_w.

According to formula (3) and (4), we define the *Evolution Algorithm* as below.
   Evolution Algorithm

Input: $\left(H(X_{K_{T_w},n}), H(X_{K_{T_w},a}) \right)$, an email service message flow pair is including two Entropy value for normal packets and abnormal packets during in sampling time duration T_w, where $T_w = 1, 2, \ldots, t$.

Output: Result cost values S of the email server status in T_w. The cost values $C_{K_{T_w}} \in \{Critical\ High,\ Very\ High,\ High,\ Normal,\ Low,\ Very\ Low\} = \{CH,\ VH,\ H,\ N,\ L,\ VL\}$ are the evolution in order to support the further judgements for the email whether under the flooding attacks or not. Also, the result will show what kind email protocol if the system is under flooding attacks.

Begin

$$x_{K,T_w} \leftarrow \max\left(-\sum_{i=1}^{t} \frac{P_{(K_{T_w},n)}}{P_{K_{T_w}}} \log_2 \frac{P_{(K_{T_w},n)}}{P_{K_{T_w}}} \right),\ T_w = 1, 2, \ldots, t;$$

$$y_{K,T_w} \leftarrow \max\left(-\sum_{i=1}^{t} \frac{P_{(K_{T_w},a)}}{P_{K_{T_w}}} \log_2 \frac{P_{(K_{T_w},a)}}{P_{K_{T_w}}} \right),\ T_w = 1, 2, \ldots, t;$$

$if\ \left(y_{K,T_w} - x_{K,T_w} \right) \gg 0\ then\ return\ \ C_{K_{T_w}} = CH;$

$else\ if\ \left(y_{K,T_w} - x_{K,T_w} \right) > 0\ AND\ \left(y_{K,T_w} - \frac{y_{K,T_{w-1}} + y_{K,T_{w-2}} + y_{K,T_{w-3}}}{3} \right) > 0,\ then\ return$

$VH;$

$return\ H;$

$if\ \left(y_{K,T_w} - x_{K,T_w} \right) \ll 0\ then\ return\ \ C_{K_{T_w}} = VL;$

$else\ if\ \left(y_{K,T_w} - x_{K,T_w} \right) < 0\ AND\ \left(x_{K,T_w} - \frac{x_{K,T_{w-1}} + x_{K,T_{w-2}} + x_{K,T_{w-3}}}{3} \right) > 0\ then\ return\ L;$

$return\ N;$

End;

$\square$

Finally, the algorithm returns the cost values for the current sampling time duration T_w. The cost values can indicate the evolution cost in order to support the judgements for the email server whether under the flooding attacks or not. Also, the result will show what kind email protocol if the system is under flooding attacks.

Conclusions

In this paper, we propose a new approach for detecting flooding attack based on Integrated Entropy Measurement in email server. Our approach can reduce the misjudge rate compared to conventional approaches. By using this approach, we can quickly analyse the current status of the email server, and determine whether the server is attacked by flooding attacks or not. Finally, according to the evolution cost value of email server by using the integrated entropy measurement we proposed, it can detect flooding attacks easily and quickly.

Acknowledgments This work was supported in part by Asia University, Taiwan, under Grant 101-asia-28, and by the National Science Council, Taiwan, Republic of China, under Grant NSC99-2221-E-468-011.

References

1. Postel JB (1982) A simple mail transfer protocol. RFC821
2. Klensin J (2008) A simple mail transfer protocol. RFC5321
3. Myers J, Rose M (1996) Post office protocol—Version 3. RFC 1939
4. Crispin M (1996) Request for comments: 2060. Standards Track, Network Working Group, Dec 1996
5. Cripsin M (2003) Internet message access protocol—version 4rev1. RFC3501
6. Karlton P (2011) Request for comments: 6101. Standards Track, Network Working Group, Aug 2011
7. Wikipedia (2013) HTTP secure. http://en.wikipedia.org/wiki/HTTP_Secure
8. Chen H-C, Sun J-Z, Wu Z-D (2010) Dynamic forensics system with intrusion tolerance based on hierarchical colour petri-nets. In: BWCCA 2010: international conference on broadband and wireless computing, communication and applications, also NGWMN-2010: the third international workshop on next generation of wireless and mobile networks, , Fukuoka, Japan, 4–6 Nov, pp 660–665
9. O'Donnell AJ (2007) The evolutionary microcosm of stock spam. Sec Priv IEEE 5:70–75
10. Bass T, Watt G (1997) A simple framework for filtering queued SMTP email. In: MILCOM 97 proceedings, vol. 3, pp 1140–1144
11. Bass T, Freyre A, Gruber D, Watt G (1998) Email bombs and countermeasure: cyber attack on availability and brand integrity. IEEE Network 12(2):10–17
12. Wang X, Chellappan S, Boyer P, Xuan D (2006) On the effectiveness of secure overlay forwarding systems under intelligent distributed DoS attacks. IEEE Trans Parallel Distrib Syst 17:619–632
13. Shannon CE (1948) A mathematical theory of communication. Bell Syst Tech J 27:379–423, 623–656
14. Absolute Astronomy (2012) Information entropy. Available from: http://www.absoluteastronomy.com/topics/Information_entropy
15. Weaver W, Shannon CE (1963) The mathematical theory of communication, 1949, republished in paperback
16. Chen H-C, Sun J-Z, Tseng S-S, Weng C-E (2012) A new approach for detecting smtpfa based on entropy measurement. In: The 9th IFIP international conference on network and parallel computing (NPC 2012), Gwangju, Korea, 6–8 Sept 2012

A Forecast Method for Network Security Situation Based on Fuzzy Markov Chain

Yicun Wang, Weijie Li and Yun Liu

Abstract In order to solve some problems associated with network security situation forecast, this study proposed a new forecast method based on fuzzy Markov chain. In this work, we focus on forecasting the threat value of network combines historical data of safe behavior with the level of threat. We establish unified information database based on multi-source log data mining techniques. By using text categorization and the threat level division, it is capable of calculating the threat value of a period of time. Due to the discrete nature of the threat values of each time and its unfollow-up effect property, considering the fuzziness of safety state, we use fuzzy Markov chain to predict the threat value in next period of time.

Keywords Network security situation · Log audit · Text mining · Fuzzy Markov chain · Forecast

Introduction

In recent years, how to grasp the network security situation and forecast the development trend has become one of the hottest spots of researches in network emergency response at home and abroad [1, 2]. Traditional network security

Y. Wang
School of Communication and Information Engineering, Beijing Jiaotong University, Beijing 100044, China
e-mail: 12120209@bjtu.edu.cn

W. Li
Key Laboratory of Communication and Information Systems, Beijing Municipal Commission of Education, Beijing Jiaotong University, Beijing 100044, China
e-mail: liwj@itsec.gov.cn

Y. Liu (✉)
China Information Technology Security Evaluation Center, Beijing, China
e-mail: liuyun@bjtu.edu.cn

Y.-M. Huang et al. (eds.), *Advanced Technologies, Embedded and Multimedia for Human-centric Computing*, Lecture Notes in Electrical Engineering 260, DOI: 10.1007/978-94-007-7262-5_108, © Springer Science+Business Media Dordrecht 2014

situation prediction algorithms are based on Naïve Bayes and grey incidence model [3], these methods can only provide network managers past and current network security situation, but cannot predict the next stage of network security status. Markov prediction model is suitable for the data prediction which meets the condition of large fluctuations, many uncertain factors with complex correlation and enough data [4]. Considering the correlation between the security states and the fuzziness of network security status, we use fuzzy Markov chain to predict the network security situation [5].

In this paper, we firstly preprocess the large amount of logs generated by network equipment, convert heterogeneous data formats, establish a unified database structure, and then use the method of text mining to classify all kinds of attacks. We will introduce it in "Log Audit". In "Network Situational Awareness", we will rank the level of threat based on past experience and evaluate the current situation of network security, so as to arrive threaten values in each period. Predication is covered in "Prediction", we use the threaten values of each period to predict the network security situation in next period based on fuzzy Markov chain.

Log Audit

Security situational awareness refers to awareness and access to the secure element in a certain time and space, integrating and analyzing the obtained data, and predicting the future trend based on the results of analysis. In recent years, this technology is gradually applied to computer network. In this paper, we use the log audit to analyze the correlation between various logs, get security event information, and calculate the theoretical security threat value, in order to predict the network security status.

Text Preprocessing

In a computer network, there are a large number of hosts, servers and network equipments, which will generate numerous logs when the system running. There is a certain correlation between these logs, which includes the security event information. However, in the process of information construction, the use of network devices becomes increasingly complicated. This generated a large number of heterogeneous, distributed security audit data, which makes it difficult for data analysis and decision supporting. The solution is preprocessing the large amount of logs by converting heterogeneous data formats and establishing a unified database structure. Because there exist considerable research efforts in this subtask, we will not detail or discuss it in this paper.

Text Mining

Security audit can be seen as a classification problem: we hope to classify each audit records to the possible categories, such as the normal or a particular intrusion or an abnormal operation. The audit log is usually based on the rule base, mathematical statistics, or data mining. In this paper, we use text mining to classify the log entries according to the type of attacks. Typical classification algorithms include Naïve Bayes, KNN and TFIDF algorithm. SVM owns the highest accuracy, but the computation time is too much; Naïve Bayes is the fastest in classifying time as well as the most efficient in terms of system memory usage although it not owns the highest accuracy [6]. As there are huge amount of logs and network security forecast has a strong real-time characteristic, we use Naïve Bayes to classify the log entry in this paper.

According to the Bayes formula $P(c|x) = \dfrac{P(x|c)P(c)}{P(x)}$, the task of Naïve Bayes text classification is to classify the text vector $X(x_1, x_2, \ldots, x_n)$ to its most closely correlated categories $C(c_1, c_2, \ldots, c_j)$, where $X(x_1, x_2, \ldots, x_n)$ is the feature vector of X_q, and $C(c_1, c_2, \ldots, c_j)$ is the given category system. The assumption of Naïve Bayes is that any two terms from $X(x_1, x_2, \ldots, x_n)$ representing a text T and classified under category c are statistically independent of each other. This can be expressed by:

$$P(X|c) = \prod_{i=1}^{n} P(x_i|c) \tag{1}$$

The category predicted for T is based on the highest probability given by:

$$c_{NB} = \arg\max p(x_1, x_2, \ldots, x_n|c_j)P(c_j) = \arg\max P(c_j) \prod_{i=1}^{n} P(x_i|c_j) \quad c_j \in C \tag{2}$$

Network Situational Awareness

There are many types of system attacks, and different security events have different threat level. From the angle of attack level, this paper will divide all attacks into four levels [7] (Fig. 1).

Level 1 includes the denial-of-service attacks and the mail bombs, and DoS includes distributed denial of service attacks, reflective distributed denial of service attacks, the DNS distributed denial of service attack and FTP attacks. Level 2 includes such events, such as local users illegally obtain the read and write permissions. Level 3 is the problems related to that external access to internal documents. Due to the server configuration error, harmful CGI and overflow problem, related vulnerabilities will appear in large numbers. Level 4 occurs in the

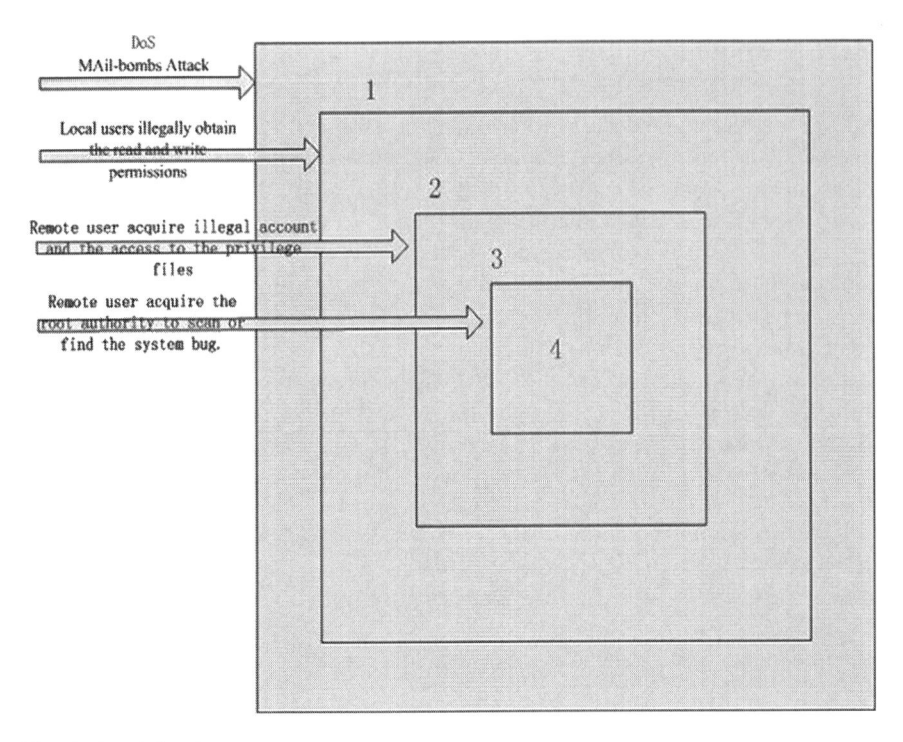

Fig. 1 Attack levels

Table 1 Threaten values of the attack levels

Attack level	0	1	2	3	4
Threaten values	0	2	4	8	10

environment that what happened should never be allowed. Main attacks include TCP/IP continuous theft, the passive channel eavesdrop, packet interception, and attackers gain root privileges.

According to the four attack levels, we can divide different security events into different grades, and assign a weight to each of the attack levels (Table 1).

Where 0 stands for the security incidents, and threaten values of 0 to 10 correspond to the above attack classification. We can use the following formula to normalize the threat level:

$$p(j) = \frac{M(j) - MIN}{MAX - MIN} \tag{3}$$

Where MAX is the supremum of threat of attacks, and MIN is the infimum of the threat of attacks, M(i) is the current threat.

We denote c_{ij} as the security event which belongs to the category i and the attack level j. Such as in Naïve Bayes classification, $c_1, c_2, \ldots, c_m$ are different attacks, their attack level are all 1, then we denote them as $c_{11}, c_{12}, \ldots, c_{1m}$. Thus, the threaten value of network in a period of time can be calculated by the following formula:

$$A = \sum_{j=0}^{4} [p(j) \cdot \sum_{i=1}^{m} c_{ij}] \qquad (4)$$

Thus, we get the threaten value of the network in a period of time.

Prediction

In Part 3, we combined the threaten values of different attacks with the number of attacks, and get the threat value of the network in a period of time, which reflects the network security situation. Network security state is not a specific subset, and we cannot make a clear division to the network security situation according to A-value. So we have to use the prediction method based on fuzzy Markov chain.

Several Concepts of Markov Chain

Markov chain is a mathematical system that undergoes state transitions from one state to another, between a finite or countable number of possible states. It is a random process characterized as memoryless: the next state depends only on the current state and not on the sequence of events that preceded it. This specific kind of "memorylessness" is called the Markov property.

The changes of state of the system are called transitions, and the probabilities associated with various state-changes are called transition probabilities: $p_{ij} = P(X_{n+1} = j | X_n = i)$. The set of all states and the transition probabilities completely characterizes a Markov chain. A Markov chain is a sequence of random variables $x_1, x_2, x_3 \ldots$ with the Markov property, namely that, given the present state, the future and past states are independent.

$$P(X_{n+1} = x | X_1 = x_1, X_2 = x_2, \ldots, X_n = x_n) = P(X_{n+1} = x | X_n = x_n) \qquad (5)$$

The possible values of X_i form a countable set S called the state space of the chain. And transition Matrix can be stated as:

$$P = \begin{bmatrix} p_{00} & p_{01} & p_{02} & \cdots & p_{0j} & \cdots \\ p_{10} & p_{11} & p_{12} & \cdots & p_{1j} & \cdots \\ \vdots & \vdots & \vdots & & \vdots & \cdots \\ p_{i0} & p_{i1} & p_{i2} & \cdots & p_{ij} & \cdots \\ \vdots & \vdots & \vdots & & \vdots & \cdots \end{bmatrix}$$

Fuzzy Markov Chain Model

1. We set random variable X_i as the threaten value A of a period of time of the network security situation, the range is set to U. Set up fuzzy state $\tilde{N}_1, \tilde{N}_2, \ldots, \tilde{N}_r$ on U, in the condition of $\forall x \in U, \sum_{i=1}^{r} \tilde{N}_i(x) = 1$.
2. In period K to period $K + 1$, the fuzzy state transition frequency number of state $\tilde{N}_i$ transfer to state $\tilde{N}_j$ can be expressed as:

$$\tilde{S}_{ij} = \bigvee_{n-1}^{k=1} \left[\tilde{N}_i(x_k) \wedge \tilde{N}_j(x_{k+1}) \right] \tag{6}$$

Fuzzy state transition frequency can be calculated by:

$$\tilde{P}_{ij} = \tilde{S}_{ij} \Big/ \sum_{j=1}^{r} \tilde{S}_{ij}, \ i = 1, 2, \ldots, r \tag{7}$$

Experimental Validation and Analysis

The experimental data set is KDD99 [8], which includes numerous varied invasions simulated in a military network environment and have more than 4,000,000 event items. Each event item contains the feature value extracted from the original network data collected during the simulated invasion stage. We use the data set as the text set which has been classified during the log preprocessing process. We take the top 20 % of KDD99 as our training set and the 1 % of the followed data of KDD99 as our test set. Attacks and their categories are shown in Table 2:

1. DoS(Denial of Service), useslarge amount of legal requests to take up too much service resource.
2. U2R(User to Root), refers to the unauthorized access attacks that local non-privileged user instead of local super-user launched.

Table 2 Attacks and their categories

Category	Attack
DoS	Back, land, neptune, pod, smurf, teardrop
U2R	Buffer_overflow, loadmodule, perl
R2L	ftp_write, guess_passwd, imap, multihop, phf, spy,
PROBE	warezclient, warzmaster, lpsweep, nmap, portsweep

Table 3 Threaten value of each period of time

Period	τ_1	τ_2	τ_3	τ_4	τ_5	τ_6	τ_7	τ_8	τ_9	τ_{10}
Threat Value	1,926	6,859	7,745	8,930	10,000	10,000	7,968	9,270	9,865	7,428
Period	τ_{11}	τ_{12}	τ_{13}	τ_{14}	τ_{15}	τ_{16}	τ_{17}	τ_{18}	τ_{19}	τ_{20}
Threat Value	5,621	4,215	6,098	9,707	13,359	14,695	17,520	11,593	8,596	5,207

3. R2L(Remote to Local) refers to the unauthorized access of the remote host.
4. Probing, means that the attackers scan computers on the network to collect information or to find the known system vulnerabilities.

We assume that the records are evenly distributed in time τ, and the total time of collecting training data set. We divide the total time into 20 equal sections which are stated by $\tau_1, \tau_2, \ldots, \tau_{20}$ respectively, and calculate the network threats of each section, classify threats according to the security level stated in part 3 and compute the threaten value of each period of time. The computed results are shown in the Table 3.

According to the needs of different systems, fuzzy state level setting and membership degree equations are different. The most suitable membership degree equation can be derived from training. In order to illustrate our forecast method, in this paper, we set up fuzzy states on the set of real numbers R as follows, where x is the network threaten value of each period of time.

$$\tilde{N}_1 = \begin{cases} 1 & x \in [0, 1000) \\ \frac{3000-X}{2000} & x \in [1000, 3000) \\ 0 & else \end{cases}$$

$$\tilde{N}_3 = \begin{cases} \frac{X-6000}{2000} & x \in [6000, 8000) \\ 1 & x \in [8000, 11000) \\ \frac{13000-X}{2000} & x \in [11000, 13000) \\ 0 & else \end{cases}$$

$$\tilde{N}_2 = \begin{cases} \frac{X-1000}{2000} & x \in [1000, 3000) \\ 1 & x \in [3000, 6000) \\ \frac{8000-X}{2000} & x \in [6000, 8000) \\ 0 & else \end{cases}$$

$$\tilde{N}_4 = \begin{cases} \frac{X-11000}{2000} & x \in [11000, 13000) \\ 1 & x \in [13000, 16000) \\ \frac{18000-X}{2000} & x \in [16000, 18000) \\ 0 & else \end{cases}$$

$$\tilde{N}_5 = \begin{cases} \frac{X-16000}{2000} & x \in [16000, 18000) \\ 1 & x \in [18000, 50000) \end{cases}$$

Easy to calculate the threat of fuzzy state membership in each period as follows (Table 4):

Based on formula (6) and (7), it is easy to get Tables 5 and 6:

Thus, we can get the fuzzy state transition Markov chain:

Table 4 Fuzzy state membership in each period

	τ_1	τ_2	τ_3	τ_4	τ_5	τ_6	τ_7	τ_8	τ_9	τ_{10}
$\tilde{N}_1$	0.537	0	0	0	0	0	0	0	0	0
$\tilde{N}_2$	0.463	0.5705	0.1275	0	0	0	0.016	0	0	0.286
$\tilde{N}_3$	0	0.4295	0.8725	1	1	1	0.984	1	1	0.714
$\tilde{N}_4$	0	0	0	0	0	0	0	0	0	0
$\tilde{N}_5$	0	0	0	0	0	0	0	0	0	0
	τ_{11}	τ_{12}	τ_{13}	τ_{14}	τ_{15}	τ_{16}	τ_{17}	τ_{18}	τ_{19}	τ_{20}
$\tilde{N}_1$	0	0	0	0	0	0	0	0	0	0
$\tilde{N}_2$	1	1	0.951	0	0	0	0	0	0	1
$\tilde{N}_3$	0	0	0.049	1	0	0	0	0.7035	1	0
$\tilde{N}_4$	0	0	0	0	1	1	0.24	0.2965	0	0
$\tilde{N}_5$	0	0	0	0	0	0	0.76	0	0	0

Table 5 Fuzzy state transition frequency number

$\tilde{s}_{11} = 0$	$\tilde{s}_{12} = 0.537$	$\tilde{s}_{13} = 0.4295$	$\tilde{s}_{14} = 0$	$\tilde{s}_{15} = 0$
$\tilde{s}_{21} = 0$	$\tilde{s}_{22} = 0.714$	$\tilde{s}_{23} = 0.951$	$\tilde{s}_{24} = 0$	$\tilde{s}_{25} = 0$
$\tilde{s}_{31} = 0$	$\tilde{s}_{32} = 1$	$\tilde{s}_{33} = 1$	$\tilde{s}_{34} = 1$	$\tilde{s}_{35} = 0$
$\tilde{s}_{41} = 0$	$\tilde{s}_{42} = 0$	$\tilde{s}_{43} = 0.24$	$\tilde{s}_{44} = 1$	$\tilde{s}_{45} = 0.76$
$\tilde{s}_{51} = 0$	$\tilde{s}_{52} = 0$	$\tilde{s}_{53} = 0.7035$	$\tilde{s}_{54} = 0.2965$	$\tilde{s}_{55} = 0$

Table 6 Fuzzy state transition frequency

$\tilde{p}_{11} = 0$	$\tilde{p}_{12} = 0.5556$	$\tilde{p}_{13} = 0.4444$	$\tilde{p}_{14} = 0$	$\tilde{p}_{15} = 0$
$\tilde{p}_{21} = 0$	$\tilde{p}_{22} = 0.4183$	$\tilde{p}_{23} = 0.5817$	$\tilde{p}_{24} = 0$	$\tilde{p}_{25} = 0$
$\tilde{p}_{31} = 0.1252$	$\tilde{p}_{32} = 0.2916$	$\tilde{p}_{33} = 0.2916$	$\tilde{p}_{34} = 0.2916$	$\tilde{p}_{35} = 0$
$\tilde{p}_{41} = 0$	$\tilde{p}_{42} = 0$	$\tilde{p}_{43} = 0.12$	$\tilde{p}_{44} = 0.5$	$\tilde{p}_{45} = 0.38$
$\tilde{p}_{51} = 0$	$\tilde{p}_{52} = 0$	$\tilde{p}_{53} = 0.7035$	$\tilde{p}_{54} = 0.2965$	$\tilde{p}_{55} = 0$

$$\tilde{P} = \begin{pmatrix} 0 & 0.5556 & 0.4444 & 0 & 0 \\ 0 & 0.4183 & 0.5817 & 0 & 0 \\ 0.1252 & 0.2916 & 0.2916 & 0.2916 & 0 \\ 0 & 0 & 0.12 & 0.5 & 0.38 \\ 0 & 0 & 0.7035 & 0.2965 & 0 \end{pmatrix}$$

The threat value of the twentieth section is 5,207 and the subordination allocation vector of its fuzzy state is $(0, 1, 0, 0, 0)$, so we can presume the probabilities of the network threat value of the next time section belong to the five states respectively according to the transition probability matrix $\tilde{P}$, as shown in the formula:

$$(0, 1, 0, 0, 0)\tilde{P} = (0, 0.4183, 0.5817, 0, 0) \tag{8}$$

According to the maximum subordination principle, it is easily to be seen that the threat value of the next time section most possibly belongs to the fuzzy state $\tilde{N}_3$.

In the test set, the actual value of the next period is 7041 and the security situation of the experiment value and the actual value of the next period both belong to the same fuzzy state $\tilde{N}_3$.

Conclusions and Future Work

We have introduced a method for forecasting the network security situation based on fuzzy Markov chain. We firstly preprocessed the large amount of logs and established a unified database structure, and used the text mining to classify all kinds of attacks. Based on the past experience, we assigned to each attack a different weight according to their danger degrees. Through the calculation, we obtained the Markov state transition matrix, and realized the prediction of the network security situation. Because the attacks are correlative, this method showed a good predictive performance.

Different systems have different requirements, and the fuzzy membership function is different from each other. We can always train the best membership function according to the training set. Another interesting aspect or further study is how to deal with the unknown threats. Our current approach is based on the classification of known threats. However, it would be desirable to explore other more sophisticated approaches.

Acknowledgments This work has been supported by the National Natural Science Foundation of China under Grant 61172072, 61271308, the Beijing Natural Science Foundation under Grant 4112045, the Research Fund for the Doctoral Program of Higher Education of China under Grant W11C100030, the Beijing Science and Technology Program under Grant Z121100000312024.

References

1. Onwubiko C (2009) Functional requirements of situational awareness in computer network security. In: IEEE international conference on on intelligence and security informatics ISI'09, pp 209–213
2. Wei-wei X, Hai-feng W (2010) Prediction model of network security situation based on regression analysis. In: IEEE international conference on wireless communications, networking and information security (WCNIS 2010), pp 616–619
3. Jibao L, Huiqiang W, Liang Z (2006) Study of network security situation awareness model based on simple additive weight and grey theory. in: International conference on computational intelligence and security, pp 1545–1548
4. Zhang Y, Tan XB, Cui XL et al (2011) Network security situation awareness approach based on Markov game model. J Software 22(3):495–508

5. Kuang GC, Wang XF, Yin LR (2012) A fuzzy forecast method for network security situation based on Markov. In: International conference on computer science and information processing (CSIP 2012), pp 785–789
6. García Adeva JJ, Pikatza Atxa JM (2007) Intrusion detection in web applications using text mining. Eng Appl Art Intell 20(4):555–566
7. http://www.ibm.com/developerworks/cn/security/se-parseatt/
8. http://www.sigkdd.org/kdd-cup-1999-computer-network-intrusion-detection

IT Architecture of Multiple Heterogeneous Data

Yun Liu, Qi Wang and Haiqiang Chen

Abstract This paper studies the multiple heterogeneous data IT infrastructure issues in Internet, focusing on the data storage management, the integration analysis and the efficient data transmission. On this basis, we propose architectural solutions and key technology programs which can effectively solve and handle the problems of large data trend analysis and risk assessment. Based on the characteristics of big data, considering the flexibility and scalability of the framework and taking in-depth mining the potential value of big data as the future goal, this paper studies and designs data storage and processing framework which can be adapted to developing more application functionality and meet future demand for Internet services and new business for big data processing requirements.

Keywords Big data · Data mining · Multiple heterogeneous data · Data warehouse

Y. Liu (✉) · Q. Wang
School of Communication and Information Engineering, Beijing Jiaotong University,
Beijing 100044, China
e-mail: liuyun@bjtu.edu.cn

Q. Wang
e-mail: 12125042@bjtu.edu.cn

Y. Liu · Q. Wang
Key Laboratory of Communication and Information Systems, Beijing Municipal
Commission of Education, Beijing Jiaotong University, Beijing 100044, China

H. Chen
China Information Technology Security Evaluation Center, Beijing, China
e-mail: chenhq@itsec.gov.cn

Y.-M. Huang et al. (eds.), *Advanced Technologies, Embedded and Multimedia for Human-centric Computing*, Lecture Notes in Electrical Engineering 260, DOI: 10.1007/978-94-007-7262-5_109, © Springer Science+Business Media Dordrecht 2014

Introduction

The rapid growth of the number of data, to a large extent, is promoted by the revolutionary network applications, such as micro-blog, e-commerce, messaging, social networking and location-related information search and aggregation. Such kind of big data not only has mass and high-speed characteristics, but also the diversity and variability. The diversity also known as heterogeneity, is considered to be generated for the widely using of Internet search and another reason is the new multi- structured data, including the data types of blogs, social media, Internet, phone call records and sensor networks. Especially sensors, each sensor increases the diversity of the data, due to the different installation locations and functions. Variability of big data is reflected in the multi-layer structure and varied forms of the data.

The larger scale data, the processing and storage is more difficult, but its mining potential value is greater. Large data contains the value, but there are still many difficulties and challenges in the use of big data technology. Due to the potential value of big data and the features of specific business and data source, this paper studied and proposed a multiple heterogeneous data IT infrastructure which is appropriate to the specific business needs. We focus on studying the key technologies such as the data storage management, the integration of analysis and the efficient data transmission, and present technical architecture level solutions and key technical solutions based on the data trend analysis and the risk assessment. Based on the characteristics of big data, considering the flexibility and scalability of the framework and taking in-depth mining the potential value of big data as the future goal, this paper studies and designs data storage and processing framework which can be adapted to developing more application functionality and meet future demand for Internet services and new business for big data processing requirements.

Related Work

Google launched its own distributed file system GFS [1] in 2003. GFS is a scalable distributed file system. It is a dedicated file system designed to store massive amounts of search data. GFS is used for large-scale, distributed, large amounts of data access applications and runs on inexpensive commodity hardware, but can provide fault tolerance, and higher overall performance of the services provided to a large number of users. Dean et.al. [2], proposed a parallel computing software framework for large data sets (more than 1 TB)—MapReduce. Chang et.al. [3] proposed a framework Bigtable. Bigtable is a distributed structured data storage system, which is designed to handle massive amounts of data: Usually the PB level data which is distributed in the thousands of ordinary servers. Current most of mainstream data processing frameworks are customized and improved according to Hadoop [4, 5] which is developed based on the early Google's GFS/MapReduce core idea. These customizations and improvements some focus on improving system reliability and manageability, and some focus on improving the processing speed of the system and reducing the delay, while others focus on considering the data processing capabilities of the system, as well as the specific hardware or service—specific optimization. The difficulties faced by a big data processing focus on three aspects of large-capacity, heterogeneity and high-speed, and thus the core technologies focus to address these issues. In addition, big data usually has the characteristic of multi-source diversity, the effective integration of multivariate data is the key link of fully mining the data value.

Contributions

In this paper, we study the technologies of large data storage, integration of heterogeneous data, real-time computing architecture and efficient data transmission mode based on the characteristics of big data and the existing big data technologies. Then we presented key technologies and key considerations of big data system and proposed referenced big data system architecture.

Multiple Heterogeneous Data IT Infrastructure

Multiple heterogeneous big data IT infrastructure mainly consists of four parts: the heterogeneous big data tiered storage and storage management, the integration of heterogeneous data, the off-line and real-time computing architecture, the efficient transmission mode of big data. Typical big data frameworks are based on MapReduce and Hadoop and combined with the specific services to be designed. In this paper, we designed a complete technical architecture according to the multiple heterogeneous data processing and storage requirements of the upper services. This technical structure consists of the underlying file system, structured data storage system, semi- structured data and unstructured data storage system, the distributed storage of data, the unified data access interface, data indexing and positioning, processing task decomposition and scheduling management, e distributed execution of processing tasks, e service interface of the system and the secondary development interface and system manageability and security and completely defined a data storage and processing platform meeting the specific needs. In the following paper, we will present the four aspects of our big data framework respectively.

Multiple Heterogeneous Data Tiered Storage and Storage Management

Multiple heterogeneous big data have the characteristics of multiple dimensions, multiple sources, structured + semi-structured + unstructured heterogeneous data and the huge amount of data, thus the traditional relational database has been very difficult to store such data. In addition, the demand for the processing of such data is not limited to the mode supported by SQL, but shows the features of the connection and interweaving between the data on the Internet. There are a lot of storage methods for heterogeneous data, such as serial method, graph method, tree methods, file system method, database field methods and object methods. Due to the difference between the services of application layer, the characteristics of multi-source heterogeneous data also are different with different services, so our data storage and management must use corresponding methods for differing services, for example, Bigtable, the Keyvalue of NoSQL and so on. We use the Multiple heterogeneous data tiered storage as our storage method.

Big data needs to mass storage, so we cannot storage and manage them like the traditional solutions treating them as equally important. However, Hierarchical storage is using different storage methods corresponding to the characteristics and values of different data and utilize the hierarchical storage management software to achieve automatic sorting, automatic storage, which greatly increases the effectiveness and speed of data storage and meets the storage needs of different kinds of data. In our system, there are four kinds of data which are static data, dynamic data, offline data and online data respectively. The framework of hierarchical storage is shown in Fig. 1.

In order to achieve the storage of big data, especially the Heterogeneous data storage of the structured data, semi-structured data and unstructured data, we superimpose a relational database and a distributed non-relational database on a distributed file system to store huge amounts of data. In the file system level, data are managed by the master + multi-slave physical structure. The master node is used to store metadata, and slave nodes are used to store the actual data. Logical data level is also constituted by the same master + multi-slave structure, where the master node stores metadata, making the access to RDBMS and NoSQL is consistent and also playing the role of slave control. Slave nodes in the logical data level are used to store logical data. Distributed file system is responsible for the management and maintenance of the persistence and positioning of the data on the storage medium and provides read and write physical media I/O for the logical data layer. The logical data layer is responsible for the management of the logical structure of the data and provides services of access to the logical data for the high level. Figure 2 shows us the storage framework of big data.

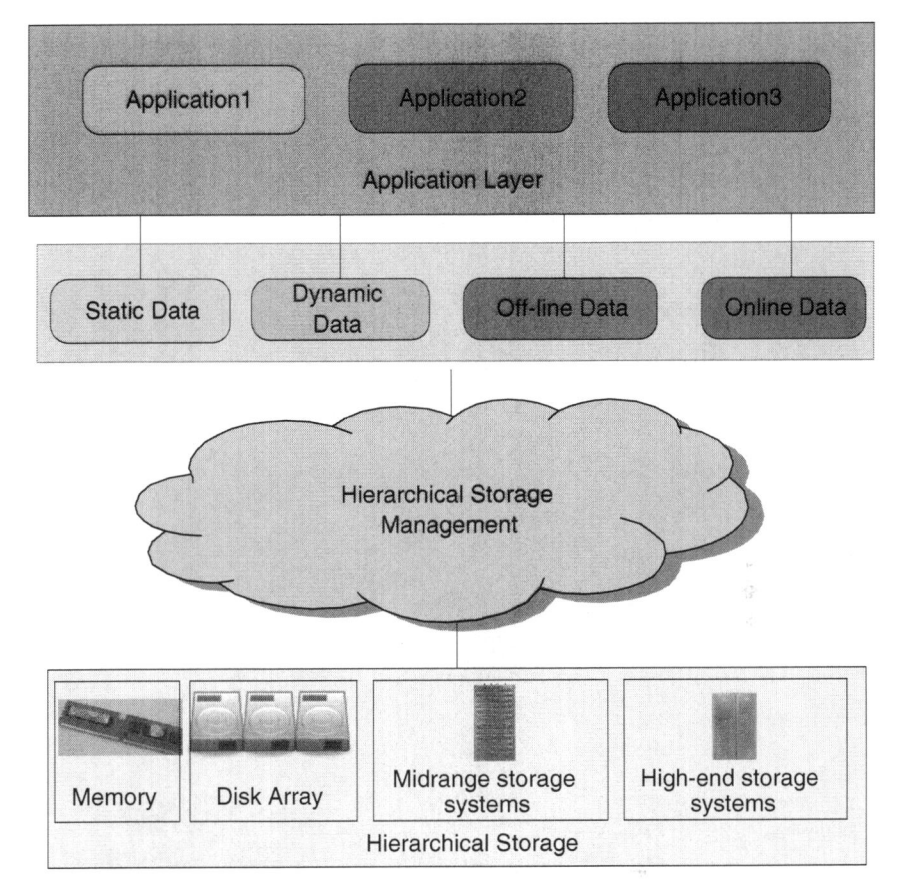

Fig. 1 Hierarchical storage diagram

Heterogeneous Data Fusion

Heterogeneous data fusion is the core part of the deal with heterogeneous data. Fusion of the data can be divided into three levels of pixel-based fusion, feature-level fusion and decision fusion. The pixel-based fusion also has a function similar to the data de-noising and the cleaning process in addition to the significance of fusion. Feature-level fusion classifies data based on the features and data attributes. Decision Fusion uses data to implement the trend assessment and the macro perspective service processing at the highest level. According to the specific services and the specific characteristics of the data, the corresponding fusion algorithms and computational structures are used to treat on different data. Utilizing data fusion technology, we can remove redundant information to reduce the amount of data increase the access efficiency. Moreover, we can extract useful

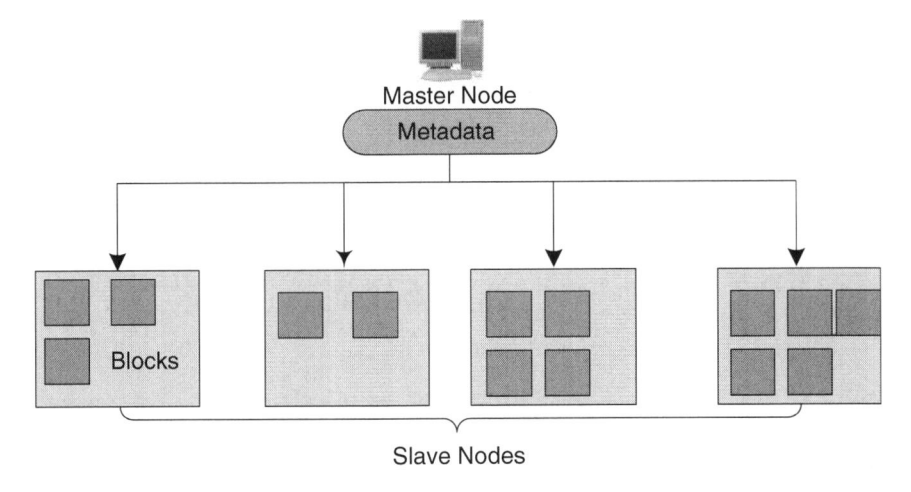

Fig. 2 Distributed file system structure

information from a large number of heterogeneous data, providing services and support for the business of the upper services. Figure 3) shows the functional model of heterogeneous data fusion.

Fig. 3 Heterogeneous data fusion diagram

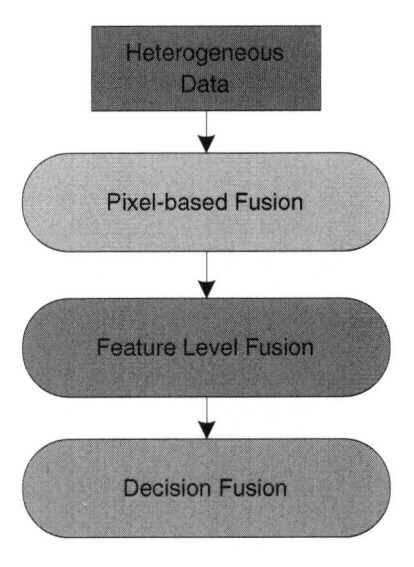

Off-line and Online Computation Framework

There are off-line and online computing demands in a large data environment. Off-line computation is more relaxed in the computing time requirement. However, under the conditions of mass data, there are still computing time limit and the balance between recalculation and the incremental calculation becomes an inevitable problem with the increasing of data. Online calculation has a higher requirement on the computing time, which is a quit hard problem for the mass data and needs to a variety of methods to guarantee the calculation real-time.

In this paper, we class the data into static data and dynamic data. Static data are the historical read-only data, but dynamic data are the read and write data including intermediate result. For static data, when we implement off-line computation, we must design a reasonable storage structure and create effective index to improve the processing efficiency according to the demands of specific services. Hadoop use the idea of Mapreduce. It slices the data to deal with large amount of off-line data and then assign computing tasks to more than one computer, which will greatly improve the speed and efficiency of computation. Online computation needs to use reasonable caching mechanism to solve the massive data processing problem. For dynamic data, we will combine data structure designing, index designing and caching mechanism designing together to complete the data processing. Storm is a common stream computing framework which has been widely used for real-time log processing, real-time statistics, real-time risk control. Storm also is used to process the data preliminarily and it can store the data in a distributed database like HBase in order to facilitate the subsequent queries. Storm is a scalable, low-latency, reliable and fault-tolerant distributed computing platform. In addition, we will employ the distributed computing mode for data computation and processing and the assignment, scheming and management of the computing tasks. The off-line and online computation framework is shown in Fig. 4.

Fig. 4 Off-line and online computation

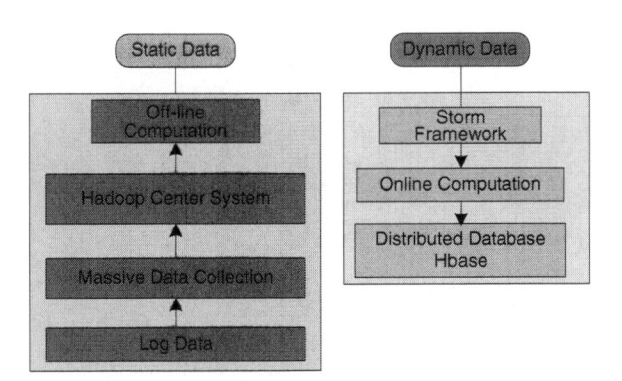

Efficient Interactive Transmission Mode of Big Data

Storage and processing of big data necessarily involves the network-based distributed storage and computation, the collection of multi-source data, the remote data management on the basis of network and data sharing and exchange. So Data exchange and interactive data service is an indispensable supporting technology.

In this paper, we designed the data exchange framework according to two modes of data service and heterogeneous data exchange. Data service is established data access function for large-scale predefined data transferring and migration and heterogeneous data exchange refers to the data exchange between the different modules within a framework and clients. Data service mode uses connection-oriented channel exchange. However, Heterogeneous Data uses connectionless datagram exchange and defines the structure of datagram by xml or json.

The framework defines three types of service interfaces for the data interactive interface: the interfaces between the modules within the system, which include two modes of service and API; the external interfaces of the system functions, which appear in the form of API and are used for the developing and data access of the service applications; interactive service interfaces, which have two modes of service and REST API. System interfaces can support each other to form the overall function. The external interfaces in the form of API can be used for the developing the service applications and REST API can be used for extending the service applications of the system and create a more convenient remote data access. Data exchange modes are shown in Fig. 5.

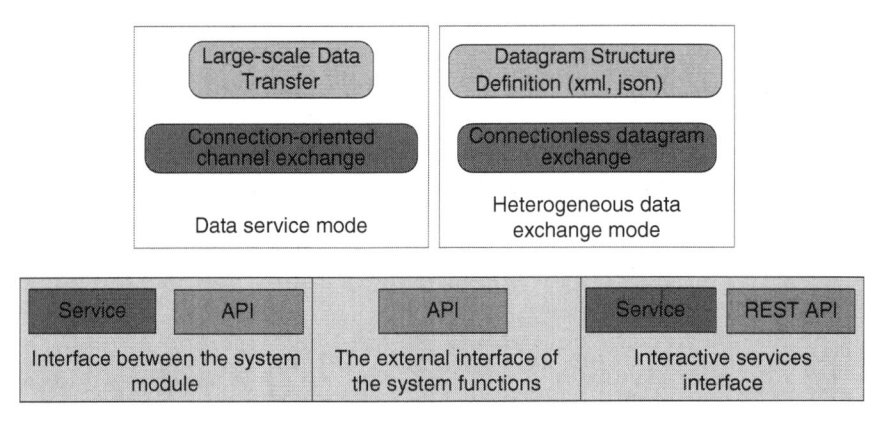

Fig. 5 Data exchange module structure

Conclusion

In this paper we talked about the framework of multiple heterogeneous big data from four aspects of the storage management, the data fusion, the online and offline computation and the efficient interactive transmission mode of big data. We then proposed a framework of multiple heterogeneous data which can be adapted to developing more application functionality and meet future demand for Internet services and new business for big data processing requirements.

Acknowledgments This work has been supported by the National Natural Science Foundation of China under Grant 61172072, 61271308, the Beijing Natural Science Foundation under Grant 4112045, the Research Fund for the Doctoral Program of Higher Education of China under Grant W11C100030, the Beijing Science and Technology Program under Grant Z121100000312024.

References

1. Ghemawat S, Gobioff H, Leung ST (2003) The google file system. ACM SIGOPS Oper Syst Rev 37(5):29–43
2. Dean J, Ghemawat S (2008) MapReduce: simplified data processing on large clusters. Commun ACM 51(1):107–113
3. Chang F, Dean J, Ghemawat S et al (2008) Bigtable: a distributed storage system for structured data. ACM Trans Comput Syst (TOCS) 26(2):4
4. Hadoop: the definitive guide. O'Reilly Media, California
5. White T (2012) Hadoop: the definitive guide. O'Reilly Media, California

An Implementation Scheme of BB84-Protocol-Based Quantum Key Distribution System

Shao-sheng Jiang, Rui-nan Chi, Xiao-jun Wen and Junbin Fang

Abstract The BB84 protocol is a secure communication solution that has been proved to be unconditionally secure. The quantum key distribution system that is based on this protocol has been gradually developed towards the application stage. This paper proposes a scheme to realize quantum key distribution system based on BB84 protocol using weak coherent light source, polarization controller and delay multiplexing detection technology to prepare single photon, to design photon polarization states orthogonal to each other, and to detect single photon respectively.

Keywords Quantum key distribution · Single photon signal · Photon polarization states · Delay multiplexing detection

Introduction

Quantum key distribution system is formed through the operation of quantum bits on the basis of the inherent properties of such quantum bits. Because the quantum bits comply with the no-cloning theorem and the uncertainty principle, the quantum key distribution system theoretically has the capability of unconditional security protection, and thus the technology and scheme for its realization are considered as an important direction in the research of present information security.

S. Jiang · R. Chi · X. Wen (✉)
School of Computer Engineering, ShenZhen POLYTECHNIC, Shenzhen 518055, China
e-mail: szwxjun@sina.com

J. Fang (✉)
Department of Optoelectronic Engineering, Jinan University, Guangzhou 510632, China
e-mail: junbinfang@gmail.com

Y.-M. Huang et al. (eds.), *Advanced Technologies, Embedded and Multimedia for Human-centric Computing*, Lecture Notes in Electrical Engineering 260, DOI: 10.1007/978-94-007-7262-5_110, © Springer Science+Business Media Dordrecht 2014

BB84 protocol is the first quantum key distribution protocol proposed by Bennett and Brassard. It is a four-state scheme based on two conjugate bases, which uses complementarity of linearly polarized photon state and circularly polarized photon state to satisfy the uncertainty principle in single photon quantum channel so as to ensure the unconditional security of quantum channel, and then complete the distribution of quantum key through the quantum transmission channel, quantum measurement channel and classic auxiliary channel [1]. Thus, the realization of quantum key distribution system mainly depends on the relevant technology design in the preparation, transformation and detection of quantum bits.

At present, the existing literatures revealed the research on the security of BB84 protocol of quantum key distribution system [2]. For system realization, Capraro et al. used Java to simulate the BB84 system [3], and Chen et al. used the method of polarization feedback compensation to achieve the BB84 protocol polarization encoding key distribution experiments [4]. This paper mainly studies the realization scheme of BB84-protocol-based quantum key distribution system; that is, by fully taking physical features on which the key system is based into account in the design, the key distribution system is ultimately achieved as a basis for the completion of all kinds of application based on such a system.

Generation of Quantum Signal

Realization of quantum key is based on the physical laws of quantum bits. According to quantum bits attribute, classical signals obviously do not have conditions to carry the quantum bits, so only quantum signals are used to complete [1] The quantum signal is the evolution of quantum system with a particular parameter, and can be described by Schrodinger equations as follows:

$$p = \frac{h}{j}$$ (1)

General, considering the quantum system's characteristics of physics value A satisfies the following relations:

$$A \sim h$$ (2)

Quantum efficiency cannot be ignored, and then A can be used as quantum signal. Where h is the normalized Planck constant. If $A \gg h$, A is described as the classical signal. Micro-systems all have the obvious quantum effects, so all these systems can be used as a carrier of quantum signals.

Because laser signal is relatively excellent in fiber transmission, a single photon is inseparable, so a single photon is an ideal carrier of quantum signals. The evolution of photons can be described by a state vector $|\Psi\rangle$, so $|\Psi\rangle$ can be used to describe the quantum signals that are carried, referred to as the single photon sign.

But the preparation of single photon signal in the technical realization is difficult, so considering realizing single photon signal with weak laser pulse to be emitted by common lasers. The average photon number in a single laser pulse shall satisfy the following conditions:

$$\overline{N} \leq 0.1 \tag{3}$$

In the above formula $\overline{N}$ represents the average photon number in a single laser pulse. Calculation shows that, the laser pulse signal in this case has obvious quantum effect, with features approximated to the real single photon quantum signal.

The specific realization method is: design a quantum signal transmitter that can randomly emit four pulse lasers at the frequency of 1 MHz, and the center wave length is 850 nm, so each photon energy is

$$E_p = \frac{hc}{\lambda} = 2.337 \times 10^{-19} W \tag{4}$$

The number of photons in each pulse is

$$n = \frac{E}{E_p} = 100 \times 10^{12} / (2.337 \times 10^{-19}) = 4.279 \times 10^8 \tag{5}$$

In order to ensure the photon number per pulse satisfying $N \leq 0.1$, attenuation is required. When using an attenuator with multiple attenuation of 96 dB, the average photon number per pulse becomes

$$\overline{N} = n \times 10^{-10} = 0.043 < 0.1 \tag{6}$$

Therefore, the single photons signals are generated.

Transformation of Quantum Signals

According to the protocol of BB84, four single photon signals produced by quantum signal transmitting mechanism must be transformed into two orthogonal polarization states of photons, including one pair of linear polarized photon and another pair for circularly polarized photons, which meets the uncertainty principle [5]. However, concerning that the preparation and transformation of circularly polarized photons in implementation is difficult, we design two pairs of linearly polarized photons, as long as it meets two pairs of photon state that are not mutually orthogonal and undistinguishable. The designed diagram is shown in Fig. 1:

Fig. 1 Schematic diagram of
the quantum signal transform

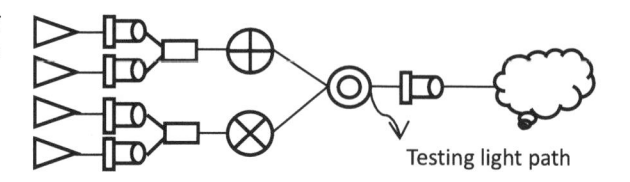

In the above diagram:

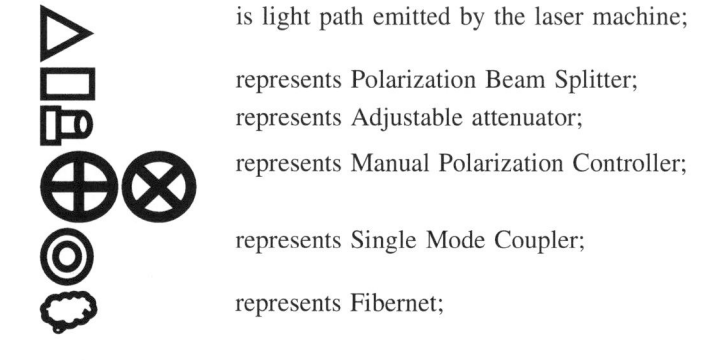

is light path emitted by the laser machine;

represents Polarization Beam Splitter;

represents Adjustable attenuator;

represents Manual Polarization Controller;

represents Single Mode Coupler;

represents Fibernet;

First of all, we set two Polarization Beam Splitters (hereinafter referred to as PBS), which will be used to provide two photonic inputs for beam, with each polarization direction perpendicular to each other, thus forming the two orthogonal polarization states of photons.

Then we design two Manual Polarization Controllers (hereinafter referred to as MPC), which will be used to design two photon states mutually non-orthogonal. Here only let one pair of photons go through MPC rotation at the angle of 45° and then transfer, and another pair of the photons is transferred with the original mutual vertical states. Two MPC can also make two pairs of photons by adjusting the polarization feedback to well compensate the light's polarization changes in the path.

Finally, we use the Single Mode Coupler (hereinafter referred to as SMC) to form four kinds of polarized light into a beam, as a sender of quantum key distribution system.

Another important thing is to form a single photon pulse in the implementation. As mentioned before, the attenuation of polarized light into 96 dB is required. The direct attenuation with large coefficient is neither safe nor conducive to regulate the system, so generally we use 2 ∼ 3 concatenated attenuators for attenuation, and an optical power meter for step-by-step measurement, until it finally reaches the total attenuation of 96 dB. Usually after the emissions of four paths of light, the first attenuation is about 50 dB, and afterwards the attenuation is about 46 dB, and the error can be allowed within the range of 1 dB.

The testing light path in the diagram is used to help calibrate the polarization direction of MPC. The two properly-set MPCs can make the beam of two mutually perpendicular polarized light being prepared into $|0°>$, $|90°>$, $|+45°>$, $|−45°>$ four kinds of polarized light (later we referred to these four kinds of light condition for $\leftrightarrow$, $\updownarrow$, $\nearrow$, $\searrow$) and complement each other, so as to meet the requirements of the implementation of BB84 protocol.

Detection of Quantum Bits

The receiver of quantum key must be a photoelectric detection system, which can detect very weak quantum signals, and identify the quantum key carried by quantum signals. In the system only a single photon detector is set, by using avalanche effect to probe photons. Electric pulse signal output is processed by the main control plate, and after processing the information is transferred to a computer, with the sender completing basis vectors alignment by classical network. The realization diagram of the receiver is designed as Fig. 2:

Above $\leftharpoondown$ indicates a 2 m-long jumper, we use the jumper with multiple SMCs to make four polarization-state photon line to form a beam into a single photon detector, then use delay multiplexing technology to separately detect photons of various polarization states. As shown in Fig. 2, through the optical fiber transmission, a single photon will randomly select a path in the first SMC transmission. (For example, the photon of $\leftrightarrow$ photon polarization state had a 50 % probability of going to the wrong path, and by choosing $\otimes$ state MPC, it is equivalent to the receiver choosing a wrong measurement basis vector, this part of detection signals will be discarded directly. When the photons of polarization states select correct measurement basis vectors, the MPC is mainly used for polarization mode dispersion for feedback compensating polarized light in the system, which makes it consistent with the polarization beam direction.)

In Fig. 2, the main function of PBS is to split polarized light of the two mutually perpendicular states into beams. For the detection of $\leftrightarrow$, $\updownarrow$, $\nearrow$, $\searrow$ four states of polarized light, the optical receiver increases 0, 2, 4, 6 m jumpers

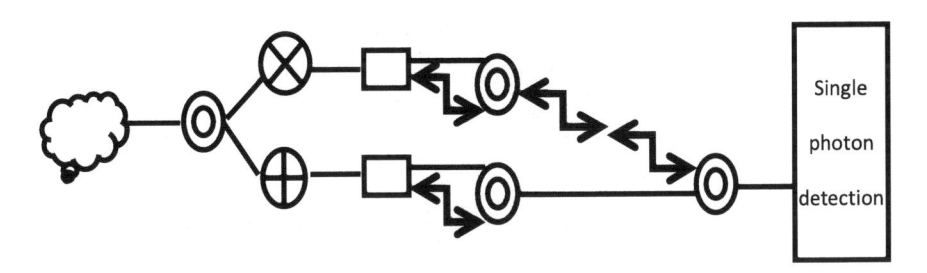

Fig. 2 Schematic diagram of the realization

respectively. If adopting the conversion of 1 m to 5 ns, i.e., delay of X + 0 ns for detecting $\leftrightarrow$ photons, delay of X + 10 ns for detecting $\updownarrow$ photons, delay of X + 20 ns for detecting $\diagup$ photons, and delay of X + 30 ns for detecting $\diagdown$ photon. The X value is the value acquired by the laser pulse detector in roughly scanning at the delays.

Conclusion

The BB84 protocol is the first quantum cryptography communication agreement on the basis of single photon of a four-state scheme based on two conjugate bases. This paper proposes an implementation technology and realization circuit schematic diagram based on the protocol, with the key components including the preparation of single photon signals, the design of two pairs of orthogonal polarization of photons, the delay multiplexing detection technology of single photon, which has been realized in the laboratory. It has been proved to be unconditionally secure [5, 6]. At present, many applications have been realized on the basis of the key distribution system, which has the very good application and practical value.

References

1. Zeng GH (2006) Quantum cryptography. Science Press, Beijing
2. Chen ZX, Tang ZL, Liao CJ et al (2006) Analysis of BB84 protocol security eavesdropping extended practical quantum key distribution. Acta Photonica Sinica 35:126–129
3. Ivan C, Tommaso O (2007) Implementation of real time high level protocol software for quantum key distribution. In: IEEE international conference on signal processing and communications, Dubai, United Arab Emirates, pp 24–27
4. Chen J, Li Y, Wu G et al (2007) Quantum key distribution in polarization stability control. J Phys 56:5243–5247
5. Wen XJ, Chen YZ (2012) Quantum signature and its application. Aviation Industry Press, Beijing
6. Liu JF, Liang RS, Tang ZL, et al (2004) Eavesdropping of practical QKD system based on BB84 protocol. Acta Photonica Sinica 33:1356–1359

Precise Abdominal Aortic Aneurysm Tracking and Segmentation

Shwu-Huey Yen, Hung-Zhi Wang and Hao-Yu Yeh

Abstract In this paper we propose a mean-shift based technique for a precise tracking and segmentation of abdominal aortic aneurysm (AAA) from computed tomography (CT) angiography images. Output data from the proposed method can be used for measurement of aortic shape and dimensions. Knowledge of aortic shape and size is very important for selection of appropriate stent graft device for treatment of AAA. Comparing to conventional approaches, our method is very efficient and can save a lot of manual labors.

Keywords Aneurysm · Abdominal aortic aneurysm · Mean-shift · Computerized tomorgraphy

S.-H. Yen (✉) · H.-Z. Wang · H.-Y. Yeh
Department of Computer Science and Information Engineering, Tamkang University, New Taipei, Taiwan, Republic of China
e-mail: 105390@mail.tku.edu.tw

H.-Z. Wang
e-mail: 600410640@s00.tku.edu.tw

H.-Y. Yeh
e-mail: 601411191@s01.tku.edu.tw

Y.-M. Huang et al. (eds.), *Advanced Technologies, Embedded and Multimedia for Human-centric Computing*, Lecture Notes in Electrical Engineering 260, DOI: 10.1007/978-94-007-7262-5_111, © Springer Science+Business Media Dordrecht 2014

Part X
Advances in Multimedia Algorithms, Architectures, and Applications

Mining High Utility Itemsets Based on Transaction Deletion

Chun-Wei Lin, Guo-Cheng Lan, Tzung-Pei Hong and Linping Kong

Abstract In the past, an incremental algorithm for mining high utility itemsets was proposed to derive high utility itemsets in an incrementally inserted way. In real-world applications, transactions are not only inserted into but also deleted from a database. In this paper, a maintenance algorithm is thus proposed for reducing the execution time of maintaining high utility itemsets due to transaction deletion. Experimental results also show that the proposed maintenance algorithm runs much faster than the batch approach.

Keywords Utility mining · Maintenance · Transaction deletion · Two-phase approach · FUP concept

C.-W. Lin (✉) · L. Kong
IIIRC, School of Computer Science and Technology, Institute of Technology Shenzhen
Graduate School, Xili, Shenzhen, People's Republic of China
e-mail: jerrylin@ieee.org

L. Kong
e-mail: konglingping@utsz.edu.cn

C.-W. Lin
Shenzhen Key Laboratory of Internet Information Collaboration Harbin, Institute of
Technology Shenzhen Graduate School, HIT Campus Shenzhen University Town, Xili,
Shenzhen, People's Republic of China

G.-C. Lan
Department of Computer Science and Information Engineering, National Cheng Kung
University, Tainan, Taiwan, Republic of China
e-mail: rrfoheiay@gmail.com

T.-P. Hong
Department of Computer Science and Information Engineering, National University of
Kaohsiung, Kaohsiung, Taiwan, Republic of China
e-mail: tphong@nuk.edu.tw

T.-P. Hong
Department of Computer Science and Engineering, National Sun Yat-sen University,
Kaohsiung, Taiwan, Republic of China

Y.-M. Huang et al. (eds.), *Advanced Technologies, Embedded and Multimedia
for Human-centric Computing*, Lecture Notes in Electrical Engineering 260,
DOI: 10.1007/978-94-007-7262-5_112, © Springer Science+Business Media Dordrecht 2014

Introduction

Association rules mining from binary database is the fundamental approach for knowledge discovery [1–4]. In some applications, frequent itemsets may only contribute a small portion to the overall profit, while non-frequent ones may contribute a large portion to the profit [5]. In the past, utility mining [6] was thus proposed to partially solve the above limitations. It may be thought of as an extension of frequent itemset mining with the sold quantities as a quantitative database with the item profits considered [5, 7–10].

In real-world applications, transactions in the database do not usually remain in a stable condition. Some transactions may be inserted or deleted from an original database. The discovered frequent itemsets may become invalid, or some new implicit information may emerge in the whole updated database. Lin et al. thus proposed an incremental algorithm for transaction insertion in high utility mining [11] based on the Fast UPdated (FUP) concepts [12]. In addition to transaction insertion, transaction deletion is also commonly seen in real-world applications. In this paper, a maintenance algorithm is thus proposed to update the mined utility itemsets for deleted transactions. The FUP2 (Fast UPdated) algorithm [13] is then adopted in the proposed algorithm to reduce the time of re-processing the whole updated database. In the experiments, it shows that the proposed algorithm has the better performance than the two-phase algorithm [9].

Review of Related Works

FUP Concepts

Data mining can be divided into many specific areas due to its applications [14–16], but the most common approach is to extract patterns or rules from data sets in a particular representation [1, 2, 4]. In real-world applications, transaction databases usually grow over time and the mined association rules must be re-evaluated. Considering an original database and some transactions to be deleted, four cases may arise based on the FUP2 concepts [13] showing in Fig. 1. Each case is then executed by its designed procedure.

Mining High Utility Itemsets

Yao et al. proposed the utility model by considering quantities and profits of items [10]. Chan et al. proposed the topic of utility mining to discover high utility itemsets [7]. Liu et al. then presented a two-phase algorithm for fast discovering high utility itemsets by adopting the downward-closure property [9] and named it

Fig. 1 Four cases when
transactions are deleted from
an existing database

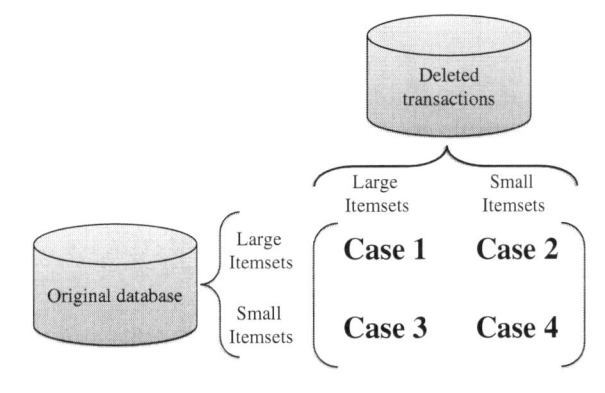

Table 1 An illustrated database

TID	A	B	C	D	E	tu
1	6	0	0	0	0	36
2	0	1	0	5	0	37
3	0	6	0	0	0	12
4	0	0	5	0	0	75
5	0	0	0	0	8	80
6	2	0	2	0	4	82
7	0	1	0	4	0	30
8	0	4	0	0	0	8
9	0	2	3	7	0	98
10	0	0	0	0	1	10
11	0	5	2	5	0	75
12	0	3	0	3	0	27
twu	118	287	330	267	172	Total utility $= 570$

as the transaction-weighted-utilization (TWU) model. Lin et al. then proposed an
incremental approach [11] for efficiently mining the high utility itemsets based on
TWU mode [9]. An example is given below to briefly demonstrate the TWU
model. Assume the original database is shown in Table 1 consisting of 12 transactions with 5 items, denoted by A to E. The profits for items are {A:6, B:2, C:15,
D:7, E:10}.

The *transaction utility* (*tu*) in Table 1 is the summation of the item utilities in
the transaction. The *transaction-weighted utilization* (*twu*) value of each item is
the summation of the *tu* value where the item existing in the transaction. Suppose
the upper (minimum high utility) threshold is set at 30 %. The minimum high
utility count is thus calculated as (570×0.3) $(= 171)$. The final results for the high
(large) transaction-weighted utilization itemsets and their *actual utilities* (AU) are
{$twu(B) = 287$, $AU(B) = 44$, $twu(C) = 330$, $AU(C) = 180$, $twu(D) = 267$,
$AU(D) = 168$, $twu(E) = 172$, $AU(E) = 130$, $twu(B, C) = 173$, $AU(B, C) = 104$,

$twu(B, D) = 267, AU(B, D) = 192, twu(C, D) = 173, AU(C, D) = 159, twu(B, C, D) = 173, AU(B, C, D) = 174\}.$

The Proposed Maintenance Algorithm

The transaction-weighted utilization itemsets and its actual utility values are firstly derived from the original database before transaction deletion. The details of the proposed maintenance algorithm for transaction deletion are described below.

The maintenance algorithm for transaction deletion:

INPUT: A profit table P of items, an original database D, a minimum high utility threshold λ, the total utility TU^D of D, the large (high) transaction-weighted utilization itemsets $HTWU^D$ with their transaction-weighted utilization values and actual utility values discovered from D, and a set of deleted transactions d extracted from the original database D.

OUTPUT: A set of high utility itemsets (HU^U) for the updated database $U (= D - d)$.

STEP 1: Calculate the utility value u_{jk} of each item i_j in each deleted transaction t_k as $u_{jk} = q_{jk} \times p_j$, where q_{jk} is the quantity of i_j in t_k and p_j is the profit of i_j. Accumulate the utility values all items in each deleted transaction t_k as the transaction utility tu_k. That is:

$$tu_k = \sum_{j=1}^{m} u_{jk}.$$

The total utility in the deleted transactions is the summation of all transaction utilities in the deleted transactions as:

$$TU^d = \sum_{k=1}^{n} tu_k.$$

STEP 2: Calculate the updated minimum high utility count muc^U $(= TU^D - TU^d) \times \lambda$.

STEP 3: Set $k = 1$, where k records the number of items in the itemset s currently being processed.

STEP 4: Generate the candidate k-itemsets and calculate their transaction-weighted utility $twu^d(s)$ from the deleted transactions as the summation of the utilities of the deleted transactions which include i_j. That is:

$$twu^d(s) = \sum_{i_j \in t_k} tu_k.$$

STEP 5: Check whether the $twu^d(s)$ of each k-itemset in the deleted transactions is larger than or equal to the minimum high utility count muc^d $(= TU^d \times \lambda)$.

If s satisfied the condition, put it in the set of high transaction-weighted utilization k-itemsets for the deleted transactions, $HTWU_k^d$.

STEP 6: For each k-itemset s in the set of high transaction-weighted utilization k-itemsets ($HTWU_k^D$) from the original database, if it also appears in the set of high transaction-weighted utilization k-itemsets ($HTWU_k^d$) in the deleted transactions, do the following substeps (**Case 1**):

Substep 6-1:Set the updated transaction-weighted utility of itemset s as:

$$twu^U(s) = twu^D(s) - twu^d(s),$$

where $twu^D(s)$ is the high transaction-weighted utility of itemset s in the original database ($HTWU_k^D$) and $twu^d(s)$ is the high transaction-weighted utility of itemset s in the deleted transactions ($HTWU_k^d$), respectively.

Substep 6-2: Check whether the updated transaction-weighted utility $twu^U(s)$ is larger than or equal to the updated minimum high utility count muc^U. If s satisfied the above condition, put it in the set of updated high transaction-weighted utilization k-itemsets, $HTWU_k^U$. Otherwise, remove s from the set of high transaction-weighted utilization k-itemsets ($HTWU_k^D$) in the original database.

STEP 7: For each k-itemset s in the set of high transaction-weighted utilization k-itemsets ($HTWU_k^D$) from the original database, if it does not appear in the set of high transaction-weighted utilization k-itemsets ($HTWU_k^d$) in the deleted transactions, do the following substeps (**Case 2**):

Substep 7-1: Set the updated transaction-weighted utility of itemset s as:

$$twu^U(s) = twu^D(s) - twu^d(s).$$

Substep 7-2: Put s in the set of updated high transaction-weighted utilization k-itemsets, $HTWU_k^U$.

STEP 8: For each k-itemset s, if it does not appear both in the set of high transaction-weighted utilization k-itemsets ($HTWU_k^D$) from the original database and in the set of high transaction-weighted utilization k-itemsets ($HTWU_k^d$) in the deleted transactions, do the following substeps (**Case 4**):

Substep 8-1: Rescan the original database to determine the transaction-weighted utility $twu^D(s)$ of itemset s.

Substep 8-2: Set the updated transaction-weighted utility of itemset s as:

$$twu^U(S) = twu^D(S) - twu^d(S).$$

Substep 8-3: Check whether the updated transaction-weighted utility $twu^U(s)$ is larger than or equal to the updated minimum high utility count muc^U. If s satisfied the above condition, put it in the set of updated high transaction-weighted utilization k-itemsets, $HTWU_k^U$.

STEP 9: Generate the candidate $(k + 1)$-itemsets from the set of high transaction-weighted utilization k-itemsets ($HTWU_k^U$) in the updated database.

STEP 10: Set $k = k + 1$;

STEP 11: Repeat STEPs 4 to 10 until no new candidate itemsets are generated.

STEP 12: Process each itemset s in the set of $HTWU^U$; if $au^U(s) \geq muc^U$, itemset s is a high utility itemset. Put itemset s into the set of HU^U.

STEP 13: Set $HTWU^D = HTWU^U$ as the set of the original large (high) transaction-weighted utilization itemsets for the next transaction deletion in maintenance mining.

After STEP 13, the final high utility itemsets for the updated database can then be updated.

Experimental Results

In the experimental evaluation, the TP-HUI algorithm [9] and the proposed FUP-HUI-DEL algorithm are then compared to respectively show the performance. When transactions are deleted from the original database, the TP-HUI algorithm has to re-scan the updated database for extracting the updated high utility itemsets in a batch way. The proposed FUP-HUI-DEL algorithm divides the itemsets into four parts according to whether their transaction-weighted utilizations are large or small in the original database and in the deleted transactions. Each part is then processed on its own way to update the discovered knowledge.

The IBM data generator [17] is used to generate the simulation dataset called T10I4N4KD200K. Firstly, 200,000 transactions are used to initially mine the large (high) and pre-large transaction-weighted utilization itemsets with their actual utility values. Each 2,000 transactions are then sequential deleted from the original database bottom up at each time. The minimum high utility threshold is set at 0.2 % to evaluate the performance comparing to the TP-HUI and the proposed FUP-HUI-DEL algorithms. The results are then shown in Fig. 2.

In Fig. 2, the TP-HUI algorithm has to process the updated database in a batch way whenever transactions are deleted. The proposed FUP-HUI-DEL algorithm requires re-scanning the whole database if it is necessary to re-calculate the itemsets appearing in in case 4. Experiments are also made to evaluate the

Fig. 2 The comparisons of the execution time for transaction deletion

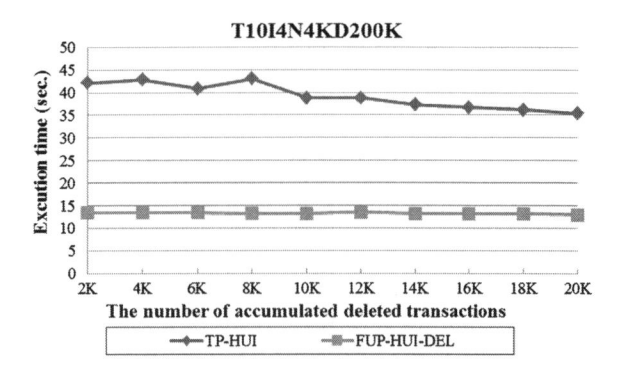

Fig. 3 The comparisons of the execution time in different minimum utility thresholds

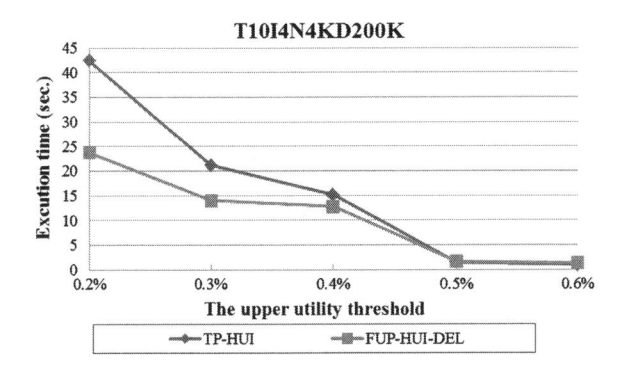

efficiency of the proposed FUP-HUI-DEL algorithm in different minimum high utility threshold values shown in Fig. 3. The minimum high utility threshold is then set from 0.2 to 0.6 %, increases 0.1 % each time.

It can easily be observed from Fig. 3 that the execution time of the proposed FUP-HUI-DEL is much less than that by the TP-HUI algorithm for handling transaction deletion in different minimum high utility thresholds.

Conclusions

In this paper, a maintenance mining algorithm is thus proposed for transaction deletion based on the two-phase and FUP2 approaches. When transactions are deleted from the original database, the proposed maintenance algorithm processes the itemsets into four parts according to whether they are high (large) transaction-weighted utilization itemsets or small in the original database and in the deleted transactions to update the discovered high utility itemsets. Each part is then processed by its own procedure. Experimental results also show that the proposed maintenance algorithm runs faster than the batch approach for updating the high utility itemsets.

Acknowledgments This research was partially supported by Shenzhen peacock project, China, under contract No. KQC201109020055A, and Shenzhen Strategic Emerging Industries Program under Grants No. ZDSY20120613125016389.

References

1. Agrawal R, Imielinski T, Swami A (1993) Database Mining: a Performance Perspective. IEEE Trans Knowl Data Eng 5:914–925
2. Agrawal R, Srikant R (1994) Fast algorithms for mining association rules in large databases. In: Proceedings of the international conference on very large data bases, pp 487–499

3. Hong TP, Lin CW, Wu YL (2008) Incrementally fast updated frequent pattern trees. Expert Syst Appl 34:2424–2435
4. Park JS, Chen MS, Yu PS (1997) Using a hash-based method with transaction trimming for mining association rules. IEEE Trans Knowl Data Eng 9(5):813–825
5. Liu Y, Liao WK, Choudhary A (2005) A fast high utility itemsets mining algorithm. In: Proceedings of the international workshop on utility-based data mining, pp 90–99
6. Yao H, Hamilton HJ, Butz CJ (2004) A foundational approach to mining itemset utilities from databases. In: Proceedings of the siam international conference on data mining. pp 211–225
7. Chan R, Yang Q, Shen YD (2003) Mining high utility itemsets. In: Proceedings of IEEE international conference on data mining, pp 19–26
8. Lan GC, Hong TP, Tseng VS (2011) Discovery of high utility itemsets from on-shelf time periods of products. Expert Syst Appl 38(5):5851–5857
9. Liu Y, Liao WK, Choudhary A (2005) A two-phase algorithm for fast discovery of high utility itemsets. Lect Notes Comput Sci 3518:689–695
10. Yao H, Hamilton HJ (2006) Mining itemset utilities from transaction databases. Data Knowl Eng 59(3):603–626
11. Lin CW, Lan GC, Hong TP (2012) An incremental mining algorithm for high utility itemsets. Expert Syst Appl 39(8):7173–7180
12. Cheung DW, Jiawei H, Ng VT, Wong CY (1996) Maintenance of discovered association rules in large databases: an incremental updating technique. In: Proceedings of the international conference on data engineering, pp 106–114
13. Cheung DW, Lee SD, Kao B (1997) A general incremental technique for maintaining discovered association rules. In: Proceedings of the international conference on database systems for advanced applications, pp 185–194
14. Hong TP, Wu CH (2010) An improved weighted clustering algorithm for determination of application nodes in heterogeneous sensor networks. J Inf Hiding Multimedia Sig Process 2(2):173–184
15. Lin CW, Hong TP, Chang CC, Wang SL (2013) A greedy-based approach for hiding sensitive itemsets by transaction insertion. J Inf Hiding Multimedia Sig Process 4(4):201–227
16. Lin CW, Hong TP (2013) A survey of fuzzy web mining. Wiley Interdisc Rev: Data Min Knowl Discovery 3:190–199
17. IBM: quest synthetic data generation code (1996) Available: http://www.almaden.ibm.com/cs/quest/syndata.html

Design and Implementation of a LBS General Website Content Extract System for Android

Yongbo Chen, Xin Zhou and Yun Liu

Abstract Every day more than 1 million new Android devices are activated worldwide [1]. This paper researched on the topic of design and implementation of a location based public opinion information system on android system, which provides personal or group users, such as people, enterprise, and government an easy access to information what they care about in specified location area with specified focus and interest. Those latest information would extract from some API-opened websites, such as a microblog website, a specified social network website, and take advantage of Google Geocoding API, we can extends the ability of the system to an infrastructure for other applications.

Keywords Content extract · LBS · Android

Introduction

Every day more than 1 million new Android devices are activated worldwide [1]. This paper researched on the topic of design and implementation of a location based public opinion information system on android system, which provides

Y. Chen · Y. Liu (✉)
School of Electronic and Information Engineering, Beijing Jiaotong University, Beijing 100044, China
e-mail: liuyun@bjtu.edu.cn

Y. Chen
e-mail: 12120065@bjtu.edu.cn

Y. Chen · Y. Liu
Key Laboratory of Communication and Information Systems, Beijing Municipal Commission of Education, Beijing Jiaotong University, Beijing 100044, China

X. Zhou
China Information Technology Security Evaluation Center, Beijing, China
e-mail: zhoux@itsec.gov.cn

Y.-M. Huang et al. (eds.), *Advanced Technologies, Embedded and Multimedia for Human-centric Computing*, Lecture Notes in Electrical Engineering 260, DOI: 10.1007/978-94-007-7262-5_113, © Springer Science+Business Media Dordrecht 2014

personal or group users, such as people, enterprise, and government an easy access to information what they care about in specified location area with specified focus and interest. Those latest information would extract from some API-opened websites, such as a microblog website, a specified social network website, and take advantage of Google Geocoding API, we can extends the ability of the system to an infrastructure for other applications.

Related Work

OAuth

OAuth is an open standard for authorization [2]. OAuth provides a method for clients and third-party applications to access server resources on behalf of a resource owner (such as a different client or an end-user). It takes an advantage of making resource sharing easily without provide the resource owner's, credentials (e.g. username and password pair), using user-agent redirections, which make end-users risk less security problem [3].

OAuth is required to take advantage of Open API [4]. Before OAuth standard, there are generally three traditional methods to verify an internet user:

a) Using username and password;
b) Using a temporary secret link address, such as a link in an account activation email;
c) Using cell phone verification code.

Those methods are either vulnerable structure from security point of view or cost extra resources, and would bring some security risk. OAuth was begun in November 2006 when Blaine Cook was developing the Twitter OpenID implementation [3]. By now OAuth 2.0 is the next evolution of the OAuth protocol.

By using OAuth protocol, the system would provide users a more secure permission control and data storage service. Generally, Open API system would take a validation with OAuth protocol for applications. So the system is designed with an OAuth Configuration Manage component, to get the corresponding OAuth access token when accessing the variety Open API system.

LBS

Location-based services are a general class of computer program-level services used to include specific controls for location and time data as control features in computer programs [5].

LBS is widely used in social network websites, IM software, indoor object search, entertainment and so on, and plays an important role in many people's daily life. Android system provides a well-organized location service. With Android Location package, it's efficient to use the location service, and by communicate with server side; the location information could be used to find out nearby events on internet or some spot activity on target location area. There are examples that are common LBS application type recommending social events in a city, requesting the nearest business or service, such as an ATM or restaurant, locating people on a map displayed on the mobile phone etc.

Android Location Listener interface provides an access to the device's location changing.

Webpage Content Extracting

Web pages often contain clutter (such as pop-up ads, unnecessary images and extraneous links) around the body of an article that distracts a user from actual content [6].

To make large scale of data fetching from website possible, automatically extract the content is an essential solution. By researching on extract the useful content of a website, further more application are given, such as make small screen devices efficient to visited content wrapped by order less advertisement in webpages.

Opera re-organized the page content with content analyzing, by extracting the main content of websites and make them user-friendly on small screen devices.

Currently, most used methods to extract web content is xpath-extract and regex-extract, both of them need a hand-tailored extractor. Suhit Gupta advised a method to generate content extractors based on DOM information of the html file [6].

Rahman et al. propose another technique that uses structural analysis, contextual analysis, and summarization. The structure of an HTML document is first analyzed with a definition of 'zones' and then properly decomposed into smaller subsections. The content of the each section well be extracted then summarized afterwards. However, this proposal still remains to be implemented yet [7]. Design and Prototype

Design of the System

When designing the system, there are two factors that mainly decided the architecture of it:

- As android system provides a variety of UI elements, it's convenient to build friendly GUI programmatically.
- The technical improvements of mobile devices and the growing mobile network bandwidth.

Mobile Client Structure and Responsibility

- Provides current location of the device;
- Calculate the distance between the device and a specified location;
- Maintenance of connection to server;
- Receive notification of server with an acceptable delay;
- Maintenance of location data synchronization;

Server Structure and Responsibility

- A OAuth Configuration Manage module, providing requests sent to different Open API websites with OAuth access tokens;
- A permission control module, between different users;
- A data providing and persistence module, containing client synchronizing location data. Composed of such sub modules:

  - A synchronizing configuration maintenance module;
  - A notification module;
- A Website information synchronizing module, which implemented by two cooperation sub-modules with different mechanisms:

  - A high performance multi-thread API-based information extract module which extracts abstract and partial of contents;
  - An aiding information extract sub module by analyzing html content with xpath and regex;
- A configuration maintenance module, manages Auto information of different API-opened website that requests OAuth;

Prototype of the System

OAuth Configuration Manage

OAuth access token should be applied and maintained manually, this module doesn't generate OAuth for client, but take responsibility of make each request sent to Open API website acceptable on target, and won't be denied due to OAuth token exception or error.

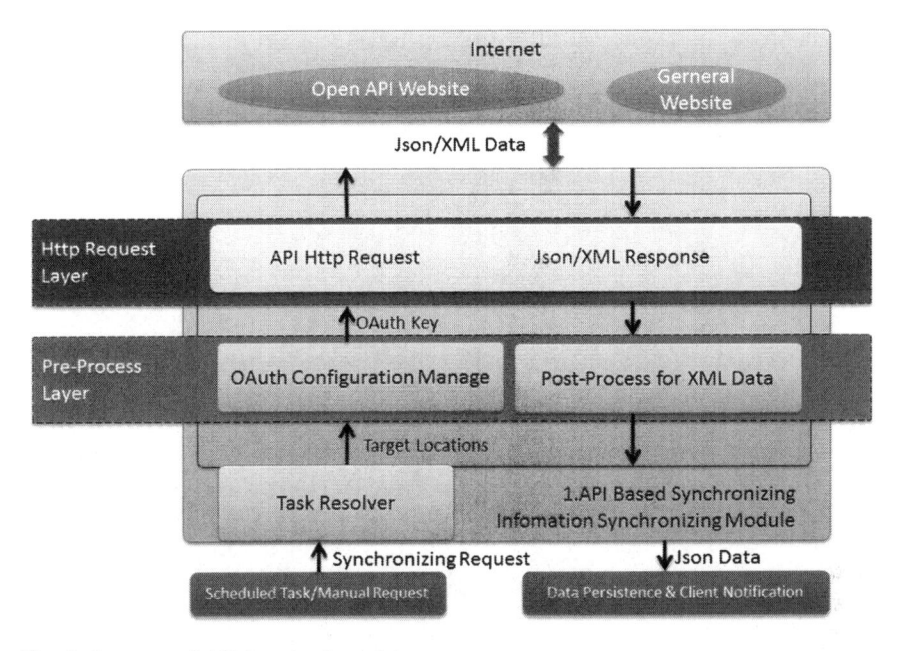

Fig. 1 Structure of API based sub-module

Information Synchronizing Module

In order to take use of the latest information on websites effectively, we need to make the system able to automatically synchronize content from some target websites. Reasonably, the system had been designed to be capable for both executing scheduled task and manual request once or repeat in a specified frequency.

There are two different methods to execute the information synchronizing, API-based and content analyzing based. This module composed of two sub-modules according to those two methods, shown in the following graph (Figs. 1, 2):

In the API-based case, an OAuth key is generally needed, the OAuth configuration manage component manages a variety keys that manually applied from target websites that provides Open API. The Http Request Layer uses a restful web service to proxy scheduled task or request, transform the synchronizing work detail into concrete URL request that can be easily processed through third party tools such as Apache Http Commons. The request result should be in Json or XML format, to make data consistent, we transform the XML data into Json data and processed in following up procedures.

In a synchronizing progress The API based sub-module provides more accurate results, while due to API restriction, less in amount and has a limited usage in rate. Meanwhile, the content analyzing sub-module support an unrestrained usage by mocking a common user requesting the website automatically, the actions containing searching keywords, retrieving the content, and automatically extract the

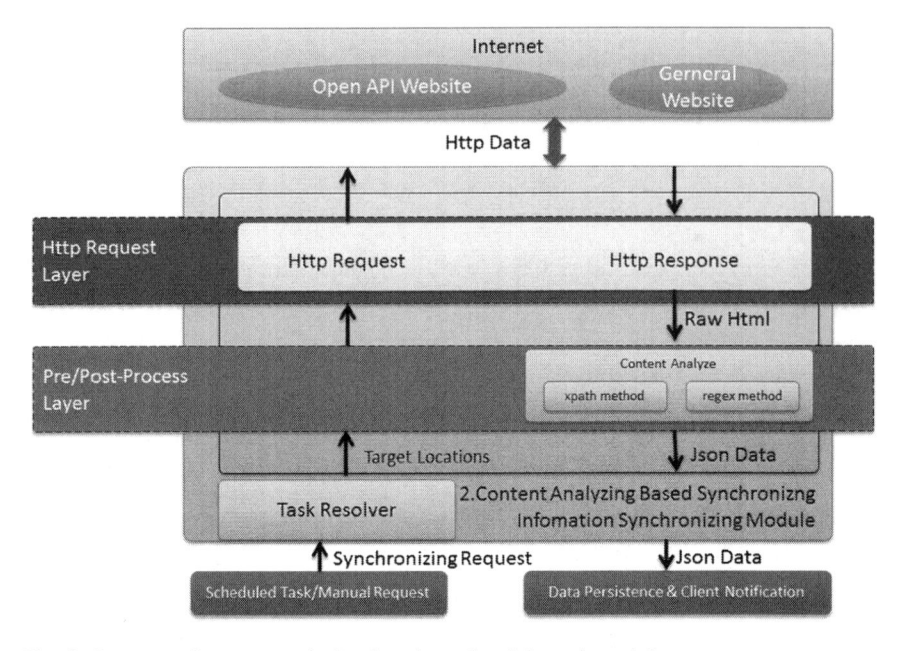

Fig. 2 Structure of content analyzing based synchronizing sub-module

key opinion of result page. Both of the sub-modules making the results into a consistency data structure presented as well formatted Json and they would be persisted into database for further usage such as client notification.

Notification Module

The server would notification the client with messages. Androidpn is used here to provide powerful and comprehensive notification service. As the server side sends a short message, target client or target client group would immediately response with a short ring or some vibration due to the cell phone hint mode configuration.

Screenshots

Figure 3 shows a graph of manually sending a notification, while running the system, this procedure would be called by notification module automatically by sending an http request posting the messages as parameters, and the procedure would be without GUI.

Fig. 3 A screen shot of sending manually notification

Figure 4 shows the client receiving the notification which was sent by server before. We use XMPP (Extensible Messaging and Presence Protocol) to implementation the notification. The XMPP tool here used is Androidpn which implements the XMPP based on asmack. The Androidpn framework contains a client side project and a sever side project. On the client side, it establishes a listener on port 5222, the XMPPConnection class of Androidpn is responsible for communication with the server side. On the server side, it's in fact a light weight Http container, which would send notification and handle the request sent by client. The top level of the server side consist of Session Manager that manages session between client and server, Auth Manager that authenticate client users, Presence Manager that manages the login state of client users, and Notification Manager that send notifications to client side. The scheme guarantees security, simplify, make the client side independent from Android operation system version. It also keeps the system extensible for further design.

Figure 5 shows that the application displayed notification location on map with an icon of broadcast.

Summary and Prospects

As price keep being lower and functions keep enhancing, smart cell phone popularize more every day, meanwhile, mobile communication technology is developing. In such an environment, LBS are greatly increasing. The combination of

Fig. 4 Client receiving notification

Fig. 5 Display location on map

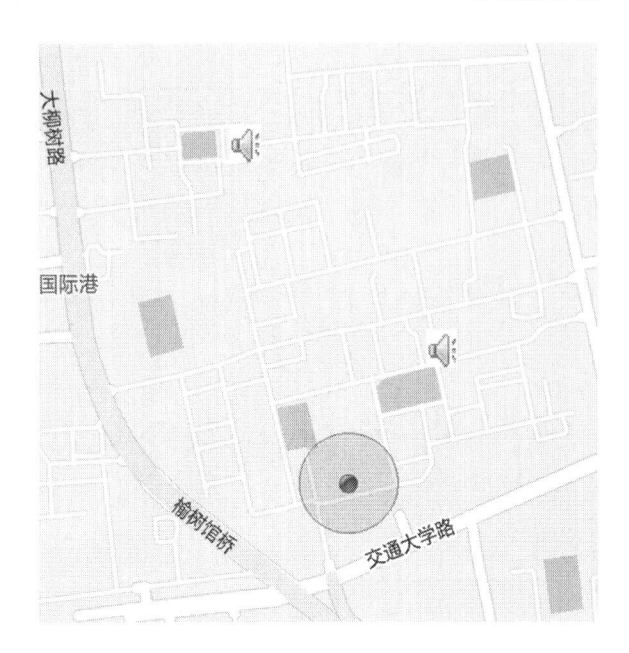

LBS and Website information would be a promising trend. Some applications on smart phones already benefit from the combination of LBS and some social network, in future, there would be more applications take advantage of LBS and the combination of LBS with other service would provide powerful enhancement to daily life.

Acknowledgments This work has been supported by the National Natural Science Foundation of China under Grant 61172072, 61271308, the Beijing Natural Science Foundation under Grant 4112045, the Research Fund for the Doctoral Program of Higher Education of China under Grant W11C100030, the Beijing Science and Technology Program under Grant Z121100000312024.

References

1. http://developer.android.com/about/index.html
2. http://en.wikipedia.org/wiki/OAuth
3. http://tools.ietf.org/html/rfc5849
4. Lee D, Shin J, Lee S (2012) Design and implementation of user information sharing system using location-based services for social network services. J Meas Sci Instrum 3(2)
5. http://en.wikipedia.org/wiki/Location-based_service
6. Gupta S et al (2003) DOM-based content extraction of HTML documents. In: Proceedings of the 12th international conference on world wide web. ACM, 2003

MR. Eye: A Multi-hop Real-time Wireless Multimedia Sensor Network

Yanliang Jin, Yingxiong Song, Jian Chen, Yingchun Li, Junjie Zhang and Junni Zou

Abstract Simple data, such as temperature, humidity and luminary, gathered by nodes in wireless sensor networks (WSNs), can hardly meet the growing needs for real-time information. The availability of low-cost hardware is enabling the development of wireless multimedia sensor networks (WMSNs), i.e., networks of resource-constrained wireless devices that can retrieve multimedia content such as video and audio streams, still images, and scalar sensor data from the environment. However, multi-function means complex design, effective hardware, robust protocol and many other considerations. Comparing some of those current mainstream WMSN test beds, limits and shortcomings can be easily found. In this paper, MR. Eye is designed. Nodes, within MR. Eye, built on OMAP3530, are capable of coping with video, images and other real-time information. Moreover, the load balanced airtime with energy awareness (LBA-EA) is applied in this system and nodes can be easily controlled by SNMP gateway. The last but not the least, real-time video data can be transmitted fluently in multi-hop scenarios.

Keywords WMSNs · Mesh protocol · Sensor nodes · Multi-hop · Real-time

Introduction

Wireless sensor network (WSN) is currently a frontier research field, involving a high degree of interdisciplinary and highly integrated knowledge, drawing intense international attentions. The development of technologies, such as sensor technology, micro electro-mechanical systems, modern network and wireless communication, push forward the generation and development of modern wireless sensor

Y. Jin (✉) · Y. Song · J. Chen · Y. Li · J. Zhang · J. Zou
Key Laboratory of Specialty Fiber Optics, Shanghai University, Shanghai, China
e-mail: baizhsh@126.com

Y.-M. Huang et al. (eds.), *Advanced Technologies, Embedded and Multimedia for Human-centric Computing*, Lecture Notes in Electrical Engineering 260, DOI: 10.1007/978-94-007-7262-5_114, © Springer Science+Business Media Dordrecht 2014

networks. Nonetheless, along with the increasingly complex of environment, simple data gathered from conventional node devices cannot meet inspection needs, which require fine-gained and accurate information. Consequently, wireless multimedia sensor networks (WMSNs) come into being.

Comparing with conventional WSNs, WMSNs possesses apparent attributes in energy distribution, QoS requirements and sensor model, confronting huge challenges both in fundamental theory and on technical implementation level. As far as our best knowledge, those test beds listed above have their own limits more or less. Some of those have low power consumption, but they can only transmit still pictures with low resolution, such as Mica, MicaZ, Yale XYZ, Sun SPOT and Stargate in [1–5] respectively. Others, however, with high power consumption, can merely transmit real-time video within single-hop, such as Stargate and Embedded PC in [2] and [3] respectively. In this paper, we propose the MR. Eye, which incorporates wireless multimedia sensor nodes and a network management system. It possesses functions such as video and data acquisition, network traffic control, system management and configuration of wireless sensor node devices. Network management system mainly fulfill following tasks: real-time topology monitoring, viewing performance data, viewing alarm information, real-time video monitoring, user configuration management and device configuration management. Wireless multimedia sensor nodes are mainly in charge of video capturing, temperature and humidity data acquisition, GPS data acquisition, extracting data of battery state and current state. Nodes can also send data to network management system through agent-side software with UDP protocol. What's more, node devices have high-performance power management function.

Hardware Architecture

The hardware architecture of our proposed sensor node adopts next-generation mobile applications processor OMAP3530 as host processor, which is produced by TI Corporation for low-power portable applications such as smart phones, GPS system and notebook. OMAP3530 integrates an ARM Cortex-A8 core, a TMS320C64x+ DSP core, a graphics engine, a video accelerator and other multimedia peripherals on one single chip. We also realize USB extension in MR. Eye, which enables connection with 802.11 wireless modules, Ethernet ports and other USB devices. Each node device carries a digital and analog video capture port, connecting to a 2 GB SD memory card and various types of common sensor nodes through local bus and I^2C bus respectively. Besides, two separate power modules and clock modules provide the system with stable power supply and clock signal. Figure 1 plots the structure of a whole node device.

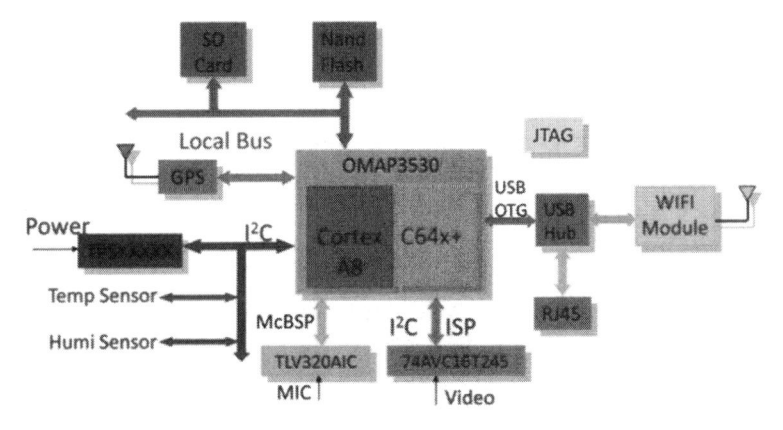

Fig. 1 Hardware architecture

Software Architecture

GStreamer

GStreamer, a framework used to build audio and video applications, is applied in our system. Video processing task is assigned on the DSP core by applying relevant video codec plug-in. The main tasks of GStreamer in our designed node devices are video acquisition, adjustment, coding, RTP package and transmission. Video is displayed in PC control system after the received video data is unpacked and decoded respectively.

Basing on the framework discussed above, we can build up multi-media tunnel basing on GStreamer. Real-time video data acquired by 'v4l2src' plug-in is adjusted by 'video scale' plug-in. Video data, after adjustment, is buffered and compressed through 'TIVidenc1', a plug-in where algorithm, bit rate and frame rate can be set. Thereafter, data stream flows into 'rtpmp4vpay', where data is packed with RTP to guarantee real-time video transmission. Finally, data enters 'udpsink' for transmitting data to destination. In PC system, the tunnel is just the opposite from the one in the node. PC asks for video data from nodes through 'udpsrc' and receives data packages. The header information of RTP package and payload data is unpacked by 'rtpmp4vdepay'. Then payload data is decoded by 'ffdec_mpeg4' to be played by 'dshowvideosink'. Figure 2 plots the structure of the GStreamer multi-media tunnel.

SNMP Gateway Control

SNMP gateway control [6] can be divided into two parts: network management and network agent. Network management part can be controlled through graphical user interface conveniently. What users need to do is just configuring the management

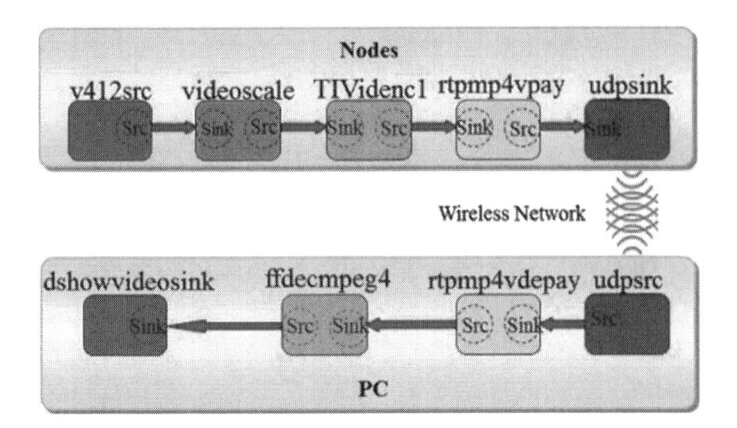

Fig. 2 GStreamer multi-media tunnel

interface. Configuration information is sent to network management module to be analyzed and displayed consequently. Network agent part mainly deals with exception handling, agent management and maintenance, MIB library management and message handling. Management part and agent part are connected by two communication modules. Figure 3 plots the SNMP gateway control framework.

LBA-EA and Mesh Protocol

In MR. Eye, we adopt Mesh topology and HWMP (hybrid wireless mesh protocol) [7] as routing protocol. We propose a new routing metric, load balanced airtime

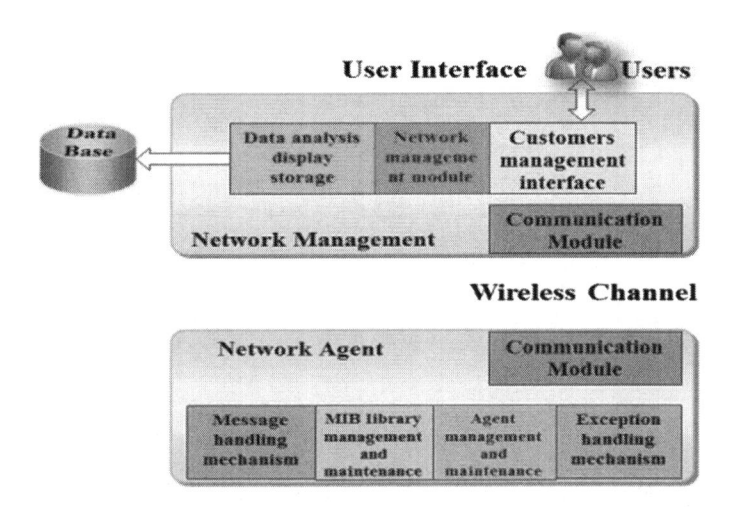

Fig. 3 SNMP gateway control framework

with energy awareness (LBA-EA) [8], which is based on airtime link metric defined in IEEE 802.11s draft. The proposed metric is composed of two parts. One is a load-aware airtime factor, and the other is an energy-aware factor, with a tunable parameter. The weight of a path is calculated as follows:

$$W = (1 - \beta) \times \sum_{i=1}^{n_L} C_{lba}^i + \beta \times \sum_{j=1}^{n_N} \frac{E_{init}^j}{E_r^j} \tag{1}$$

where C_{lba}^i is an enhancement of airtime and represents a more accurate expected transmission time of link i. n_L is the number of links along the selected path. E_r^j is the residual value of node j's energy with its initial energy E_{init}^j, and n_N is the number of nodes along the selected path.

In (1),

$$C_{lba}^i = \left(T_o + \frac{L_t}{r_i} + d_q^i\right) \times \frac{f(\rho_i)}{1 - \rho_i} \times \frac{1}{1 - e_f^i} \tag{2}$$

where these parameters T_o, L_t, r_i and e_f^i are similar to those defined in airtime and d_q^i denotes the queuing delay of a forwarding node in the ith link, which is calculated as:

$$d_q^i = \frac{E\left(L_q^i\right)}{r_i} \tag{3}$$

where $E\left(L_q^i\right)$ is an average value of queue length in a given period of time.

And ρ_i denotes the factor of inter-flow interference which caused by contending neighbors, while $f(\rho_i)$ is a weight function that depends on ρ_i.

$$f(\rho_i) = \begin{cases} 1 & \rho_i < \rho_0 \\ K & \rho_0 \leq \rho_i \leq \rho_{max} \\ \infty & \rho_i > \rho_{max} \end{cases} \tag{4}$$

IEEE 802.11s protocol [9] is the most part of MAC80211 protocol framework, which contains network mgt, QoS rqt, security mgt, neighbor mgt, routing control and data forwarding. MAC80211 provides underlying drivers and user programs with uniform interface, facilitating the development.

Results

To evaluate MR. Eye, several node devices are manufactured to form a certain network on the campus. Figure 4 plots the hardware of the node and Fig. 5 shows the user interfaces in MR. Eye system. The details of hardware architecture have been elaborated in section Hardware Architecture. In this section, the simulation and the experiments of both indoor and outdoor scenarios are done.

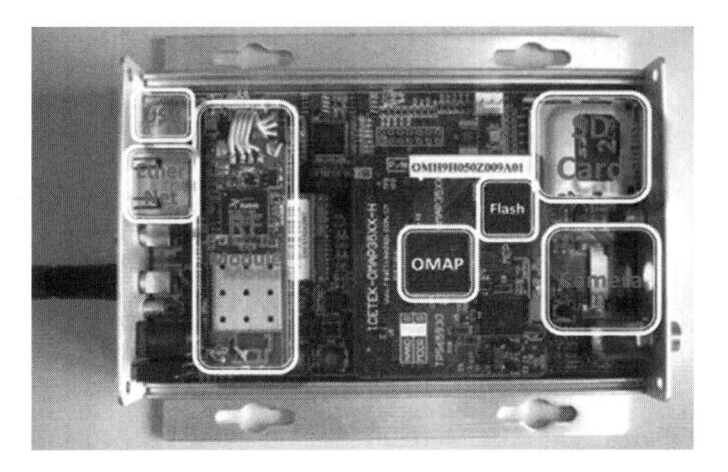

Fig. 4 Internal structure of a node

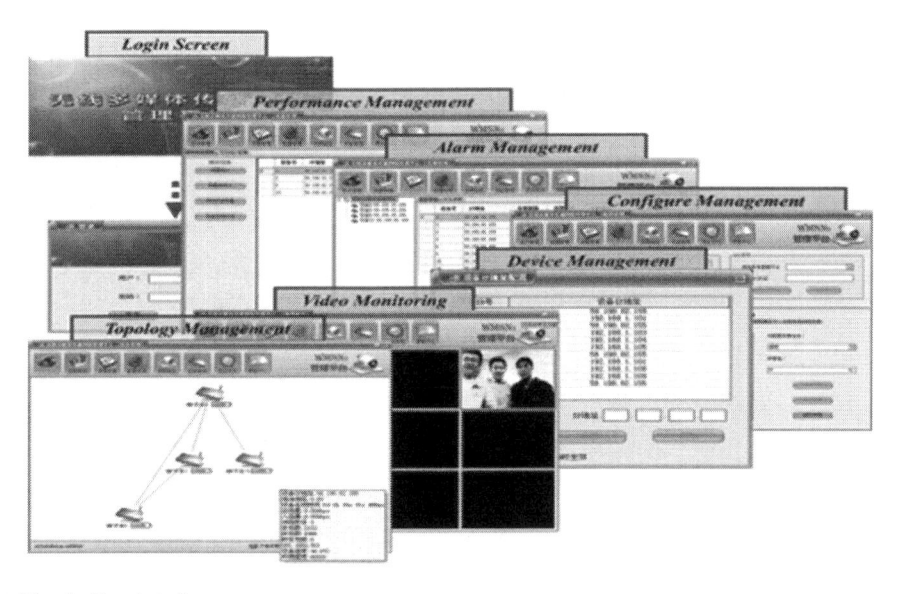

Fig. 5 User interfaces

Simulation

In this section, we implement our simulation in NS3 [10] to evaluate the performance of LBA-EA. we place 50 nodes randomly in an area for 600×600 m and the node 0 is the sink. We use 802.11e EDCA in Data/ACK mode for medium access and use RM-AODV defined in 802.11s as path selection protocol. We consider UDP to be the transport layer and assume that all 10 flows generate data at a constant rate. We consider the packet size of 1,024 bytes. The sources of the flows are randomly chosen and the destination is the sink node. Each simulation run has been executed 100 times and the average results are plotted in the graphs.

As shown in Fig. 6, LBA-EA HWMP has higher delivery rate than HWMP because LBA-EA can disperse flows into different paths thus reduce the probability of collision while Airtime introduces flows into a single path with high data rate. As is shown in Fig. 7, we can see that LBA-EA HWMP has less normalized routing load than HWMP when the number of source nodes increases from 5 to 30. The new scheme uses the load balanced to control the broadcasting information and decrease the overhead.

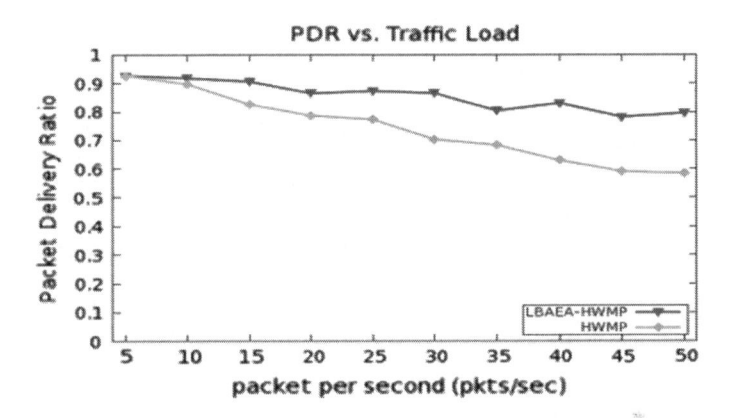

Fig. 6 Comparation of packet delivery ratio

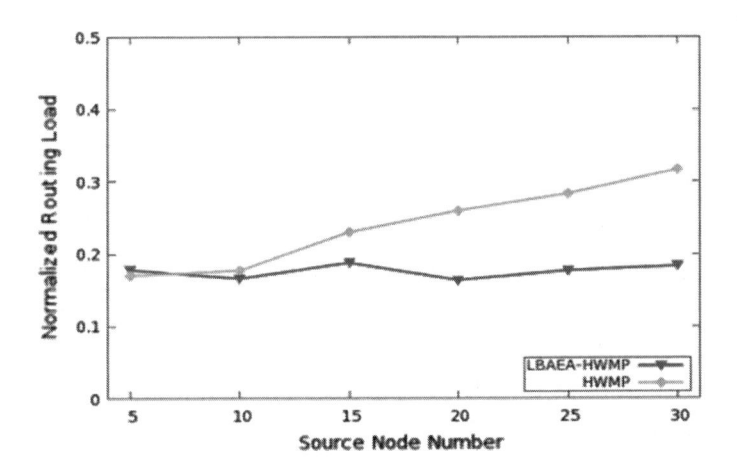

Fig. 7 Comparation of normalized routing load

Indoor Scenario

For indoor experiment, three major aspects are tested: generating dynamic topology, bandwidth and time-jitter respectively.

In bandwidth experiment, to form a three-hop chain topology, four nodes are put in linear shape in the corridor of the lab. The bandwidth of this scenario is plotted in Fig. 8. Same experiment condition with the bandwidth experiment, the time-jitter between the two ends of the chain topology is measured. Figure 9 plots the result of time-jitter. Clearly, the average time-jitter stays around 10 ms, which is acceptable for real-time video data.

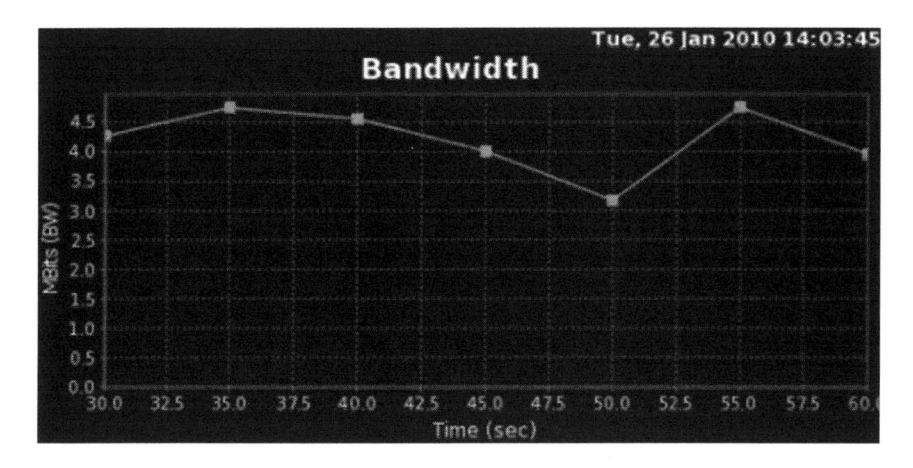

Fig. 8 Bandwidth experiment

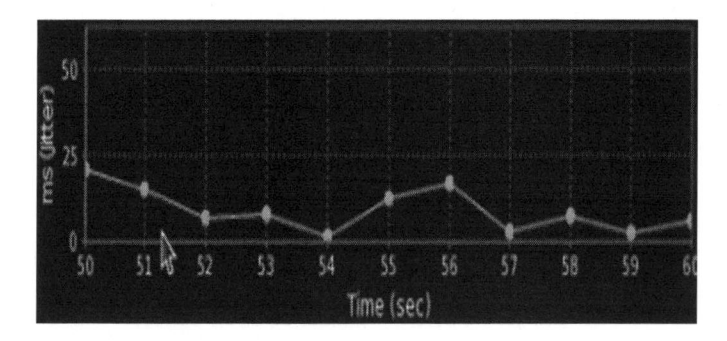

Fig. 9 Time-jitter experiment

Outdoor Scenario

As for outdoor scenario, we locate 8 nodes on the campus. In Fig. 10, we set node 4 as the sink node, which is connected to the SNMP gateway control system. Through the gateway system, real-time monitoring video from other distributed nodes can be easily displayed on the screen.

Figure 11 shows the video data gathered from node 8 and node 7 respectively. In both two-hop and three-hop scenarios, video can be played frequently.

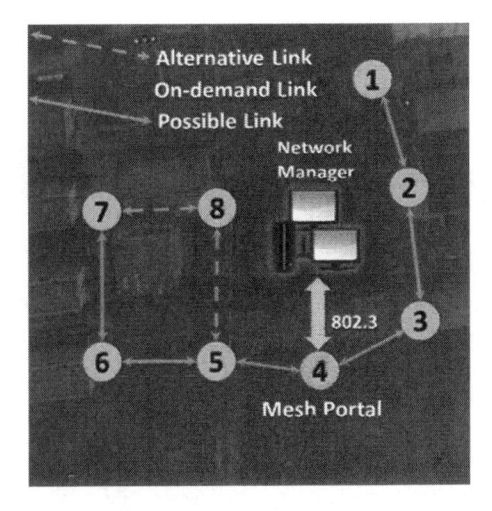

Fig. 10 Practical outdoor experiment

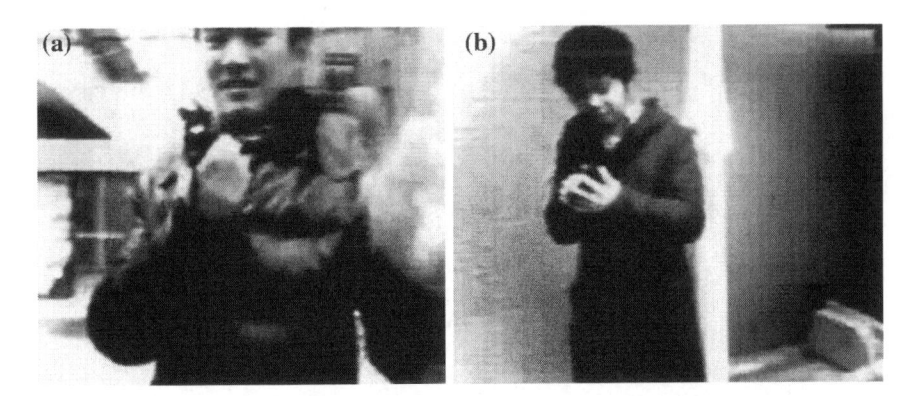

Fig. 11 a Video from a two-hop node 8. **b** Video from a three-hop node 7

Conclusion

A high performance MR. Eye is demonstrated in this paper. In MR. Eye, nodes with low power consumption have been designed. Meanwhile, highly effective mesh protocol, powerful SNMP gateway control system and friendly interface are developed. Multi-media nodes are able to collect and process signals like location, temperature, humidity, audio and video. Adopting HWMP routing protocol is suitable for the attributes of mesh topology. SNMP gateway control system facilitates the management of data gathered from nodes. Amiable graphical user interface realizes an easy access to configuration and monitoring of nodes. Indoor and outdoor experiments have proven the fact that MR. Eye is quite competent in practical applications. We will further expand the node number of the test-bed and enhance the system performance in the future.

Acknowledgments The authors would like to thank the reviewers for their detailed comments on earlier versions of this paper. This work was supported by Shanghai leading academic discipline project (S30108, 08DZ2231100), funding of key laboratory of wireless sensor network and communication, shanghai institute of micro system and information technology, Chinese academy of sciences (2012002) and Shanghai science committee (12511503303).

References

1. Gerla M, Xu K (2003) Multimedia streaming in large-scale sensor networks with mobile swarms. In: Papakonstantinou Y (ed) Proceedings of the ACM SIGMOD 2003. ACM Press, New York, pp 72–76
2. Kahn JM, Katz RH, Pister KSJ (1999) Next century challenges: mobile networking for "Smart dust". In: Proceedings of the 5th annual ACM/IEEE Int'l conference on mobile computing and networking. ACM Press, New York, pp 271–278
3. Kulkarni P, Ganesan D, Shenoy P, Lu QF (2005) SensEye: a multi-tier camera sensor network. In: Zhang HJ, Chua T-S (eds) Proceedings of the 13th annual ACM international conference on multimedia'05. ACM Press, New York, pp 229–238
4. Turning vision into reality (2005) http://www.research.sun.com/spotlight/SunSPOTSJune30.pdf/
5. Feng W, Code B et al (2003) Panoptes: a scalable architecture for video sensor networking applications. In: Rowe L, Vin H (eds) Proceedings of the ACM Int'l conference on multimedia. ACM Press, New York, pp 151–167
6. Mauro D, Schmidt K (2001) Essential SNMP. O'Reilly Media, California
7. Cornils, M, Bahr M, Gamer T (2010) Simulative analysis of the hybrid wireless mesh protocol (HWMP), wireless conference (EW), European, 12–15, pp 536–543
8. Sun W-Z, Song Y-X, Chen M (2010) A load-balanced and energy-aware routing metric for wireless multimedia sensor network, the third IET international conference on wireless, mobile and multimedia networks, pp 21–24
9. Hiertz GR, Max S, Zhao R, Denteneer D, Lars B (2007) Principles of IEEE 802.11s. In: Computer communications and networks, ICCCN 2007. Proceedings of 16th international conference on Honolulu, HI, pp 1002–1007
10. Zhu Z, Wang P (2010) A multimedia system based on OMAP3530. In: applied mechanics and materials, 40–41, pp 506–509

Effects of the Online VOD Self-learning on English Ability of Taiwanese College Students: The ARCS Approach

Da-Fu Huang

Abstract This study investigated the effect of the Live DVD self-learning system on English learning motivation of students of a Taiwanese technological university. Modeled on the ARCS motivation model of Keller (1987), this study validated the ARCS model, examined the effects of Majors and English Levels on motivation, and tested the proposed structural model involving the effect of the ARCS model on self-assessed English skills. Five hundred and twenty-seven students of three English ability groups completed a survey questionnaire. Statistical procedures including confirmatory factor analysis, factorial ANOVA, and structural equation modelling were performed to address the research questions. The participants were found to have overall positive motivation toward the Live DVD System in terms of the ARCS model. English Levels had significant influence on Attention, with students of Mid and Low-level groups having higher attention to the Live DVD system than High-level group students. For Satisfaction, aside from the main effect of English Levels, the English Levels*Majors interaction effect was also found to be statistical; for non-engineering majors, students of mid and low English abilities were found to have higher satisfaction with the Live DVD System than high proficiency students, while for Level B groups, non-engineering had higher satisfaction with the self-learning system than engineering students. Only Confidence of the ARCS in the proposed structural model was found to have significant effect on Self-assessed Skills, students with stronger confidence tending therefore to have higher self-rated English abilities.

Keywords Live DVD · ARCS motivation model

D.-F. Huang (✉)
Southern Taiwan University of Science and Technology, Tainan, Taiwan, Republic of China
e-mail: dfjhuang@mail.stust.edu.tw

Y.-M. Huang et al. (eds.), *Advanced Technologies, Embedded and Multimedia for Human-centric Computing*, Lecture Notes in Electrical Engineering 260, DOI: 10.1007/978-94-007-7262-5_115, © Springer Science+Business Media Dordrecht 2014

Introduction

Lack of learning motivation of students, especially those of vocational and technological universities, has long been claimed to cause ineffectiveness of English education in Taiwan. Teaching approaches geared towards initiating self-directed learning via technology-mediated support with a view to enhancing learning motivation have hence been gaining ascendancy in English learning settings across different educational levels in Taiwan. Multimedia provides variety and excitement to a computer-supported teaching and learning environment, adapting instruction to the diverse learning preferences of students [1]. Advanced multimedia instruction heightens visual aspects of communication, provides dynamic learning experiences and raises learning results [2]. Multimedia materials such as DVDs and VODs as effective self-directed learning aids enhances students' comprehension and memory, increases their motivation and promotes their concentration on the content in a near natural environment [3–5]. Best multimedia, however, will be rendered useless unless they are utilized by students to the utmost effect, which should be ensured through some mechanism. This consideration made Southern Taiwan University of Science and Technology create an online VOD Self-learning System, brand named Live DVD and the first of its kind in Taiwan, where students viewed designated films as a requirement. Students were monitored of their use behaviors and tested on their learning achievement after viewing the target films and using the system's embedded language learning features, including repeated listening, vocabulary collection, dictionary, target expression searching, multiple language display mode, etc. This study aimed to understand to what extent the Live DVD System affected learning motivation, which in turn would affect self-assessed English abilities, and whether this relationship depends on majors and English levels. Specifically, this study intended to test a proposed structural equation model involving Keller's ARCS motivation model [6] as a framework to investigate student perceptions of the Live DVD System and its effects on self-assessed English abilities in terms of whether the system draws their *attention*, shows *relevance* to their learning goals, helps build *confidence* in realistic expectations and learning outcomes, and makes the learning *satisfying* (see Fig. 1).

Method

Twelve classes of first-year non-English major undergraduate students participated in the study by completing the motivation survey. The students were placed at classes of three English levels: high, mid, and low for better learning of the required English class. As a class requirement, the students did English learning by viewing the films at the Live DVD Platform, and were invited to complete the questionnaire at the semester end, with a return of 527 questionnaires. The

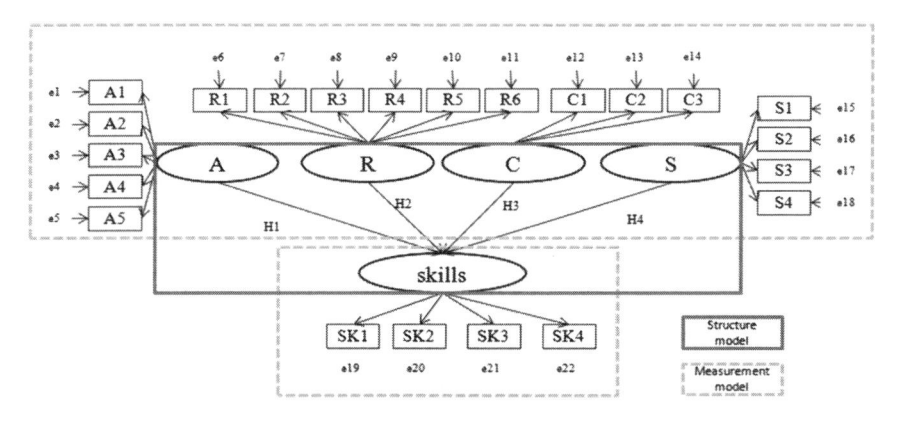

Fig. 1 The proposed structural equation model

questionnaire instrument on the Lickert scale was piloted on two classes of students and modified to insure a high reliability of the instrument for the formal data collection. After completion of data collection of 527 valid samples, confirmatory factor analysis, aside from reliability analysis, was conducted using SPSS v.18 and AMOS v.18 to check the convergent and discriminant validity of the instrument to validate the scale construct and the measurement model. Factorial ANOVA was then performed via SPSS v.18.0 to obtain the effects of Majors and English Levels on the ARCS motivation as well as the interaction effects. Structural equation modeling via AMOS was also conducted to test the hypotheses of the posited model featuring the structural model involving the effect of the ARCS motivation latent variables on Self-assessed Skills.

Result

Factor analysis employing principal component and Varimax extraction methods extracted five components including Attraction, Relevance, Confidence, Satisfaction, and Self-assessed Skills, cumutively accounting for ca. 72 % of total variation. A total of 22 question items were loaded on the 5 factors: Attraction (5 items), Relevance (6 items), Confidence (3 items), Satisfaction (4 items), and English Skills (4 items).

The reliability of the five factors turned out to be acceptably high, with Cronbach Alpha values of 0.91, 0.90, 0.74, 0.88, and 0.90 respectively for the preceding five factors. Confirmatory factor analysis supported the acceptable convergent validity of the scale [7–9]. In ARCS motivation model, the RMR was 0.039 (<0.05), the GFI, NFI, CFI being 0.904, 0.930, 0.947 respectively. In addition, the factor loading of each variable was significant and higher than 0.5, the CR of each dimension was over 0.7, and the AVE of each dimension was larger

Table 1 Two-way ANOVA for attention of ARCS model

Source	Type III sum of squares	df	Mean square	F	Sig.
Corrected model	10.362	5	2.072	3.014	0.011
Intercept	10,144.505	1	10,144.505	14,751.916	0.000
Major	0.511	1	0.511	0.744	0.389
English level	8.084	2	4.042	5.878	0.003**
Major * level	1.567	2	0.783	1.139	0.321
Error	358.278	521	0.688		
Total	10,571.360	527			
Corrected total	368.640	526			

than 0.5. As to the Self-assessed Skills measurement model, the RMR was 0.019 (<0.05), the GFI, NFI, CFI being 0.987, 0.988, 0.990 respectively. In addition, the variables' factor loadings were significant and higher than 0.5, the CR value over 0.7, and the AVE value larger than 0.5, indicating sufficient convergent validity and reliability of the model. The five components were also shown to have adequate discriminant validity complying with the rule that the correlation coefficient of two dimensions was less than the Cronbach's alpha reliability coefficients [10], and that the correlation coefficient of two dimensions was less than the square root of AVE [11].

The descriptive statistics of the ARCS model indicated a grand mean of 4.34, with separate means/SDs of 4.40/0.98, 4.29/0.84, 4.16/0.95, 4.49/0.94, respectively for each of the components, indicating a fairly positive motivation toward Live DVD self learning. Two-way ANOVA was performed to examine the effects of the variables of Group (of Majors) and Level (of English Ability) on the ARCS Motivation. As shown in Table 1, main effect of English Level ($F = 5.88$, $p < 0.01$) was found for Attention of ARCS Model. Post-hoc comparison showed Level A to be significantly lower in attention ($p < 0.05$) than Level B and Level C. Table 2 showed that main effect of English Level ($F = 7.96$, $p < 0.01$) and interaction effect ($F = 3.33$, $p < 0.05$) between Major and English Level were also found for Satisfcation of ARCS Model. In Table 3, the simple main effect of Major showed a significant difference ($F = 10.346$, $p = 0.000$) among non-engineering majors, where both Level B and Level C significantly surpassed Level

Table 2 Two-way ANOVA for satisfaction of ARCS model

Source	Type III sum of squares	df	Mean square	F	Sig.
Corrected model	14.716	5	2.943	4.652	0.000
Intercept	10,562.019	1	10,562.019	16,691.946	0.000
Major	0.313	1	0.313	0.495	0.482
English level	10.075	2	5.037	7.961	0.000**
Major * level	4.216	2	2.108	3.332	0.036*
Error	329.669	521	0.633		
Total	10,968.938	527			
Corrected total	344.385	526			

Table 3 Test of simple main effect of major and english level on satisfaction of ARCS model

	Sum of squares	df	Mean square	F	Sig	Post hoc
Majors						
Engr	1.092	2	0.546	0.878	0.417	
Non-Engr	13.318	2	6.659	10.346	0.000**	Level B > Level A
						Level C > Level A
Levels						
A	1.156	1	1.156	1.531	0.218	
B	3.418	1	3.418	6.453	0.012*	Non-Engr > Engr
C	0.080	1	0.080	0.128	0.721	

A students in satisfaction. The simple main effect of English Level showed a significant difference (F = 6.453, p < 0.05) between non-engineering and engineering majors of Level B, the former significantly surpassing the latter in satisfaction.

Structural equation modelling was conducted to test the proposed structural model. The model fit analysis includes preliminary fit criteria, fit of internal structure of model and overall model fit [8]. The analysis of absolute fit measures (χ^2/d.f. = 2.282, GFI = 0.929, RMR = 0.041, RMSEA = 0.049), incremental fit measures (AGFI = 0.907, CFI = 0.97, NFI = 0.947) and parsimonious fit measures (PNFI = 0.791, PGFI = 0.709) suggested that the overall model fit of the proposal model after modest modification was good [12]. The regression coefficients of path analyses were used for testing the hypotheses of the proposed structural model, yielding the results as represented in Table 4 and Fig. 2.

Discussion

The participants' fairly positive attitude toward Live DVD System suggested that the online self-learning system is a worthwhile resource for students. However, Levels B and C tended to have higher attention to the system than Level A students, implying that while boosting the learning motivation of lower end students, the system needs to be modified in the designing of learning features to appeal to the interests and motivation of higher proficiency students. The

Table 4 Summary of hypotheses results

Hypothesized path		Standardized regression weights estimate	t-Value	Results
H1:	Attention → Eng skills	0.006	0.055	NH not rejected
H2:	Relevance → Eng skills	−0.098	−0.905	NH not rejected
H3:	Confidence → Eng skills	0.536***	6.045	NH rejected
H4:	Satisfaction → Eng skills	0.126	1.26	NH not rejected

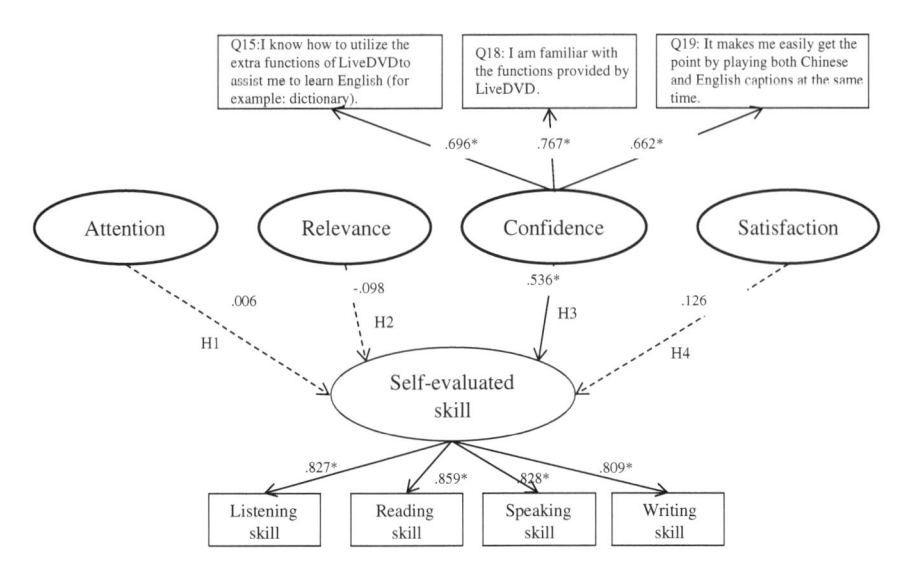

Fig. 2 The structural equation model with parameter estimates

precedence of Levels B and C over Level A in attention also raised the issue of the credibility of the placement test, which might have placed medium proficiency students into Group A, and reversely high proficiency students into Group B. The validity and reliability of the placement test for the incoming undergraduates needs therefore to be reviewed. Group A teachers are better advised to highlight the uses of the learning system and guide them to more advanced learning results. This implication seems to be particularly relevant for Non-engineering students given another result showing Levels B and C Non-engineering students to have higher satisfaction with the learning system than Level A counterparts.

In the ARCS model only Confidence was found to have direct effect on students' self-assessed skills, reflecting the important role of self-confidence in the learning process of college students, particularly of technological and vocational higher education, who tended to have lower learning motivation and English proficiency and hence lower self-confidence as well. Relative to the constructs of Attention, Relevance, and Satisfaction, Confidence, or students' perceived proficiency in using available learning features, linked to motivation most prominently in the EFL learning contexts. This research thus called attention to reinforcing of self-directed learning resources with user-friendly and practical features to build students' self-confidence which would directly affect their perceived English learning progress, and indirectly their learning motivation.

Conclusion

This study investigated the technological university students' perceptions of the Live DVD Self-learning System by testing a hypothetical structural equation model involving the ARCS measurement model and a structural model comprising the latent variables of ARCS and Self-assessed Skills. The results of the study suggested an English Level effect on Attention, and a Major effect as well as a Major*English Level interaction effect on Satisfaction, engendering implications for improvement of placement test validity and reliability and teaching practice of Level A classes. More important, Confidence was validated to be the sole construct of the ARCS motivations that affected student self-assessed skills, a result that highlights the critical need of constructing self-directed learning resources in a way that would facilitate student utilization and best build self-confidence. The results of this research are significant and applicable to technological and vocational higher education institutions in Taiwan with similar student backgrounds for purposes of English education reform and curriculum improvement.

References

1. Zaidel M, Luo XH (2010) Effectiveness of multimedia elements in computer supported instruction. J College Teach Learn 7(2):11–16
2. Wang L (2008) Developing and evaluating an interactive multimedia instructional tool. J Edu Multimedia Hypermedia 17(1):43–54
3. Astleitner H, Wiesner C (2004) An integrated model of multimedia learning and motivation. J Edu Multimedia and Hypermedia 13(1):3–22
4. Deimann M, Keller JM (2006) Volitional aspects of multimedia learning. J Edu Multimedia and Hypermedia 15(2):137
5. Guariento W, Morley J (2001) Text and task authenticity in the EFL classroom. ELT J 55(4):347–354
6. Keller JM (1987) Development and use of the ARCS model of motivational design. J Inst Dev 10(3):2–10
7. Anderson JC, Gerbing DW (1988) Structural equation modeling in practice. Psychol Bull 103:411–423
8. Bagozzi RP, Yi Y (1988) On the evaluation of structural equation models. J Acad Market Sci 16:74–94
9. Gefen D, Straub DW, Boudreau MC (2000) Structural equation modeling and regression. Commun Assoc Inf Syst 7(7):1–78
10. Gaski JF, Nevin JR (1985) The differential effects of exercised and unexercised power sources in a marketing channel. J Mark Res 22(5):130–142
11. Fornell C, Larcker D (1981) Structural equation models with unobservable variables and measurement error. J Mark Res 18(1):39–50
12. Hair JF, Black WC, Babin BJ, Anderson RE, Tatham RL (2006) Multivariate data analysis, 6th edn. Prentice-Hall, New Jersey

Applying Firefly Synchronization Algorithm to Slot Synchronization

Yanliang Jin, Zhishu Bai, Lina Xu, Wei Ma, Xuqin Zhou and Muxin Wang

Abstract Firefly synchronization algorithm, in a bio-inspired way, has been proposed to replace conventional methods of solving the synchronization problem in wireless sensor networks. For the well-known M&S model and the RFA, a direct application of them can hardly lead to considerable results. This paper presents several vital improvements in order to make the RFA more applicable for a certain wireless sensor network. To better evaluate the results, the algorithm is modulated in a scenario with realistic parameters of a network. Moreover, differing network topologies have also been taken into consideration.

Keywords Firefly synchronization · RFA · WSNs

Introduction

In order to schedule a wireless sensor network in a sufficient way, a common notion of time slot should be the fundament. It is like the clock of a whole operating system that scheduling can only be achieved when every components work on a certain rhythm. To gain this goal, an accurate clock on wall would be necessary no more. Besides, dealing with a distributed network consists of sensor nodes, with limited capability of power and computation, conventional mechanism would be inappropriate and a nonhierarchical distributed algorithm may be preferable. Fireflies in South-East Asia, however, also distributed in trees with low intelligence, can flash in a perfect synchrony within a huge range from a chaotic situation. This striking phenomenon has been delved for a long run. Buck et al.

Y. Jin · Z. Bai (✉) · L. Xu · W. Ma · X. Zhou · M. Wang
Key Laboratory of Special Fiber Optics and Optical Access Networks of Ministry of Education, Shanghai University, Shanghai 200072, China
e-mail: baizhsh@126.com

Y.-M. Huang et al. (eds.), *Advanced Technologies, Embedded and Multimedia for Human-centric Computing*, Lecture Notes in Electrical Engineering 260, DOI: 10.1007/978-94-007-7262-5_116, © Springer Science+Business Media Dordrecht 2014

investigated fireflies' reaction to external flashings and studied their behaviors in more details.

Then, Mirollo and Strogatz thoroughly concluded a mathematical model for firefly synchronization based on pulse-coupled oscillators [1]. The rule governing this model is a quite simple one: (1) Each oscillator maintains a periodic time-phase function that acts as a build-in timer and when the phase counts to a certain threshold, the phase turns back to zero. (2) Once an oscillator receives a "flashing" from one of its neighbors, the function of whom updates its phase ϕ to a certain value ϕ' immediately. They have also proven that for an arbitrary number of entities and independent of the initial condition, the network shall always synchronize [2], so long as the constraints on the coupling between entities are met. The RFA is an algorithm based on the M&S model proposed in [3]. It differs slightly from the M&S model: the function is only adjusted once at the next beginning of cycle just after the firing instant. This difference, although small, makes the RFA more practical compared with the M&S model, for it alleviates the burden on computation ability of sensor nodes and neutralizes the impact of delays. Both the M&S model and the RFA serve as the way to force all sensor nodes in a network to achieve a rhythm, based on which scheduling can be implemented in order to gather or spread data in a well-organized way. However, direct application of the M&S model and the RFA in wireless communications can hardly achieve a satisfactory result. Either Topology impacts or realistic radio effects, such as propagation delays, channel attenuation and noise, place a huge effect on the performance of synchronization. What's more, in real wireless communication scenario, synchronization flash is composed of a sequence of pulses with several bytes or even more, which means that the reception and parse of synchronization words shall also bring in delays. Half-duplex transmission in the physical layer and random packet arrivals in the MAC layer also lead to a huge difference from experimental situations. As for the M&S model, lots of works have been done [4] to make it more applicable in WSNs.

This paper investigates how realistic radio effects, topology difference, half-duplex transmission and random packet arrivals affect the RFA. For practical applications of the RFA, several improvements were proposed in this paper. Simulation results show that with these improvements the RFA works well in a wireless sensor network. However, network topology variance still place a huge impact on the performance of the RFA. This paper is structured as follows: section Classic Mathematical Models of Firefly Synchronization reviews the M&S model and investigates two different situations of the RFA. Section Synchronization Metric and Blind Area gives the metric of synchronization and the principle of blind area. In section The RFA in Slot Synchronization, intertwining with CC1100, several details that may impact the RFA are pinpointed. Analysis of topology variance is given in section Topology Simulation and Analysis. Finally, section Conclusion and Future Work draws the conclusions.

Classic Mathematical Models of Firefly Synchronization

In this section, both the M&S model and the RFA are reviewed. Then, a divergence of the RFA is settled.

The M&S Model

According to the M&S model, pulse-coupled oscillators are governed by one simple rule:

1. Each oscillator maintains a periodic time-phase function that acts as a build-in timer. When the phase counts to a certain threshold, the phase turns back to zero.
2. Once an oscillator receives a "flashing" from one of its neighbors, the function of whom updates its phase ϕ to a certain value ϕ' immediately:

$$\phi' = f^{-1}(f(\phi) + \varepsilon) \tag{1}$$

where

$$f(\phi) = \frac{1}{b}\ln\left(1 + [e^b - 1] \cdot \phi\right) \tag{2}$$

is the phase-state function and ε is the coupling strength. The phase-state function describes a curve that is smooth, monolithic increasing and concave down. The term b is the dissipation factor that decides the radian of the curve. Therefore, the later a flashing arrives, the larger the phase adjustment is.

The Reachback Firefly Algorithm

By the theory of the RFA [5], unlike the M&S model, oscillators use timestamps to record and queue up the firing messages from neighbors without responding in phase adjustment immediately. At the beginning of the next period, oscillators calculate the overall phase adjustment from the firing messages in queue and update the phase. The phase adjustment is also decided by phase-state function mentioned in section The M&S Model. Compared with the M&S model, the RFA avoids frequent phase adjustment to alleviate the computation burden. MAC-layer time stamping is also used in the RFA to avert two unpleasant conditions caused by propagation delays in WSNs: (1) The firing message of node A arrives just right at the firing instant of node B. (2) Unpredictable delays may cause disordered

Fig. 1 The RFA

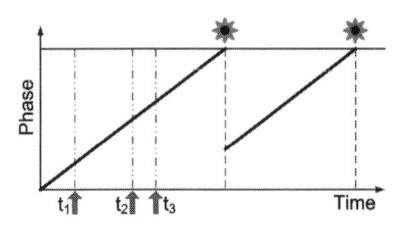

firing message arrivals. Time stamping is a valid method that makes the RFA more applicable in a realistic situation (Fig. 1).

As for the RFA, there is still a cardinal divergence unsettled. For those messages arrive more close to the firing instant, it is more likely that the phase would jump to a value that exceeds the threshold according to function (1). How to deal with those irregular firing messages has not been settled yet. Moreover, how to determine the boundary has not been mentioned anywhere else so far as our best knowledge. Apparently, discarding all those irregular firing messages would be a better choice, for it at most prolongs the time to synchronization without causing instability. However, there still exists one particular situation. Suppose there are two sensor node clusters A and B, whose phases are almost the same and just meet the discarding condition. Consequently, the states of A and B are locked and shall never merge together until other nodes join into break this deadlock. To avoid this peculiar situation, nodes should keep the first firing message which leads to an over-jump and record the phase adjustment as $1 - \phi$. Then, the boundary is set at this phase, after which no more firing messages shall be recorded.

Synchronization Metric and Blind Area

Synchronization Metric

To evaluate the performance of the RFA, the Kuramoto synchronization metric [6] is used in this paper. Kuramoto metric is described as the following:

$$r \cdot \exp(j2\pi\bar{\phi}) = \frac{1}{N}\sum_{n=1}^{N}\exp(j2\pi\phi_n)$$

$$\Rightarrow r = \frac{1}{N} \cdot \exp(-j2\pi\bar{\phi}) \cdot \sum_{n=1}^{N}\exp(j2\pi\phi_n) \tag{3}$$

where r is the synchronization index, ϕ_n is the instant phase of oscillator n and $\bar{\phi}$ is the mean phase of all N oscillators. The synchronization index r equals to 1 when all nodes in a network are synchronized. When the system is in disorder, i.e. firing instants are randomly distributed within [0, T], the index r approaches 0.

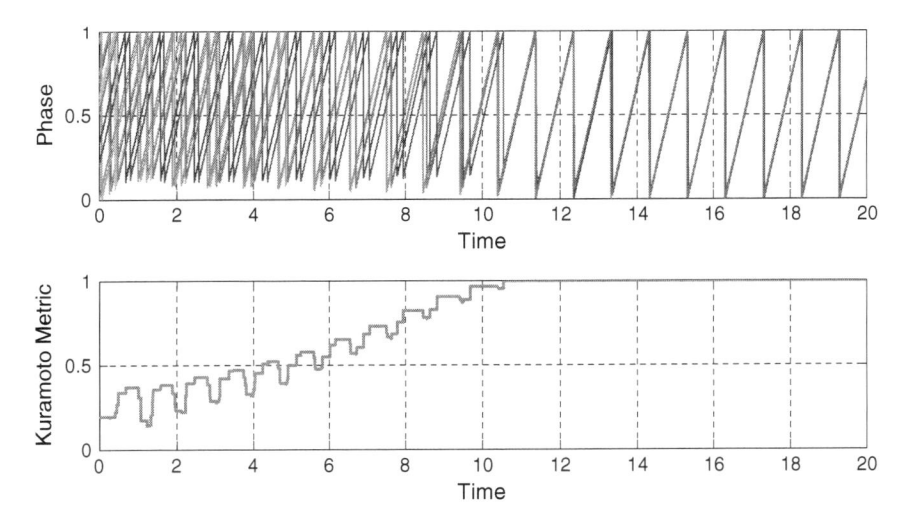

Fig. 2 Kuramoto Metric applied to the RFA

Figure 2 plots how the Kuramoto Metric evaluates the RFA, where 20 nodes are considered. At first, when the initial phases are randomly allocated, the Metric is pretty low. At about the 10th cycle, as the network achieves synchronization and finally stays steady, the Metric tends to 1.

Blind Area

Blind area is a fixed phase, below which nodes do not react to any firing messages received from neighbors. This specific phase is chosen as a value that is half of the phase threshold. The essence of blind area is similar with refractory period [7], proposed for the M&S model to avoid "echoes" caused by delays in networks. However, blind area still differs from refractory period for two aspects: (1) Blind area is a constant for a certain network. (2) Blind area is proposed to bring about a wait-and-chase situation in order to achieve the most efficient synchronization.

Refractory period T_{refr} is defined as the duration which satisfies that $T_{refr} \geq 2 \cdot v_{ij}$, where v_{ij} is the largest transmission delay between two nodes in a network. Apparently, for a wireless network consists of a number of nodes with multi-hop, finding out this delay is not an easy work. In addition, a refractory period is a wide range to choose from and it could vary a lot depending on the topology. Blind area is set as $\frac{1}{2}\Phi$th regardless of the topology which is easy to be located. Moreover, blind area always divides the nodes into two groups: a chasing group and a waiting group, where the chasing group always goes for the waiting group. $\frac{1}{2}\Phi$th is a peculiar value wiping off the possibility that two nodes chasing

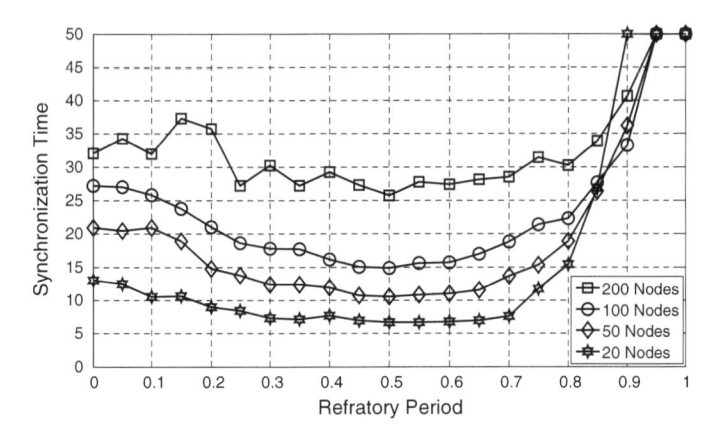

Fig. 3 Synchronization time with varying refractory period

for each other. Theoretically, blind area ensures both the efficiency and the stability of synchronization.

As for a network with certain node number, synchronization time varies with different refractory periods. Simulation result plotted in Fig. 3 shows that the shortest synchronization time can always be achieved when refractory period equals 0.5, which is a half of the phase threshold, for almost all situations. This has been explained above as the theory of blind area.

The RFA in Slot Synchronization

Figure 4 presents a wireless communication block diagram of realizing the RFA with chip CC1100. The diagram is divided into three parts: transmission, reception and arbitration. For every payload data $x(t)$ that is about to be transmitted shall be added with a header consists of several preamble bits and a sync-word $s(t)$ with 16 or 32 bits by CC1100. In the reception part, a sync-word is detected by CC1100 as the indicator of a valid payload data. The RFA is inserted as an arbitrator between the transmission part and the reception part to decide which part shall preempt the antenna. Node i can only transmit the data $y_i(t)$ when the result of its arbitration part is true. The RFA receives signals from the sync-word detector to queue up valid flashings and finally adjust the phase, which is elaborated in section Classic Mathematical Models of Firefly Synchronization. Once a sync-word is detected, the arbitration part should be informed no matter the whole data frame is successfully received. Because any two or more overlapping data frames could lead to a fail reception, namely collision which happens quite frequently when a network is in a chaotic state. From the transmitter side of node j, it takes T_{Tx} for a sync-word to be sent out completely, v_{ij} to go through the air to reach the receiver side of node i. Finally, sync-word detector takes T_{detec} to detect whether it is a valid

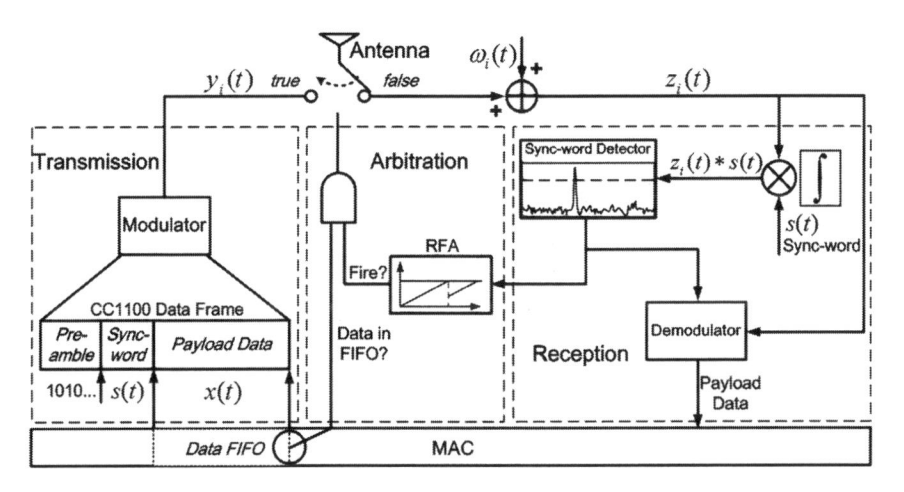

Fig. 4 Block diagram of wireless communication with the RFA applied as the arbitration part

sync-word or not. The total delay is written as $D = T_{Tx} + v_{ij} + T_{detec}$. Consider an additive white Gaussian noise $\omega_i(t)$ with zero-mean and variance σ^2, the incoming signal can be written as

$$z_i(t) = \sum_{j \in Z_i} A_{ij} \cdot y_j(t - \hat{D}) + \omega_i(t) \tag{4}$$

where Z_i is the set of nodes whose transmissions are observed by node i within its reception slot, $\hat{D}$ represents the delay that

$$\hat{D} = T_{Tx} + v_{ij} \tag{5}$$

and A_{ij} donates the amplitude

$$A_{ij} = \sqrt{d_{ij}^{-x}P} \tag{6}$$

in which d_{ij} is the distance between node i and node j, x is the path loss exponent and P is the transmit power. The incoming signal $z_i(t)$ directed to the reception part is scanned for the sync-word to decide the beginning of the data frame. The sync-word is detected by cross-correlating a known sync-word with $z_i(t)$:

$$z_i(t) * s_i(t) = \int z_i(t - \tau)s_i(\tau)d\tau$$

$$= \int \left[\sum_{j \in Z_i} A_{ij} \cdot y_j(t - \hat{D} - \tau) + \omega_i(t - \tau) \right] s_i(\tau)d\tau \tag{7}$$

$$= \sum_{j \in Z_i} A_{ij} \int y_j(t - \hat{D} - \tau)s_i(\tau)d\tau + \omega_i(t) * s_i(t).$$

Only a sync-word that is fully covered in the reception slot of node i can generate a wave peak with magnitude $A_{ij}S$, in which S is the length of sync-word. This wave peak is decided by the sync-word detector that whether it is a valid one by comparing the peak with a given threshold. Because of the existence of noise in the channel, false alarm and/or omission of a sync-word could happen. The likelihood of a successful detection is decided by both the threshold set by CC1100 and the length of sync-word.

Comparing with the ideal synchronization model, several cardinal differences should be taken into account:

1. The RFA alone cannot decide whether or when shall a data frame be transmitted. As a matter of the MAC strategy, nodes may not always transmit a data frame at their firing instant.
2. Multiple delays exist in the entire process of transmission and reception as the sync-word is not just an ideal pulse without length and processing delay.
3. To a certain node, at one time, the reaction to one node or a cluster of synchronized nodes is the same, for the sync-word detector can only detect a wave crest at one time no matter high or low.
4. Half-duplex mode makes the utilization of the RFA even more complicated, because nodes are "deaf" while transmitting and nodes which are receiving can only be scheduled to transmit mode after a whole data frame is completely received.

Apparently, although the MAC strategy macroscopically decrease the amount of firing messages a node could receive from its neighbor in one listen slot, it prolongs the time to synchronization without affecting the accuracy. Delays, however, would drastically compromise the attainable slot synchronization accuracy. The maximum transmission distance of CC1100 is about 150 meter for an indoor scenario. Then $v_{ij} \leq \frac{d}{c} = \frac{150}{3 \cdot 10^8} = 0.5\,\mu s$, where d is the distance between two nodes and c is the speed of light. The maximum transmission data rate of CC1100 is 500 kbps, when preamble is set as 8 bits and sync-word is set as 32 bits, then $T_{Tx} \geq \frac{32+8}{500 \cdot 10^3} = 80\,\mu s$, which is far greater than v_{ij}. Consequently, $D \approx T_{Tx} + T_{detec}$. For a certain hardware platform, D is a constant that can be easily obtained.

Figure 5 plots a general process of how the RFA works in a sensor node A. In the listen state, it takes $D \approx T_{Tx} + T_D$ for node A to completely receive and parse the sync-word ever since one of its neighbors has reached the firing instant. Once node A detects a valid sync-word for node B, it could calculate the right time at which node B fired by subtracting the constant delay D. For nodes which fires at $t1$, $t2$ and $t3$, node A only react to the node firing at $t3$ for the firing messages of the other two nodes are blocked by blind area. At the firing instant of node A, only when there is payload data in the MAC layer can it turn to the transmit mode to emit the data frame. It is should be mentioned that there may exist nodes which fire at the previous cycle of node A, but their sync-words could not be detected until the next cycle. This specific situation is like the node C which fires at $t4$ in Fig. 5.

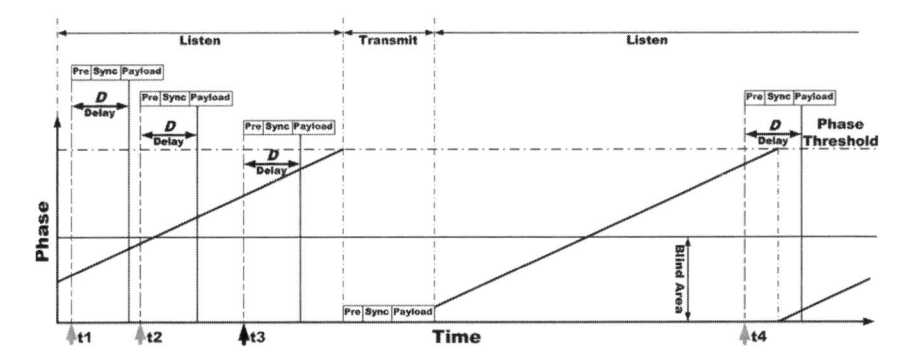

Fig. 5 Listen and transmit slot with the RFA applied

Even though the right firing instant of node C can be obtained, it is too late to add it to the previous queue. Given this kind of situation, the firing message at $t4$ should not be discarded and shall be added into the next queue to be calculated. In the diagram that Fig. 4 plots, the randomly scheduled data transmission at the firing instant could vastly compromise the synchronization rhythm, for the sensor nodes can only work in half-duplex mode. In order to regain the stability of synchronization, a vital modification is proposed to the diagram to overcome the drawback of half-duplex, shown in Fig. 6.

Now that nodes do not react to any flashings received within a blind area slot, the slot could be assigned to transmit data frames. During the transmission period, the time-phase function is still maintained and synchronization rhythm is neither violated. To make sure that the data frame could be sent out completely within the blind area T_{BA}, it should fulfill:

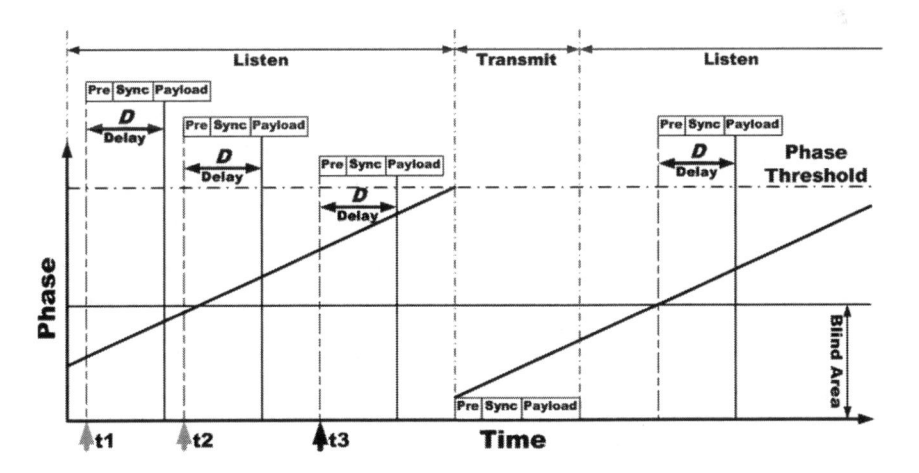

Fig. 6 Modified listen and transmit slot

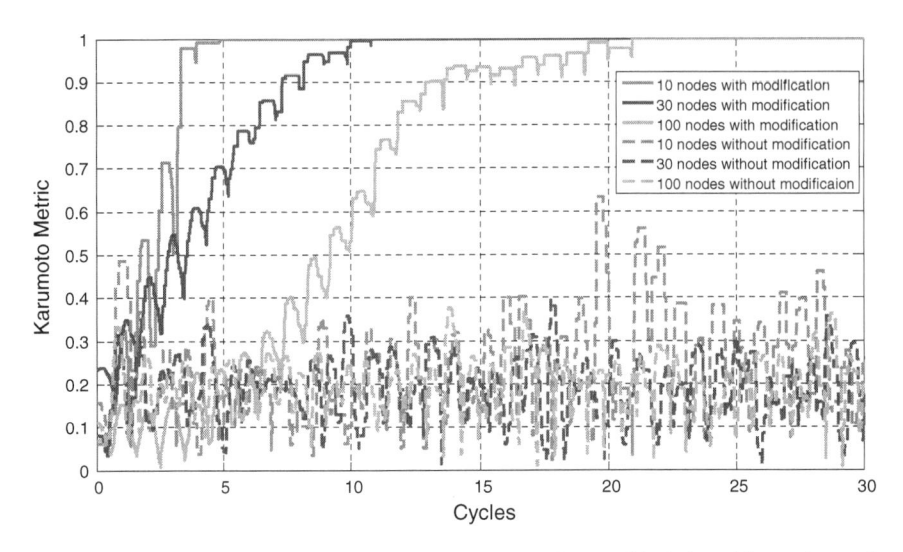

Fig. 7 The performances of the RFA in modified slot and unmodified slot with varying node number

$$T_{BA} > (N - 1) \cdot \varepsilon + T_{DF} \tag{8}$$

where T_{DF} is the time it takes to completely transmit a data frame and the term N is number of sensor nodes in a network. Figure 7 shows the simulation results of the RFA applied in slot synchronization within a fully-connected network. Simulation parameters are given in Table 1. It is clear that the RFA works well with the modification but the dash lines show that networks can hardly get synchronized with the effect of half-duplex mode. Compared Fig. 3, the synchronization time is apparently prolonged, for the influence from MAC strategy.

Topology Simulation and Analysis

In this section, the performance of the RFA in slot synchronization, elaborated in section The RFA in Slot Synchronization, is analyzed when different topologies

Table 1 Slot synchronization simulation parameters		
	Data frame length	25 bytes
	Sync-word length	32 bits
	Preamble length	32 bits
	Cycle duration	1 s
	Total delay D	0.66 ms
	Baud rate	400 Kbps
	Repeat	100 times

Table 2 Topology simulation parameters	Node number	30
	Coupling strength	0.00239
	Dissipation factor	10
	Data frame length	25 bytes
	Sync-word length	32 bits
	Preamble length	16 bits
	Cycle duration	1 s
	Total delay D	0.12 ms
	Baud rate	400 Kbps
	Repeat	100 times

are applied. Similar works have been done based on the M&S model. However, a crucial discrepancy between this section and the result in [8] is found and a sound explanation is drawn finally.

In this paper, a concept named connection rate is utilized to characterize topological properties of a network. Given a certain graphic:

$$G = G(V, E).$$

Connection rate L_G is defined with the number of existing links E and the number of nodes N in a network:

$$L_G = \frac{2|E|}{|V| \cdot (|V| - 1)} \tag{9}$$

where $|E|$ and $|V|$ is the number of edges and nodes respectively in the graphic. Connection rate is a simple way to characterize the density of a network. Moreover, with a given L_G and certain number of nodes, it is easy to generate and handle a network. Especially, $L_G = 1$ means a fully-connected network.

The simulation is evaluated with MATLAB accounting for topologies like line, ring, star and mesh with varying connection rate. Simulation parameters are given in Table 2 and the result is shown in Fig. 8. Mesh-n means this topology is generated with a connection rate that equals n and Mesh-fully means the connection rate is 1.

Mesh Topology. For mesh topology, three different connection rates are taken into consideration. It is easy to tell from Fig. 8 that the synchronization time is inversely proportional to the connection rate, which is reasonable. A higher connection rate means that a node could have a larger neighbor set and phase adjustment could happen more frequently in one cycle. In mesh topology, therefore, the performance of the RFA, applied to slot synchronization, is mainly affected by the connection rate. However, the same connection rate does not lead to the same topology shape, and there still exist some special topology shapes that worth the attention, such as star, line and ring.

Star Topology. Star topology, in which every node is connected to a center node, may lead to the worst performance when the RFA is applied, shown in

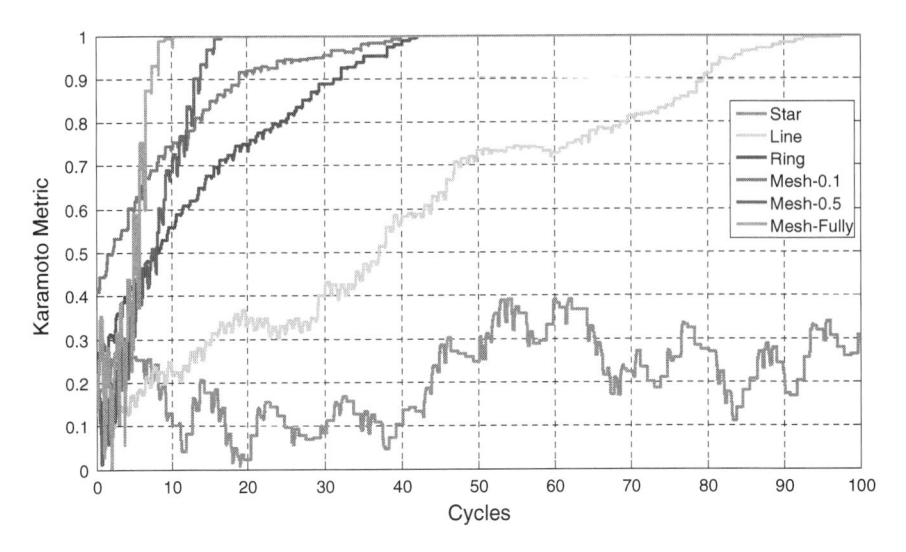

Fig. 8 Simulation results for different topologies

Fig. 8. The reason that causes such a bad condition is that the center node is always dragged around by its neighbors and its neighbors cannot communicate each other. This situation, however, is not permanent. After a number of experiments, it is interesting to find out that a network with star topology could be synchronized when the number of nodes in it decreases below about 10. The conclusion is just opposite to [9], in which star topology leads to the best performance.

Line Topology. In a line topology, nodes are connected one by one but the head and tail are not connected. Except for the head and tail, each node has only two neighbors. Clearly, this kind of shape does no good to synchronization. A longer line makes it harder to achieve an overall rhythm. Given a certain connection rate, when the number of nodes if greater than 3, a mesh topology always leads to a better performance than what a line topology does.

Ring Topology. By connecting the head and the tail of a line topology a ring topology is generated. Although the connection rate increases a little, as shown in Fig. 8, the performance is improved vastly comparing with the line topology.

Conclusion and Future Work

In this paper the RFA has been modified to be more feasible in wireless communication with realistic radio effects from delay, half-duplex mode and so on. Simulation results show that blind area works well in slot synchronization. Besides, half-duplex mode effect is neutralized by inserting the transmit mode in the blind area. Different Topologies have also been taken into account. For certain connection rates, a topology that is randomly generated is always more suitable than those special ones, such as line, ring and star especially. In future work, it will be interesting to test the theory of this paper on a realistic test bed. Other factors that may affect synchronization accuracy shall also be studied.

Acknowledgments The authors would like to thank the reviewers for their detailed comments on earlier versions of this paper. This work was supported by Shanghai leading academic discipline project (S30108, 08DZ2231100), funding of key laboratory of wireless sensor network and communication, shanghai institute of micro system and information technology, Chinese academy of sciences (2012002) and Shanghai science committee (12511503303).

References

1. Mirollo R, Strogatz S (1990) Synchronization of pulse-coupled biological oscillators. SIAM 50(6):1645–1662
2. Leidenfrost R, Elmenreich W, Bettstetter C (2010) Fault-tolerant averaging for self-organizing synchronization in wireless ad hoc networks. Proceedings of the international symposium on wireless communication systems (ISWCS), New York
3. Wieser S, Montessoro PL, Loghi M (2013) Firefly-inspired synchronization of sensor networks with variable period lengths. In: Adaptive and natural computing algorithms, pp 376–385, Springer, Berlin Heidelberg
4. Tyrell A, Auer G, Bettstetter C (2007) Biologically inspired synchronization for wireless networks. In: Dressler F, Carreras I (eds) Studies in computational intelligence, 69:47–62, Springer, Berlin, Germany
5. Tyrrell A, Auer G, Bettstetter C (2010) Emergent slot synchronization in wireless networks. IEEE Trans Mob Comput 9:719–732
6. Kuramoto Y (1984) Chemical oscillations, waves, and turbulence. Springer, Verlag
7. Mathar R, Mattfleldt J (1996) Pulsed-coupled decentral synchronization. SIAM J on applied math 56(4):1094–1106
8. Leidentfrost R, Elmenreich W (2009) Firefly clock synchronization in an 802.15.4 wireless network. EURASIP J Embed Syst, 17
9. Moreno Y, Pacheco AF (2004) Synchronization of Kuramoto oscillators in scale-free networks

Novel Mutual Information Analysis of Attentive Motion Entropy Algorithm for Sports Video Summarization

Bo-Wei Chen, Karunanithi Bharanitharan, Jia-Ching Wang, Zhounghua Fu and Jhing-Fa Wang

Abstract This study presents a novel summarization method, which utilizes attentive motion analysis, mutual information, and segmental spectro-temporal subtraction, for generating sports video abstracts. The proposed attentive motion entropy and mutual information are both based on an attentive model. To capture and detect significant segments among a video, this work uses color contrast, intensity contrast, and orientation contrast of frames to calculate saliency maps. Regional histograms of oriented gradients based on human shapes are also adopted at the preliminary stage. In the next step, a new algorithm based on mutual information is proposed to improve the smoothness problem when the system selects the boundaries of motion segments. Meanwhile, differential salient motions and oriented gradients are merged to mutual information analysis, subsequently generating an attentive curve. Furthermore, to remove non-motion boundaries, a smoothing technique based on segmental spectro-temporal subtraction is also used for selecting favorable event boundaries. The experiment results show that our proposed algorithm can detect highlights effectively and generate smooth playable clips. Compared with existing systems, the precision and recall rates of our system outperform their results by 8.6 and 11.1 %, respectively. Besides, smoothness is enhanced by 0.7 on average, which also verified feasibility of our system.

B.-W. Chen (✉) · J.-F. Wang
Department of Electrical Engineering, National Cheng Kung University, Tainan, Taiwan, Republic of China
e-mail: dennisbwc@gmail.com

K. Bharanitharan
Department of Electrical Engineering, Feng Chia University, Taichung, Taiwan, Republic of China

J.-C. Wang
Department of Computer Science and Information Engineering, National Central University, Jhongli, Taiwan, Republic of China

Z. Fu
School of Computer Science, Northwestern Polytechnical University, Xi'an, China

Y.-M. Huang et al. (eds.), *Advanced Technologies, Embedded and Multimedia for Human-centric Computing*, Lecture Notes in Electrical Engineering 260, DOI: 10.1007/978-94-007-7262-5_117, © Springer Science+Business Media Dordrecht 2014

Keywords Video summarization · Attentive motion entropy · Mutual information analysis · Segmental spectro-temporal subtraction

Introduction

Video summarization techniques have been proposed for years to offer people
comprehensive understanding of the whole story in a video. To date, a great deal of
effort has been devoted to providing people with more friendly interfaces and better
concept interpretation [1–13]. Traditional video summarization approaches can be
roughly classified into two categories: One is the static storyboard [1, 10, 13], which
is composed of still images extracted from the original video; the other is the dynamic
skimming [2–9, 11, 12], which concatenates several shorter clips. Both of them aim
to offer users a compact view of a video. This work mainly focuses on the study of
dynamic skimming approaches because generating playable clips is suitable for users
to navigate sports videos. So far, a large amount of related research has been proposed to analyze sports videos, including colors, motion vectors [6, 7, 12, 14],
saliency maps [15, 16], information theory-based features [8, 11], and so forth.

Although these systems are capable of capturing salient shots among videos, it
is still difficult to detect genuine events and remove false alarms. For example,
Walther et al. [15, 16] attempted to highlight objects in an image by using low-
level visual features and generating saliency maps. Like many other color-based
approaches, their method required heuristic rules for filtering unnecessary frames.
On the other hand, [7] focused on another features and balanced this problem via
estimating motion vectors. Cernekova et al. [8] adopted an information theory-
based model and estimated motion events by detecting mutual information
between frames. Nevertheless, such methods were susceptible to global motion
changes, such as camera panning and zooming, which may cause false positive
results. Duan et al. [3] then developed an enhanced video skimming system by
extracting video objects and converting them into motion vector fields. Although
precision rates increased, Duan's approach needed to map quite many visual
features to the semantic level. A more sophistic classifier was required in the
training phase.

To solve such problems, a compromised way, which combines merits of the
above algorithms for summarizing sports video events, is proposed in this study.
Therefore, this work provides:

1. A novel approach based on attentive motion entropy analysis, which joins
   salient motions and regional histograms of oriented gradients;
2. A new mutual information estimator that calculates differential salient motions
   and differential oriented gradients for detecting coherent frames and the
   boundaries of a motion event;

3. A boundary-smoothing scheme based on segmental spectro-temporal subtraction for removing obscure motion segments.

With the use of the proposed techniques, the system can decrease false positives caused by conventional motion analyses and create a friendly navigation interface for users.

The rest of this paper is organized as follows. Section Related Work introduces the related approaches about dynamic video skimming. Section Proposed Video Summarization System then describes details of the proposed video summarization method. Next, section Experimental Results summarizes the performance of the proposed method and the analysis results. Conclusions are finally drawn in section Conclusion.

Related Work

To date, a great amount of research has devoted to generating dynamic video abstracts, using various approaches. In [12], the authors employed two types of models, "global motion model" and "local motion model," to distinguish camera events from sports events. Ma et al. [6, 17] made use of different features, such as motions, color, and other media descriptors, and fuse them into a curve. Such curves are often referred to as attentive curves; they can offer developers numeric conversion of images. In 2004, Cernekova et al. [8] adopted another approach, modeling shot and scene changes by detecting mutual information between frames. There were also some approaches [5, 6] that exploited a more sophisticated way to extract highlights. For example, the authors [6] utilized content-based parsing techniques to detect face regions. However, the major drawback is that it requires a recognition system and involves a training phase. In addition to the aforementioned methods, heuristic rules are often used as a criterion for identifying event patterns. In [5], the authors presented several feature extraction methods, including wavelet-based motion-trend analyses, hybrid field-color models, and prior knowledge-driven line detection. Liu et al. [2] has devised a perceived motion model for key-frame extraction. They came up with a triangular modeling rule to reshape the generated attentive curve. Duan et al. [3] developed a semantic video skimming system by extracting video objects and converting them into motion vector fields. Instead of proposing visual features as the aforementioned research did, Lu et al. [18] made use of a transition graph model to calculate skimming length of shots. Like Lu's approach, Ngo et al. [19] modeled a temporal graph for scene classification using N-Cut algorithm. Although video abstracts can be presented to users in forms of scenes and shots, both of Lu's and Ngo's approaches required scene and shot detection, which are inapplicable in sports videos due to obscure scene changes.

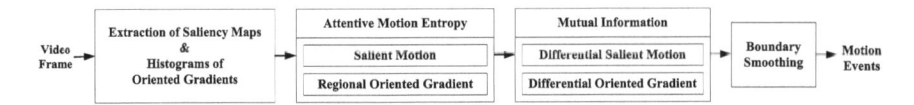

Fig. 1 System Overview

Generally speaking, finding an effective way to transform video data into a numeric curve is the major challenge in dynamic skimming. Accordingly, this study aims to provide another feasible solution to this issue.

Proposed Video Summarization System

The proposed system consists of four stages, as shown in Fig. 1. The first stage of the workflow is the saliency map extraction along with computation of histograms of oriented gradients. The second calculates attentive motion entropy according to salient motions and oriented gradients; potential motion events are preliminarily marked in this step. Then, the third stage calculates mutual information between frames using differential salient motions and differential oriented gradients; the objective is to find out the boundaries of a motion event. Finally, the last stage is to detect sports activities from mutual information curves and refine the results using boundary smoothing. The outputs of the system are concatenated playable clips.

Extraction of Saliency Maps and Histograms of Oriented Gradients

The following description gives a short introduction to saliency map extraction. A saliency map is an image that can point out the visual regions, which have the most perceptual impact on human brains. For these years, this technique has been intensively and broadly studied, and it has a wide range of applications [6, 15, 16]. The idea behind the algorithm is to provide visual perception analysis by using color contrast, intensity contrast, and orientation contrast in an image. In this study, the main reason to adopt such techniques in our system is that unnecessary global motions could be removed by employing saliency map extraction. Besides, such techniques can avoid detecting false highlights in videos.

The following are the key-points of the saliency map according to the authors' research [15]: After the system extracts each frame, the image is sub sampled into a Gaussian pyramid and then decomposed into several channels for red (R), green (G), blue (B), yellow (Y), intensity (I), and local orientation (O_θ). From these channels, two feature maps, "center c" and "surround s" are constructed and normalized. A saliency map can be obtained by averaging the functions proposed in [15] at this stage.

To capture moving athletes, this study follows [20], which extracted intensity-based and gradient-based features, to model human shapes. An Adaboost classifier is also used to detect humans in video frames. After the classifier finds human contours or silhouettes, a regional histogram of oriented gradients (HOGs) based on the labeled areas is created. Let $f(x,y)$ represent the pixel of coordinate x and y, ∇ denote gradients, W refer to the weight of a coordinate, and ϕ be an edge direction. The histogram of oriented gradients can be expressed as follows.

$$\nabla_{x,y}^{\text{Horizon}} = 2f(x + 1, y) - 2f(x - 1, y) + f(x + 1, y + 1)$$
$$- f(x - 1, y + 1) + f(x + 1, y - 1) - f(x - 1, y - 1) \tag{1}$$

$$\nabla_{x,y}^{\text{Vertical}} = 2f(x, y + 1) - 2f(x, y - 1) + f(x + 1, y + 1)$$
$$- f(x + 1, y - 1) + f(x - 1, y + 1) - f(x - 1, y - 1) \tag{2}$$

$$W_{x,y} = \left(\left(\nabla_{x,y}^{\text{Horizon}} \right)^2 + \left(\nabla_{x,y}^{\text{Vertical}} \right)^2 \right)^{1/2} \tag{3}$$

$$\phi_{x,y} = \tan^{-1} \left(\nabla_{x,y}^{\text{Horizon}} \Big/ \nabla_{x,y}^{\text{Vertical}} \right). \tag{4}$$

When gradients are computed, a histogram of edge directions is then created to record the number of pixels that belongs a direction.

Attentive Motion Entropy

After extracting saliency maps and regional gradients of video frames, this step uses those data to calculate motion entropy. As mentioned in the previous section, although dominant motion features were widely used for key-frame identification in sports videos, we observed that such techniques were subject to global camera motion effects. Therefore, motion entropy based on saliency maps and regional gradients is proposed herein to alleviate those problems in video skimming. Motion entropy [11] has been proved effective in rejecting false alarms and efficient computation. The major difference between motion entropy and attentive motion entropy is that the latter is operated in the attentive regions of frames, including saliency maps and regional gradients. Motion entropy of saliency maps is given as follows:

$$H_{\text{SaliencyMap}} = \sum_k w_k \times h_k \tag{5}$$

where k is the index of the sector after the polar axis is partitioned into equal segments; h_k (motion directivity entropy) is the total entropy that belongs to the corresponding sector, and w_k is the total weights in the sector. Let p_k represent the proportion of motion vectors, whose angles fall into the kth sector, and h_k can be defined as

$$h_k = -p_k \times \log p_k. \tag{6}$$

It describes the activity rate of a frame when it is converted to a saliency map; the strength of directivity entropy is calculated by

$$w_k = \sum_i r_{i,k} \Big/ \sum_i r_i \tag{7}$$

where $\sum_i r_{i,k}$ is the sum of the lengths of the motion vectors in the kth region, and the denominator means the sum of the total lengths.

Based on the same concept in [11], motion entropy of regional gradients can be obtained by using histograms of oriented gradients. Thus, attentive motion entropy can be expressed by

$$H_{\text{Attentive}} = H_{\text{SaliencyMap}} \times H_{\text{RegionalHOG}} \tag{8}$$

where $H_{\text{SaliencyMap}}$ calculates inter frame saliency, and $H_{\text{RegionalHOG}}$ represents intra frame saliency. The following figures compare the proposed attentive motion entropy and the other approaches.

Mutual Information Based on Differential Salient Motions and Oriented Gradients

Once a highlighted frame is detected in a video, the system then determines the boundaries of an event. Thus, playback would become smooth when users watch it. A simple way to estimate the boundaries of an event involves using color information between frames, such as [8]. The authors presented a novel concept of applying mutual information to detecting similar images. Although it may work in movies, some problems still arise in analyzing sports videos. This is because color features in different scenes among a sports video slightly change. Therefore, this work modifies the original equation to compute motion data.

$$\begin{aligned} MI_{t,t+1}^{\text{SaliencyMap}} &= MI_{t,t+1}^{\text{Active}} - MI_{t,t+1}^{\text{Inactive}} \\ &= \mathbb{C}_{t,t+1} \times \log \frac{\mathbb{C}_{t,t+1}}{\mathbb{C}_t^{\text{Active}} \mathbb{C}_{t+1}^{\text{Active}}} - \mathbb{C}_{t,t+1}' \times \log \frac{\mathbb{C}_{t,t+1}'}{\mathbb{C}_t^{\text{Inactive}} \mathbb{C}_{t+1}^{\text{Inactive}}} \end{aligned} \tag{9}$$

where $\mathbb{C}_{t,t+1}$ denotes the proportion of the motion vectors, which remain as active (directional) states from one frame to the next, whereas $\mathbb{C}_{t,t+1}'$ represents inactive ones; $\mathbb{C}_t^{\text{Active}}$ or $\mathbb{C}_t^{\text{Inactive}}$ calculates the percentage of the directional or unidirectional vectors in a single frame, respectively. When multiplied with $MI_{t,t+1}^{\text{RegionalHOG}}$

which calculates mutual information of differential oriented gradients (i.e., $MI_{t,t+1}^{\text{Active}} - MI_{t,t+1}^{\text{Inactive}}$), [21] becomes

$$MI_{t,t+1}^{\text{Saliency}} = MI_{t,t+1}^{\text{SaliencyMap}} \times MI_{t,t+1}^{\text{RegionalHOG}}. \tag{10}$$

Equation 10 measures the coherent status between consecutive frames. The higher mutual information is, the more coherent two frames are. With the use of 10, the system can find the start and the end of a motional event efficiently because a sports event usually comes along with significant motion activity changes.

Boundary Smoothing Based on Segmental Spectro-Temporal Subtraction

After mutual information is computed, the browsing system needs to output meaningful parts of a video by choosing "salient peaks" from the curve. A simple way is to set up a predefined threshold, and then the system can determine sports events based on the value. Nevertheless, many local minimums appear in a segment, and they may cause the system to misclassify false boundaries. Consequently, removing those local minimums becomes an important issue.

Activity detection technology has been widely used in signal processing. It involves detecting silence segments and recognizing whether an audio or visual signal of interest begins or not. Zero-crossing rates [22] are considered as the simplest way to separate silence from an unknown signal. However, in our case, the value of mutual information is always larger than zero, which makes zero-crossing rates inapplicable. Other research, such as spectral entropy [23] and spectral flatness [24], also offers the same functionality

In this work, segmental spectro-temporal subtraction, which is derived from spectral subtraction [21], is proposed at this stage. Spectral subtraction was originally developed to suppress noise from voice signals in the frequency domain. Let X represent the noiseless mutual information and N denote noise signals in a video. The spectro-temporal representation of MI, X, and N can be given as follows.

$$MI(f,t) = X(f,t) + N(f,t) \tag{11}$$

Taking power spectrum on both sides yields

$$|MI(f,t)|^2 = |X(f,t)|^2 + |N(f,t)|^2. \tag{12}$$

The above function can be rewritten in terms of segments, where η is the index of segments. The estimated X is calculated by subtracting averaged local minimum of noise near η.

$$\left|\hat{X}_\lambda(f,t)\right|^2 = \left|MI_\lambda(f,t)\right|^2 - \left|N_\lambda(f,t)\right|^2 = \left|MI_\lambda(f,t)\right|^2 - E\left(\sum_\tau \left|N_{\lambda-\tau}(f,t)\right|^2\right).$$

$$(13)$$

Experimental Results

To assess the performance of our system, six soccer videos (videos 1–4) were used in the test. These video dataset contained 152 sports highlights, including corner kicks, goalkeeper shots, block tackles, goal-and-cheer shots, etc., which were manually labeled and segmented by humans. All of these sports events were motion events. Besides, unnecessary segments, such as commercial parts, anchor shots, audience scenes, and captions, were removed from the videos. Motion events in the same video were concatenated together, each of which is separated with at least one non motion event. To evaluate whether the proposed system can distinguish motion events from non highlights in videos, average precision and recall rates are used as the criteria in the experiment.

Performance test between the parameters of the proposed method is listed in Table 1. Table 1 compares average precision and recall rates of the parameters, where the first column of the table is the parameter number, the second denotes the parameter type, and the rest columns list the test results. Closely examining precision and recall rates revealed that motion entropy-based parameters (parameters 1 and 2) could detect more accurate motion events, compared with mutual information-based ones (parameters 3 and 4). However, the recall rates of mutual information-based parameters rose to more than 75 %, owing to selecting longer segments. Clearly, when combined with parameters 1–4, the performance could achieve as high a precision rate as 82.1 %. Also, the recall rate was as high as 83.7 %. Such experimental results indicate that integration of the two types of parameters can offer higher discriminability than that of individual one.

Table 2 lists the experimental results of different approaches. As shown in the table, our results outperformed the other methods. The precision and recall scores were respectively increased to 85.9 and 85.5 % on average. In comparison with the other baselines, the precision difference between our methods and the other

Table 1 Performance test between the parameters of the proposed method

Number	Parameter	Precision (%)	Recall (%)
1	Motion entropy based on saliency maps	80.3	72.2
2	Motion entropy based on histograms of oriented gradients	80.0	71.3
3	Mutual information based on differential salient motions	72.0	80.5
4	Mutual information based on differential oriented gradients	70.8	75.5
5	Motion entropy (1 and 2) +mutual information (3 and 4)	82.1	83.7

Table 2 Evaluation results of different approaches

Method	Precision (%)	Recall (%)
Saliency map [16]	69.5	62.2
Motion activity [7]	78.0	74.9
Motion entropy [11]	81.9	78.5
Color-based mutual information [8]	79.6	81.7
Proposed method without boundary smoothing	82.1	83.7
Proposed method with boundary smoothing	85.5	86.1

Table 3 Smoothness test. After watching video summaries, each user can give a score (integer), which ranges from 1 (worse) to 3 (better), to each system

Video number	Motion activity	Motion entropy	Color-based mutual information	Proposed
I	1.2	1.7	2.0	2.1
II	1.4	2.2	2.0	2.0
III	1.4	1.6	2.1	2.8
IV	1.0	2.1	1.8	2.6
V	1.4	2.2	2.0	2.2
VI	1.4	1.5	1.1	2.5
Average	1.3	1.9	1.8	2.4

approaches was enhanced by 8.6 %. The recall difference of our results also found an average increase of 11.1 %. Notably, when the proposed boundary smoothing technique was added to the system, the precision and recall rates can reach 86.5 and 86.1 % because inappropriate motions were filtered. In contrast to the other methods, the proposed one has demonstrated its efficacy of detecting highlights.

In order to benchmark smoothness of selected segments, a common subjective measurement, mean opinion scores (MOS), was employed in the test. Total ten persons were invited to the test, and each of them was asked to give a score ranging from 1 (worse) to 3 (better). The evaluation results are shown in Table 3. We observed that the average scores of our system were above 2.0, whereas those of the baselines were smaller than 2.0. Due to lack of mutual information for event boundary checking, the average score of Yeo and Liu's method reached only 1.3. In contrast to Yeo and Liu's system, the MOS score of Cernekova's approach, which focuses on only color mutual information, was increased to 1.8 although Cernekova's approach could determine motion boundaries. The score of the proposed method could reach as high as 2.4, which was larger than that of motion entropy system by 0.6. Such results imply that our algorithm has better ability to generate smooth segments than the baselines.

The following figures list the snapshots of the testing results for demonstration. Notably, for clarity of presentation, the total testing frames are 39 (13 different keyframes), and each snapshot is manually extracted. As shown in Fig. 2a, some undesired frames (all the bottom snapshots) are selected due to their color saliency. Like the result of Cernekova et al., although motion entropy (Fig. 2b) can remove

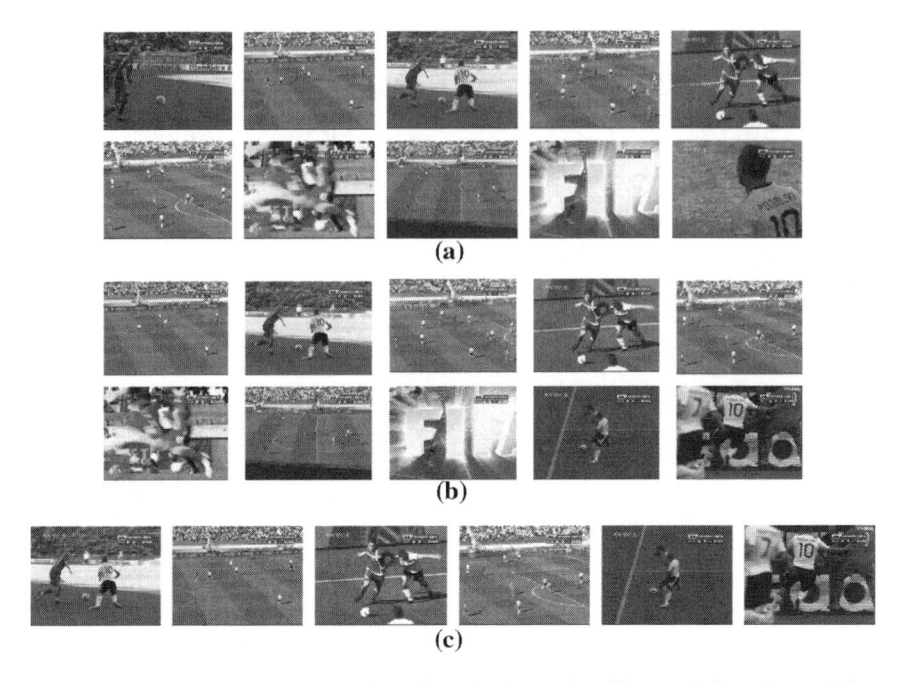

Fig. 2 Snapshots of the video summaries, where the time order of images is from the top-left to the bottom-right. **a** Color-based mutual information [8]; **b** Motion entropy [11]; **c** Proposed system. The duration of sample video is around one minute. Notably, for clarity of presentation, the total testing frames are 39 (13 different key frames), and each snapshot is manually extracted

frames that contain slight motions, such as the first and last frames in Fig. 2a, it still detects false highlights (the bottom-left four frames). Compared with the above results, the proposed system (Fig. 2c) is less susceptible to sudden movement and color saliency.

Conclusion

This work introduces a new approach for summarizing sports videos. In order to detect the highlights in a video, attentive motion entropy is employed by combining salient motions with regional histograms of oriented gradients. Another event-boundary detection algorithm is also proposed in this work to search similar frames in a video, which is based on mutual information analysis. To identify the boundaries of motion events more accurately, differential salient motions and differential oriented gradients are analyzed according to the directivity changing rates between frames. Moreover, segmental spectro-temporal subtraction is used to smooth mutual information curves and to remove unnecessary highlight boundaries that have smaller mutual information. With the above-mentioned methods, our system is capable of generating video abstracts efficiently. Experiments on a 6-video database indicate that the proposed approach can achieve a precision and recall rate of 85.5 and 86.1 %, respectively. Besides, smoothness is also enhanced by 0.7 on average. Furthermore, a comparison reveals that the proposed approach is superior to the other baselines.

References

1. Bagga A, Hu J, Zhong J, Ramesh G (2002) Multi-source combined-media video tracking for summarization. In Proceedings of the 16th IEEE international conference pattern recognition, Quebec, Canada, Aug 11–15. IEEE computer society, Washington, pp 818–821
2. Liu T, Zhang H-J, Qi F (2003) A novel video key-frame-extraction algorithm based on perceived motion energy model. IEEE trans. circuits and systems for video technology 13(10):1006–1013
3. Duan L-Y, Xu M, Tian Q, Xu C-S, Jin JS (2005) A unified framework for semantic shot classification in sports video. IEEE Trans Multimedia 7(6):1066–1083
4. Li Z, Schuster GM, Katsaggelos AK (2005) MINMAX optimal video summarization. IEEE trans. circuits and systems for video technology, 15(10):1245–1256
5. Liu T-Y, Ma W-Y, Zhang H-J (2005) Effective feature extraction for play detection in American football video. In: Proceedings of the 11th international multimedia modeling conference (Melbourne, Australia, Jan. 12–14). IEEE computer society, Washington, pp 164–171
6. Ma Y-F, Hua X-S, Lu L, Zhang H-J (2005) A generic framework of user attention model and its application in video summarization. IEEE Trans Multimedia 7(5):907–919
7. Yeo B-L, Liu B (2005) Rapid scene analysis on compressed video. IEEE trans circuits and systems for video technology, 5(6):533–544
8. Cernekova Z, Pitas I, Nikou C (2006) Information theory-based shot cut/fade detection and video summarization. IEEE transactions circuits and systems for video technology, 16(1):82–91
9. Li Y, Lee S-H, Yeh C-H, Kuo C-CJ (2006) Techniques for movie content analysis and skimming: tutorial and overview on video abstraction techniques. IEEE Signal Process Mag 23(2):79–89
10. Taskiran CM, Pizlo Z, Amir A, Ponceleon D, Delp EJ (2006) Automated video program summarization using speech transcripts. IEEE Trans Multimedia 8(4):775–791
11. Chen C-Y, Wang J-C, Wang J-F, Hu Y-H (2007) Event-based segmentation of sports video using motion entropy. In: Proceedings of the 9th IEEE international symposium multimedia (Taichung, Taiwan, 10–12). IEEE computer society, Washington, pp 107–111
12. You J, Liu G, Sun L, Li H (2007) A multiple visual models based perceptive analysis framework for multilevel video summarization. IEEE trans. circuits and systems for video technology, 17(3):273–285
13. Chen B-W, Wang J-C, Wang J-F (2009) A novel video summarization based on mining the story-structure and semantic relations among concept entities. IEEE Trans Multimedia 11(2):295–312
14. Black MJ (1996) The robust estimation of multiple motions: parametric and piecewise-smooth flow fields. Comput Vis Image Underst 63(1):75–104
15. Walther D, Rutishauser U, Koch C, Perona P (2005) Selective visual attention enables learning and recognition of multiple objects in cluttered scenes. Comput Vis Image Underst 100(1–2):41–63
16. Walther D, Koch C (2006) Modeling attention to salient proto-objects. Neural Networks 19(9):1395–1407

17. Ma Y-F, Lu L, Zhang H-J, Li M (2002) A user attention model for video summarization. In: Proceedings of the 10th ACM international conference multimedia (Juan-les-Pins, France, Dec. 01–06). ACM Press, New York, pp 533–542

18. Lu S, King I, Lyu MR (2005) A novel video summarization framework for document preparation and archival applications. In: Proceedings of the 2005 IEEE aerospace conference (Big Sky, Montana, United States, Mar. 05–12). IEEE computer society, Washington, 1–10

19. Ngo C-W, Ma Y-F, Zhang H-J (2005) Video summarization and scene detection by graph modeling. IEEE transactions circuits and systems for video technology, 15(2):296–305

20. Chen Y-T, Chen C-S (2008) Fast human detection using a novel boosted cascading structure with meta stages. IEEE Trans Image Proc 17(8):1452–1464

21. Kamath SD, Loizou PC (2002) A multi-band spectral subtraction method for enhancing speech corrupted by colored noise. In: Proceedings of the IEEE international conference acoustics, speech, and signal processing (Orlando, Florida, United States, May 13–17). IEEE computer society, Washington, pp 4164–4167

22. Zhang T, Kuo C-CJ (2001) Audio content analysis for online audiovisual data segmentation and classification. IEEE Trans Speech Audio Proc 9(4):441–457

23. Misra H, Vepa J, Bourlard H (2006) Multi-stream ASR: an oracle perspective. In: Proceedings of the ISCA international conference spoken language processing (Pittsburgh, Pennsylvania, United States, Sep. 17–21)

24. Gray AH, Markel JD (1974) A spectral-flatness measure for studying the autocorrelation method of linear prediction of speech analysis. IEEE Trans Acoustics, Speech and Signal Processing 22(3):207–217

A Framework Design for Human-Robot Interaction

Yu-Hao Chin, Hsiao-Ping Lee, Chih-Wei Su, Jyun-Hong Li, Chang-Hong Lin, Jhing-Fa Wang and Jia-Ching Wang

Abstract Multimodal human-robot interaction integrates various physical communication channels for face-to-face interaction. However, face-to-face interaction is not the only communication method. The proposed framework achieves more flexible communications such as communication to others at a distance. Intimate and loose interactions are categorized as ubiquitous multimodal human-robot interaction. Therefore, this work presents a framework design for human-robot interaction using ubiquitous voice control and face localization and authentication implemented for intimate and loose interactions. The simulation results demonstrate the practicality of the proposed framework.

Introduction

Humanoid and human-friendly robots are useful in daily life owing to their ability to assist individuals in executing some tasks by easily communicating with each other. Face-to-face interaction is almost adopted for human-robot interaction. Additionally, individuals can simply, naturally, and efficiently interact with robots without location and time constrains. Such communication is called "ubiquitous human-robot interaction".

Many investigations on multimodal human-robot interaction focus on various communication channels in social interaction like speech and visual communication [1–4]. Unfortunately, the only communication in most of these systems is face-to-face interaction. Ubiquitous human-robot interaction has been proposed [5]

Y.-H. Chin · C.-W. Su · J.-H. Li · Chang-HongLin · J.-C. Wang (✉)
Department of Computer Science and Information Engineering,
National Central University, Jhongli, Taiwan, Republic of China
e-mail: jcw@csie.ncu.edu.tw

H.-P. Lee · J.-F. Wang
Department of Electrical Engineering, National Cheng-Kung University,
Tainan, Taiwan, Republic of China

Y.-M. Huang et al. (eds.), *Advanced Technologies, Embedded and Multimedia for Human-centric Computing*, Lecture Notes in Electrical Engineering 260, DOI: 10.1007/978-94-007-7262-5_118, © Springer Science+Business Media Dordrecht 2014

to exploit gesture and voice to realize different interactions such as intimate and loose interactions. For loose interaction, author of [5] employed a camera to look over the room to detect whether someone is requesting interaction through gesturing. However, such an approach is not the most natural and efficient means of interacting with robots because individuals need to wave their hands to call robots to come close to them in order for them to instruct the robots. Therefore, speech may be a better choice for interaction between human and robots. The proposed framework design for human-robot interaction developed ubiquitous voice control subsystem to construct six microphones in the room to detect whether someone is giving a request by voice commands for loose interaction. Furthermore, one microphone installed in each robot is utilized to record close talking for intimate interaction. As well as speech, face localization and authentication subsystem is applied to recognize a user face for intimate interaction, and to localize a user in a different distance for loose interaction.

The target scenario of the proposed framework is broad. For example, the framework can be employed in a ward or a hotel room, in which robots recognize user faces, then pay attention to their voice command like 'take tea to me' everywhere, and localize people to accomplish this task.

The remainder of this paper is organized as follows. Section The Framework Design for Human-Robot Interaction describes the framework design for human-robot interaction. Section Ubiquitous Voice Control then presents the ubiquitous voice control subsystem in detail. Section Experimental Results summarizes the experimental results. Conclusions are finally drawn in Section Conclusion.

The Framework Design for Human-Robot Interaction

The Framework Design for Intimate Interaction

Figure 1 describes the framework design. For the close interaction between the user and the robots, the face is first recorded by a camera on a robot and then authenticated by the face recognition module. If the authentication result is acceptance, then the audio signal recorded by a microphone embedded on a robot is processed. The recorded audio signal is usually distorted by distributed background noises in the real world. Thus, the speech enhancement module has to improve the noisy recorded audio signal. The enhanced audio signals are then recognized via a recognition module. If the enhanced audio signals are recognized as voice commands like "take tea to me", then the robot executes the associate actions. The robot needs to execute two actions, one is taking tea and the other is localizing and recognizing the user. For the face localization and authentication subsystem, six audio signals recorded by six microphones embedded on ceiling are processed by the preprocessing module, and the strongest signal is selected to determine the region in which the user is localized. Additionally, the face detection module detects the user from the region size and position of the face. The face

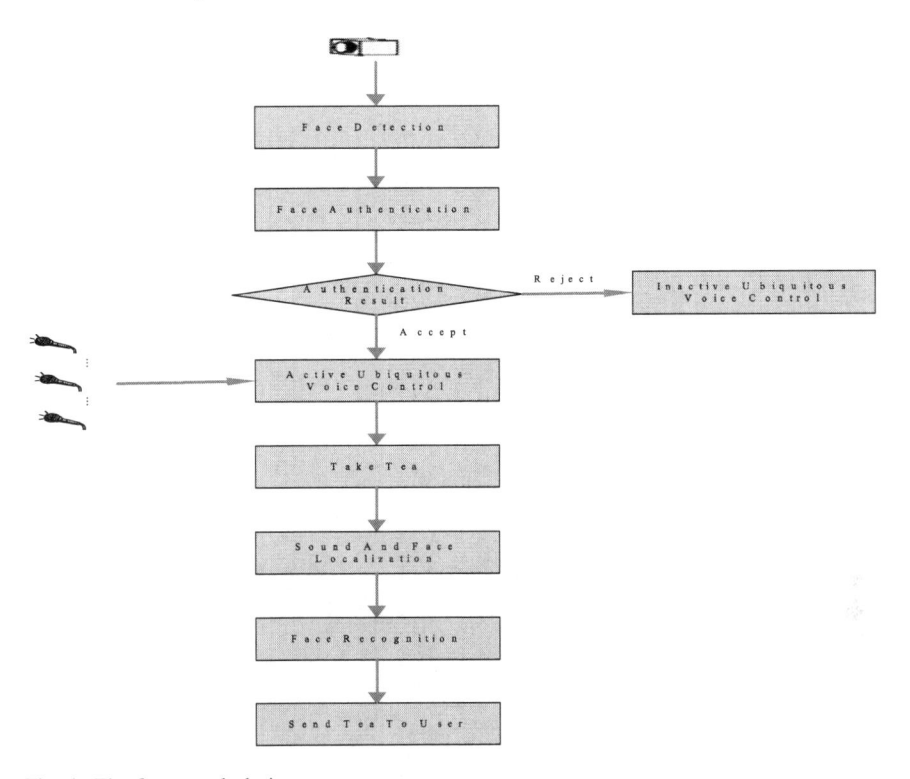

Fig. 1 The framework design

recognition module then recognizes the facial image. If the result is acceptance, then the robot sends tea to user.

The Framework Design for Loose Interaction

If users and robots interact loosely, then the robots locate the user and turn toward the user's location to detect and recognize the user's face. If the recognition result is acceptance, then the audio signals recorded by six microphones is transmitted to the preprocessing module which normalizes the signals and selects some stronger signals for enlarging the signals to be efficiently recognized and reducing computation time in the speech enhancement and speech recognition modules. The speech enhancement module is then run to enhance some recorded audio signals. Finally, the enhanced audio signals are then recognized by the speech recognition module. If the enhanced audio signals are recognized as voice commands, then the robots will execute the associate actions. The robots react as described in the previous section.

In the next section, we will address on the ubiquitous voice control subsystem.

Ubiquitous Voice Control

Our ubiquitous voice control subsystem mainly includes a speech enhancement module and a speech recognition module. To apply speech recognition in ubiquitous computing environment, a speech enhancement module is essential.

A well-known difficulty in speech recognition is recording audio signal with remote microphones at varying distances. Therefore, in many cases, head-mounted close-taking microphones such as Bluetooth headset are often used for speech recognition. This approach is called a wearable computing environment [6, 7]. However, this approach is not very comfortable and convenient for users.

Therefore, we need to overcome this limitation by using ubiquitous computing environment. The ubiquitous computing environment [6, 7] which utilizes multiple microphones provides an intuitive solution to this limitation. In a ubiquitous computing environment, microphones are embedded everywhere in the environment. Figure 2 illustrates the ubiquitous computing environment in the proposed design. Microphones are installed on the ceiling, and arranged separately by unique-distance. Therefore, microphones can record any audio signal irrespective of its origin. However, due to the restriction of the microphone number and distance, the recorded audio sources away from the microphones are noisy and weak. To overcome this problem, besides using sensitive and anti-noise microphones, speech enhancement and normalization are adopted. In the proposed

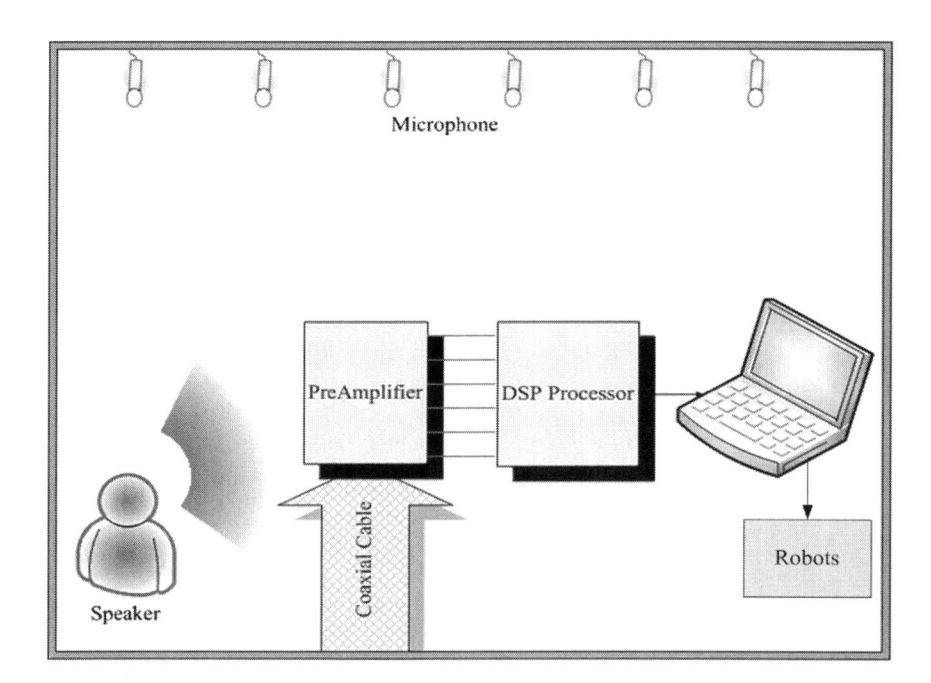

Fig. 2 Illustration of the ubiquitous computing environment in our design

ubiquitous computing environment, six noisy audio signals recorded by six microphones have to be pre-processed to reduce the number of noisy audio signals and speed up the computing of the system. This work uses voting strategy to select the most accurate recognition result. Furthermore, the maximum amplitude of the sounds also gives the hints to robots for sound localization.

Experimental Results

Six microphones were placed on the ceiling of an room for recording pervasive audio signals. Passing through the silver link microphone cables, the audio signals were delivered to a preamplifier card with six input ports. The parallel amplified streams were converted to a serial stream by the 16-bit PCMCIA analog input card, and then received by a notebook.

In our experiments, 19 keywords were adopted as voice commands. The testing database contained 57 spoken sentences. The recognition rate in the proposed ubiquitous computing environment was around 70 %. We are currently working on adapting the placement of microphones and survey the appropriate microphone suit for far distance recording.

Conclusion

This work presents a framework design for human-robot interaction which is being developed by building technologies for ubiquitous voice control subsystem and face localization and authentication subsystem. This multimodal human-robot interaction consists of two communication methods, intimate and loose interaction. These interactions are realized by combining ubiquitous voice control subsystem, including preprocessing, speech enhancement and speech recognition modules with face localization and authentication subsystem, including face detection and face recognition modules. Hence, the proposed framework design demonstrates a ubiquitous, convenient and natural environment, which is appropriate for human-robot interaction.

References

1. Breazeal C, Aryananda L (2002) Recognizing affective intent in robot directed speech. Auton Robots 12(1):83–104
2. Sidner C, Kidd C, Lee C, Lesh N (2004) Where to look: a study of human-robot engagement. In: Proceedings of intelligent user interfaces Madeira, Island of Funchal, Portugal, pp 78–84
3. Hanafiah ZM, Yamazaki C, Nakamura A, Kuno Y (2004) Human-robot speech interface understanding inexplicit utterances using vision. In: Proceedings conference on human factors in computing systems, Vienna, Austria, April pp 1321–1324

4. Yoshizaki M, Kuno Y, Nakamura A (2001) Human-robot interface based on the mutual assistance between speech and vision. In: Proceedings workshop on perceptive user interfaces, CD–ROM
5. Takeda H, Kobayashi N, Matsubara Y, Nishida T (1997) Towards ubiquitous human-robot interaction. In: Workiing Notes for *IJCAI*-97 Workshop Intell Multimodal Syst, pp 1–8
6. Furui S (2000) Speech recognition technology in the ubiquitous/wearable computing environment. In: Proceedings of IEEE international conference on acoustics, speech, and signal processing Istanbul, vol 6, pp 3735–3738
7. Rhodes BJ, Minar N, Weaver J (1999) Wearable computing meets ubiquitous computing: reaping the best of both worlds. In: Proceedings of 3rd international symposium on wearable computers (ISWC'99), Oct

Blind Signal Separation with Speech Enhancement

Chang-Hong Lin, Hsiao-Ping Lee, Jyun-Hong Li, Chih-Wei Su, Yu-Hao Chin, Jhing-Fa Wang and Jia-Ching Wang

Abstract A new speech enhancement architecture using convolutive blind signal separation (CBSS) and subspace-based speech enhancement is presented. The spatial and spectral information are integrated to enhance the target speech signal and suppress both interference noise and background noise. Real-world experiments were carried out in a noisy office room. Experimental results demonstrate the superiority of the proposed architecture.

Introduction

Many multiple- microphone speech enhancement methods have been proposed to exploit spatial information to extract the single source signal [1–4]. To significantly reduce the number of microphones and do not require a priori information about the sources, blind signal separation (BSS) methods are adopted to effectively separate interfering noise signals from the desired source signal. The second-order decorrelation based convolutive blind signal separation (CBSS) algorithm was recently developed [5]. Since estimating second-order statistics is numerically robust and the criteria leads to simple algorithms [6].

We proposed a novel architecture, in which critical-band filterbank is utilized as a preprocessor to provide improved performance and further savings on convergence time and computational cost. However, only critical-band CBSS does not work for removing background noise, which originates from a complex combination of a large number of spatially distributed sources. Therefore, a subspace-

C.-H. Lin · J.-H. Li · C.-W. Su · Y.-H. Chin · J.-C. Wang (✉)
Department of Computer Science and Information Engineering,
National Central University, Jhongli, Taiwan, Republic of China
e-mail: jcw@csie.ncu.edu.tw

H.-P. Lee · J.-F. Wang
Department of Electrical Engineering, National Cheng-Kung University,
Tainan, Taiwan, Republic of China

Y.-M. Huang et al. (eds.), *Advanced Technologies, Embedded and Multimedia for Human-centric Computing*, Lecture Notes in Electrical Engineering 260, DOI: 10.1007/978-94-007-7262-5_119, © Springer Science+Business Media Dordrecht 2014

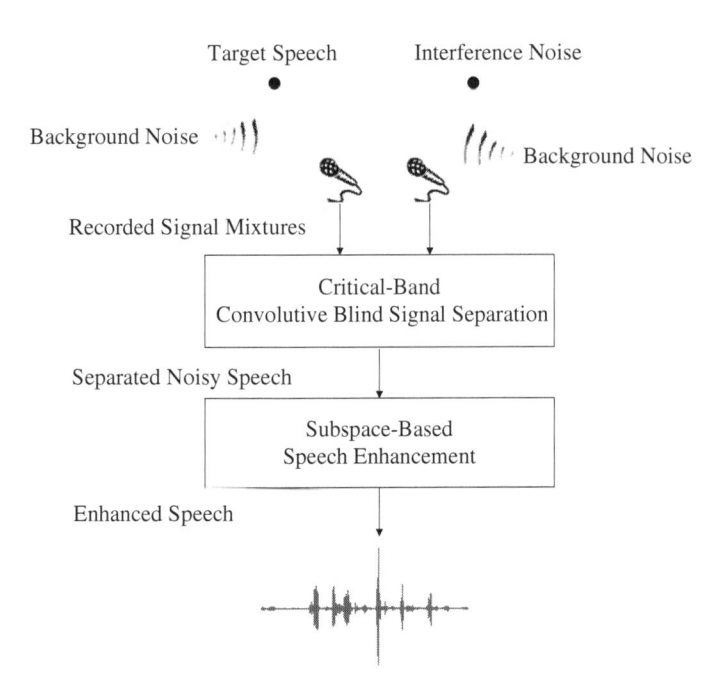

Fig. 1 Block diagram of proposed speech enhancement system

based speech enhancement method is utilized to reduce the background noise by exploiting additional spectral information [7].

Proposed Architecture

Figure 1 schematically depicts the block diagram of the proposed speech enhancement system. This architecture comprises a critical-band CBSS module and a subspace-based speech enhancement module. The input mixed signals are first processed by using the critical-band CBSS to separate the target speech from the interference noise. Next, the extracted target speech is fed into the subspace-based speech enhancement module to reduce the residual interferences and background noise. The proposed architecture adopts both spatial and spectral processing, and needs only two microphones.

Critical-Band Convolutive Blind Signal Separation

First, a critical-band filterbank based on the perceptual wavelet transform (PWT) is built from the psycho-acoustic model. The recorded signal mixtures are decimated into critical band time series by PWT. The CBSS is performed to separate the noisy speech and the interference noise in each critical band. A signal selection strategy based on high order statistics is then adopted to extract the target speeches. Finally, the inverse perceptual wavelet transform (IPWT) is applied to the critical-band extracted speeches to reconstruct the full-band separated noisy speech.

Perceptual auditory modeling is very popular for speech analysis and recognition. The wavelet packet decomposition is designed to adjust the partitioning of the frequency axis into critical bands which are widely used in perceptual auditory modeling. Within the 4 kHz bandwidth, this work uses 5-level wavelet tree structure to approximate the 17 critical bands derived based on the measurement [8, 9].

Convolutive Blind Signal Separation

This work assumes that two mixture signals $\bar{x}(t) = [x_1(t), x_2(t)]^{\mathrm{T}}$ composed of two point source signals $\bar{s}(t) = [s_1(t), s_2(t)]^{\mathrm{T}}$ and additive background noise $\bar{n}(t)$ are recorded at two different microphone locations:

$$\bar{x}(t) = \sum_{\tau=0}^{P} \mathbf{A}(\tau)\bar{s}(t - \tau) + \bar{n}(\tau). \tag{1}$$

The mixing matrix $\mathbf{A}$ is a 2×2 matrix and P represents the convolution order. Passing through the critical-band filterbank, PWT separates mixture signals into 17 critical-band wavelet packet coefficients. In each critical band, using an M-point windowed discrete Fourier transformation (DFT), the time-domain equation (1) can be converted into frequency-domain. The convolutive BSS is then performed in each critical band.

Signal Selection

In each critical-band, the CBSS has separated the mixed signals as speech dominant and interference dominant signals. Next, we should identify the target speech from the two separated outputs. Nongaussianity can be considered as a measure for discriminating the target speech and the interference noise by using kurtosis [3].

The last stage simply synthesize the enhanced speech using the inverse perceptual wavelet transform (IPWT).

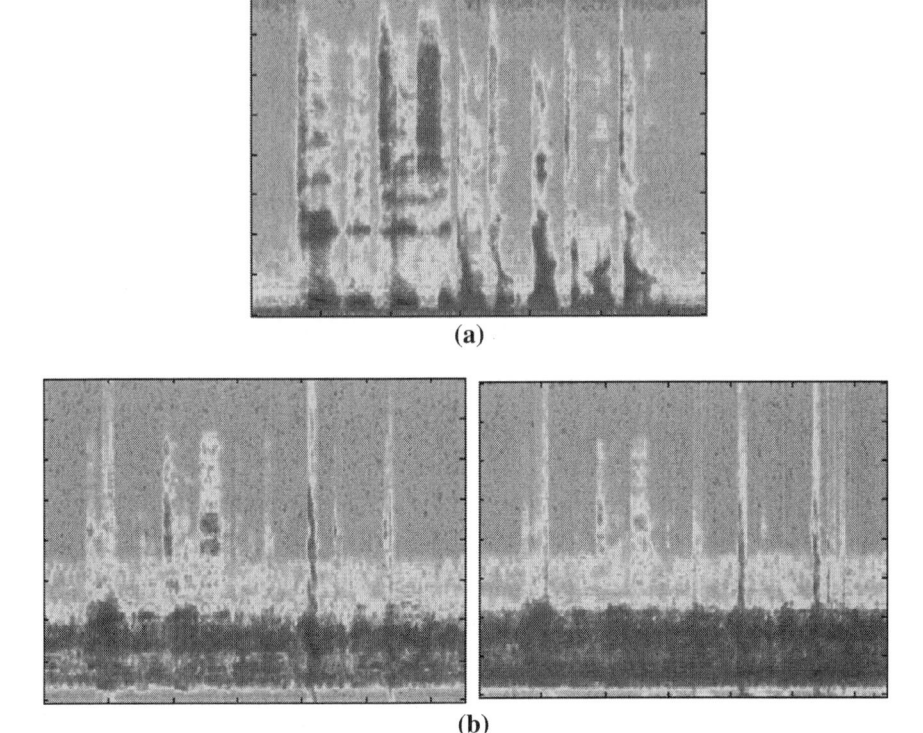

(a)

(b)

Fig. 2 a The original clean speech signals; **b** the 2-channel corrupted speeches under babble noise

Subspace-Based Speech Enhancement

The subspace-based speech enhancement is used to enhance the separated noisy speech by minimizing the background noise. The additive noise removal problem can be described as a clean signal $\bar{s}$ being corrupted by additive noise $\bar{n}$. The resulting noisy signal $\bar{u}$ can be expressed as

$$\bar{u} = \bar{s} + \bar{n}, \tag{2}$$

where $\bar{s} = [s(1), s(2), \ldots, s(L)]^{\mathrm{T}}$, $\bar{n} = [n(1), n(2), \ldots, n(L)]^{\mathrm{T}}$, and $\bar{u} = [u(1), u(2), \ldots, u(L)]^{\mathrm{T}}$. The observation period has been denoted as L. Henceforth, the vectors $\bar{s}$, $\bar{n}$, and $\bar{u}$ will be considered as part of real space R^{L}.

Ephraim and Van Trees proposed a subspace-based speech enhancement method [7]. The goal of this method is to find an optimal estimator that would minimize the speech distortion by adopting the constraint that the residual noise fell below a preset threshold.

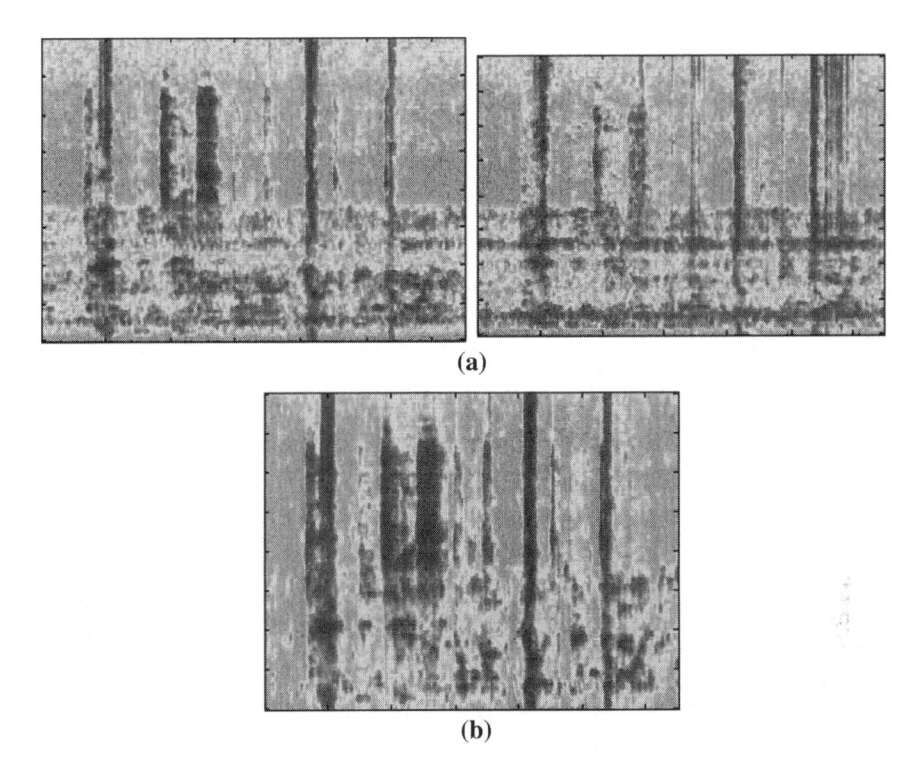

(a)

(b)

Fig. 3 **a** The two-channel critical-band CBSS outputs; **b** the selected enhanced speech

Experiment Results

The experiment was performed with a speech source and a babble interference noise at an angle of 150° and a distance of 40 cm from the center of the microphone array. Twenty different spoken sentences were played, each with about 50,000 samples and babble noise in AURORA database was employed as interference noise.

For objective evaluation, the SNR measure was adopted to evaluate these speech enhancement algorithms. Additionally, the modified Bark spectral distortion (MBSD) was also applied to assess speech quality. Since MBSD measure, presented by Yang et al. [10], is a perceptually motivated objective measure for mimicking human performance in speech quality rating. In both measures, the proposed architecture significantly outperforms conventional subspace enhancement method.

Figure 2 shows the spectrograms of original clean speech and two speeches corrupted by babble noise. Figure 3 illustrates the spectrograms of the critical-based CBSS outputs and the enhanced result. Figure 3a clearly reveals that one output is target speech dominant, while the other is interference dominant.

Conclusion

This work develops a spatio-spectral architecture for speech enhancement. The architecture consists of a critical-band CBSS module and a subspace-based speech enhancement module. The spatial and spectral information are exploited to enhance the target speech, and to suppress strong interference noise and background noise using two microphones. Kurtosis analysis is then adopted to select the target CBSS output. The enhancement performance is improved significantly.

References

1. VanVeen BD, Buckley KM (1988) Beamforming: a versatile approach to spatial filtering. IEEE Acoust, Speech Sig Process Mag 5:4–24
2. Kellermann W (1991) A self-steering digital microphone array. In: Proceedings of IEEE international conference on acoustics, speech, and signal processing, vol 5, April pp 3581–3584
3. Low SY, Nordholm S, Togneri R (2004) Convolutive blind signal separation with post-processing. IEEE Trans Speech Audio Process 12(5):539–548
4. Visser E, Otsuka M, Lee TW (2003) A spatio-temporal speech enhancement scheme for robust speech recognition in noisy environments. Speech Commun 41(2):393–407
5. Parra L, Spence C (2000) Convolutive blind source separation of nonstationary sources. IEEE Trans Speech Audio Process 8(3):320–327
6. Parra L, Fancourt C (2002) An adaptive beamforming perspective on convolutive blind source separation. Davis G (ed) In: noise reduction in speech applications, CRC Press LLC
7. Ephraim Y, Van Trees HL (1995) A signal subspace approach for speech enhancement. IEEE Trans Speech Audio Process 3(4):251–266
8. Zwicker E, Terhardt E (1980) Analytical expressions for critical-band rate and critical bandwidth as a function of frequency. J Acoust Soc Am 68:1523–1525
9. Chen SH, Wang JF (2004) Speech enhancement using perceptual wavelet packet decomposition and Teager energy operator. J VLSI Sig Process Syst 36(2–3):125–139
10. Yang W, Benbouchta M, Yantorno R (1998) Performance of the modified bark spectral distortion as an objective speech quality measure. In: Proceedings of IEEE international conference on acoust, speech, signal process pp 541–544

A Robust Face Detection System for 3D Display System

Yu Zhang and Yuanqing Wang

Abstract Face detection is a kind of extremely useful technology in many areas, such as security surveillance, electronic commerce and human–computer interaction and so on. Face detection can be viewed as a two-class classification problem in which an image region can be classified as being a "face" or "non-face". Detection and locating the position of the observers' face exactly play a critical role in stereoscopic display system. Accuracy, speed and stability are some main standards to evaluate an object-tracking system. The face detection system presented in the paper with classifiers trained by AdaBoost algorithm can meet the specific requirements of stereoscopic display in detecting speed, accuracy and stability. After accurate face detection, we utilize a certain method to detect the pupil in the area of face which is obtained in the above process. At last, the active 3D display equipment will project corresponding images of the same scene to users' pupil respectively to make sure the viewer can obtain the sense of depth. According to the experimental results, this system is highly accurate, stable and users can get well experience through this 3D display system.

Keywords Face detection · 3D display · AdaBoost

Introduction

Face detection is a kind of extremely useful technology in many areas, such as security surveillance, video search, electronic commerce, human–computer interaction and so on [1–3]. Face detection can be viewed as a two-class classification problem in which an image region could be classified as being a "face" or "non-face." Face detection also provides interesting challenges to the pattern recognition and machine learning area.

Y. Zhang (✉) · Y. Wang
Department of Electronic Science and Engineering, NanJing University, NanJing, China
e-mail: yxu828@sina.com

Y.-M. Huang et al. (eds.), *Advanced Technologies, Embedded and Multimedia for Human-centric Computing*, Lecture Notes in Electrical Engineering 260, DOI: 10.1007/978-94-007-7262-5_120, © Springer Science+Business Media Dordrecht 2014

After several decades of research, the existing face detection algorithms can be generally divided into four categories: knowledge-based methods [4], feature invariant approaches [5], template matching methods [6] and appearance-based methods [7, 8]. Accuracy, speed and stability are some main standards of evaluating an object-tracking system. The face detection system presented in the paper with classifiers trained by AdaBoost algorithm [7, 8] can meet the specific requirements of stereoscopic display in detecting speed, accuracy and stability. Afterwards, the accurate location of pupil will be detected in the next process with reference to the face area in the image which could reduce the computational complexity effectively and improve the detection accuracy in pupil tracking process. Subsequently, the stereoscopic display system will accurately project the corresponding images to the pupils of observers to achieve good user stereoscopic experiences without any accessory equipment.

Haar Features

Haar-like features [7, 8] are a kind of simple rectangle features which has five basic feature templates to describe an area feature in an image as shown in Fig. 1. The object can be detected efficiently and quickly through these features than basic pixels. The value of haar feature can be defined as the sum of the white pixels subtracts the sum of the black pixels in grayscale area which reflects the local gray level change in an image. Some selected haar feature with well performance to distinguish face and non-face are shown in Fig. 2 and each haar feature is described collectively by the parameters of feature type, size and location in the image.

The five basic haar features could be placed at any position of the images in any size, so there will be lots of features in the training process and it will cost considerable time if we calculate these feature values directly. In order to enhance the computing speed, we utilized integral image [7, 8] to avoid plenty of repetitive computation. The value of every point in integral image could be defined as the sum of grayscale value of the pixels located on the left and upper the point which is shown in (1). The integral image of a whole image can be quickly calculated by some simple certain rules. Subsequently, we can calculate the haar feature value by some ordinary addition and subtraction operation with the integral image only.

Fig. 1 Some haar features

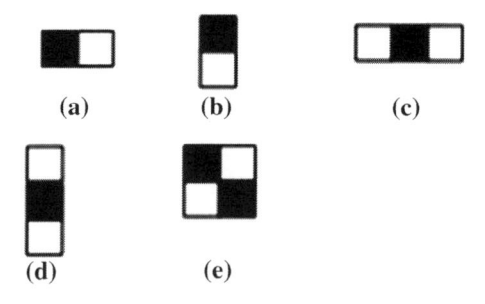

Fig. 2 Some haar features with well performance to distinguish face and non-face

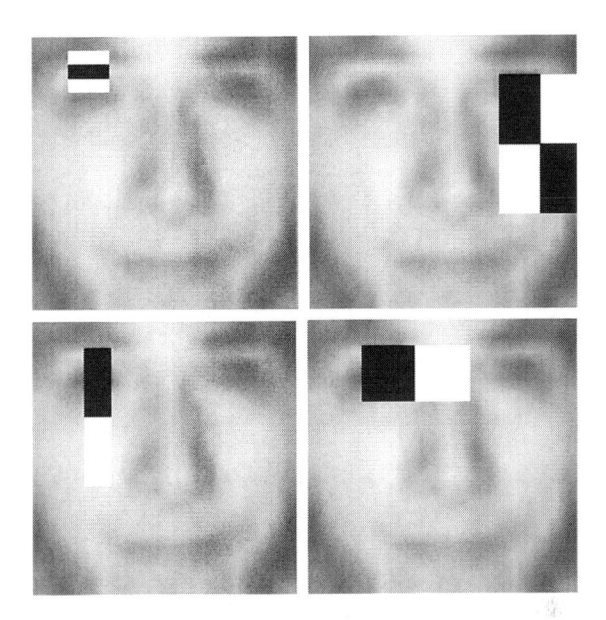

Fig. 3 Integral image

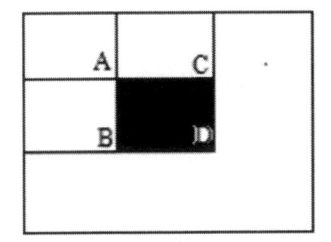

For instance, the sum of grayscales of black area in Fig. 3 could be calculated quickly only with reference to the values of the four points A, B, C, D in the integral image and the specific operation can be seen in (2). By this means, the haar feature can be calculated in constant time which could greatly reduce the computation complexity.

$$I(x, y) = \sum_{\substack{x\prime \le x \\ y\prime \le y}} i(x\prime, y\prime) \tag{1}$$

$$I_{area} = I_A + I_D - I_C - I_B \tag{2}$$

AdaBoost Algorithm

The AdaBoost algorithm [7, 8] is one kind of machine learning algorithm, which greatly improves the original boosting algorithm by dynamically adjusting the weight of weak classifiers to automatically meet the basic algorithm requirement.

The basic idea of Adaboost algorithm is to select some weak classifiers from weak classification space and integrate these weak classifiers to form a strong classifier according to certain rules. AdaBoost algorithm usually could be divided into two categories: discrete AdaBoost and real AdaBoost. In early discrete Adaboost algorithm, the classifier output are only two cases no matter it is a weak classifier or a strong classifier, which limits the performance of the weak classifier. Afterwards, in real AdaBoost, the weak classifier is improved to output real number which preferably depicts the confidence level meantime. The real Adaboost algorithm can make better performance than the discrete AdaBoost algorithm with the consideration from the experimental result so we choose the real Adaboost algorithm as our core method in the face detection system.

Experimental Results and Conclusions

The face detection system can work well in the test environment with single user and multiple observers. After accurate face detection, we utilize corresponding method to detect the pupil in the area of face which is obtained in the above process. After obtaining the coordinate of the pupils' location in the face, the active 3D display equipment will project a pair of images obtained in the same scene from different perspective to users' pupil respectively to make sure the viewer can obtain the sense of depth in order to form vivid stereoscopic feelings. Now this technology is utilized in our active 3D display system as a module to provide well stereoscopic experience for the observers without any attachment such as glasses and so on (Fig. 4).

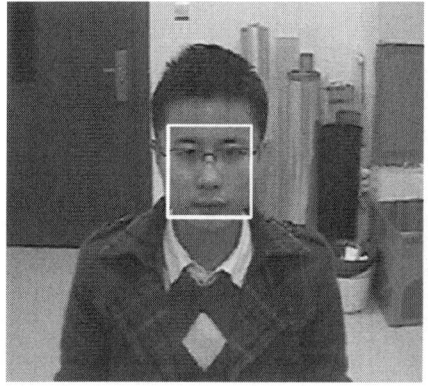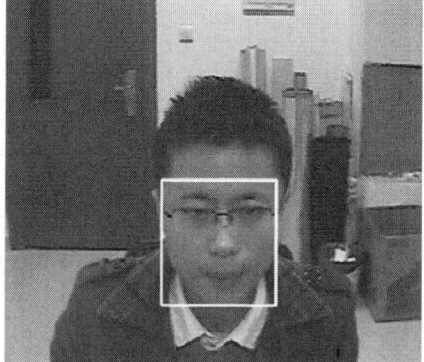

Fig. 4 some experimental results

Acknowledgments The project was supported by the program (Grant No. AHJ2011Z001).

References

1. Hsu RL, Abdel-Mottaleb M, Jain AK (2002) Face detection in color images. Pattern Analysis Mach Intell 24(5), 696–706. IEEE Press
2. Gorbenko A, Popov V (2013) Face detection and visual landmarks approach to monitoring of the environment. Int J Math Anal 7:213–217
3. Chen ZX, Liu CY, Chang FL, Han XZ (2013) Fast face detection algorithm based on improved skin-color model. Arabian J Sci Eng 38(3):629–635
4. Lin C, Fan KC (2001) Triangle-based approach to the detection of human face. Pattern Recogn 34(6):1271–1284
5. Li G, Xu Y, Wang J (2010) An improved adaboost face detection algorithm based on optimizing skin color model. In: 2010 sixth international conference on natural computation, vol 4, pp 2013–2015
6. Aiping C, Lian P, Yaobin T, Ning N (2010) Face detection technology based on skin color segmentation and template matching. In: 2010 second international workshop on education technology and computer science, vol 2, pp 708–711. IEEE
7. Viola P, Jones MJ (2004) Robust real-time face detection. Int J Comput Vision 57(2):137–154
8. Viola P, Jones M (2001) Rapid object detection using a boosted cascade of simple features. In: Computer Vision and Pattern Recognition, vol 1, pp 511–518 IEEE

Part XI
Virtual Reality for Medical Applications

Using Inertia Measurement Unit to Develop Assessment Instrument for Self-Measurement of the Mobility of Shoulder Joint and to Analyze Its Reliability and Validity

Shih-Ching Yeh, Si-Huei Lee and Yi-Hang Gong

Abstract Frozen shoulder is one type of shoulder disease that is commonly seen clinically, and its main symptom is the limit in the mobility of the shoulder joint of the patient and the shoulder pain. Since its therapeutic process of rehabilitation takes a long period of time, the patient usually abandons the therapy. In addition, under the current medical situation, physical therapist usually does not have extra time and space to help each patient densely to measure the mobility of the joint and to evaluate the progress of rehabilitation, hence, the patient usually is lack of in-time feedback to understand clearly the current rehabilitation progress, which in turn results in future low willingness of the patient to take rehabilitation, eventually, the goal of early and continuous therapy cannot be reached. Therefore, the objective of this research is to develop a set of "Assessment instrument for self-measurement of the mobility of shoulder joint", meanwhile, its reliability and validity is tested too. The system has associated wireless sensor technology and virtual reality, and the patient only has to follow the instruction and teaching on the screen to finish all kinds of standard shoulder joint actions, the progress in shoulder joint can then be evaluated at any time, eventually, the goal of real-time and self-assessment of the effectiveness of rehabilitation can be achieved.

Keywords Wireless IMU sensor · Virtual reality · Frozen shoulder

S.-C. Yeh · Y.-H. Gong (✉)
National Central University, Zhongli, Taiwan, Republic of China
e-mail: sam77917sam@hotmail.com

S.-C. Yeh
e-mail: shihchiy@csie.ncu.edu.tw

S.-H. Lee
Taipei Veterans General Hospital, Taipei, Taiwan, Republic of China
e-mail: sihuei.lee@gmail.com

Y.-M. Huang et al. (eds.), *Advanced Technologies, Embedded and Multimedia for Human-centric Computing*, Lecture Notes in Electrical Engineering 260, DOI: 10.1007/978-94-007-7262-5_121, © Springer Science+Business Media Dordrecht 2014

Introduction

Frozen shoulder is a general name for the injury of the soft tissue of shoulder and articular capsule, which is commonly seen in people around the age of 50. It is a commonly seen shoulder disease clinically, and its symptom includes the limit of active or passive mobility of shoulder joint and shoulder pain. Codman [1] was the first one to use the term frozen shoulder to describe such patient, later on, Neviaser [2] has observed the change of glenohumeral joint synovium and proposed adhesive capsulitis. The general population prevalence of frozen shoulder is usually in the range of 2–5 %, and the incidence among females is higher than that among males (with ratio of 58:42) [3]. The incidence for people with diabetes or hyperthyroidism is also higher than that of general people.

Lundberg [4] was the first one to propose the classification of frozen shoulder, namely, idiopathic or primary and secondary frozen shoulder, and the cause of disease of secondary frozen shoulder can also be divided into systemic, extrinsic and intrinsic. The systematic cause of disease includes diabetes, thyroid disease, and the extrinsic cause of disease includes stroke, humerus fracture and Parkinson's disease, etc. The intrinsic cause of disease includes rotator cuff injury, tendinitis or tendon calcination.

According to the disease process of frozen shoulder and histological check, Nevaiser and Nevaiser [5] has proposed four stages for frozen shoulder: (1) Pre-adhesive stage: It usually has only slight capsulitis, clinically, only slight pain can be felt at the end angle of the joint, and there is almost no problem such as the limit of mobility of the joint. (2) Freezing stage: Histologically, red and thickened joint capsule can be seen, and clinically, serious pain and limit on the joint is seen. (3) Frozen stage: Histologically, only slight inflammation can be seen in bursa synovialis, but the adhesive tissue becomes mature and thickening, meanwhile, the clinical pain symptom is also greatly reduced, and joint ankylosis becomes more significant, furthermore, active and passive mobility of joint becomes close. (4) Thawing stage: In this stage, almost no pain will be felt, the mobility of the joint will be significantly due to the completion of remodeling reaction.

Two major goals for the therapy of frozen shoulder are: (1) Reduce the pain in the disease part so that the patient can cooperate with the subsequent therapy; (2) Keep and improve the mobility of the joint so that the patient can get back to normal life as soon as possible. The most important thing in therapy is to use appropriate joint exercise, for example, stretching exercise (Including Codman's exercise, pulley therapy, front and side finger wall-climbing exercise and towel exercise, etc.) and joint mobilization, to extend the adhesive joint capsule so as to improve the mobility and function of the joint. In addition, appropriate muscle strengthening exercise should also be given to prevent muscle atrophy due to long term exercise reduction.

For the rehabilitation effectiveness of frozen shoulder, in addition to patient's subjective reduction in pain and progress in function, objectively, it has to rely on the test on active and passive mobility of joint implemented by the therapist, however, the implementation from the therapist is very time-consuming, and it takes an appropriate space too. Hence, if a good and simple joint assessment method for the patient to make self-test at home can be set up, then not only the patient can make self-test on the mobility of the joint, the home-stay rehabilitation patient can also get real time feedback too. In addition, since the patient can make the test at home, hence, only through appropriate learning, even without the existence of the therapist, the patient can, without spending too much time, complete the assessment of the active mobility of the shoulder joint.

Related Work

Virtual reality technology has fast development in recent years. Due to the progress in computer software and hardware technology and great enhancement in computer graphic capability, its application field becomes also much larger than before. Virtual reality is a technique using computer graphic and image composition technology in association with voice signal processing to simulate real environment in existence. The user, through all kinds of stimulations on the sense organs provided by the virtual reality system, can own the feeling like he is personally on the scene. Another feature is that the virtual reality environment allows the user to interact with the environment and to provide real time feedback response.

Today, there are lots of researches in the physical therapy industry using virtual reality in association with medical rehabilitation, and currently, most of the researches are focusing on stroke patient. Fischer et al. [6] has announced a rehabilitation system for the patient with upper extremity stroke. In the system, the specifically designed gloves developed by them together with sensor is used to let the patient make finger stretching rehabilitation, meanwhile, helmet display is also used to display the entire virtual reality scene. The user has to use the glove, and through the detection by the sensor, to catch and release all kinds of objects in the virtual reality environment to achieve the task, and the final data analysis result also proves that the patient gets significant progress after using this rehabilitation system. Kuttuva et al. [7] proposed another type of rehabilitation system for upper extremity stroke patient. The user sits beside a chair of the same height as his waist and faces the computer, and the hand on the disease side is put on the table, then a detector device is installed in the neighborhood of the wrist, through the detector, the moving track of the wrist of the user can then be detected, meanwhile, together with two games developed by them, rehabilitation exercise for upper extremity can be carried out. Camerao et al. [8] similarly used the stroke patient as the main target, together with the self-developed upper extremity sensor, the traditional rehabilitation therapeutic process can be put on virtual reality, and information such as hand moving speed and time lag of seizing the object can be more accurately assessed.

Hauschild et al. [9], in 2007 has published a paper regarding research using virtual reality to help patient with spinal cord injury for rehabilitation. In the research, electric stimulus and the power-supplied artificial limb for upper arm developed by them are used to put the traditional rehabilitation method on virtual reality. Through the helmet display, the patient can receive rehabilitation training using virtual reality to simulate the catching of object by upper extremity.

In 2000, scholars such as Rose et al. [10] has proposed a project for 210 persons under test to receive training in different groups, it was found that in the group tested with sense action training by stability tester in virtual reality shows significant progress in the within-the-group performance after therapy. After finishing the therapy, these people under test show significant progress when the same action is made by them in the real environment.

The above researches have explained the feasibility of applying virtual reality technology in medical rehabilitation, and they have also shown that virtual reality technology has become one very useful assisted tool in the medical rehabilitation field.

Goal

The main objective of this research is to apply wireless sensor and virtual reality technology to develop a set of system which can objectively assess the active mobility of the shoulder joint of the patient, meanwhile, the correlation between the measurement result of "Assessment instrument for self-measurement of the mobility of the shoulder joint" and the result of the traditional measurement is measured to assess the reliability and validity of the system.

System Architecture

This system architecture mainly associates wireless sensor IMU and virtual reality technology (Fig. 1). Wireless sensor is mainly used to detect the posture of joint on human body, through the installation of wireless sensor on limb and body part corresponding to each rehabilitation action, the acquisition of joint angle of that part is then performed, meanwhile, the acquired posture value is sent into the system for further processing. The system will, based on the received sensor posture value, display it on the screen according to human body animation reconstruction way so that the patient can clearly know the current body and joint posture; meanwhile, the system has also designed multi-representations, which include text, graph, image and audio, to remind the patient the current angle measurement result and the historical information, through these methods, real time information can be provided to the patient.

Fig. 1 System architecture
drawing

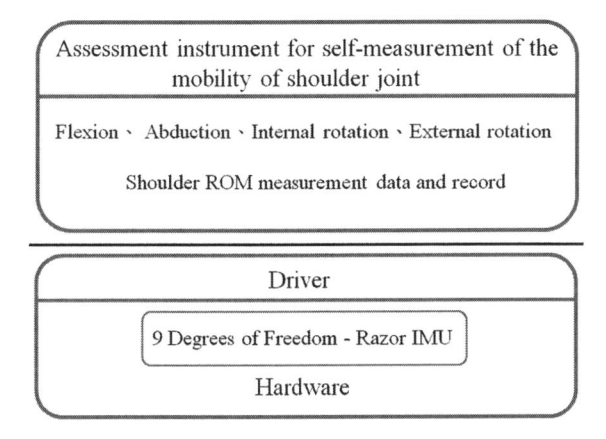

Wireless IMU Sensor

This system mainly uses wireless sensor technology to measure the angle of the shoulder joint of the patient. Wireless sensor can be free from the binding of cable mainly through low power wireless transmission, meanwhile, using the IMU (Inertial measurement unit), the posture values of an object can be measured (pitch, yaw and row). 9DOF Razor IMU has associated three sensors such as gyroscope, accelerator and magnetic meter to provide three-axis and nine degrees of freedom of measurement, meanwhile, after measuring the posture value of an object, the wireless transfer module will be used to transfer the measurement data, and the angle received by the sensor will be converted and transferred to PC in the form of 0–255 digital signals through Zigbee.

Software

The main development environment of the system software is in game engine 3D Unity, and four actions are mainly measured in the angle measurement of the shoulder joint: shoulder abduction, shoulder flexion, shoulder internal rotation and shoulder external rotation. In the formal measurement process, on the right side of the screen, there will be angle bar chart to show the angle measured by the sensor so that the patient can know at any time the current bending angle of the shoulder joint, meanwhile, three recorded angles and average values made by the system will all be listed as reference used by the patient and the physical therapist. In addition, the system will also compare this record with the last record, finally, whether progress can be seen when this measurement result is compared with the previous measurement result (Fig. 2) will be listed. During the measurement process, the system will process the received posture value, and the posture of

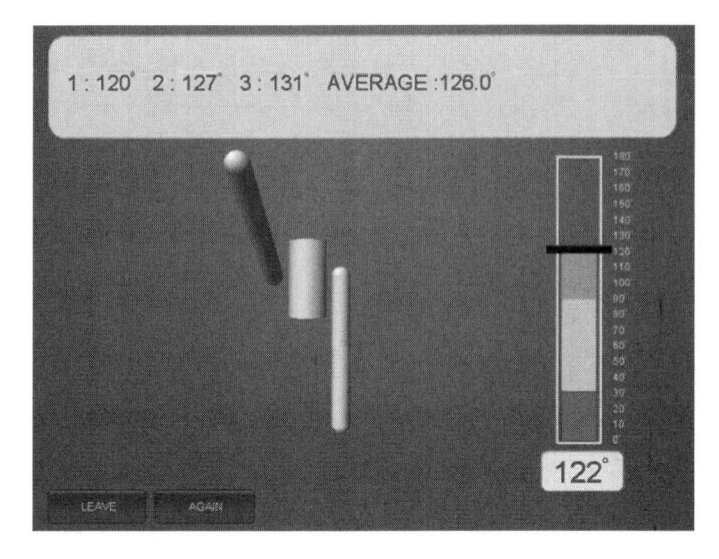

Fig. 2 Angle measurement screen

human body can then be displayed through animation, in other words, real time visual feedback can be sent to the patient.

Experiment Design

Participants

This research is foresighted, interfering, random-distributed and single-blind clinical test. Person under test meeting the condition will be assigned randomly according to the random number table generated by the computer into the experiment group and reference group. There will be 25 persons under test in each group.

Experimental Procedure

The shoulder joint angle to be measured in the experiment can be mainly divided into four actions: shoulder flexion, shoulder abduction, shoulder internal rotation and shoulder external rotation. Person under test in the experiment group and reference group will finish sequentially angle measurement for shoulder joint for four actions, and each action will be measured three times respectively, finally, the

Fig. 3 The situation of real measurement of the angle of shoulder joint of person under test in the experiment group

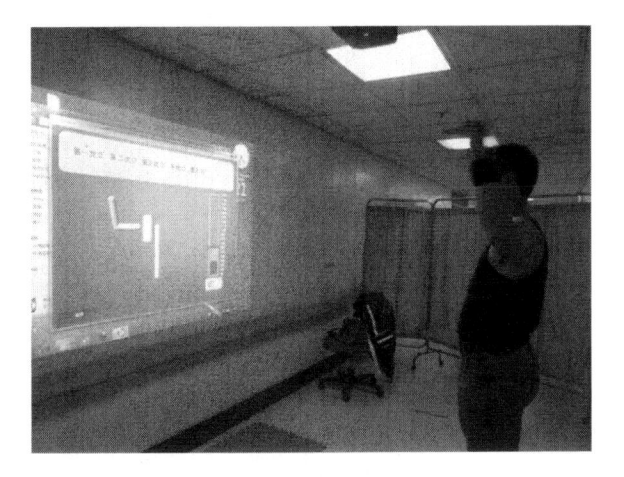

results of three shoulder joint angle measurements are averaged, and then the values are recorded by the physical therapist.

In the experiment group, the person under test will follow the sequence of the action to attach the sensor on specific location, before the formal angle measurement, the person under test can, through the teaching film provided by the system, get to understand how to perform the action and the steps to be watched during the angle measurement stage, during the measurement process, the person under test can, through the system, know the real time information of the exercise of his own shoulder joint. In addition, through the display of animation, person under test can judge correctly the current exercise posture of the shoulder joint, hence, the goal of real time feedback can be achieved (Fig. 3). In addition, the system will also display each measurement result and the finished average value on the screen, not only it can draw the attention of the person under test on the entire experimental process, but also it can save lots of time for the entire set of experiment.

In the reference group, manual measurement method is used. The measurement of then angle of shoulder joint is mainly performed by the physical therapist using traditional angle measurement tool. During the experimental process, there is no assistance from any software system (see Fig. 4), and the experimental step is the same as that of the experimental group, that is, for each action, the physical therapist will use traditional measurement method to do three measurements, then the average value will be calculated and recorded.

Measurement and Analysis

In the experiment group, wireless sensor is mainly used for the measurement, but in the reference group, the physical therapist will use traditional angle measurement tool to do the measurement, for each action in both groups, three times of

angle measurement have to be done and the values then averaged, finally, the data will be analyzed by SPSS. The analysis is done through a comparison of correlation coefficient for the difference between the experiment group and reference group, meanwhile, the reliability and validity of the system is inspected too.

In addition, survey questionnaire is used to evaluate patient's feeling on five items such as Presence, Usefulness, Playfulness, Intention to Use and Flow Theory. In the scoring of the questionnaire, 7 selection items of each question are given with scores, very disagree is given with a score of 1, disagree is given with a score of 2, slightly disagree is given with a score of 3, no comment is given with a score of 4, slightly agree is given with a score of 5, agree is given with a score of 6, very agree is given with a score of 7, and the calculated average result is called susceptibility.

Discussion

In this study, correlation coefficient is mainly used to compare the correlation between experiment group and reference group, and the main task of correlation coefficient is to, aiming at two variables suspicious of the existence of linear correlation (in random), calculate a number to reflect the strength of the linear correlation, hence, when the absolute value of overall correlation coefficient get closer to 1, the linear correlation between two variables will get stronger.

The analysis result is as shown in Table 1. It can be seen that the correlation coefficient values of four actions are very close to 1, which means that the

Table 1 Comparison of correlation coefficient between experiment group and reference group

Flexion	Abduction	External rotation	Internal rotation
0.997**	0.978**	0.897**	0.984**

Table 2 The average susceptibilities of five scopes

Presence	Usefulness	Playfulness	Intention to use	Flow theory
5.8	5.7	6.1	5.4	5.7

measurement results between experiment group and reference group do not have significant difference but have very high correlation, and it means that the results measured by the system are very close to the results measured using traditional measurement method.

In addition, in the feedback of the survey questionnaire, 17 questionnaires have been collected. Table 2 shows the average susceptibility of five scopes of the survey questionnaire.

The result shows that the average susceptibilities of five scopes are all close to or even exceed a score of 5.5, which means that person under test has very good susceptibility on this system.

Conclusion

In this research, a new set of system has been proposed to measure the mobility of the shoulder joint of the patient with frozen shoulder. The system has mainly associated wireless sensor IMU and virtual reality technology. The final experiment result shows that the result measured by system sensor has very high correlation to the result measured by traditional angle measurement tool, moreover, there is no significant difference too. In addition, through questionnaire survey, the patient has very good susceptibility to the system, hence, it is proved that this system can effectively and reliably measure the mobility of the shoulder joint. Through the assistance of this system, in the future, it will save a lot of time and labor when the patient is going to measure the mobility of shoulder joint, meanwhile, the patient will be able to self-measure and self-assess the progress of his own rehabilitation.

References

1. Codman EA (1934) The shoulder: rupture of the supraspinatus tendon and other lesions in or about the subacromial bursa. T. Todd Co, Boston, MA
2. Neviaser JS (1945) Adhesive capsulitis of the shoulder, a study of the pathological findings in periarthritis of the shoulder. J Bone Joint Surg Am 27(2):211–222
3. Thierry D (2005) Adhesive capsulitis. Emedicine 11:7
4. Lundberg BJ (1969) The frozen shoulder. Clinical and radiographical observations. The effect of manipulation under general anesthesia. Structure and glycosaminoglycan content of the joint capsule. Local bone metabolism. Acta Orthop Scand Suppl 119:1–59

5. Neviaser RJ, Neviaser TJ (1987) The frozen shoulder. Diagnosis and management. Clin Orthop Relat Res 223:59–64

6. Fischer HC, Stubblefield K, Kline T, Luo X, Kenyon RV, Kamper DG (2007) Hand rehabilitation following stroke: a pilot study of assisted finger extension training in a virtual environment. Top Stroke Rehabil 14(1):1–12

7. Kuttuva M et al (2006) The Rutgers Arm, a rehabilitation system in virtual reality: a pilot study. Cyberpsychol Behav 9(2):148–151

8. Cameirao MS, i Badia SB, Zimmerli L (2007) The rehabilitation gaming system: a virtual reality based system for the evaluation and rehabilitation of motor deficits. IEEE of Virtual Rehabilitation, Venice, Italy, pp 29–33

9. Hauschild M, Davoodi R, Loeb GE (2007) A virtual reality environment for designing and fitting neural prosthetic limbs. IEEE Trans Neural Syst Rehabil Eng 15(1):9–15

10. Rose FD et al (2000) Training in virtual environments: transfer to real world tasks and equivalence to real task training. Ergonomics 43(4):494–511

The Development of Interactive Shoulder Joint Rehabilitation System Using Virtual Reality in Association with Motion-Sensing Technology

Shih-Ching Yeh, Si-Huei Lee and Yao-Chung Fan

Abstract Virtual Reality combined with medical rehabilitation has become the modern trend. In the clinical therapy, frozen shoulder is a common shoulder disease. They usually use proper rehabilitation exercise to pole adhesive joint capsule. In these long time process, patients usually can't continue. Our research is using virtual reality to combine with motion tracking in order to develop interactive shoulder joint rehabilitation system. We decide to use interactive game-based oriented way and to give real-time visual and auditory feedback to improve user's motivation and willingness of doing rehabilitation. In this way, patients can achieve better result of rehabilitation compared with traditional therapy.

Keywords Frozen shoulder · Virtual reality · Microsoft kinect

Introduction

Shoulder joint is the joint with largest mobility among all human's joints. It is supported by complicated soft tissue, hence, once shoulder joint is injured, lots of daily life functions will be limited. Frozen shoulder is a general name for the injury of the soft tissue of shoulder and articular capsule, which is commonly seen in people around the age of 50. It is a commonly seen shoulder disease clinically,

S.-C. Yeh (✉) · Y.-C. Fan
National Central University, Zhongli, Taiwan, Republic of China
e-mail: shihchiy@csie.ncu.edu.tw

Y.-C. Fan
e-mail: 100522070@cc.ncu.edu.twl

S.-H. Lee
Taipei Veterans General Hospital, Taipei, Taiwan, Republic of China
e-mail: sihuei.lee@gmail.com

Y.-M. Huang et al. (eds.), *Advanced Technologies, Embedded and Multimedia for Human-centric Computing*, Lecture Notes in Electrical Engineering 260, DOI: 10.1007/978-94-007-7262-5_122, © Springer Science+Business Media Dordrecht 2014

and its symptom includes the limit of active or passive mobility of shoulder joint and shoulder pain. Codman [1] was the first one to use the term frozen shoulder to describe such patient, later on, Neviaser [2] has observed the change of gleno-humeral joint synovium and proposed adhesive capsulitis. The general population prevalence of frozen shoulder is usually in the range of 2–5 %, and the incidence among females is higher than that among males (with ratio of 58:42) [3]. The incidence for people with diabetes or hyperthyroidism is also higher than that of general people.

Lundberg [4] was the first one to propose the classification of frozen shoulder, namely, idiopathic or primary and secondary frozen shoulder, and the cause of disease of secondary frozen shoulder can also be divided into systemic, extrinsic and intrinsic. The systematic cause of disease includes diabetes, thyroid disease, and the extrinsic cause of disease includes stroke, humerus fracture and Parkinson's disease, etc. The intrinsic cause of disease includes rotator cuff injury, tendinitis or tendon calcination.

Two major goals for the therapy of frozen shoulder are: (1) Reduce the pain in the disease part so that the patient can cooperate with the subsequent therapy; (2) Keep and improve the mobility of the joint so that the patient can get back to normal life as soon as possible. The most important thing in therapy is to use appropriate joint exercise, for example, stretching exercise (Including Codman's exercise, pulley therapy, front and side finger wall-climbing exercise and towel exercise, etc.) and joint mobilization, to extend the adhesive joint capsule so as to improve the mobility and function of the joint. In addition, appropriate muscle strengthening exercise should also be given to prevent muscle atrophy due to long term exercise reduction. Before implementing exercise therapy, hot compression, short wave, ultrasonic wave or interference wave, etc. can be used first to improve the extensibility of joint capsule, muscle, tendon and ligament, etc.

For the rehabilitation of frozen shoulder, the patient has to go between the home and hospital in a long term, physical and psychological depletion usually lets the patient give up, and the effectiveness of rehabilitation is thus reduced. In recent years, virtual reality is gradually applied in the rehabilitation therapy, and computer aided virtual reality can accurately create task of different strength and difficulty, through digital interface control, the doctor or therapist can easily set up and adjust the environment stimulations. This makes the design of the therapeutic process of rehabilitation richer, not only the patient can be provided with more chances to practice, but also real time sound and image feedback can be provided so that the patient can have better limb and body practice result within real-time and interactive operation environment design. Together with the game design, the rehabilitation therapy will become more interesting, and the patient's motivation for rehabilitation will be enhanced, in other words, through the giving of certain feedback and feeling of achievement to the patient, the maximal effectiveness of the rehabilitation can be achieved.

Related Work

Along with technological progress, we gradually are capable of using some high tech instrument and technology to help patients. Today, lots of scholars performing medical rehabilitation related researches and first-line doctor and therapist around the world start to use these technologies to make clinical therapy, in other words, they perform all kinds of new therapeutic process of rehabilitation and development of rehabilitation technology based on Virtual Reality and Augmented Reality, the design concept of user centered, consideration of the perception of the user on the system, usability and immersion, use of the interactive mode and strategy provided by the human machine interface [5, 6].

Virtual reality does not take too much expense to get rehabilitation situation close to the real world. Moreover, virtual reality can be adjusted on the schedule table according to the level of injury and rehabilitation of the patient. Currently, researches associating virtual reality with medical rehabilitation are mostly on patients with stroke [7, 8].

There is a research [9] used virtual reality to construct interactive game for rehabilitation training. In the research, Microsoft Kinect Sensor is also used to make action control of the full body, and the user, under the sensing of the sensor, has completed a series of rehabilitation balance tasks.

These researches have pointed out that the association of virtual reality with medical rehabilitation is feasible, in addition to that, motion-sensing detection can also be added to increase interactive element and to enhance the effectiveness of rehabilitation.

Goal

The major goals of this research are:

1. The situation design and system analysis of interactive shoulder joint rehabilitation system.

This research has associated the application of Kinect sensor to design a set of interactive virtual reality rehabilitation system, and innovative technology has been applied to construct new and interactive rehabilitation system. For the selected rehabilitation exercise mode, life situation simulation is performed, and all kinds of perceptual stimulations are designed. In addition, since the assisted equipment for today's rehabilitation action training only has single direction functional purpose, there is lack of signal feedback design, hence, for the rehabilitation effectiveness aspect, in addition to the subjective assessment from the rehabilitation doctor and the patient himself, there is still no quantified objective analysis. The interactive virtual reality rehabilitation system designed in this research not only provides interactive, virtual and real rehabilitation assisted tool, but also can provide quantified index of the effectiveness of rehabilitation to the patient, and all these can enhance the motivation of continuous rehabilitation of the patient.

2. Initial investigation was made on the effectiveness of interactive shoulder joint rehabilitation system for the improvement of frozen shoulder.

Interactive and signal feedback medical assisted training simulator in association with Kinect sensor is used together with perspective and interventional research method to cure frozen shoulder patient and to enhance therapeutic motivation and to quantify therapeutic effectiveness. Empirical medical method is used to assess the subjective and objective effectiveness difference before and after the therapy. Moreover, the overall effectiveness is investigated so as to set up the clinical rehabilitation new model for frozen shoulder patient.

System Architecture

This system uses Unity3D as the development platform to develop five traditional actions of rehabilitation such as flexion, abduction, internal rotation, external rotation and circumduction. The association of virtual reality and medical rehabilitation has resulted in the design of seven rehabilitation games. Kinect is used together to prepare interactive shoulder joint rehabilitation system so that during the rehabilitation process, training and assessment can be performed. Based on the requirement, system construction is done by aiming at the following two goals:

1. Different training posture is given according to the requirement of severity at different quadrant for the patient's frozen shoulder:

For example, if the patient has serious flexion angle problem, then the patient should be made to face the screen to train the flexion extension of the shoulder joint;
If the patient has more serious flexion abduction angle problem, then the patient should be made to face screen to train the abduction extension of the shoulder joint.

2. Task of different level of difficulty should be given according to different lifting capability of the arm of the patient:

Since the severity and speed of progress of each frozen shoulder patient is quite different, hence, each patient will have different capability during each time of rehabilitation (Fig. 1).

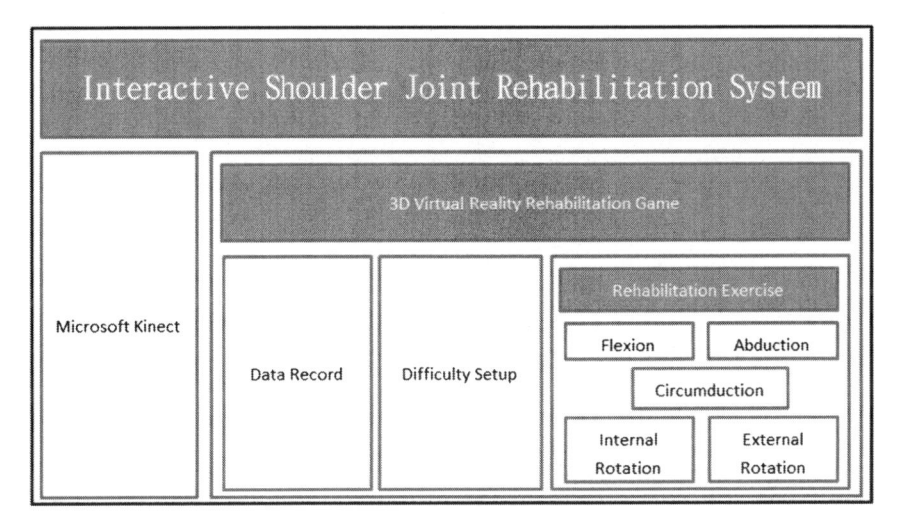

Fig. 1 System architecture

Hardware

In this interactive shoulder joint rehabilitation system, Kinect is used to acquire node coordinate of the skeleton of human body, then these information are transferred back to the computer, and in Unity3D, program is written to receive these skeleton information so as to achieve the interaction among scenes.

Kinect itself has two lenses and one infrared transmitter, and a total of three cameras (RGB color camera in the middle, infrared CMOS camera in the right side and infrared transmitter in the left side). Through the camera, object moving in front of the lens is caught; in addition, there is also one set of array type microphones. In the data acquisition part, Kinect has provided more than 20 skeleton nodes for the entire body to record the location information of the body skeleton in the space when the user is moving.

Software

The main development environment of this system software is based on game engine Unity3D. Unity3D is interactive game engine for development, after simple installation, it can be easily operated by the user. It has advantages such as user friendly interface and simple operation; hence, it plays a very critical role in the development of virtual reality scenes. The software has been used to develop task goal, and these game based tasks have brought brand new feeling to the patient in the rehabilitation process. In addition to that, using the image, voice and video to provide instruction for rehabilitation action for the patient can bring deeper impression and memory to the patient as compared to that by oral explanation. Such task oriented rehabilitation method not only can enhance patient's usage motivation but also can enhance the rehabilitation efficiency. In addition, some interactive feedback designs are added to transfer patient's attention on the pain during the rehabilitation process and to remove their discomfort in their mind. Software design can be divided into three directions:

1. **Situational rehabilitation task design**

In the traditional therapeutic process of rehabilitation, real task is used to lead the patient to complete the task goal and to achieve the goal of exercise, training and rehabilitation. Each game of this system has very clear task instruction to tell the patient to achieve the goal of the game as much as possible.

2. **Human machine interactive design**

The interaction between the user and virtual reality environment includes the tracking of hand location and the visual and audio feedback.

The tracking technology of hand location uses Kinect to acquire the skeleton information of the upper extremity, in other words, the program will read these information into the game scene of Unity3D, and skeleton information will be corresponded to the setup target object, hence, during the game process, the patient can operate on scene interactive object to achieve the task goal.

Visual feedback part is displayed in score. During the game process, the upper side of the screen will display the accumulated score or the remaining time; after the game is ended, the upper side of the screen will display total score or total time spent; in the entire process, the edge of the screen will display bar chart, that is, it will display the goal set up this time and the highest angle arm can be lifted at that moment. Moreover, sound feedback is displayed in different sound effect, in the beginning of the game, there will be instruction voice, beside, during the game process, when the patient has achieved the goal, feedback will be given immediately too. Such design can reflect the effectiveness of rehabilitation of the patient in real time, in other words, the patient can see real performance and level of progress exactly, and the rehabilitation motivation can then be greatly enhanced.

3. **Task performance**

The system itself, after the finish of the game, will display the total score acquired and total time spent by the patient so that the patient can know the game result.

Such situational rehabilitation task design process needs a professional therapist to lead and instruct 40 patients to try for several times, and the feedback and suggestion need to be recorded at any time so as to revise the software into a situational mode that can fit to most of the patients (Fig. 2).

Game Design :

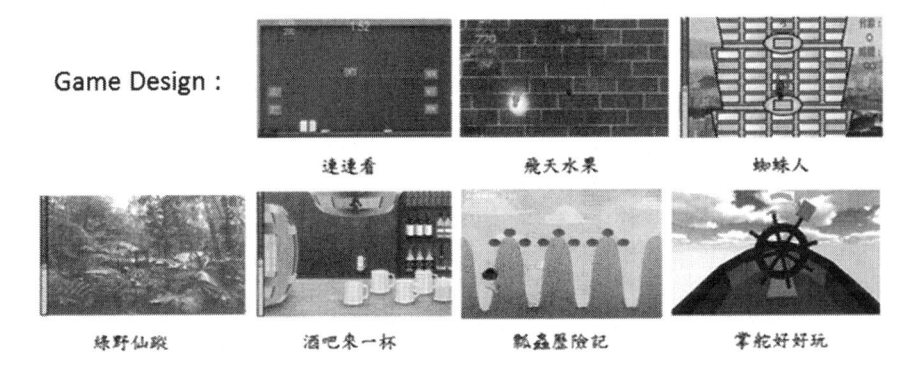

連連看　　　　　飛天水果　　　　　蜘蛛人

綠野仙蹤　　　酒吧來一杯　　　瓢盆歷險記　　　掌舵好好玩

Fig. 2 Unity3D game design

Experiment Design

Participants

This research is perspective, interventional, randomized controlled and single-blind clinical test. Persons meeting the condition will be assigned randomly according to random table into experiment group and reference group. 40 persons are expected to be used in this research.

Experimental Procedure

The shoulder joint angle to be measured in the experiment can be mainly divided into four actions: shoulder flexion, shoulder abduction, shoulder internal rotation and shoulder external rotation. Person under test in the experiment group and reference group will finish sequentially angle measurement for shoulder joint for four actions, and each action will be measured three times respectively, finally, the results of three shoulder joint angle measurements are averaged, and then the values are recorded by the physical therapist.

Before the start of the experiment, the experiment process and purpose of this experiment will be explained first to the person under test, and letter of consent will be singed under the agreed condition. This experiment is designed into two times each week for a total period of four weeks. Before and after the experiment, professional medical personnel will make Constant-Murley Score (CMS) assessment, then the experiment group and reference group will follow the following flow to do rehabilitation:

Experiment group: The shoulder joint of the disease side will receive hot compression and electric therapy of interfering wave, and each time will last 20 min, for two times a week and a total of 4 weeks. Moreover, it will receive rehabilitation from interactive shoulder joint rehabilitation system, for 20 min each time, two times a week and a total of 4 weeks.

Reference group: The shoulder joint at the disease side will receive hot compression and electric therapy from interfering wave, 20 min each time, two times a week and a total of four weeks. Person under test in this group will receive homestray rehabilitation exercise for general frozen shoulder, and the rehabilitation action is similar to that in the experiment group, but there is no situational design (Codman's exercise and front and side finger wall-climbing exercise).

When rehabilitation from interactive shoulder joint rehabilitation system is received, the implementation sequence each time is fixed, and each person each time will finish these seven rehabilitation games.

Measurement and Analysis

If division is made based on the measurement time, then one respective measurement will be done before the start of the therapeutic process and at the end of the entire therapeutic process, and the measurement will be done aiming at the shoulder joint angle and the overall function of the upper extremity of the patient. The part-time assistant, without knowing the experiment group or reference group, will be responsible for the measurement of the angle of shoulder joint of the patient before and after the therapy.

1. Active and passive shoulder joint angle: Standardized angle measurement tool is used to measure the angle of shoulder joint at the disease side using standard posture, and it includes flexion, abduction, internal rotation and external rotation.
2. The overall function of upper extremity: Using Constant-Murley Score (CMS) for the assessment.

Before and after the experiment on the patient, professional medical and nursing personnel should ask the patient the severity of pain, then together with the extendable angle of the patient, it will be quantified into scores with full score of 100.

For the comparison of differences among numerical variables, non-parameteric statistics (Kruskal–Wallis Test) is used to carry out statistical analysis. In the statistics, $p < 0.05$ is used to represent that there is significant statistical difference. Moreover, survey questionnaire is prepared to assess patient's perception on Presence, Playfulness, Intention to Use and Flow Theory, etc. In the scoring of the survey questionnaire, 7 selection items of each question are given with score respectively, when it is very disagree, score of 1 will be given, when it is disagree, score of 2 will be given, when it is slightly disagree, score of 3 will be given, when it is no comment, score of 4 will be given, when it is slightly agree, score of 5 will be given, when it is agree, score of 6 will be given, when it is very agree, score of 7 will be given, and the calculated average result is called perception.

Table 1 *P*-value comparison between experiment group and reference group

Median(IQR) (degree)	Experiment group (before)	Reference group (before)	P-value	Experiment group (after)	Reference group (after)	P-value
Flexion	149.1	149.0	0.904	171.6	166.0	0.050
	(129.2–162.7)	(131.7–157.5)		(153.2–174.0)	(154.7–171.2)	
Abduction	145.8	144.0	0.947	168.0	157.7	0.040
	(104.4–163.5)	(100.7–163.9)		(150.3–171.3)	(127.7–169.9)	
External rotation	62.1	60.0	0.076	83.65	74.3	0.017
	(55.7–83.9)	(39.7–71.6)		(71.5–88.2)	(56.5–81.2)	
Internal rotation	40.5	37.1	0.051	65.5	50.65	0.003
	(36.4–64.2)	(21.9–53.7)		(54.0–71.9)	(40.2–59.0)	

Discussion

Moreover, non-parametric statistics of Wilcoxon Rank Sum Test is used for statistical analysis, and it was found that P value <0.05, which means that significant difference does exist.

The analysis results are as shown in Tables 1 and 2. The P-values of four actions are all <0.05 and the result shows that there is significant difference between the experiment group and the reference group. It means that our designed rehabilitation system can make greater progress to the angle of shoulder joint of the patient as compared to that of the traditional therapy (Table 3).

Table 2 Comparison of *P*-value of CMS score between experiment group and reference group

Constarit-Muriey score(CMS)	Experiment group (before)	Reference group (before)	P-value	Experiment group (after)	Reference group (after)	P-value
Median(IQE)	63.5	63.0	0.738	85.0	76	0.046
	(43.5–70.7)	(50.0–71.2)		(72.5–89.0)	(68.2–84.7)	

Table 3 Average perception of four scopes

Presence	Playfulness	Intention to use	Flow theory
5.4	5.7	5.5	5.7

Conclusion

The therapeutic process of rehabilitation is very tedious. As technology advances continuously, we can associate virtual reality with medical rehabilitation model to bring brand new feeling and perception to the patient, and we can also create the experience of playing game and performing rehabilitation at the same time.

Through the turning of the patient's attention onto playing games, the patient can forget the pain perceived in doing rehabilitation and the boring during tedious rehabilitation process. Interactive shoulder joint rehabilitation system indeed shows its effectiveness in therapy from the clinical test result, when it is compared with the current frozen shoulder rehabilitation, during the entire rehabilitation process, the patient not only has objective and quantified data to be referred to, but also can use interesting content of situational task as well as visual and audio real time feedback, hence, the rehabilitation motivation of the patient is greatly enhanced. Eventually, optimal therapeutic effectiveness can be reached continuously.

This study is a pilot study, which has used very new technology and tool to design and develop "interactive shoulder joint rehabilitation system" that has put together computer aided tool virtual reality, and the clinical therapeutic effectiveness is only an initial test. In the future, it is hoped that it can be developed and extended into software with more shoulder joint rehabilitation games, and it is hoped that deeper, more perfect and larger scale of clinical test can be performed. It is hoped that cloud technology can be associated in the future to create a new model of remote health care.

References

1. Codman EA (1934) The shoulder: rupture of the supraspinatus tendon and other lesions in or about the subacromial bursa. T. Todd Co, Boston, MA
2. Neviaser JS (1945) Adhesive capsulitis of the shoulder, a study of the pathological findings in periarthritis of the shoulder. J Bone Joint Surg Am 27:211–222
3. Thierry D (2005) Adhesive capsulitis. Emedicine 11:7
4. Lundberg BJ (1969) The frozen shoulder. Clinical and radiographical observations. The effect of manipulation under general anesthesia. Structure and glycosaminoglycan content of the joint capsule. Local bone metabolism. Acta Orthop Scand Suppl 119:1–59
5. Rheingold H (1991) Virtual reality (1st edn). Simon and Schuster, New York
6. Rizzo AA et al (2006) A virtual reality scenario for all seasons: the virtual classroom. CNS Spectr 11(1):35–44
7. Sveistrup H (2004) Motor rehabilitation using virtual reality. J Neuroeng Rehabil 1(1):10
8. Meldrum D et al (2012) Virtual reality rehabilitation of balance: assessment of the usability of the Nintendo Wii((R)) Fit Plus. Disabil Rehabil Assist Technol 7(3):205–210
9. Lange B et al (2011) Development and evaluation of low cost game-based balance rehabilitation tool using the Microsoft Kinect sensor. Conf Proc IEEE Eng Med Biol Soc 2011:1831–1834

Development of a Virtual Reality-Based Pinch Task for Rehabilitation in Chronic Hemiparesis

Shuya Chen, Shih-Ching Yeh, Margaret McLaughlin, Albert Rizzo and Carolee Winstein

Abstract Impaired pinch performance affects dexterity function after stroke. Virtual reality-based training may be beneficial for improving dexterity function. This study aimed to develop a virtual reality-based pinch task and to investigate its feasibility for chronic hemiparesis. The pinch task in the virtual environment was accomplished by coordinating two PHANTOM devices that provide haptic feedback. Participants grasped and lifted a virtual cube with 30-sec time limit for 10 trials. Cube size, cube mass and lift height were systematically varied. The participant poststroke attempted an average of 38 trials per session with a 60 % success rate and without complaint of fatigue or pain. After the training, the participant poststroke decreased the total time. However, the peak pinch force did not change. The results suggest that the virtual reality-based pinch task was

S. Chen (✉)
Department of Physical Therapy, China Medical University, Taichung, Taiwan
e-mail: sychen@mail.cmu.edu.tw

S. Chen · C. Winstein
Division of Biokinesiology and Physical Therapy at the School of Dentistry, University of Southern California, Los Angeles, CA, USA
e-mail: winstein@usc.edu

S.-C. Yeh
Department of Computer Science and Information Engineering, National Central University, Taoyuan, Taiwan
e-mail: shihchiy@csie.ncu.edu.tw

M. McLaughlin
Annenburg School for Communication and Integrated Media Systems Center, University of Southern California, Los Angeles, CA, USA
e-mail: mmclaugh@usc.edu

A. Rizzo
Institute for Creative Technologies, University of Southern California, Los Angeles, CA, USA
e-mail: arizzo@usc.edu

Y.-M. Huang et al. (eds.), *Advanced Technologies, Embedded and Multimedia for Human-centric Computing*, Lecture Notes in Electrical Engineering 260, DOI: 10.1007/978-94-007-7262-5_123, © Springer Science+Business Media Dordrecht 2014

feasible for chronic hemiparesis. Further investigation is warranted to better understand the effect of pinch force regulation using hepatic feedback.

Keywords Virtual reality · Pinch · Rehabilitation · Hemiparesis

Introduction

Stroke is a leading cause of long-term disability [1]. Recently, evidence show that stroke survivors have not used their affected arm to full extent in daily life even the arms are able to do so [2, 3]. While many patients regain most of the reaching and grasp capabilities for their upper extremity function, recovery of the pinch skill remains incomplete for most of the patients. Pinch movement is one the important skills of upper extremity [4, 5]. Impaired pinch skill significantly affects dexterity function after stroke. With practicing various interesting exercise tasks for considerable amount of time, the regained pinch capability would significantly improve the upper extremity function after stroke.

Virtual reality-based training has been shown to be beneficial for the enhancement of hand function poststroke [6–9], because game-like exercises in virtual environment are motivating and allow therapists to easily modify training parameters [8, 10]. This study aimed to develop a new virtual reality-based task for pinch function and to investigate its feasibility for chronic hemiparesis.

Methods and Materials

Participants

One participant with poststroke hemiparesis and one age-matched control participant were recruited. The inclusion criteria for the participants poststroke were: (1) stroke at least 1 month prior, (2) more than 18 years of age, (3) Mini-Mental Status Exam score $\geq$ 24, (4) no significant range of motion limitations in the hemiparetic upper extremity, (5) ability to perform the VR-based pinch task. Tested hand and the total time on pinch task were also matched for both participants.

Virtual Reality-Based Pinch Task

The pinch task was one of four upper extremity virtual reality-based tasks from our pilot project [8]. To develop the pinch task (specifically for the opposite movement

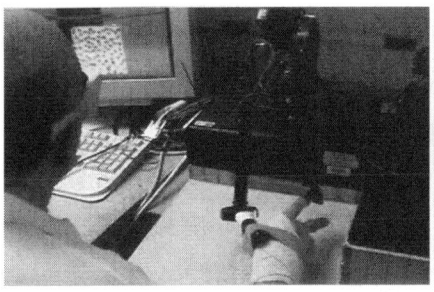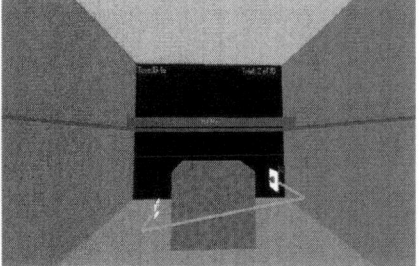

Fig. 1 Actual training with two PHANToM devices (*left*) and the display (*right*)

of thumb and index finger) in virtual environment, there were two PHANToM devices (SensAble Technologies) placed perpendicular to each other for the pinch movement (Fig. 1) and reconfigured to provide hepatic feedback for the pinch task. The goal of pinch task was to successfully pick the virtual cube up from the floor, lift it to a specified height, and place it back on the floor without dropping. Haptic feedback was provided for the thumb and index finger through the PHANToM devices, so that the participants felt they were lifting a real cube with mass. There were 10 trials per block. Each trial was configured using eight parameters: cube width (20–40 mm), cube height (20–40 mm), cube length (20–40 mm), mass (50–150 gw), dynamic friction (0.5–1.0), static friction (0.5–1.0), stiffness (0.5–1.0), and lift height (20–80 mm). Cube size, cube mass and lift height could be systematically varied according to the participant's capability level. A maximum of 30 s was allowed for each trial. The number of successful hit and total time on the task were provided as performance feedback after each block.

Outcome Measures

A preset benchmark block of the pinch task (Table 1) and the behavioral assessments including the Fugl-Meyer Assessment (FMA) [11] and the Box and Block test (B&B) [12] were conducted pre-and post- training to evaluate the effectiveness of the virtual reality system. Total time on the task and the peak pinch force during the benchmark block were further calculated.

Testing Procedure

Both participants practiced the pinch task for 15 min over three weeks for 12 training sessions. An experienced physical or occupational therapist worked with the participants. The responsibility of the therapist was to choose appropriate

Table 1 Task configuration of the benchmark block

Trial #	Size	Mass	Lift height
1	30	100	50
2	40	100	20
3	20	50	80
4	20	150	80
5	30	150	50
6	40	50	20
7	30	100	20
8	20	50	50
9	40	150	50
10	30	100	80

practice blocks and task parameters for participants. The practice blocks were chosen based on the participants' capability with some challenge. Participants sat in front of the computer screen with their tested thumb and index finger attached to the PHANTOM devices (Fig. 1, left). Before the trials, the 7-cm distance between the thumb and index finger were positioned and calibrated to reconfigure the metrics in the virtual environment. Participants were asked to grasp and lift a virtual cube within 30 s for 10 trials per block. The cube size, cube mass, and lifting height were systematically varied in order to keep a moderate level of difficulty. Performance summary feedback including successful hit and total time was provided after each 10-trial block.

Results and Discussion

The demographic data is shown in Table 2.

Training Performance

The participant poststroke completed 11 training sessions with the virtual reality-based pinch task (Table 3). The participant poststroke attempted an average of 38 trials per session with a 60 % success rate and without complaint of fatigue or

Table 2 Demographic data of study participants

	Age (y)	Sex	Dominant hand	Tested hand	Time after stroke (m)
Stroke	59	Male	Right	Right	24
Control	60	Female	Right	Right	

Table 3 Overall training performance

	Training days	Training time (hrs)	Blocks completed	Trials completed	Successful trials	Successful rate (%)
Stroke	11	2.28	48	415	250	60.24
Control	8	2.12	134	1,260	1,252	99.37

pain. Within the same block, time per trial and total time per block were longer for the participant poststroke compared to the control participant (Fig. 2).

Pre-Post Training Effect

After training, the participant poststroke decreased total time per block (Fig. 2) but peak pinch force did not change (Fig. 3). Peak pinch force was significantly greater for the participant poststroke across trials. The findings suggest that the participant poststroke could perform a virtual reality-based pinch task and improve speed of performance with practice without a change in peak force modulation. As for the behavior performance, the participant poststroke improved on the score of the Fugl-Meyer Assessment (FMA) and the total time on the benchmark block (Table 4). The results suggest the effectiveness of the virtual reality-based pinch task for stroke rehabilitation.

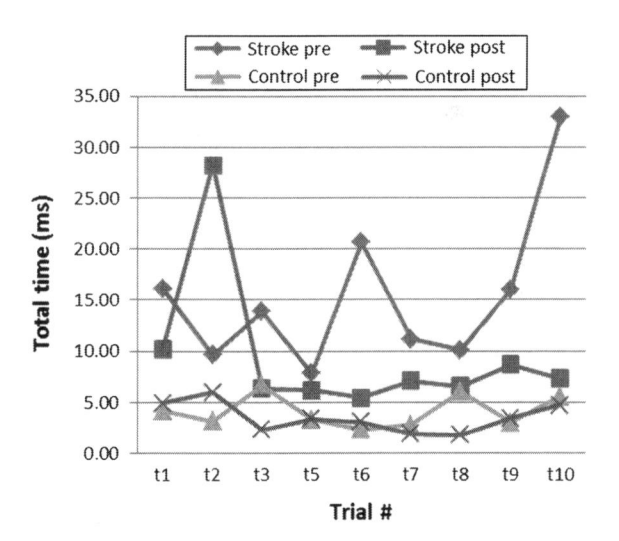

Fig. 2 Total time for each trial on the benchmark block pre- and post- training

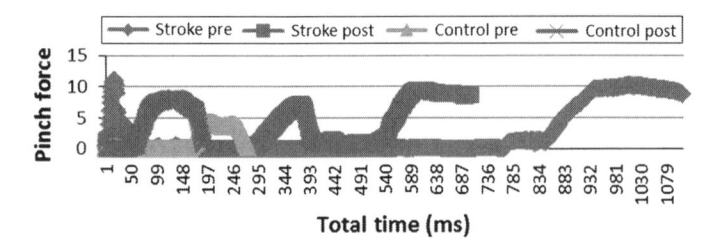

Fig. 3 An example (t3) of peak pinch force pre and post training for both participants

Table 4 Behavior and virtual reality performance data pre-post training

	FMA		B&B		Total time on BM (ms) (9 trials)*	
	Pre	Post	Pre	Post	Pre	Post
Stroke	48	55	17	17	138.37	85.88
Control					36.94	31.42

FMA Fugl-Meyer Assessment; *B&B* Box and Block test; *BM* Benchmark block
*Trial 4 was not included in calculating total time on BM

Table 5 Peak pinch force for each trial in the benchmark block pre- and post- training

		t1	t2	t3	t5	t6	t7	t8	t9	t10	mean
Stroke	Pre	6.27	10.40	8.75	9.97	12.43	11.07	9.76	11.68		9.64
	Post	9.62	8.94	10.81	10.48	12.00	9.48	10.28	12.41	11.23	10.58
Control	Pre	3.52	4.05	2.90	3.39	5.02	4.27	3.70	3.49	5.01	3.93
	Post	7.35	7.47	7.48	8.44	8.04	8.10	7.10	8.92	8.96	7.89

*Trial 4 was eliminated due to programming problems

Pinch Force pre- and post- Training

Peak pinch force in the benchmark block shows in Table 5 and Fig. 3. Pre-training peak pinch force was greater for the participant poststroke than the control participant. Both participants used a higher grip force after training. This may be associated with the concomitant increased lifting speed.

Conclusions

Post-stroke participant could perform a virtual reality-based pinch task and improve speed of performance with practice. A lack of explicit feedback regarding force during task performance may explain the similar peak force after training. Further investigation is warranted to better understand the effect of pinch force regulation during virtual reality-based training for patients poststroke.

Acknowledgments This study was supported by the Interdisciplinary Study of Neuroplasticity and Stroke Rehabilitation (ISNSR), an NIH Exploratory Center for Interdisciplinary Research (Grant # P20 RR20700-01) and the Integrated Media Systems Center, an NSF Engineering Research Center (Cooperative Agreement # EEC-9529152), both at the University of Southern California.

References

1. Rosamond W et al (2008) Heart disease and stroke statistics–2008 update: a report from the American heart association statistics committee and stroke statistics subcommittee. Circulation 117(4):e25–146
2. Andrews K, Stewart J (1979) Stroke recovery: he can but does he? Rheumatol Rehabil 18(1):43–48
3. Sterr A, Freivogel S, Schmalohr D (2002) Neurobehavioral aspects of recovery: assessment of the learned nonuse phenomenon in hemiparetic adolescents. Arch Phys Med Rehabil 83(12):1726–1731
4. Blennerhassett JM, Carey LM, Matyas TA (2006) Grip force regulation during pinch grip lifts under somatosensory guidance: comparison between people with stroke and healthy controls. Arch Phys Med Rehabil 87(3):418–429
5. Blennerhassett JM, Matyas TA, Carey LM (2007) Impaired discrimination of surface friction contributes to pinch grip deficit after stroke. Neurorehabil Neural Repair 21(3):263–272
6. Merians AS et al (2002) Virtual reality-augmented rehabilitation for patients following stroke. Phys Ther 82(9):898–915
7. Crosbie J et al (2005) Development of a virtual reality system for the rehabilitation of the upper limb after stroke. Stud Health Technol Inform 117:218–222
8. Stewart JC et al (2007) Intervention to enhance skilled arm and hand movements after stroke: a feasibility study using a new virtual reality system. J Neuroeng Rehabil 4:21
9. Wade E, Winstein CJ (2011) Virtual reality and robotics for stroke rehabilitation: where do we go from here? Top Stroke Rehabil 18(6):685–700
10. Stewart JC et al (2007) Training reach movements in individuals with hemiparesis: effect of a virtual environment. J Neurol Phys Ther 31(4):192
11. Fugl-Meyer AR et al (1975) The post-stroke hemiplegic patient. 1. a method for evaluation of physical performance. Scand J Rehabil Med 7(1):13–31
12. Desrosiers J et al (1994) Validation of the Box and Block Test as a measure of dexterity of elderly people: reliability, validity, and norms studies. Arch Phys Med Rehabil 75(7):751–755

Developing Kinect Games Integrated with Virtual Reality on Activities of Daily Living for Children with Developmental Delay

I-Ching Chung, Chien-Yu Huang, Shyh-Ching Yeh, Wei-Chi Chiang and Mei-Hui Tseng

Abstract Children with developmental delay (DD) often have difficulty in executing activities of daily living (ADL). Although independence in ADL is one of the ultimate goals of rehabilitation for children with DD, ADL training is challenging in the contexts of hospitals and clinics due to a lack of natural settings, and the difficulty in transferring the skills learned in hospitals and clinics to home environment. Kinect games integrated with virtual reality (VR) simulating a home environment can provide a natural environment for effective ADL training on children with DD. Thus, the aim of the study is to develop game-based ADL training tasks using Kinect integrated with VR for children with DD. Kinect games for training purposes are developed. In addition, two pilot studies are conducted for typically developing children and children with DD aged 3–5.9 years respectively to test the applicability of Kinect games. Kinect games integrated with VR designed for ADL training provide opportunities for children with DD to practice ADL in simulated real-life situations, which reinforces the effectiveness of training at clinics and decreases the burn of caregivers in training the child. The efficiency and feasibility of ADL training could thus be improved.

Keywords Kinect games · Virtual realty · Children

I.-C. Chung · C.-Y. Huang · W.-C. Chiang · M.-H. Tseng (✉)
School of Occupational Therapy, College of Medicine, National Taiwan University, Taipei, Taiwan
e-mail: mhtsengster@gmail.com

S.-C. Yeh
Department of Computer Science and Information Engineering, National Central University, Jhongli, Republic of China

Y.-M. Huang et al. (eds.), *Advanced Technologies, Embedded and Multimedia for Human-centric Computing*, Lecture Notes in Electrical Engineering 260, DOI: 10.1007/978-94-007-7262-5_124, © Springer Science+Business Media Dordrecht 2014

Introduction

The child's independence in activities of daily living (ADL), including basic ADL and instrumental ADL, is crucial for children to gain self-control and could decrease caregiver burden. Children with developmental delay (DD), however, often have difficulty executing ADL, such as toileting and shopping. ADL training is thus important for children with DD and is one of primary goals of rehabilitation. Nevertheless, ADL training is challenging in the contexts of hospitals and clinics. The ADL training setting in hospitals and clinics is different from home setting, resulting in weak generalization to the real world environment. In addition, ADL training at home often causes the tension between children and caregivers.

Kinect games allow players using their whole body to engage in games and interact with scenes and characters in the games. Virtual reality (VR) allows operators perceiving sensory inputs close to the real world so that operators could completely blend into the simulated environment. When ADL training designed in the form of Kinect games integrated with VR, children could use their whole body to practice activities in the real-life context simulated by the computer. The problem of weak generalizability could hence be solved. Furthermore, Kinect games integrated with VR provide interactive-interface, the animation stories with VR, and visual and auditory feedbacks. These elements increase the appealingness of ADL training to children, with resulting enhancement of children's motivation for participating in training activities, and consequently promote the training effectiveness.

Besides, as regards equipment, only a few items are needed for Kinect games, i.e., a computer and a sensor. Such simple equipment exempts caregivers from the burden of preparing ADL appliances in a real-life situation, thereby maintaining the fidelity of caregivers to implement the ADL training at home.

Kinect games integrated with VR have been widely applied to rehabilitation, including stroke rehabilitation [1–5], balance training [3, 6–8], and motor training [1, 9]. All the aforementioned studies reported that participants had significant improvements, indicating that Kinect games integrated with VR are potentially useful tools for ADL training and benefit children with DD. However, no Kinect games integrated with VR to date have been developed for ADL training in children with DD.

The aim of the study is to develop Kinect games integrated with VR for ADL training for children with DD in order to provide opportunities for children with DD to practice ADL in the simulated real-life situations. Since the Kinect games integrated with VR require only simple equipment, the feasibility and efficiency as well as caregivers' fidelity of implementing the training program at home are improved.

Proposed Method

This study has two phases: (1) the phase of developing Kinect games, and (2) the tryout phase of Kinect games.

The Phase of Developing Kinect Games

Equipment

Kinect

The present study utilizes Kinect, a type of interactive motion-sensing systems, as the main device to catch sensory inputs such as body motions and facial expressions and sounds. Kinect sensor detects the human body as the main controller without the needs of additional operational devices. During the operating process, Kinect sensor could record the patterns and the accuracy of the body movement which could be used to record the progress of the child with DD.

Kinect Games

Training purposed Kinect games integrated with VR for ADL training purposes are developed for children with DD. One example of the program is described as follows. For the purpose of shopping in the community, a 360-degree spherical image of a virtual convenience store is provided in the game. Children are asked to walk around the store, search for the assigned goods, pay appropriate amount of money to clerks or get accurate exchange back, finally take the goods and walk out from the store. Other ADL training activities are also developed in the present study, including answering the phone at home and expressing the need for toileting, etc.

Procedure

1. Applying Kinect Sensor and Unity
The program applies Kinect sensor to capture three information, including the colorful image, 3D depth alignment and auditory information. The main tracking information from players is the skeletal information, including the position and the direction. The unity is a powerful rendering game engine integrated with complete set of intuitive tools, including technique of light mapping, a type of rendering path. Users could create interactive 3D content and publish games for multiplatform easily.

2. Writing Scripts for Kinect Games and Developing Kinect Games
Researchers first collect common difficulties or problems encountered by children with DD when they executing activities of daily living (ADL) from clinicians, parents, and teachers. Then, researchers write up stories containing the ADL training tasks targeted at those problems. Besides, each activity is graded by levels of difficulty in order to fit children with DD functioning levels. Scripts are designed for Kinect games through 3D animation and computer programs. Kinect sensor could then be used to play Kinect games.

The Tryout Phase of Kinect Games

A Pilot Study of Kinect Games on Typically Developing Children

Participants

The present study recruit twenty children aged 3–5.9 years and their caregivers. In addition, caregivers should be able to read Mandarin Chinese in order to fill out the Survey.

Instrument

The Survey of Feedback for Kinect games (Appendix A.) is used to evaluate the applicability of the Kinect games. Four areas are accessed in the survey, including difficulties, interestingness, fluency, and comments for improvement. Caregivers complete the survey through observing the child's performance and asking the child's opinions. Researchers then revise the Kinect games based on the survey results. The survey takes about 10 min.

Procedures

1. Cover letters explaining the research project and consent letters are mailed to kindergartens in the greater Taipei area to recruit participants. After caregivers returned the signed consent letters, researchers either set up the computer along with the Kinect sensor in children's home or ask participants to come to the research lab.
2. Children are asked to play Kinect games, and caregivers fill out the Survey of Feedback for Kinect Games.
3. Researchers revise the scripts and program of Kinect games according to opinions from children and caregivers.
4. Repeat step 2 and step 3 until no problems are reported from caregivers or children.

A Pilot Study of Kinect Games: Children with DD

Participants

The study recruits 20 children with DD aged 3–6.9 years and their caregivers. In addition, caregivers should be able to read Mandarin Chinese in order to fill out the Survey.

Instruments

The Survey of Feedback for Kinect Games is also used in this study.

Procedures

1. Cover letters explaining the research project and consent letters are mailed to pediatrics clinics of Department of Physical Medicine and Rehabilitation at hospitals or medical centers, and child developmental centers in the greater Taipei area to recruit participants. After caregivers returned the signed consent letters, researchers either set up the computer along with the Kinect sensor in children's home or ask participants to come to the research lab.
2. The revised Kinect games are provided to children with DD. In addition, the provided Kinect games are just-right challenge activities according to children's abilities and their intervention goals.
3. The following steps are following the same with the procedures on typically developing children.

Discussion

Enhancing the Feasibility of ADL Training Programs and Promoting the Effects of Rehabilitation Therapy

It is easier for caregivers and clinicians to train real-life ADL skills using Kinect games that create simulated real-life situations. The Kinect games can be played both in clinics and at home so that children have sufficient practice opportunities to ensure learning new ADL skills and achieving the optimal rehabilitation effects. Moreover, caregivers are usually unable to fully understand the home program dictated by therapists right after treatment sessions. With Kinect games, it becomes easier for caregivers to carry out home programs for children. Furthermore, with Kinect games caregivers can obtain the progress of children with objective data.

Enhancing Children's Motivation in Participation

Kinect games have been gaining more attention and become more prevalent as a new tool in the field of rehabilitation for children with cerebral palsy due to its fascination. The interactive interface, the animation stories with VR, and the feedback of audio and video effects could enhance the children's motivation, attention, and vitality in rehabilitation therapy to further promote the therapy effects and achieve functional independence.

Recording and Monitoring Progress of ADL Training for Children with DD

The Kinect ADL training system allows therapists selecting appropriate training program according to evaluation results. In addition, children could practice home-based ADL training activities whenever they like to such that therapeutic effects can be enhanced. Besides, with online training program, children's performance on ADL tasks and their progress are automatically recorded in the system, which allows the rehabilitation professionals to obtain the profile of children's ADL skills thus, effective ADL training program will be established.

Conclusion

This is the first study using the Kinect games integrated with VR to design ADL training program for children with DD. There are three advantages of applying Kinect games on ADL training: (1) improve the feasibility of ADL training, (2) enhance the child's motivation on the training, and (3) record and Monitor progress of ADL training. Thus, effective ADL training program will be established by clinicians as soon as possible result in three-win situation of the medical personnel, the caregivers, and the children.

Appendix A. Examples of The Survey of Feedback for Kinect games:

A. Operation

    1. Is it easy to find the icons of each Kinect games?
    ☐ Yes, ☐ No, Suggestion: _____
    2. Is the plot interesting?
    ☐ Yes, ☐ No, Suggestion: _____

B. Satisfaction

Items	Strongly agree	Agree	Neutral	Disagree	Strongly disagree
1. I felt satisfied with the content of the training programs.	SA	A	N	D	SD
The reason is:					
2. My child felt interesting about the Kinect games.	SA	A	N	D	SD
The reason is:					

References

1. Gallo L (2011) Controller-free exploration of medical image data: experiencing the Kinect. In: Paper presented at the 24th international symposium on computer-based medical system (CBMS)
2. Mouawad MR, Doust CG, Max MD et al (2011) Wii-based movement therapy to promote improved upper extremity function post-stroke: a pilot study. J Rehabil Med 43:527–533
3. Padala KP, Padala PR, Burke WJ (2011) Wii-Fit as an adjunct for mild cognitive impairment: clinical perspectives. J Am Geriatr Soc 59:923–932
4. Saposnik G, Manmdani M, Bayley M et al (2010) Effectiveness of virtual reality exercises in stroke rehabilitation (EVREST): Rationale, design, and protocol of a pilot randomized clinical trial assessing the Wii gaming system. Int J Stroke 5:47–51
5. Yavuzer G, Senel A, Atay MB et al (2008) "Playstation Eyetoy Titles" improve upper extremity-related motor functioning in subacute stroke: a randomized controlled clinical trial. Eur J Phys Rehabil Med 44:237–244
6. Clark R, Kraemer T (2009) Clinical use of Nintendo Wii bowling simulation to decrease fall risk in an elderly resident of a nursing home: a case report. J Geriatr Phys Ther 32:174–180
7. Kliem A, Wiemeyer J (2010) Comparison of a traditional and a video game based balance training program. Int J Comput Sci Sport 9:80–91
8. Smith ST, Sherrington C, Studenski S et al (2009) A novel dance dance revolution (DDR) system for in-home training of stepping ability: basic parameters of system use by older adults. Br J Sports Med 45:441–445
9. Bryanton C, Bosse J, Brien M, McLean J, McCormick A, Sveistrup H (2006) Feasibility, motivation, and selective motor control: virtual reality compared to conventional home exercise in children with cerebral palsy. CyberPsychology Behavior 9(2):123–128
10. Kerrebrock N, Lewit EM (1999) Children in self-care. The future of children pp 151–160
11. Case-Smith J (1994) Defining the specialization of pediatric occupational therapy. Am J Occup Ther 48(9):791–802
12. Dougherty J, Kancel A, Ramer C et al (2011) The effects of a multiaxis balance board intervention program in an elderly population. Mo Med 108:128–132
13. Flynn S, Palma P, Bender A (2007) Feasibility of using the Sony playstation 2 gaming platform for an individual poststroke: a case report. J Neurol Phys Ther 31:180–189
14. Kerrebrock N, Lewit EM (1999) Children in self-care. The Future of Children, 151–160
15. Lange B, Flynn S, Rizzo A (2009) Initial usability assessment of off the-shelf video title consoles for clinical title-based motor rehabilitation. Phys Ther Rev 14:355–363
16. Rand D, Kizony R, Weiss P (2008) The Sony playstation II EyeToy: low-cost virtual reality for use in rehabilitation. J Neurol Phys Ther 32:155–163

Automate Virtual Reality Rehabilitation Evaluation for Chronic Imbalance and Vestibular Dysfunction Patients

Ming-Chun Huang, Shuya Chen, Pa-Chun Wang, Mu-Chun Su, Yen-Po Hung, Chia-Huang Chang and Shih-Ching Yeh

Abstract Dizziness is a major consequence of chronic imbalance and vestibular dysfunction, which prevents people performing their routine tasks and affects their quality of life. It may lead to severe injuries caused by unexpected falling. Medicine treatment can alleviate the syndrome of dizziness and past research shows that dizziness can be further reduced if appropriate vestibular function rehabilitation exercises are practiced regularly. Nevertheless, these exercises are usually time-consuming and tedious because of repetitive motions. Most of patients ceases practicing accordingly and reduce the effectiveness of rehabilitation. In order to encourage patients to be involved in the rehabilitation process, interactive rehabilitative gaming systems are introduced in the recent research. Virtual reality technology is used to enhance the gaming experiences and vision sensors are served as gaming inputs. This paper proposes a series of novel virtual reality games adapted by Cawthorne-Cooksey exercises, which are extensively used in clinical for chronic imbalance and vestibular dysfunction rehabilitation. 32 patients participate in the rehabilitation processes within two months period and their gaming parameters and quantified balance indices are analyzed by the

M.-C. Huang (✉)
Computer Science Department, University of California, Los Angeles, USA
e-mail: mingchuh@cs.ucla.edu

M.-C. Su · Y.-P. Hung · S.-C. Yeh
Computer Science and Information Engineering, National Central University, Chungli, Taiwan, Republic of China

S. Chen
Physical Therapy, College of Health Care, China Medical University, Taichung, Taiwan, Republic of China

P.-C. Wang
Otolaryngology, Cathay General Hospital, Fu-Jen Catholic University School of Medicine, Taipei, Taiwan, Republic of China

C.-H. Chang
Outcomes Research Unit, Cathay Medical Research Institute, Hsichih, Taiwan, Republic of China

Y.-M. Huang et al. (eds.), *Advanced Technologies, Embedded and Multimedia for Human-centric Computing*, Lecture Notes in Electrical Engineering 260, DOI: 10.1007/978-94-007-7262-5_125, © Springer Science+Business Media Dordrecht 2014

supported vector machine (SVM) classifier. It shows that ∼81 % patients improve their game parameters and balance indices after undertaking the dizziness rehabilitation training compared to the measurement they had. Our clinical observations also reveal that our patients have higher willing and motivation to regularly perform rehabilitation with the proposed system.

Introduction

Chronic imbalance and vestibular dysfunction usually lead to dizziness [1]. Common dizzy types are Meniere's disease, Benign Positional Vertigo, Vertebrobasilar insufficiently, and Vestibular neuritis, and sick headache. Dizzy may persist for minutes, hours, or longer, and thus interrupts people's daily routine [2–4].

Moreover, Dizziness may cause severely dangerous while people holds glass/ knife in the kitchen or stands in a bathroom. Steady posture may be difficult to maintain, and the person may experience shaking, tilting, difficulty standing, and falls. Worst of all, the syndrome of dizziness will keep deteriorate if no treatment is applied. Common treatments are vestibular surgery, medicine treatment, and balance training exercises. Past research reveals that the combination of these treatments can effectively reduce the syndrome of dizziness [5]. Compared to surgery and medicine treatment, balance training is the more non-invasive and suitable for most of patients. Cawthorne-Cooksey is a balance training exercises designed to improve patients' ability of keeping balance by training patients' ability to control their eyeball movement, head movement, extremities stretch, and bilateral body balance. Through undertaking repeated training exercises, patients' ability of cervico-ocular reflex should be enhanced and provides better compensation for retaining balance situation during dizzy [5, 6]. Cawthorne-Cooksey vestibular rehabilitation exercises requires the subject move their eyeball without moving the orientation of the head, move the head without changing their body position, rotate and bend the full body with the force of waist, rotate head and should with the force of shoulder and throwing/catching objects in between both hands. These exercises are usually time-consuming and tedious because of repetitive motions. In addition, it is difficult to customize training exercises for individual patients because the differences among the patients are not quantifiable. To overcome the limitation of the traditional training methods, interactive game-based rehabilitation programs and sensor based recording system are introduced [2, 7–10]. Gaming and sensing systems not only motivate patients to exercise more, but also quantified the process of balance training into meaningful balance indices [11, 12]. Virtual reality technology makes gaming more playful and close to the daily life setup. Physical therapists can adjust the difficulty levels and game contents for meeting the need of individual patient. On the other hand, camera based recording system reduces the interference of the training process, but still record training processes in full. Optical tracking technology and skeleton detection technology provided by Microsoft Kinect further provides abundant information of human motion. A full rehabilitation system can provides immediate visual, auditory, and tactile feedback to make patients more involved in the training processes and gain better training effectiveness.

System Architecture

The proposed VR interactive rehabilitation system consists of game-based training system and center of pressure measuring system. 4 training tasks adopt from Cawthorn-Cooksey vestibular rehabilitation exercises: eyeball movement, head movement, extremities stretch, and bilateral body balance. These exercises are transformed into interactive 3-dimensional VR games. The difficulties and contents of these games can be adjusted by physical therapists based on the condition of the respective patient. Indeed, the prototype games strictly follow the guidance of our medical collaborators and are implemented as similar as the functionality of Cawthorn-Cooksey exercises as possible. Patients are expected to wear 3D glasses and receive real-time gaming feedback in terms of visual and audio output. System flowchart (Fig. 1) presents the work flow of the system from user interaction, data collection, feature selection, to SVM classification. Patients should follow the visual and audio instructions to complete the assigned tasks in the games. Their kinematic performance in each session is automatically recorded by the optical tracking system and Microsoft Kinect. This information are served as gaming inputs, performance indices for assisting therapists and programmers to design more appropriate gaming scenarios. Same for the pre/post-test of the center of

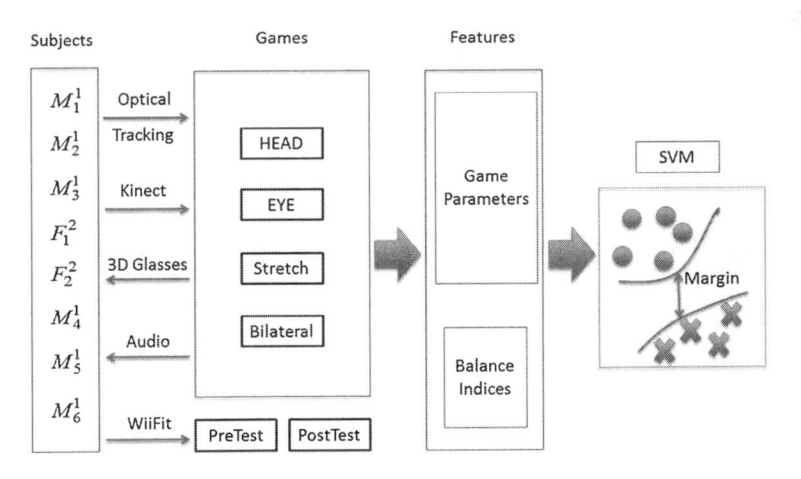

Fig. 1 System flowchart

Fig. 2 Four Cawthorn-Cooksey based game snapshots are shown and the taped corresponding
subjects' movements

pressure measuring system, center of pressure readings and time stamps are
continuously recorded by WiiFit.

Cawthorn-Cooksey Based Games

Four rehabilitative games are designed to simulate the tasks of Cawthorn-Cooksey
exercises (Fig. 2). The first game requires users to read numbers randomly generated on the screen. The numbers are shown on the margin of the screen. The goal
of this exercise is to train the eyeball movement; therefore, users should keep their
head in a static position and move their eyeball in order to read the number on the
corner. The ratio of hit or miss is recorded as performance indices. The game for
training head movement is similar to eyeball movement training game. Instead,
users are asked to gaze on a red box in the screen and read the random number
around the box by moving their heads. In order to encourage users to perform
stretch exercise, a basketball shooting game is designed. Users should grasp a
virtual basketball in the game, lift their arms to a randomly assigned location
indicated by a red ball, and the ball will be threw to the virtual basket automatically. During the exercise, users should move their shoulder and stretch their arms
to assigned levels. The ratio of hit or miss is recorded as performance indices as
well. The last game is for bilateral coordination exercise. Users are asked to
capture a ball threw from one hand to the other hand repeatedly. The throwing
process will continue until the users miss a catch.

Pre/Post-Test

Pre/Post-test center of the pressure data are collected by WiiFit. The trajectory of the center of the pressure can derive statokinesigram (SKG) [13, 14], which is an important index of balance level. The smaller envelope of the trajectory is preferred in our case. Smaller envelope area means that users have less variance in terms of their center of the pressure while they are standing. Together with SKG, maximum X and Y directional placements (MaxML and MaxAP), X and Y directional standard deviations (SDML and SDAP), and average X and Y directional speeds (meanML and MaxAP) can be calculated before and after each game-based training. These indices and the gamed based parameters form a feature for each subject and become an input data for SVM classification.

Binary Supported Vector Machine

Binary linear SVM is a classifier which attempts to partition data set into two categories with the maximum barrier in between. The barrier, formally named as a margin, stands for the width increased by before hitting a data point for both categories. Intuitively, a good separation is achieved by finding the hyper-plane that has the largest distance to the nearest training data point of any class. However, some data set may contain data points which are not linear separable. To overcome this situation with linear SVM, a kernel trick is performed to project those data points into a higher feature dimension; presumably, those data points are projected into a space where they are linear separable. Commonly used kernels are linear, quadratic, and Gaussian kernels. Process data points with SVM classifier is simple if each data point is correctly labeled, such as marked as $+1$ and -1 for respective group. The normal and offset vector of the hyper-planes can be obtained via training processes. After training, a set of supported vectors can be obtained. For binary linear SVM case, testing data should be projected on to the obtained hyper-plane and multiply with its label to determine the group it belongs to. Accuracy can be computed by analyzing the differences between the results generated by projection from the testing data labels.

Evaluation

32 subjects, including both males and females, participate in a series of VR gaming experiments. Each subject participates in at least 6 trials during approximately two month period. Subjects are required to stand on Wiifit to record their seven indices of center of pressure before and after completing all games in the training exercises. Subjects experience 4 types of Cawthorn-Cooksey based games.

	Linear	Quadratic	Gaussian
Full feature set	65.12%	79.17%	81.40%
Remove game based feature	55.26%	66.11%	70%

Fig. 3 Experimental results of binary linear SVM (*first row* uses all features including game related parameters and center of pressure indices and *second row* uses only balance indices of the center of pressure)

All the kinematics information are recorded by the optical tracking system and Microsoft Kinect. In order to make the rehabilitation process smooth, a copy of the technical manual was provided to assist the therapists in operating the interactive rehabilitation system for each subject. All subjects after completing the sixth trial are invited to join our survey, targeting on analyzing the sufficiency of gaming instructions, game appearance, system usefulness/playfulness, motivation promotion, and the easiness in terms of daily use for all game types: eye movement, head movement, hand stretching, and bilateral balance. In addition, subjects' personal information and their type of dizziness are recorded. Benign Paroxysmal Positional Vertigo, BPPV, Meniere's disease, Vestibular neuritis, vertebrobasilar insufficiency, VBI, and headache induced dizziness.

Recorded game parameters and balance indices from Wiifit are classified by SVM classifier. Linear, quadratic, and Gaussian kernels are applied to understand the structure of the collected data. From the Fig. 3, we can identify that the data are not linear separable. This is because different patients' severity does not equal and their recover speeds are not the same. Some patients get much more improvement during gamed-based rehabilitation and others may get relative less improvement. With Gaussian kernels, we still can achieve ~ 81 % recognition accuracy, which indicates that most of improvement can be identified with SVM classifier and the differences among all patients are tolerable. We are also interested in verifying that gaming related parameters actually help in classification. Hence, in the second row of the Fig. 3, we remove the features from games and run the training and testing procedure again. We can see that the recognition rate decrease to ~ 70 % accordingly. It reveals that the gaming parameters provide valuable information as well to diagnose patients' recovery.

Conclusion

This paper proposes a series of novel and interactive virtual reality games adapted by Cawthorne-Cooksey exercises for promoting chronic imbalance and vestibular dysfunction rehabilitation. In order to evaluate the performance of system, collected quantified game parameters and balance index are analyzed by the SVM classifier. It shows that ~ 81 % patients receive better game parameter balance indices after undertaking the dizziness rehabilitation training compared to the game parameter and balance indices they had. The results of the follow-up questionnaire reveal that using new VR technology for rehabilitation is helpful to promote chronic imbalance and vestibular dysfunction rehabilitation.

References

1. Monsell EM (1995) New and revised reporting guidelines from the Committee on hearing and equilibrium. American Academy of Otolaryngology-Head and Neck Surgery Foundation, Inc. Otolaryngology Head Neck Surgery 1995, vol 113, pp 176–178
2. Gauchard GC, Jeandel C, Perrin PP (2001) Physical and sporting activities improve vestibular afferent usage and balance in elderly human subjects. Gerontology 47:263–270
3. Greenspan SL, Myers ER, Maitland LA, Resnick NM, Hayes WC (1994) Fall severity and bone mineral density as risk factors for hip fracture in ambulatory elderly. JAMA 271:128–133
4. Luukinen H, Herala M, Koski K, Honkanen R, Laippala P, Kivelä SL (2000) Fracture risk associated with a fall according to type of fall among the elderly. Osteoporos Int 11:631–634
5. Schubert MC, Whitney SL (2010) From Cawthorne-Cooksey to biotechnology: where we have been and where we are headed in vestibular rehabilitation? J Neurol Phys Ther 34:62–63
6. Corna S, Nardone A, Prestinari A, Galante M, Grasso M, Schieppati M (2003) Comparison of Cawthorne-Cooksey exercises and sinusoidal support surface translations to improve balance in patients with unilateral vestibular deficit. Arch Phys Med Rehabil 84:1173–1184
7. Adamovich SV, Fluet GG, Merians AS, Mathai A, Qiu Q (2009) Incorporating haptic effects into three-dimensional virtual environments to train the hemiparetic upper extremity. IEEE Trans Neural Syst Rehabil Eng 17:512–520
8. Holden MK (2005) Virtual environments for motor rehabilitation: review. Cyberpsychol Behav 8:187–211
9. Lahiri U, Warren Z, Sarkar N (2011) Design of a gaze-sensitive virtual social interactive system for children with autism. IEEE Trans Neural Syst Rehabil Eng 19:443–452
10. Weiss PL, Katz N (2004) The potential of virtual reality for rehabilitation. J Rehabil Res Dev 41:vii–x
11. Hunag MC, Chen E, Xu W, Sarrafzadeh M (2012) Gaming for upper extremities rehabilitation. Proceeding of conference on wireless health 2012, vol 27
12. Huang MC, Xu W, Su Y, Lange B, Chang CY, Sarrafzadhe M (2012) Smart Glove for upper extremities rehabilitation gaming assessment. Proceeding of conference on pervasive technologies related to assistive environment 2012, vol 20
13. Takagi A, Fujimura E, Suehiro S (1985) A new method of statokinesigram area measurement. Application of a statistically calculated ellipse. In: Igarashi M, Black FO (eds) Vestibular and visual control on posture and locomotor equilibrium, Karger, Basel, pp 74–79
14. Paillard T, Costes-Salon C, Lafont C, Dupui P (2002) Are there differences in postural regulation according to the level of competition in judoists? Br J Sports Med 36:304–305

Part XII
Recent Advances on Video Analysis and its Applications

Machine-to-Machine Interaction Based on Remote 3D Arm Pointing Using Single RGBD Camera

Huang-Chia Shih and En-Rui Liu

Abstract In this paper, we propose a prototype of machine-to-machine interaction system using a single RGBD camera. Based on the 3D arm pointing and tracking algorithm, system enables user to select multiple objects using their two arms, without infrared remote controller and wireless transmitter. A fast gesture recognition method is used to identify the instruction assignment for the target objects. The system consists of three modules, including image capturing module, 3D arm pointing and target tracking module, and interaction module. The image capture module use the RGBD camera installed behind of user to capture the arm parameters in real-world space. The pointing and tracking algorithm is referred to the environmental information to establish the master and slave objects. For interaction module, the command gesture is determined by the gesture recognition, interacting with target devices by the decoded instruction. Finally, system generates a control signal for triggering the master object to communicate with the slave object according to the fetched the instruction.

Keywords Machine-to-machine interaction · Tracking algorithm · Human-computer interaction · 3D pointing · Particle filtering · RGBD camera · Motion capturing

Introduction

With the development of human-computer interaction (HCI) technique is rapidly increasing, we can expect that the computer system allows amounts of interactions around our environment. The fundamental requirement of HCI is to interact with

H.-C. Shih (✉) · E.-R. Liu
Human-Computer Interaction Multimedia Lab, Department of Electrical Engineering,
Yuan Ze University, Taoyuan, Taiwan, Republic of China
e-mail: hcshih@saturn.yzu.edu.tw

Y.-M. Huang et al. (eds.), *Advanced Technologies, Embedded and Multimedia for Human-centric Computing*, Lecture Notes in Electrical Engineering 260, DOI: 10.1007/978-94-007-7262-5_126, © Springer Science+Business Media Dordrecht 2014

an interface, this enables user to control the actions of machine intuitively. Nowadays, the remote controlling has been widely used as the major interface for delivering information and commanding the computerized equipment. The evolution of interactive media changes from the infrared ray (IR), wireless signal, and the natural body control. In recent years, the kinect sensor established by the Microsoft Company that has been attracted a lot of attention and used to real-time capture the human motion parameters.

The conventional control mechanism is based on a remote sensing controller via wired or wireless approach to communicate with devices located at a distance. For example, user tends to control the television, camera, or other electronic devices. In addition, when users send the control signal via remote controller, a target device will execute the corresponding action based on the received control signal such as the commands of turn on/off the power, turn up/down the voice. According to the technology evolutions, the mechanism of HCI changes from direct touching, with a small handset remote controller, and the speech-based and gesture-based manner. Basically, a camera and microphone is used to capture user's voice and gesture.

This study developed a prototype of machine-to-machine interaction based on the arm pointing and gesture recognition captured by a single RGBD camera (i.e., kinect sensor). This provides user to assign machines intuitively and command the interaction between the machines. Figure 1 shows the examples of arm pointing with a single RGBD camera.

System Design

This interaction system consists of three processes, including (1) pointing process, (2) tracking process, and (3) recognition process.

Pointing and Tracking Algorithm

In this paper, the extension line from elbow to hand is used to describe the pointing direction. The depth information provided by kinect sensor applies for initializing object location. The particle filtering [1] algorithm is used to track the target object, it avoids the imperfect depth information resulted from the unstable lighting condition. This instability is complemented using a probabilistic model.

A. *Arm pointing by extension line*

In real world coordination, a pointing unit vector $u = (u_x, u_y, u_z)$ formed by the user's skeleton with the elbow point E and hand point H. Theoretically, the target object location T satisfies a 3-dimensional line equation along with the unit

(a)

(b)

Fig. 1 The examples of pointing action with a single kinect sensor behind of user; **a** single arm pointing **b** two arms pointing

vector u. However, due to the stereovision system, it exists a visual error between optical line and pointing line. As shows in Fig. 2, the eyes specify target T which pointing from the direction of u is a constant offset higher than the real one.

Fig. 2 The conceptual illustration of 3D arm pointing based on the extension line

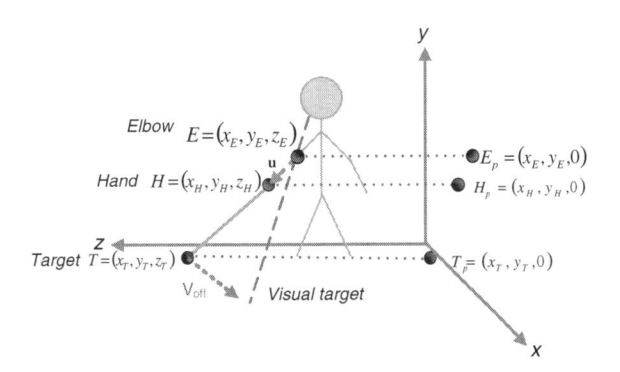

Therefore, we need to modify the pointing direction using an offset vector v_{off}. To project onto the x-y image plane, we obtain three points include E_p, H_p, and T_p which denote the corresponding feature points projected from 3-dimensional space. Therefore, we can easily to compute the offset vector from these three points. Based on the depth information, it enables the target object location more consistent with the real pointing direction.

B. *Target tracking by particle filtering*

Particle filtering (PF) [1] is based on the Bayesian theory and the Mote Carlo Sequential sampling method to handle the nonlinear and non-Gaussian problems. The probability propagation model is selected based on prior and likelihood model of every candidate.

The PF approach requires two probabilistic models: the state model and the observation model. The state model describes the evolution of the system from its past state whereas the observation model relates the current state observations to the current state of the system. Based on the observation model, the measurement of the weight for the each particle is an important process for updating the priori. Most of the PF-based visual tracking approaches which applied the color-based histogram similarity manner to achieve the measurement. Based on Bayesian sequential estimation, the prior density function can be computed by two recursive stages: *prediction stage* and *updating stage*.

Gesture Recognition by Ring Projection

We employ the human skeleton toolkit provided by Microsoft [2] to obtain the center position of hand. However it requires user with upstanding pose. In practical applications, when user shows the gesture, they are not always standup stead of siting or lying. To cope this problem, the OpenNI library [3] is used to detect and track the hand.

The goal of the gesture recognition is to recognize the form of hand gestures, supporting for inter-machine communications. Here, the centroid of tracked object is utilized for aligning the hand of commanding. A maximal bounding circle centered by the centroid of circle is obtained. Then the ring projection transformation (RPT) [3] is used to project the commanding gesture into one-dimensional histogram feature map. The cross-correlation applies for identifying the type of gesture.

The RPT is constructed along circular rings of the radius, because the one-dimensional ring-projection model is invariant to the rotation of its corresponding image plane. Figure 3b shows a template image in two distinct orientations and the plots of RPT values as functions of radius r. This template applies for the centroid of hand as shown in Fig. 3a, and computes two ring-projection plots, which are approximately identical, regardless of orientation changes.

Fig. 3 Gesture recognition using the ring projection

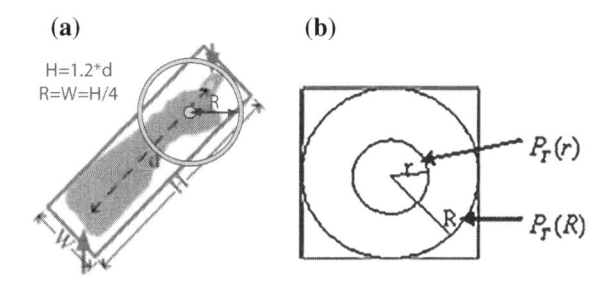

Table 1 The performance of arm pointing

Pose/arm(s)		4 Targets			8 Targets		
		Times	Hits	Accuracy (%)	Times	Hits	Accuracy (%)
Stand up	Single arm (left)	20	18	90	20	17	85
	Single arm (right)	20	19	95	20	18	90
	Two arms	20	16	80	20	15	75
Sit	Single arm (left)	20	17	85	20	17	85
	Single arm (right)	20	16	80	20	17	85
	Two arms	20	15	75	20	15	75

Experimental Results

To evaluate our system, five untrained subjects test the pointing system without any marker and remote controller. Subjects stand and sit to point 4 and 8 targets with their two arms. The results of arm pointing is shown in Table 1. Normally, this system performs the pointing results with single arm, which obtains higher accuracy than it with two arms. Based on the RGBD camera, it allows system to deal with partial occlusion problem using the depth information. In addition, the interference of lighting condition can be suppressed.

Conclusions

This paper proposed a prototype of the machine-to-machine interaction based on the visual arm pointing captured by a single RGBD camera without any remote infrared or wireless transmitter. A fast gesture recognition method used to decode the instruction assignment between the devices. The pointing and tracking algorithm is referred to the space information to establish the master and slave objects.

References

1. Isard M, Blake A (1998) Condensation—conditional density propagation for visual tracking. Int J Comput Vision 29(1):5–28
2. Shotton J, Fitzgibbon A, Cook M, Sharp T, Finocchio M, Moore R, Kipman A, Blake A (2011) Real-time human pose recognition in parts from single depth images. In: Proceeding of the computer vision and pattern recognition (CVPR), pp 1297–1304
3. [Online] OpenNI. http://www.openni.org
4. Tsai DM, Tsai YH (2002) Rotation-invariant pattern matching with color ring-projection. Pattern Recogn 35:131–141

Automatic Peak Recognition for Mountain Images

Wei-Han Liu and Chih-Wen Su

Abstract In this paper, we propose a novel method for automatically recognize the peaks of mountains based on the shape of skyline. Since the appearances of a mountain are variable due to the changes of weather, season or region, the skyline of mountains is extracted for the matching of different mountain images. We use support vector machine (SVM) to predict the possible skyline segments and the linking of incomplete fragments of skyline is formulated as a shortest path problem and solved by dynamic programming strategy. In order to resist the geometric distortion caused by view change, we perform 2D curve matching on the extracted skylines for the peak recognition task. Our experimental results demonstrate that the proposed method is our method is effective for the mountain recognition under complicated and variable circumstances.

Keywords Skyline localization · Mountain annotation

Introduction and Previous Work

In last decade, many state-of-the-art image and visual subject retrieval approaches for place instance recognition [1, 2] and scene category recognition [3, 4] have been proposed. However, the automatic image-based location recognition for

This work was supported by National Science Council, Taiwan, under the Grants NSC101-2221-E-033-062.

W.-H. Liu · C.-W. Su (✉)
Department of Information and Computer Engineering, Chung Yuan Christian University, Chungli, Taiwan, Republic of China
e-mail: lucas@cycu.edu.tw

W.-H. Liu
e-mail: g10077030@cycu.edu.tw

Y.-M. Huang et al. (eds.), *Advanced Technologies, Embedded and Multimedia for Human-centric Computing*, Lecture Notes in Electrical Engineering 260, DOI: 10.1007/978-94-007-7262-5_127, © Springer Science+Business Media Dordrecht 2014

natural scene is still a challenging task. For example, the color and texture of vegetation may be very different due to the changes of seasons, weather and sunlight. It makes appearance-based techniques [1, 2] difficult to find the robust feature descriptors of a natural scene under different circumstances.

In this paper, we consider the problem of recognizing the mountains in the image without geotagging information. Even for the geotagged photograph, only the position of the photographer is associated with the image in the most cases. To precisely annotate the names and positions of peaks in image, we extract skyline to locate the positions of mountains and then perform partial matching process on skylines to find the names of peaks.

Currently, there is a limited number of studies focus on the image-based mountain recognition in recent years. In [5], Babound et al. detected mountains silhouette by compass edge detector and some post-processing steps. They compared the silhouette with a 3D terrain model of the mountains to extract silhouette accurately. The location of each mountain peak can be assigned accurately on images. In [6], Baatz et al. proposed a system for large scale location recognition based on digital elevation models. They extracted the sky and represent the visible horizon by a set of contour words. Then, a bag-of-words like approach with integrated geometric verification was performed for looking the panorama that has a similar frequency of contour words with a consistent direction. Although the above mentioned methods are accurate and reliable, a precise 3D terrain model of the mountains on image is indispensable for matching the silhouette of mountains in their method.

The remainder of the paper is organized as follows. The next section contains the process of skyline extraction. In 2D Curve Matching for Skylines, we introduce the proposed 2D curve matching algorithm for skylines. In Experimental Results, we discuss the experiment results. Conclusion contains some concluding remarks.

Skyline Extraction

In order to extract the skyline automatically, the algorithm we proposed in [7] is adopted. First, we use Canny edge detector [8] to extract edges from an image. The neighboring pixels around the edges are then be collected as 8×8 blocks and trained by SVM for finding the possible edge segments of skyline. We extract a 210 dimensional feature descriptor for an 8×8 block. Each one contains 192 dimensions for the RGB color 8×8 neighboring pixels, 2 dimensions for the variances of illumination aside the edge and 16 dimensions for the location information of edge pixels. All of the features mentioned above are normalized within the interval of [0, 1].

Since edge extraction is always an ill-posed problem, a detected skyline edge segment may be a part of true skyline or just an irrelevant edge. We proposed an edge-linking algorithm to find the bridges for linking the skyline-like edge pixels.

We calculate an energy map by weighting a gradient image and skyline-like edge pixels, which is based on the following equation:

$$e_p = \begin{cases} 0, & \text{if } p \in S \\ \alpha_p \ (1 - G_p + \beta), & \text{otherwise} \end{cases}, \tag{1}$$

where e_p is the energy for the pixel p. S denotes the set of skyline pixels and G_p denotes the normalized gradient value of p. α_p and β are the weights of the energy function of non-skyline pixels.

In order to determine a shortest path with the lowest cumulative energy, we define the following recursive functions for solving the problem caused by steep mountain ridge:

$$E^0_{p(x,y)} = \begin{cases} e_{p(x,y)}, & \text{if } x = 1. \\ \min(E^f_{p(x-1,y-1)}, E^f_{p(x-1,y)}, E^f_{p(x-1,y+1)}) + e_{p(x,y)}, & \text{otherwise}. \end{cases} \tag{2}$$

$$E^i_{p(x,y)} = \min(E^{i-1}_{p(x,y-1)} + e_{p(x,y)}, E^{i-1}_{p(x,y)}, E^{i-1}_{p(x,y+1)} + e_{p(x,y)}). \tag{3}$$

$$E^f_{p(x,y)} = E^j_{p(x,y)}, \text{ if } E^j_{p(x,y)} = E^{j-1}_{p(x,y)} \text{ for all } x. \tag{4}$$

The idea of above recursive algorithm is similar to Seam Carving algorithm [9]. The major difference is the consideration of vertical path in Eq. (4). We incorporate vertical trace into the original Seam Carving algorithm to deal with the problem of steep mountain ridge. For the pixels at the leftmost column on image, $E^0_p = e_p$. Otherwise, E^0_p is the cumulative energy of a shortest path from the leftmost column of pixels to the pixel p. This path does not pass through any pixel at the same column of p. In contrast to E^0_p, the shortest path through the other pixels at the same column of p will be considered in E^i_p, where i denotes the ith iteration. Once the values of E^j_p becomes steady for all the pixels in the same column in jth iteration, E^j_p will be set as E^f_p which is the final lowest cumulative energy from the leftmost column of pixels to current pixel p. Some detection results in [7] are shown in Fig. 1. Our previous experimental results show that approximately 80 % of mountain images' average distance errors are lower than 3 pixels.

2D Curve Matching for Skylines

2D Curve matching is an important research issue for a long time. Mokhtarian and Mackworh proposed Curvature Scale Space (CSS) [10] to represent reliable features of a planer curve on CSS image. Afterwards, numerous CSS-based matching algorithms [11–14] have been proposed for solving the curve matching problem. In this paper, we also use curvature as the feature for the matching scheme, to resist the distortion of view change and skyline localization.

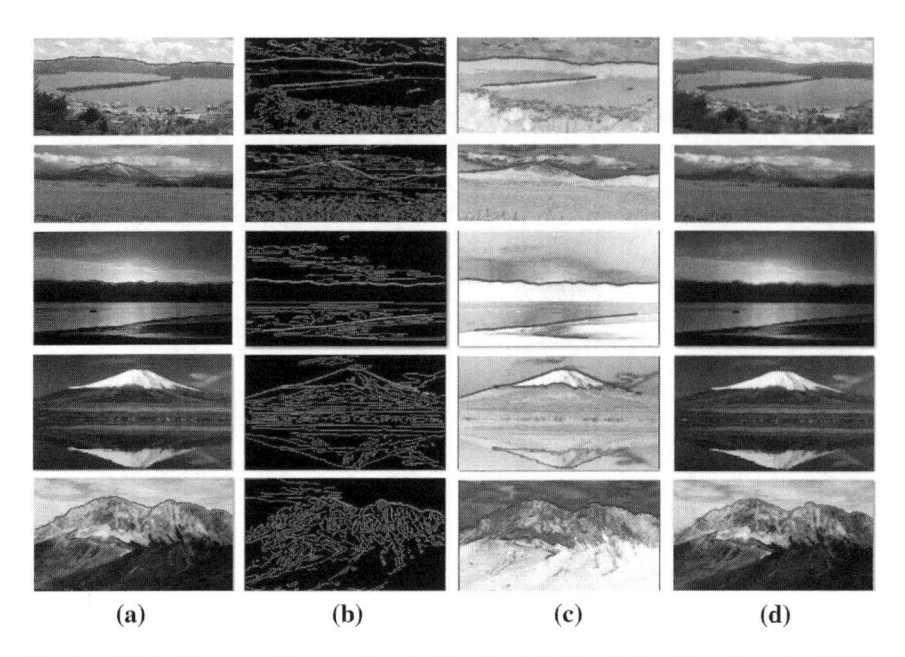

| (a) | (b) | (c) | (d) |

Fig. 1 **a** Original images with ground truth (*blue lines*). **b** The skyline-like edges classified by SVM on edge maps. **c** Energy maps. **d** The final detected skylines (*red lines*). [7]

Before we start the curve matching process, each mountain image is resized respect to the width of image and only a skyline pixel will be sampled for each image column. In other words, the number of skyline pixels is the same as the width of resized image. After the length of skyline has been normalized to N, we calculate the curvature values of the skyline pixels as follows:

$$\kappa(u, \sigma) = \frac{X_u(u, \sigma)Y_{uu}(u, \sigma) - X_{uu}(u, \sigma)Y_u(u, \sigma)}{\left(X_u(u, \sigma)^2 + Y_u(u, \sigma)^2\right)^{3/2}}, \tag{5}$$

$$\begin{aligned} X_u(u, \sigma) &= x(u) \otimes g(u, \sigma) \\ Y_u(u, \sigma) &= y(u) \otimes g(u, \sigma), \end{aligned} \tag{6}$$

where $x(u)$, $y(u)$ denote the position of a skyline pixel on the image and $g(u, \sigma)$ denotes a Gaussian filter of width σ.

Since the scale of a mountain is variable on different images, we perform brute force partial matching to find the best partial match between two skylines. Given a start position P_s and an end position P_e of skyline, we use P_s and P_e to align segments of skylines and each segment is always normalized to N pixels before matching process. On the other hand, since it is not really necessary to consider the rotation problems for mountain images, we discard the alignments with significant rotation over 15°. Then, the similarity between two skylines can be calculated based on the root-mean-square error (RMSE) of curvature value as follows.

$$\min_{i,j}\left(RMSE(A_i, B_j) + RMSE(\tilde{A}_i, \tilde{B}_j)\right), \tag{7}$$

where A_i and B_j are the aligned segments of skyline A and skyline B, respectively. $\tilde{A}_i$ and $\tilde{B}_j$ are the complement sets of A_i and B_j after alignment, respectively. Here, only the pixel that can find its corresponding pixel on the other skyline will be considered in the computation of RMSE $(\tilde{A}_i, \tilde{B}_j)$. In order to speed up the matching process, we use Fast Library for Approximate Nearest Neighbors (FLANN) [15] to perform fast approximate nearest neighbor searches in high dimensional spaces.

Experimental Results

We have collected 130 mountain images from Internet for the evaluation of our proposed method. For each query image, there are about 10 images in the dataset. Figure 2 shows two examples of the query images and top 3 retrieval results. Although the mountain images are with different fields of view, especially for the last case in Fig. 2d, our proposed method can automatically extract the complete skylines and find best alignment between two skylines. The precision versus recall curve of is shown in Fig. 3. Our method achieved 80 % average precision at 65 % recall.

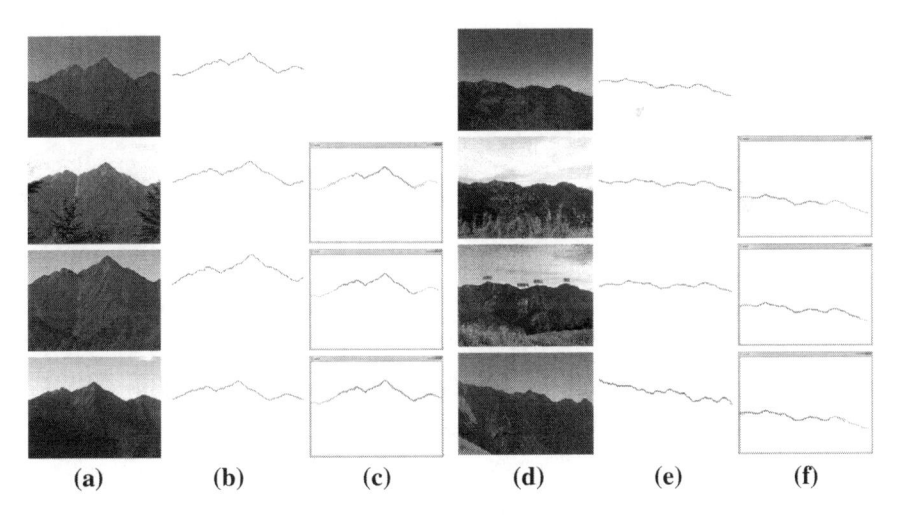

(a)   (b)   (c)   (d)   (e)   (f)

Fig. 2 Query mountain image (*first row*) and retrieved top ranked images. **a, d** Original images. **b, e** Detected skylines. **c, f** The best partial matching segments (*pink lines*)

Fig. 3 Precision versus recall *curve* (average over multiple queries)

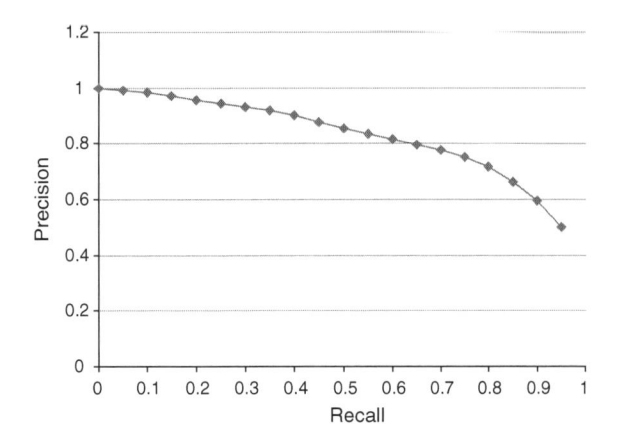

Conclusion

In this paper, we proposed a systematic method to automatically extract skyline and retrieve mountains images by curve matching. To resist the geometric distortion caused by view change, we used FLANN to find best partial alignment between two skylines effectively. The experimental results showed that our proposed method is accurate and reliable. We would like to explore the further applications of our method on the navigation and annotation of outdoor images in our future work.

References

1. Chum O, Philbin J, Sivic J, Isard M, Zisserman A (2007) Total recall: automatic query expansion with a generative feature model for object retrieval. In ICCV
2. Schindler G, Brown M, Szeliski R (2007) City-scale location recognition. In CVPR
3. Lazebnik S, Schmid C, Ponce J (2006) Beyond bags of features: spatial pyramid matching for recognizing natural scene categories. In CVPR
4. Hays J, Efros AA (2008) IM 2 GPS: estimating geographic information from a single image. In CVPR
5. Babound L, Cadik M, Eisemann E, Seidel HP (2011) Automatic photo-to-terrain alignment for the annotation of mountain pictures. In CVPR
6. Baatz G, Saurer O, Köser K, Pollefeys M (2012) Large scale visual geo-localization of images in mountainous terrain. In ECCV
7. Hung Y-L, Su C-W, Chang Y-H, Chang J-C, Tyan H-R (2013) Skyline localization for mountain images. In ICME
8. Canny J (1986) A computational approach to edge detection. IEEE Trans Pattern Anal Mach Intell 8:679–698
9. Avidan S, Shamir A (2007) Seam carving for content-aware image resizing. ACM Trans Graph 26(3):10
10. Mokhtarian F, Mackworth AK (1992) A theory of multiscale curvature-based shape representation for planar curves. IEEE Trans Pattern Anal Mach Intell 14(8):789–805

11. Abbasi S, Mokhtarian F, Kittler J (1999) Curvature scale space image in shape similarity retrieval. Multimedia Syst 7:467–476
12. Mokhtarian F, Abbasi S (2002) Shape similarity retrieval under affine transforms. Pattern Recogn 35(1):31–41
13. Pinheiro AM (2005) Identification of similar shape contours based on the curvature extremes description. In ICIP
14. Mai F, Chang CQ, Hung YS (2010) Affine-invariant shape matching and recognition under partial occlusion. In ICIP
15. Muja M, Lowe DG (2009) Fast approximate nearest neighbors with automatic algorithm configuration. In VISAPP

Gesture Recognition Based on Kinect

**Chi-Hung Chuang, Ying-Nong Chen, Ming-Sang Deng
and Kuo-Chin Fan**

Abstract In recent years, depth cameras have become a widely available sensor type that captures depth images at real-time frame rates. For example, Microsoft KINECT is a powerful but cheap device to get depth images. Even though recent approaches have shown that 3D pose estimation and recognition from monocular 2.5D depth images has become feasible, there are still some challenge problems like gesture detection and recognition. In this paper, we propose a gesture recognition method and use that to make a puzzle game with kinesthetic system. Gesture is very important for our system because instead of using some devices like keyboard or mouse, users will play the puzzle game by their own hand. We will focus on using ROI and some algorithms that we proposed to do gesture detection and recognition.

Keywords KINECT · Depth cameras · Pose estimation · Gesture detection

Introduction

Kinesthetic system is the interactive media system that has been known recently. Because of the immediateness the users gain more feedback during the operation process. By utilizing this characteristic, we integrate this system with teaching resources and make it a digital interactive platform which boost the concentration and interest of students. Hence, it becomes an excellent teaching medium. This research uses Microsoft KINECT cameras and its developed software as the foundation of kinesthetic system. KINECT cameras derive three kinds of data, which are colorful images, 3D image depth information, and audio sources. There

C.-H. Chuang (✉) · Y.-N. Chen · M.-S. Deng · K.-C. Fan
Department of Computer Engineering and EntertainmentTechnology, Tajen University,
Pingtung, Taiwan, Republic of China
e-mail: chchuang@mail.fgu.edu.tw

Y.-M. Huang et al. (eds.), *Advanced Technologies, Embedded and Multimedia for Human-centric Computing*, Lecture Notes in Electrical Engineering 260, DOI: 10.1007/978-94-007-7262-5_128, © Springer Science+Business Media Dordrecht 2014

are three cameras in KINECT: the middle one is common RGB color camera and the others are infrared launcher and 3D depth sensor imaged by infrared CMOS cameras. The data sources derived by this system detect the users' motions mainly by 3D depth sensor.

KINECT utilizes the technology of Light Coding to reach the image detecting and tracking. Light Coding [1] is a kind of technology adopted as a way to process depth information of image. This theory used coding of the measured space by Continuous light (close to infrared ray) and decodes by chips to result an image with depth. After understanding how KINECT gets images, the next step is how to process the job of recognition. The data derived from Light Coding technology are basic image information. The key point is to recognize images and transfer into order of action. KINECT can transfer 3D depth image into system of Skeleton tracking. Shotton et al. [2] is the way to integrate the colorful images with depth image to find out body node and the skeleton. Girshick et al. [3] is the procedure of Regression to increase the accuracy of the correspondence between the body pose and skeleton. This system can detect up to six people simultaneously, including recognize simultaneously the motions of two people simultaneously. It can record everyone's twenty sets of details, including body, limbs and finger. Having the approach of skeleton tracing, we can make further application, such as pose estimation [4], action recognition [5], image segmentation [6], body pose reconstruction [7, 8] Building rich 3D maps of environments etc. KINECT developed system indentified many body node from human body, but it didn't identify the detail part of hands. The system can indeed detect the position of palms. However, there is no way to effectively detect or recognize whether the palms are open or close, even the rotation of palms. In order to solve this problem, this system adopts the skeleton tracing system which can gain ROI information of image from palm node. ROI (Region Of Interest) is the area where program designers' interests are, and it is also the core of further image process and recognition. During this phrase, we use OPENNI as the foundation to detect the palm node, and take the sounding area as key palm image to process further recognition and analysis. There are other researches about KINECT palm detection, [9, 10] KINECT is used to detect palm and recognition of palm, palm and fingertip tracking. The hands were segmented using the depth vector and center of palms were detected using distance transformation on inverse image.

This research focuses on this part. We take further palm detection and discuss gesture recognition. Moreover, we can do jigsaw puzzle by corresponding this gesture recognition system with kinesthetic system.

Selection System

The kinesthetic system this research adopted structures its core program on the skeleton monitoring system as Fig. 1. The skeleton monitoring system is through the 3 M depth pictures which are detected by Light Coding method with color

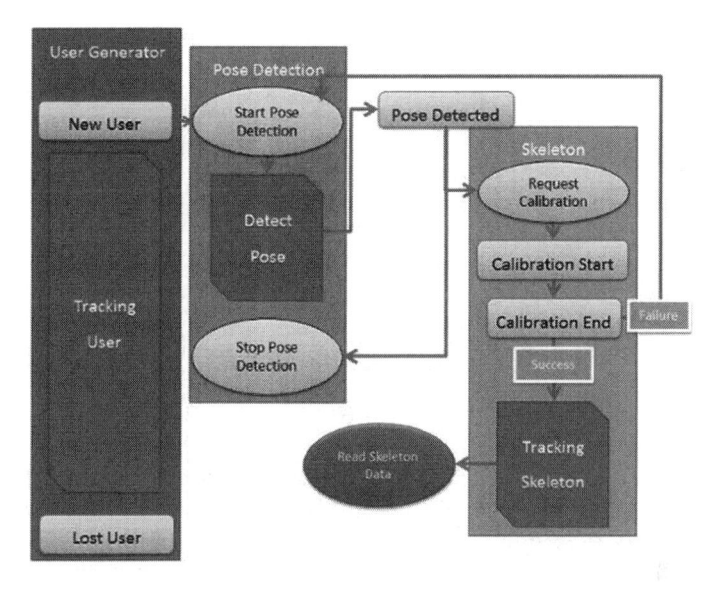

Fig. 1 Skeleton monitoring system

cameras. Once a man are detected, it will process pose detection and recognition on this man. After recognition, it will work on pose correction and track this man until he leaves the scope or the program stops. In the flow chart, the program first starts the man monitoring system. If there is a man detected in the shooting scope, it will start "New User" function and then start pose detection. "New User" and "Lost User" on the last part represent the callback function of two events: "new user detected in the scope" and "user leaving the scope for a while". When these two evens happen, these two functions will be called separately. When a new user is detected and "New User" is called, the program "New User" will call the pose detection unit "Start Pose Detection" to detect a man's pose. When the unit "Pose Detection" detects a user, it will call the callback function "Pose Detected" to process the further move. In this case, "Pose Detected" works on two jobs. One is to check if the numbers of users who is being traced reach the limit; the other is to call the function "Request Calibration", which is a unit processing the skeleton to correct and analyze the body skeleton. Once "Request Calibration" is called, the skeleton process unit will correct and analyze the skeleton. When this unit is working, the skeleton process unit will call the callback function "Calibration Start" to aware the program developer that it starts to process the skeleton correction and analysis. After the skeleton correction is done, it will call the callback function "Calibration End". However, when "Calibration End" is called, it only means the end of the skeleton correction and recognition. It is not sure the skeleton recognition succeeds or fails. If it succeeds, it will go into the next step, which is the function "Start Tracking" to make system to track the skeleton data. If it fails, it will make the pose detection unit to restart to detect the user's pose. After the

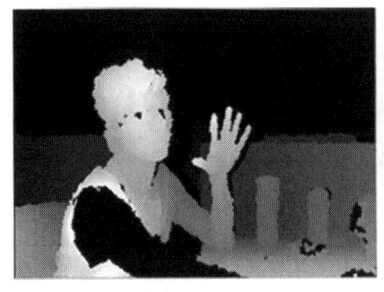 

Success case   Failure case

Fig. 2 Palm detection

success of the skeleton correction and start to track skeleton, the user only needs to call the function to read joints data, and gain the latest joint-related information to build a whole body skeleton data. Kinect development system identify many body joints, but no detail identification on palm joints. That is, when the user's palm in the shooting scope, the system can actually detect the position of palms. However, it can't detect or recognize whether the palm is open or close, or even rotates. From Fig. 2, the green dots represent the result of the palm detection. If the palms are under normal circumstances, the result will be complete and the tracking will be normal. But, if the user clenches or rotates his palms, the system will lose the tracking ability. Therefore, the probability of detection will be lower as well. This research mainly aims to make the user be able to do the jigsaw puzzle instinctively. It has to reach the stable detection and recognition on the clenched palms to allow the user to grab virtual jigsaw and move it.

Gesture Recognition

This system enlarges OPENNI development kit and apply it into the recognition of fingers and gesture. The user can play the jigsaw puzzle which this research made by hand gestures through this system. There are more applications about this hand gesture recognition system, such as using Microsoft PowerPoint. The speaker can change and operate the slides only by hand gesture. Besides Microsoft PowerPoint, it can also be integrated with other software. The more researches, the more functions.

The Concept and Procedure

Since Kinect itself is able to track the palm position, we use it to frame the scope of the palms and work on the calculation of the fingertips positions. After getting the data, we use the numbers of fingertips and the angles between them to

Fig. 3 The flowchart of palm recognition

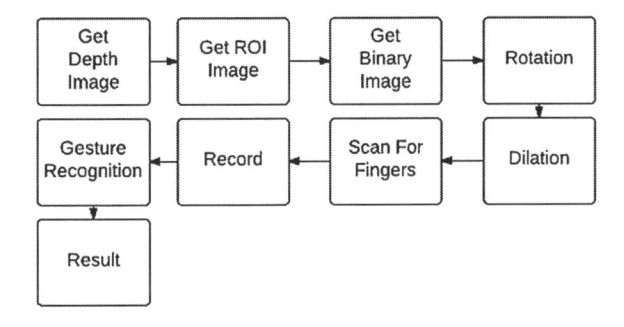

determine the movement of hand gesture. The computer will follow according to different hand gestures. Our calculation method use Kinect's original data to lower the program calculation and space cost. We rotate our palm to proper position in order to get the fingertips position and rotating angels more conveniently and we can save the direction of the palm. The detail calculation is as following Fig. 3.

In the part of image process, we describe separately each phrase according to the flow chart as following:

(a) **Getting ROI image data**: ROI (Region Of Interest) is the area the user interests. It is the core of further image process and recognition. In the phrase, using OPENNI Development environment to find palm as base, we process further recognition and analysis by getting the surrounding area as the key palm image area. However, because the position of palm and waist received by Kinect is quietly unstable, we will not adopt it. (b) **Binary image**: Each point on palm must be close to the palm Depth due to the same hand. Therefore, we frame Palm and compare each point with the palm Depth in the same time. The result like Fig. 4. (c) **Rotation**: The method we mark the finger position needs to rotate the palm to Correct position (palm's up). (d) **Dilation**: This method is to remove flaws due to some image process, such as white spots. (e) **Finding out the fingertips**: the steps

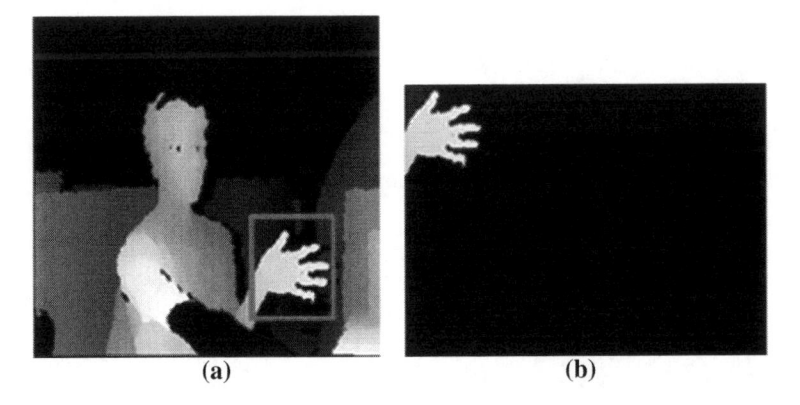

Fig. 4 Getting palm data (a palm tracking (b)palm image

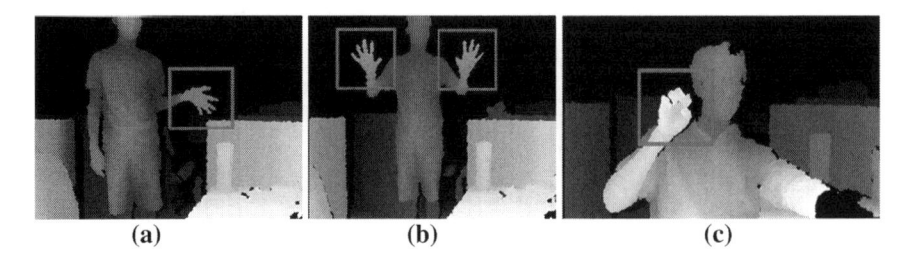

Fig. 5 Results of palm detection (a) one palm is detected (b) Two palms are detected (c) Clasped palm is detected

of finding out the fingertips as follows as: Step 1: We find the number "1", record it as fingertip (saved in the matrix) and put its position into Stack. Step 2: We take out the top position in the Stack matrix, and search five direction around Site (right, right down, down, left down, left), if the value equal $\lceil 1 \rfloor$, the value set $\lceil 0 \rfloor$, and put it into Stack. Step 3: Into recursive, we repeat Step 2 until clear Stack. Step 4: We repeat Step 1 until the end of the matrix. (f) **Hand gesture determination**: Through the palm and waist points given by OPPENNI and the fingertips found by the methods above, we determine the hand gesture by the position data of points, distance and angle. This research uses it to recognize the movement of user's palm in the jigsaw puzzle to determine status. When the user grab the virtual jigsaw, the system will recognize the gesture and connect the relative positions of the jigsaw with the coordinates of the actual palm. That will make the user feel like grabbing virtual jigsaw. When the user open his palm, the system will stop connecting jigsaw with palm, which has the effect of unclenching the virtual jigsaw.

Experimental Results

In this paper, using our proposed method of gesture recognition algorithm, With KINECT human skeleton tracking, the system can track and identify parts of the palm to achieve better results. From Fig. 5, the palm is detected very accurate.

Conclusion

The kinesthetic system is the interactive system which has risen recently because of depth image can be captured directly. The user gets more feedback during the operation. The hand gesture detection and recognition plays an important in the kinesthetic system. How to track and then recognize palms in various operation environments has been a critical issue. In this paper, the hand recognition system based on Kinect becomes practical. This method proves a successful calculation method even under various environments. That is the reason why it functions so well in the jigsaw puzzle. This kind of kinesthetic system has not only provided the user with different ways of interacting with a computer, but also drawn the attention of the user. Therefore, the user is able to operate it more fluently and more interactively.

References

1. Albitar C, Graebling P (2007) Christophe DOIGNON robust structured light coding for 3D reconstruction computer vision. International conference on computer vision (ICCV)
2. Shotton J, Fitzgibbon A, Cook M, Sharp T, Finocchio M, Moore R, Kipman A, Blake A (2011) Real-time human pose recognition in parts from single depth images. In: Proceeding of the IEEE conference on computer vision and pattern recognition (CVPR)
3. Girshick R, Shotton J, Kohli P, Criminisi A, Fitzgibbon A (2011) Efficient regression of general-activity human poses from depth images. In: Proceeding of international conference on computer vision (ICCV)
4. Grest D, Woetzel J, Koch R (2005) Nonlinear body pose estimation from depth images. In: Proceeding DAGM
5. Ye G, Liu Y, Hasler N, Ji X, Dai Q, Theobalt C (2012) Performance capture of interacting characters with handheld kinects. In: Proceeding of European conference on computer vision (ECCV)
6. Abramov A, Pauwels K, Papon J, Worgotter F, Dellen B (2012) Depth-supported real-time video segmentation with the Kinect. In: IEEE workshop on applications of computer vision (WACV)
7. Baak A, Muller M, Bharaj G, Seidel HP, Theobalt C (2011) A data-driven approach for real-time full body pose reconstruction from a depth camera. In: Proceeding of international conference on computer vision (ICCV)
8. Newcombe RA, Izadi S, Hilliges O, Molyneaux D, Kim D, Davison AJ, Kohli P, Shotton J, Hodges S, Fitzgibbon A (2011) KinectFusion: real-time dense surface mapping and tracking. In: Proceeding of IEEE international symposium on mixed and augmented reality (ISMAR), pp 127–136
9. Frati V, Prattichizzo D (2011) Using Kinect for hand tracking and rendering in wearable haptics. World haptics conference (WHC). Dipt. di Ing. dell''Inf., Univ. di Siena, Siena, Italy
10. Raheja JL (2011) Tracking of fingertips and centers of palm using KINECT. Computational intelligence, modelling and simulation (CIMSiM)

Fall Detection in Dusky Environment

Ying-Nong Chen, Chi-Hung Chuang, Chih-Chang Yu and Kuo-Chin Fan

Abstract Accidental fall is the most prominent factor that causes the accidental death of elder people due to their slow body reaction. Automatic fall detection is then an emerging technology which can assist traditional human monitoring and avoid the drawbacks suffering in health care systems especially in dusky environments. In this paper, a novel fall detection system based on coarse-to-fine strategy is proposed focusing mainly on dusky environments. Since the silhouette images of human bodies extracted from conventional CCD cameras in dusky environments are usually imperfect due to the abrupt change of illumination, our work adopts thermal imager instead to detect human bodies. In our approach, the downward optical flow features are firstly extracted from the thermal images to identify fall-like actions in the coarse stage. The horizontal projected motion history images (MHI) features of fall-like actions are then designed to verify the fall by the proposed nearest neighbor feature line embedding (NNFLE) in the fine stage. Experimental results demonstrate that the proposed method can distinguish the fall incidents with high accuracy even in dusky environments and overlapping situations.

Keywords Fall detection · Optical flow · Motion history image · Nearest neighbor feature line

Introduction

Accidental fall is the most prominent factor that causes the accidental death of elder people due to their slow body reaction. Fall accidents usually occur at night with nobody except the elder people if they live alone. It is usually too late to

Y.-N. Chen (✉) · C.-H. Chuang · C.-C. Yu · K.-C. Fan
National Central University, Jhongli, Taiwan, Republic of China
e-mail: 93542021@cc.ncu.edu.tw

Y.-M. Huang et al. (eds.), *Advanced Technologies, Embedded and Multimedia for Human-centric Computing*, Lecture Notes in Electrical Engineering 260, DOI: 10.1007/978-94-007-7262-5_129, © Springer Science+Business Media Dordrecht 2014

Fig. 1 The thermal imager

remedy the tragedy when the body is discovered hours or days after with the occurrence of accidental fall. In the occurrence of fall incident, humans usually lie flat on the ground. However, we cannot merely use the images to perceive whether this person is lying on the ground. Hence, we have to detect and avoid the risk caused by fall action before the existence of really lying on the ground. According to the survey, a sudden fainting or body imbalance is the main reason to cause a fall. No matter what kind of reasons, fall is a warning that the subject may be in danger. Moreover, if the incidents occur in a dusky and unattended environment, people usually miss the prime time for rescue because the silhouette images of human bodies extracted from conventional CCD cameras in dusky environments are usually invisible or imperfect due to the illumination constraint. To remedy this problem, a fall detection system using a thermal imager (see Fig. 1) to capture the images of human bodies is proposed in this paper. By using the thermal imager, the human bodies can be accurately located even in a dusky environment. For comparison, Fig. 2a shows the images obtained by a CCD camera in a dusky environment, whereas Fig. 2b shows the images obtained by a thermal imager in the same environment. It is obvious that the thermal imagers can extract more clear and intact human bodies in the dusky environments than CCD cameras.

Moylan [1] illustrated the gravity of falls as a health risk with abundant statistics. Larson [2] described the importance of falls in elderly. National Center for

(a)   **(b)**

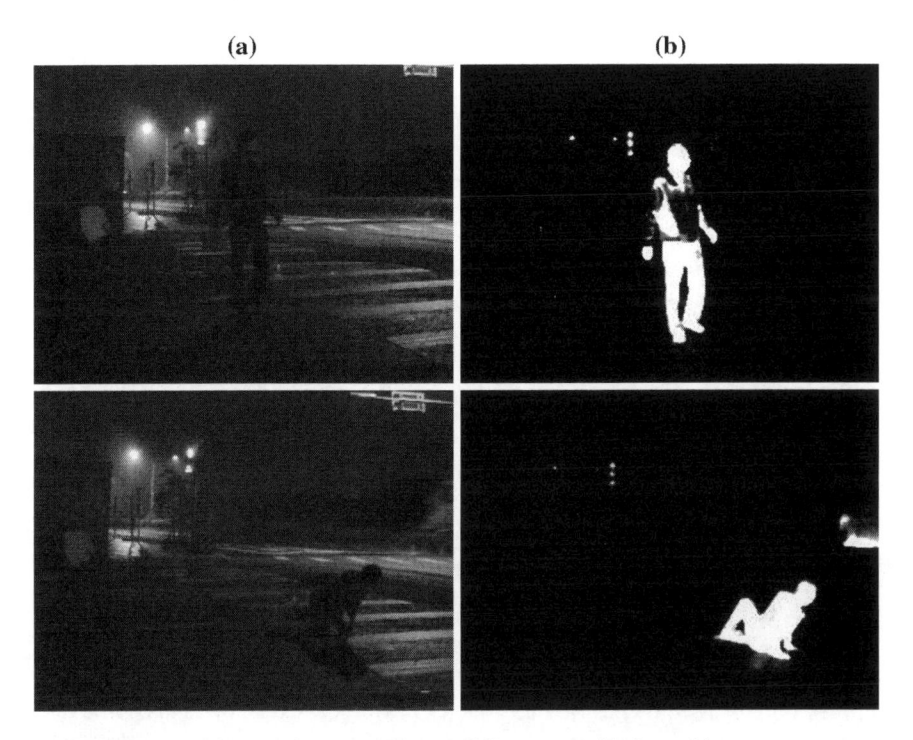

Fig. 2 Image extraction results captured by (**a**) CCD camera, (**b**) thermal imager

Health Statistics showed that more than one-third of ages 65 or older fall each year. Moreover, 60 % of lethal falls occur at home, 30 % occur in public region, and 10 % happen in health care institutions for ages 65 or older [3]. In the literatures of fall detection, Tao et al. [4] applied the aspect ratio of the foreground object to detect fall incidents. Their system firstly tracks the foreground objects and then analyzes the sequences of features for fall incidents detection. Anderson et al. [5] also applied the aspect ratio of the silhouette to detect fall incidents. The rationale based mainly on the fact that the aspect ratio of the silhouette is usually very large when the fall incidents occur. On the contrary, the aspect ratio is much smaller when the fall incidents do not occur. Juang [6] proposed a neural fuzzy network method to classify the human body postures, such as standing, bending, sitting and lying down. In [7], Foroughi et al. proposed a fall detection method using an approximated eclipse of human body silhouette and head pose as features for multi-class support vector machine (SVM). Rougier et al. [8] applied the motion history image (MHI) and variations of human body shape to detect falls. In [9], Foroughi et al. proposed a modified MHI integrating the time motion image (ITMI) as the motion feature. Then, the eigenspace technique was used for motion feature reduction and fed into individual neural network for each activity. Liu et al. [10] proposed a nearest neighbor classification method to classify the ratio of human body silhouette of fall incidents. In order to differentiate between the fall and lying, the time difference between fall and lying was used as a key feature. Liao et al. [11] proposed a slip and fall detection system based on Bayesian Belief Network (BBN). They used the integrated spatiotemporal energy (ISTE) map to obtain the motion measure. Then, the BBN model of the causality of the slip and fall was constructed for fall prevention. Olivieri et al. [12] proposed a spatio-temporal motion feature to represent activities termed motion vector flow instance (MVFI) templates. Then, a canonical eigenspace technique was used for MVFI template reduction and template matching.

In this paper, a novel fall detection mechanism based on coarse-to-fine strategy which is workable in both day and dusky environments is proposed. In the coarse stage, the downward optical flow features are extracted from the thermal images to identify fall-like actions. Then, the horizontal projected motion history images (MHI) features of fall-like actions are used in the fine stage to verify the fall by the nearest neighbor feature line embedding.

The Proposed Fall Detection Mechanism

The proposed fall detection mechanism consists of two modules including human body extraction and fall detection. In human body extraction module, temperature frames obtained from the thermal imager are processed with image processing techniques to obtain intact human body contours and silhouettes. In fall detection module, a coarse-to-fine strategy is devised to verify fall incidents. In the coarse stage, the downward optical flow features are extracted from the temperature images to identify possible fall down actions. Then, the 50-dimensional temporal based motion history images (MHI) feature vector is projected into the nearest neighbor feature line space to verify the fall down incident in the fine stage. To improve fall detection accuracy, complete silhouettes of human body must be extracted to obtain accurate bounding box of human body. To this end, the temperature images captured from a thermal imager are binarized by Otsu's method firstly. Then, the morphological closing operation is employed to obtain a complete human silhouette. Finally, a labeling process is performed to locate each human body in the image and filter out background noises. The process of human body extraction is depicted in Fig. 3. Figure 3a shows the temperature images captured from the thermal imager, Fig. 3b shows the Otsu's binarization results, and Fig. 3c shows the results of morphological closing operation. The bounding box of the human silhouette can be successfully generated after the morphological closing operations.

After the bounding box of human body has been determined, a coarse-to-fine strategy is utilized to verify fall incidents. The purpose of the coarse stage is to identify possible fall actions. Wu [13] had shown that a fall could be described by the increase in horizontal and vertical velocities. Moreover, this work observes that the histogram of vertical optical flows has also demonstrated the significant difference between walking and falling. In our work, a multi-frame optical flow

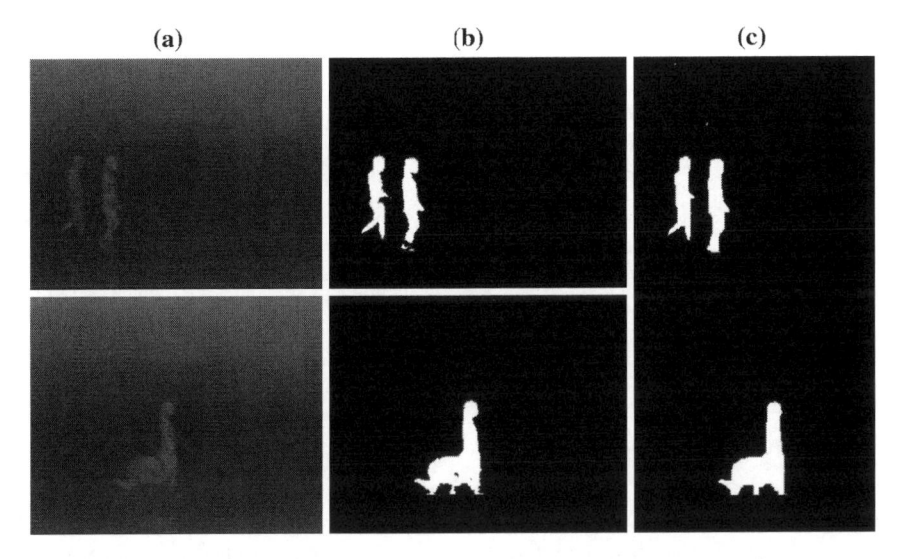

Fig. 3 Human body extraction. **a** Temperature gray level images, **b** binarization results, and **c** morphological closing operation results

method proposed by Wang [14] is adopted to extract the downward optical flow features inside the extracted bounding box in this stage. A possible fall action can be identified by two heuristic rules:

(1) Rule 1: Given 20 consecutive frames, the average vertical optical flows exhibit downward more than 75 % of frames.
(2) Rule 2: The sum of the average vertical optical flows in 20 consecutive frames is larger than a threshold, say 10 in this study.

In the coarse stage, most non-fall actions can be filtered out via the downward optical flow features. However, some fall-like actions are identified as fall incidents due to the swing of arms. To solve this problem, we devise a feature vector which is formed by projecting the MHI horizontally to verify fall incidents in the fine stage. MHI proposed by Bobick [15] is a template which condenses a determined number of silhouette sequences into a gray scale image which is capable of preserving dominant motion information. Since the main difference between fall and other actions is the vertical component changes, our work projects the MHI horizontally to obtain a 50 dimensional feature vector.

In our previous work [16], NFLE has been proven its effectiveness in pattern recognition. However, three problems of the NFLE have also been indicated in [16]. Based on the motivation of mitigating the three problems of NFLE and verify a fall-like action after the coarse stage, a modified NFLE termed Nearest Neighbor Feature Line Embedding (NNFLE) is proposed as a fall verifier in the fine stage. In this paper, we define a fine stage feature vector from a MHI as a sample, In order to overcome the extrapolation and interpolation inaccuracy, feature lines for a query point are generated from the k nearest neighborhood prototypes. The selection strategy for discriminant vectors in NNFLE is designed as follows:

(1) The within-class scatter $\mathbf{S}_W$: The NNFLs are generated from the k_1 nearest neighbor samples within the same class for the computation of the within-class scatter matrix, i.e. a set $F_{k_1}^+(x_i)$.
(2) The between-class scatter $\mathbf{S}_B$: Select k_2 nearest neighbor samples in different classes from a specified point x_i to generate the NNFLs and calculate the between-class scatter matrix, i.e. a set $F_{k_2}^-(x_i)$.

Experimental Results

In this section, experimental results conducted on fall incident detection are illustrated to demonstrate the effectiveness of the proposed method. In this paper, we compare the proposed method with two state-of-the-art methods. The results are evaluated by using the simulated video data set captured from outdoor scenes. The total number of the video data set is 320. In each video, the environment is in the dusky environments as shown in Fig. 2. The data sets used in this subsection contains only one subject in each video sequence. Two state-of-the-art methods, BBN [11] and CPL [12], are implemented for comparison. The CPL takes a sequence as a sample, whereas the BBN and our proposed method take a frame as a sample. Therefore, the performance comparison of these three methods is based on each video sequence. In the experiments, 60 video sequences of one person are used as training sets and 100 video sequences of one person are used for testing. The performance comparisons of these three methods are tabulated in Table 1. From Table 1, we can notice that the proposed coarse-to-fine strategy of fall detection outperforms the other two methods. It implies that the proposed method is much more effective than the other two methods.

Table 1 The fall detection performance on the data set (%)

Method	Classification action	Reference action (videos)	
		Fall	Walk
CPL	Fall	92.00 (46/50)	8.00 (4/50)
	Walk	10.00 (5/50)	90.00 (45/50)
BBN	Fall	80.00 (40/50)	20.00 (10/50)
	Walk	12.00 (6/50)	88.00 (44/50)
Ours	Fall	**98.00 (49/50)**	**2.00 (1/50)**
	Walk	**0.00 (0/50)**	**100.00 (50/50)**

Conclusion

In this paper, a novel fall detection mechanism based on coarse-to-fine strategy in dusky environment is proposed. The human body in dusky environment can be successfully extracted using the thermal imager and the fragments inside human body silhouette can also be significantly reduced as well. In the coarse stage, the optical flow algorithm is firstly applied on thermal images. Most of walk actions are filtered out by analyzing the downward flow features. In the fine stage, the projected MHI is used as the features followed by a nearest neighbor selection strategy adopted in the NNFLE method to verify fall incidents.

References

1. Moylan KC, Binder EF (2007) Falls in older adults: risk assessment, management and prevention. Am J Med 120(6):493–497
2. Larson L, Bergmann TF (2008) Taking on the fall: the etiology and prevention of falls in the elderly. Clin Chiropractic 11(3):148–154
3. Gs S (1988) Falls among the elderly: epidemiology and prevention. Am J Prev Med 4(5):282–288
4. Tao J, Turjo M, Wong M-F, Wang M, Tan Y-P (2005) Fall incidents detection for intelligent video surveillance. In: Proceedings of the 15th international conference on communications and signal processing 2005, p 1590–1594
5. Anderson D, Keller JM, Skubic M, Chen X, He Z (2006) Recognizing falls from silhouettes. In: Proceedings of the 28th IEEE EMBS annual international conference 2006
6. Juang CF, Chang CM (2007) Human body posture classification by neural fuzzy network and home care system applications. IEEE Trans SMC Part A 37(6):984–994
7. Foroughi H, Aabed N, Saberi A, Yazdi HS (2008) An eigenspace-based approach for human fall detection using integrated time motion image and neural networks. In: Proceedings of the IEEE international conference on signal processing (ICSP) 2008
8. Rougier C, Meunier J, ST Arnaud A, Rousseau J (2007) Fall detection from human shape and motion history using video surveillance. In: Proceedings of the 21st international conference on advanced information networking and applications workshops, vol 2, pp 875–880
9. Foroughi H, Rezvanian A, Paziraee A (2008) Robust fall detection using human shape and multi-class support vector machine. In: Proceedings of the sixth Indian conference on CVGIP 2008
10. Liu CL, Lee CH, Lin P (2010) A fall detection system using k-nearest neighbor classifier. Expert Syst Appl 37(10):7174–7181
11. Liao YT, Huang CL, Hsu SC (2012) Slip and fall event detection using Bayesian belief network. Pattern Recogn 45:24–32
12. Olivieri DN, Conde IG, Sobrino XAV (2012) Eigenspace-based fall detection and activity recognition from motion templates and machine learning. Expert Syst Appl 39(5):5935–5945
13. Wu G (2000) Distinguishing fall activities from normal activities by velocity characteristics. J Biomech 33(11):1497–1500
14. Wang CM, Fan KC, Wang CT (2008) Estimating optical flow by integrating multi-frame information. J Inf Sci Eng 24(6):1719–1731
15. Bobick AF, Davis JW (2001) The recognition of human movement using temporal templates. IEEE Trans Pattern Anal Mach Intell 23(3):257–267

16. Chen YN, Han CC, Wang CT, Fan KC (2012) Face recognition using nearest feature space embedding. IEEE Trans Pattern Anal Mach Intell 33(6):1073–1086
17. Li SZ, Lu J (1999) Face recognition using the nearest feature line method. IEEE Trans Neural Netw 10(2):439–433
18. Yan S, Xu D, Zhang B, Zhang HJ, Lin S (2007) Graph embedding and extensions: general framework for dimensionality reduction. IEEE Trans Pattern Anal Mach Intell 29(1):40–51

Mandarin Phonetic Symbol Combination Recogniztion in Visioned Based Input Systems

Chih-Chang Yu and Hsu-Yung Cheng

Abstract This paper proposes a Mandarin Phonetic Symbols combination recognition system. This work provides a prototype of a user-friendly and efficient human-centered interface. The system analyzes the fingertip trajectories to obtain the strokes of Mandarin Phonetic Symbols combinations. The trajectories are segmented into real strokes and virtual strokes. The system computes the prior probabilities of different classes of combinations and via Bayes Classifier. The likelihood that each segment is generated by a certain Mandarin Phonetic Symbols combination model is computed via Hidden Markov Models. The experiments have validated the effectiveness of the proposed system.

Keywords Mandarin phonetic symbols combination · Bayes classifier · Hidden Markov models · Human computer interface · Fingertip input

Introduction

In Chinese input systems, Mandarin Phonetic Symbol (MPS) plays a very important role. Mandarin Phonetic Symbols are extracted from Chinese characters. The importance of Mandarin Phonetic Symbols (MPS) and algorithms to recognize single MPS are discussed in [1]. Also, algorithms of locating the fingertip from the segmented foreground mask and tracking the fingertip are elaborated in [1, 2]. To develop a more user-friendly and efficient human-centered fingertip Mandarin input system, recognizing MPS combinations instead of single symbol is desired. There are total 403 different meaningful combinations of MPS. Each combination may consist of single, double, or triple symbols. For convenience, we

C.-C. Yu · H.-Y. Cheng (✉)
Department of Computer Science and Information Engineering, National Central University, Jungli, Taiwan, Republic of China
e-mail: chengsy@csie.ncu.edu.tw

Y.-M. Huang et al. (eds.), *Advanced Technologies, Embedded and Multimedia for Human-centric Computing*, Lecture Notes in Electrical Engineering 260, DOI: 10.1007/978-94-007-7262-5_130, © Springer Science+Business Media Dordrecht 2014

Fig. 1 Three different
classes of MPS combinations:
a single **b** double **c** triple

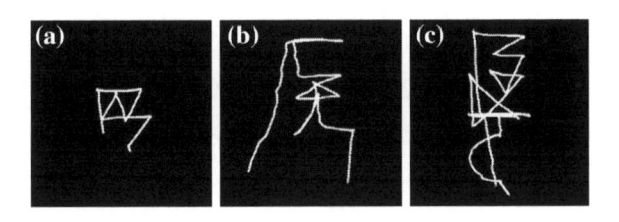

notate the single, double, and triple combinations as ω_1, ω_2 and ω_3. Figure 1a, b, and c shows an example of ω_1, ω_2 and ω_3 combination, respectively.

Mandarin Phonetic Symbol Combination Recognition

For effective MPS combination recognition, strokes must be analyzed. The strokes acquired by the trajectories of the fingertip include real strokes and virtual strokes. The input symbol combinations exhibit four different types of virtual strokes: entering stroke, leaving stroke, linking stroke, and other virtual strokes within a symbol. The entering, leaving and virtual strokes within a symbol are defined in Fig. 2. Note that in Fig. 2, starting strokes and ending strokes are real strokes. For MPS combinations, a new type of virtual stroke emerge, which is the linking virtual stroke between symbols, as illustrated in Fig. 3. Figure 3a and b show that the entering, leaving, and linking strokes may be very different for the same combination written by different users. In order to recognize the MPS combinations accurately, the entering and leaving strokes need to be eliminated, and the linking strokes need to be correctly located.

The MPS combination recognition procedure is illustrated in Fig. 4. The trajectories are encoded using 8-chain code [3] (Fig. 4a). The turning points are

Fig. 2 Entering stroke,
leaving stroke, starting
stroke, and ending stroke

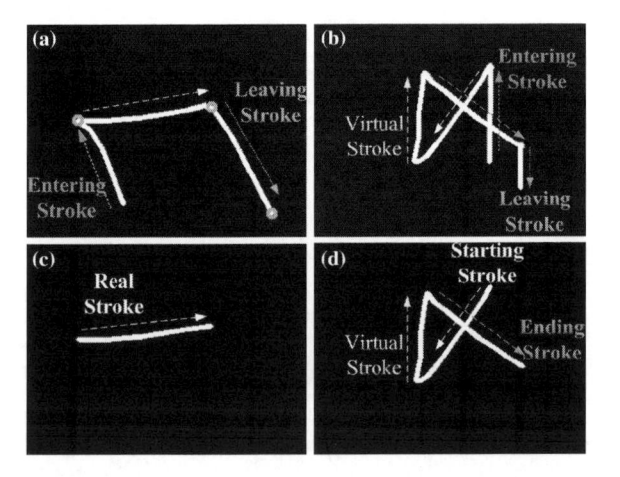

Fig. 3 Linking strokes between symbols

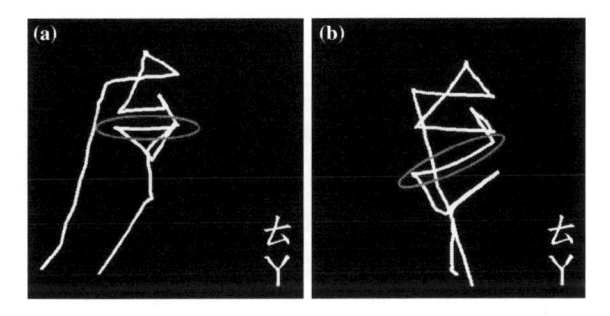

Fig. 4 MPS combination recognition procedure

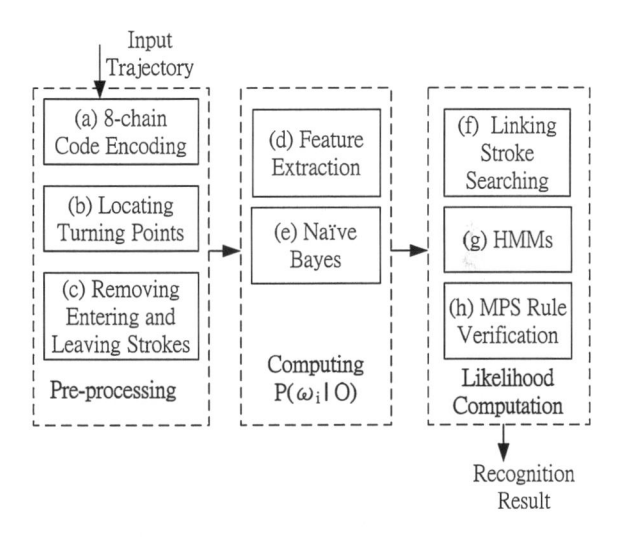

detected when the difference of two neighboring codes is larger than 2 (Fig. 4b). In other words, the neighboring strokes form an angle which is larger than 90° at a turning point. The trajectory between two turning points is regarded as a stroke. We check all the strokes and eliminate those strokes which are too short. The short strokes are usually noises and provide little contribution for recognition. Hence, we discard these short strokes for better recognition results. After obtaining the turning points, we can remove the entering and leaving strokes, which correspond to the trajectories before the first turning point and the last turning point (Fig. 4c). Some examples of encoded symbol combinations after removing entering and leaving strokes are displayed in Fig. 5. In Fig. 5, the green dots are the detected turning points between strokes.

To be able to recognize the combinations, we need to decide the number of symbols in each combination. Instead of distinguishing them into ω_1, ω_2, and ω_3 by making hard decisions, we compute the probabilities that the combination belongs to ω_i given the extracted feature vector $\mathbf{x}$, $P(\omega_i|\mathbf{x})$, $i = 1, 2, 3$. The posterior probability $P(\omega_i|\mathbf{x})$ computed by Bayes classifier (Fig. 4e) is decomposed into three terms as listed in Eq. (13). The prior probability $P(\omega_i)$ is set

Fig. 5 After removing entering and leaving strokes and encoding the trajectories using 8-chain codes

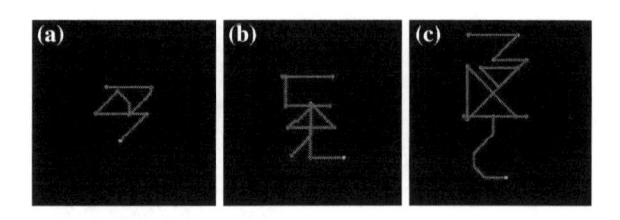

according to the appearing frequency of samples from class i. The class conditional density functions $P(\mathbf{x}|\omega_i)$ are modeled by Gaussian functions whose parameters are estimated by training samples via maximum likelihood estimation. The term $P(\mathbf{x})$ is independent of class labels and therefore does not affect the decision. We use a Naïve Bayes classifier by making the assumption that all the features are independent. A bounding box is generated for each trajectory after preprocessing. The feature vector $\mathbf{x}$ extracted from the observed trajectory includes the height of the bounding box, the width of the bounding box, the aspect ratio of the bounding box, total number of codes after encoding, number of turning points, and the percentage of each of the eight chain codes in the observed trajectory.

$$P(\omega_i|\mathbf{x}) = \frac{P(\mathbf{x}|\omega_i)P(\omega_i)}{P(\mathbf{x})} \qquad (1)$$

After computing $P(\omega_i|\mathbf{x})$, we eliminate the class with the lowest probability and perform symbol segmentation based on the two remaining classes. For ω_1, there is no need to perform symbol segmentation since there is only a single symbol. For ω_2 and ω_3, linking stroke searching is necessary (Fig. 4f). Under the assumption of ω_2, we hypothesize that the kth stroke is the linking stroke. Under the assumption of ω_3, we hypothesize that the k_1th stroke and the k_2th stroke are the linking strokes. Note that $2 \leq k \leq M - 1$, and $2 \leq k_1 < k_2 \leq M - 1$, where M denotes the length of codes after encoding. With these hypotheses, the entire trajectory is segmented into one, two, or three segments. Each segment serves as the observation to compute likelihoods $P(S|\lambda_k)$ using pre-trained single symbol Hidden Markov Models via forward–backward algorithm [4, 5] (Fig. 4g). The single symbol Hidden Markov Model λ_k is trained using the chain code of each symbol via Baum-Welch algorithm [4, 5]. Finally, MPS rules are checked to verify the recognition results (Fig. 4h).

Table 1 Three different sets of symbols

Set	Notation	Symbols
Initial	α	ㄅㄆㄇㄈㄉㄊㄋㄌㄍㄎㄏㄐㄑ ㄒㄓㄔㄕㄖㄗㄘㄙ
Medial	β	ㄧㄨㄩ
Final	γ	ㄚㄛㄜㄝㄞㄟㄠㄡㄢㄣㄤㄥㄦ

For each segment, we compute the likelihoods $P(S|\lambda_k)$ that the segment is generated by the model λ_k of symbol k using forward–backward algorithm. The MPS symbols are divided into three sets as shown in Table 1. According to the rules of MPS, ω_1 can be from the last seven symbols in α, any symbol in β, or any symbol in γ. For ω_2, the combinations of the two symbols can be $\alpha + \beta$, $\beta + \gamma$, or $\alpha + \gamma$. For ω_3, the legal combination is $\alpha + \beta + \gamma$. These rules are helpful for recognition since we can set the likelihoods of illegal combinations as zero directly.

Next, we compute the likelihood that the entire observed trajectory T is generated by a certain model combination. For ω_1, the maximum likelihood of single symbol is selected as the likelihood of the observed trajectory $P(T|\lambda_k)$, as shown in Eq. (2). For ω_2, the likelihood of the observed trajectory $P(T|\lambda_{k1}, \lambda_{k2})$ is computed using Eq. (3), where λ_i and λ_j are models from legal combinations as described above. For ω_3, the likelihood of the observed trajectory $P(T|\lambda_{k1}, \lambda_{k2}, \lambda_{k3})$ is computed using Eq. (4), where $\lambda_i \in \alpha$, $\lambda_j \in \beta$, and $\lambda_l \in \gamma$. In implementation, only top-ranked λ_i, λ_j, λ_j models need to be considered when computing Eqs. (3) and (4). The final decision is made by selecting the combination that contributes the highest weighted likelihood. The weights are the probabilities $P(\omega_i|\mathbf{x})$ computed by the Bayes classifier.

$$P(T|\lambda_k) = \max_i P(S|\lambda_i) \tag{2}$$

$$P(T|\lambda_{k1}, \lambda_{k2}) = \max_{i,j} P(S|\lambda_i)P(S|\lambda_j) \tag{3}$$

$$P(T|\lambda_{k1}, \lambda_{k2}, \lambda_{k3}) = \max_{i,j,l} P(S|\lambda_i)P(S|\lambda_j)P(S|\lambda_l) \tag{4}$$

Experimental Results

The experimental dataset for MPS combination recognition contains fingertip input trajectories contributed from 20 t esters. The features for the Naïve Bayes Classifier that computes $P(\omega_i|\mathbf{x})$ are collected from 111 ω_1 combinations, 219 ω_2 combinations, and 164 ω_3 combinations. The single symbol Hidden Markov Models are trained using 20 sets of the 37 MPS symbols. The HMMs are strictly left to right with four hidden states. Figure 6 illustrates examples of linking stroke searching for an ω_2 and an ω_3 combination. The yellow strokes in the figure are hypothesized linking strokes. The wrong hypothesis in Fig. 6a would result in a smaller likelihood than that obtained by the correct hypothesis in Fig. 6b. Similarly, Fig. 6c shows a wrong hypothesis for an ω_3 combination, and Fig. 6d shows the correct hypothesis for the combination. The MPS combination recognition rates from rank 1 to rank 10 are plotted in Fig. 7. An averaged recognition rate of more than 90 % for all possible combinations can be reached for rank 3.

Fig. 6 Examples of MPS
combination recognition

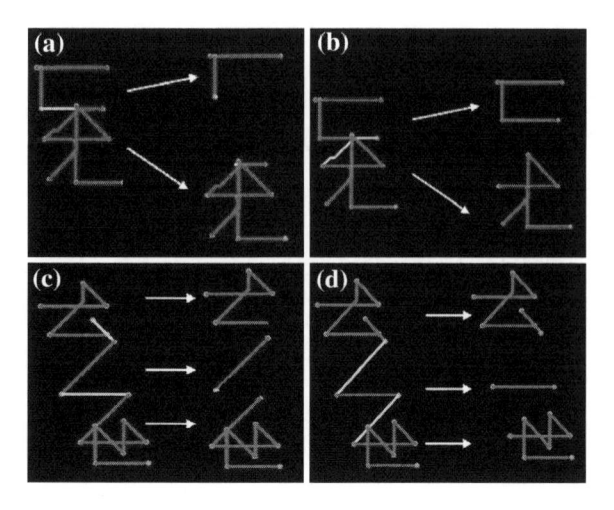

Fig. 7 MPS combination
recognition accuracy

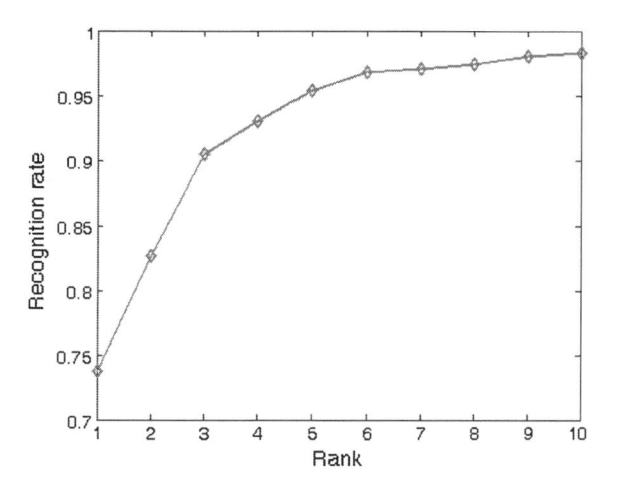

Conclusion

In this work, users employ their fingers as the input device instead of using pens or
keyboards. Unlike other languages based on alphabets, Chinese character input
systems are more challenging because of their logographic nature. The system
segments the fingertip trajectories to obtain the strokes. The strokes are further
analyzed to search for possible linking strokes between two different symbols in
the MPS combination. The system computes the prior probabilities of different
classes of combinations and via Bayes Classifier. The likelihood that each segment
is generated by a certain Mandarin Phonetic Symbols combination model is
computed via Hidden Markov Models. The experiments have validated the

effectiveness of the proposed system. The ability to recognize MPS combinations rather than single symbol makes the human computer interface system much more user-friendly.

References

1. Taele P, Hammond T (2008) Geometric-based sketch recognition approach for handwritten mandarin phonetic symbols I. International Workshop on Visual Languages and Computing, pp 270–275
2. National Taiwan Normal University (2004) Practical audio-visual Chinese 1 textbook, vol 1. Cheng Chung Book Company, Ltd, Taipei
3. Yu CC, Cheng HY, Jeng BS, Lee CC, Hong WT (2011) Human-centered fingertip Mandarin input system using single camera. In: The 17th international conference on multimedia modeling
4. Cheng HY, Hwang JN (2011) Integrated video object tracking with applications in trajectory-based event detection. J Vis Commun Image Represent 22:673–685
5. Gonzalez RC, Woods RE (2002) Digital image processing, 2nd edn. Prentice Hall, Englewood Cliffs
6. Rabiner LR, Juang BH (1986) An introduction to hidden Markov models. IEEE ASSP Mag 3:4–16
7. Rabiner LR (1989) A tutorial on hidden Markov models and selected applications in speech recognition. Proc IEEE 77:257–286

Part XIII
All-IP Platforms, Services and Internet of Things in Future

Supporting Health Informatics with Platform-as-a-Service Cloud Computing

Garrett Hayes, Khalil El-Khatib and Carolyn McGregor

Abstract Recent progression in health informatics data analysis has been impeded due to lack of hospital resources and computation power. To remedy this, some researchers have proposed a cloud-based web service patient monitoring system capable of providing offsite collection, analysis, and dissemination of remote patient physiological data. Unfortunately, some of these cloud services are not effective without utilizing next-generation hardware management techniques. In order to make cloud based patient monitoring a reality, this paper shows how leveraging an underlying platform-as-a-service (PaaS) cloud model can provide integration with web service patient monitoring systems while providing high availability, scalability, and security.

Keywords Artemis cloud · Artemis framework · Cloud computing · Health informatics · Platform as a service · Remote patient monitoring

Introduction

As technology continues to develop in our increasingly connected world, previous non-network enabled medical devices have begun to join hospital networks around the world. Thanks to recent medical device improvements, hospitals are now able to remotely collect patient data from bedside monitoring devices. The collection

G. Hayes (✉) · K. El-Khatib · C. McGregor
University of Ontario Institute of Technology, Oshawa, ON, Canada
e-mail: Garrett.Hayes@uoit.ca

K. El-Khatib
e-mail: Khalil.El-Khatib@uoit.ca

C. McGregor
e-mail: Carolyn.McGregor@uoit.ca

Y.-M. Huang et al. (eds.), *Advanced Technologies, Embedded and Multimedia for Human-centric Computing*, Lecture Notes in Electrical Engineering 260, DOI: 10.1007/978-94-007-7262-5_131, © Springer Science+Business Media Dordrecht 2014

and analysis of this data has provided many opportunities for clinicians and researchers to discover new condition onset behaviours that are evident in a collection of physiological data streams before the current methods of detection for a range of conditions in critical care. Data collected from patients worldwide gives us the ability to extrapolate observations based on populous and historical data.

In order to support researchers using data collected from bedside monitoring devices, a number of researchers have proposed a cloud-based and offsite web service framework capable of collecting and analyzing patient data worldwide [1–4]. Unfortunately, underlying traditional technical infrastructure is incapable of providing a robust underlying platform that is highly available, expandable, and secure.

In this paper we will provide an overview of cloud computing, touch on a specific cloud framework for health informatics, and show how leveraging platform-as-a-service can provide a robust underlying architecture for such a deployment.

Cloud Computing Overview

The concept of cloud computing generally refers to the delivery of one or more infrastructure components—computational, storage, or otherwise—as a scalable and reliable location-independent service. These services, often located offsite, are exposed to a consumer via pre-set client interfaces [5]. Such interfaces may include web browsers, standardized APIs, and other types of Internet-ready protocols. The location of a client's provisioned cloud resources and data, unbeknownst to them, may span multiple servers, datacenters, or even countries.

Encompassing many service models, cloud computing allows the quick provisioning of resources regardless of underlying hardware or software components. Thanks to this client-centric model, enterprises are able to deploy or request additional infrastructure resources on demand, further streamlining their already complex IT operations.

The underlying amalgamation of various servers, storage area networks, and network hardware is conformed into a single standardized service interface referred to as a cloud. The resulting cloud provides the ability to elastically provision, span, and optimize resource and storage over multiple pieces or hardware while ensuring high availability and robust security.

Although the cloud concept seems esoteric by nature, the cloud model is being used to support many services on the web. Some of these services include popular products like Gmail, Facebook, Google Drive, Dropbox, Amazon AWS, Microsoft Office Web Apps, and many more.

While the majority of well-known cloud service deployments are Internet based (i.e. public clouds), in some instances enterprises may leverage the cloud architecture to consolidate resources and improve overall datacenter efficiency. In addition, when local resources are not cost efficient or sufficient enough for large projects, hybrid cloud models may be used to provision additional computation power on the fly. The four existing cloud models are as follows [5]:

1. Private Cloud: operated by a private company.
2. Community Cloud: operated by more than one company, often working together under a common agreement or area of interest and sharing a single cloud computing instance.
3. Public Cloud: hosted by an external cloud provider whom offers cloud provisioning services to all types of customers.
4. Hybrid Cloud: when two or more clouds are merged together, often through a standardized API interface, to provide cross-datacenter provisioning services on the fly.

Regardless of the cloud model used in a cloud deployment, various types of service models can be presented to administrators. Each service model aims to provide a specific set of resources pertinent to the requirements of the cloud consumer. These cloud service models are as follows: (1) Infrastructure-as-a-Service, (2) Platform-as-a-Service, (3) Software-as-a-Service, and (4) Data-as-a-Service.

Cloud Computing Advantages in Health Informatics

Cloud computing affords many advantages over traditional data center and hardware deployments. Benefits vary from increasing service availability to even reducing data center power consumption. The various benefits of cloud computing, particularly in health informatics, are as follows:

1. Accessibility
2. Availability
3. Resource Consolidation and Multitenancy
4. Reliability
5. Scalability
6. Security.

Cloud Computing in Healthcare

The applications of cloud computing are not just limited to providing virtual machine instances for servers and other Internet facing services. Rather, the cloud model can be applied to any type of datacenter, regardless of its operational goals. One interesting application of cloud computing is in health informatics.

Current cutting-edge health informatics research projects aim to discover new condition onset behaviours that are evident in physiological data streams earlier than traditional detection of conditions in critical care data [1]. To do this, some hospitals may participate in pilot programs that aim to collect real-time patient data from network enabled monitoring devices. This collected data is then analyzed to extract relevant temporal behaviours and usually stored for future data mining and analysis operations. Naturally, not all hospitals may have the capacity—in terms of networking, computational resources, and information technology support staff—to fully support such pilot projects. In this section we will look at a case study implementing a neonatal pilot research project and show how the use of off-site platform-as-a-service cloud computing can help maximize project efficiency both inside the participating hospitals and out.

Current Cloud Applications in Health Informatics

Although cloud computing is still relatively new in the health informatics field, recently there have been a few cloud technology deployments in North American healthcare facilities. In addition, the advent of cloud-based services has created a fundamental shift in the way practitioners meet with patients and manage their electronic health record data. Below we will discuss two significant cloud deployments in the ever-growing health informatics field.

One of the most interesting recent cloud deployments is the Minnesota Tele-Health system. The Tele-Health healthcare network in Minnesota, USA leverages cloud-based medical consultation communications in their hospitals to provide secure remote video conferencing between patients and doctors [6]. This system allows patients to meet with doctors on a flexible, on-demand schedule regardless of physical location. This cloud-based Tele-Health system is capable of prioritizing calls, integrating with human translators when needed, and providing routine medical information by amalgamating existing hospital services and employees. In addition, Tele-Health conforms to the HIPAA standard thanks to the implementation of robust cryptographic controls [6].

Focusing more on health practice management, CareCloud provides a cloud-based framework for managing, analyzing, and charting patient data [7]. Their secure HIPAA compliant cloud service provides a scalable electronic healthcare record solution for healthcare practitioners, allowing them to focus on their patients instead of managing health data [7]. This system allows practitioners to add, edit, and share patient data using a web-based cloud service that provides secure, redundant, and offsite health record storage [7]. Furthermore, CareCloud provides a community connection service that allows practitioners to meet virtually with their patients while sharing selective E-Health Record (EHR) data [7].

As we can see, both Tele-Health and CareCloud have provided innovative cloud-based services capable of streamlining existing healthcare practices while still conforming to HIPAA standards. However, those research projects do not support the transmission of high frequency physiological data streams in real-time for the provision of advanced clinical decision support remotely to support the Service of Critical Care.

Artemis Framework

The Artemis framework provides multidimensional real-time analysis of high-speed physiological data and also support advanced clinical research [1]. McGregor's platform showed the plausibility of supporting online real-time monitoring and analysis of clinical data to detect onset physiological conditions [1]. Collected data may be used to provide extensive insight into novel discoveries regarding physiological data for earlier onset detection of a range of developing physiological conditions.

The Artemis framework, seen in Fig. 1, consists of five phases which, when acting together, are capable of providing in-depth online analysis of developing physiological conditions in patients at participating hospital [1].

The first phase facilitates client data acquisition. A medical data hub, located within each hospital, is responsible for collecting real-time clinical data from network-enabled medical devices onsite. In addition, medical data hubs pair real time patient data with existing e-health data from the hospital's Clinical Information System [1].

Once all relevant data has been collected—in real time- the data is pushed to the online analysis portion of McGregor's proposed Artemis framework. Patient data is streamed in real time to the Artemis software-as-a-service web interface for analysis. Collected data is then analyzed with IBM's Infosphere Streams software to apply appropriate data analysis algorithms [1]. Leveraging the Infosphere

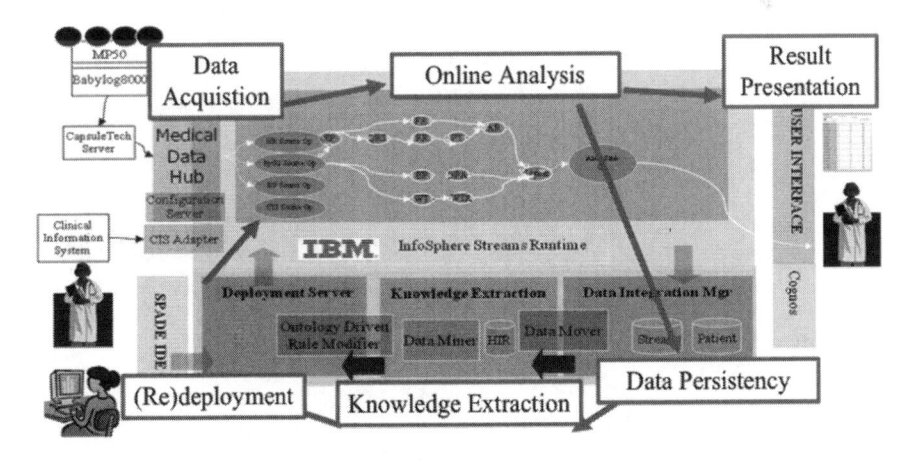

Fig. 1 The artemis framework [1]

Streams software, "enables the real-time deployment of clinical rules representing correlations between behaviours of interest in physiological streams for each condition that is being monitored" [1].

Knowledge extraction of stored data is then done via Service based Multidimensional Temporal Data Mining [3]. This can be performed on both real time and historical patient data. Extracted knowledge is then ready to be presented to researchers for further analysis. Finally, data is stored persistently for the long term to assist future knowledge extraction processes [1].

Artemis Cloud

In order to further the effectiveness of Artemis, McGregor proposed a cloud-based software-as-a-service model that would allow hospitals to interact with the Artemis framework by consuming various web services [1]. Furthermore, hospitals would have persistent access to long-term data and knowledge extraction stored in the Artemis cloud, also accessible using a data-as-a-service storage service [1]. This can be seen in Fig. 2.

Using various proposed web services, hospitals would be capable of initializing an IBM Infosphere Streams environment with custom rule sets and knowledge extraction algorithms [1]. Patient data can then be streamed into the appropriate web interface, encoded in the proposed XML protocol format [4], and coupled with existing patient data in the HL7 V3 format. Ultimately this setup would allow

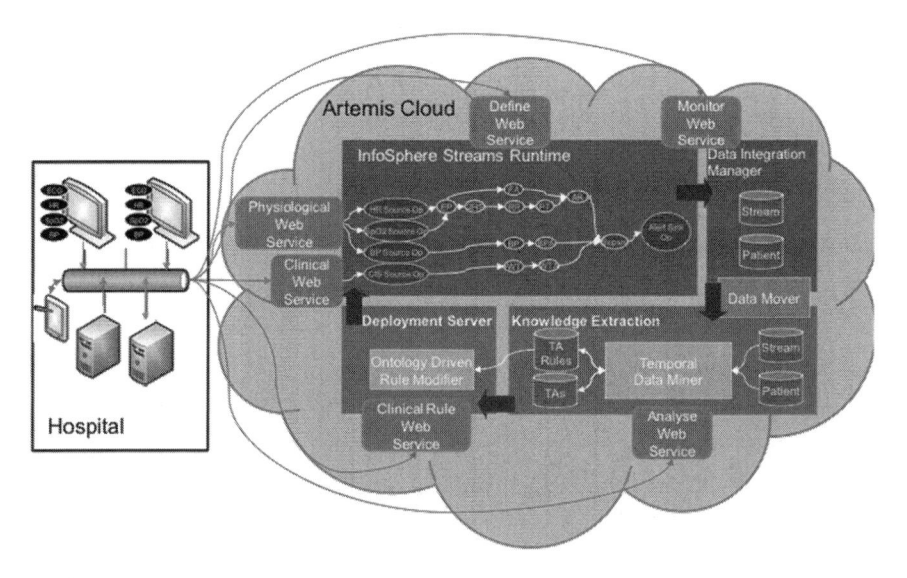

Fig. 2 The artemis cloud framework [1]

clinicians to deploy their own knowledge extraction rules to effectively monitor patient changes in real time and perform extensive clinical research.

Although this framework seems to fit the needs of both researchers and clinicians, McGregor states the "plausibility of using cloud environments for robust healthcare monitoring systems is unknown" [1].

Providing PaaS for Artemis Cloud Deployments

As we have seen in the above Artemis cloud framework, each participating hospital must be able to provision a private Infosphere Streams instance capable of conducting knowledge extraction through customized rule sets. The real challenge of successfully deploying the Artemis cloud lies in the provisioning of underlying resources needed by each hospital. If each component of the cloud architecture in Fig. 2 were run entirely on individual servers, eventually computational resources would be exceeded. In addition, the provisioning and addition of extra hardware resources would be difficult and could result in unneeded downtime. In this section we will define the requirements for an Artemis cloud deployment and show how the use of an underlying platform-as-a-service model can be effectively leveraged by McGregor's proposed web service framework.

Requirements

The underlying PaaS deployment leveraged by the Artemis cloud framework must meet various requirements to be effective. These range from high availability to stringent security policies. The most fundamental set of requirements is as follows:

1. Hospitals must be able to securely interact with the Artemis web endpoint without worrying about data leakage or plaintext data transfers. This can be done using various virtual private network (VPN) technologies. VPN security settings should be chosen based on regional data privacy requirements.
2. Each hospital must have access to its own instance of IBM's Infosphere Streams software via their Artemis web service interface.
3. Each hospital's patient data must be isolated as per regional regulatory standards and must be stored long-term for later knowledge extraction. Long-term data storage should have high redundancy and zero loss, even during hardware failure.
4. The Artemis cloud must suffer no downtime. 99.999 % availability or better should be provided at all times, regardless of hardware failure.
5. Underlying physical resources can be added or removed at will with zero downtime. This ensures the expandability of computation and data storage resources in the future.

6. The exploitation of a single hospital's web service should under no circumstances provide access to, or affect, another hospital's patient data.
7. The Artemis web service should be capable of interfacing with underlying platform-as-a-service components using a standardized API. This facilitates the deployment of Infosphere Streams and customized rule sets via the Artemis web services interface.

High Level PaaS Architecture

In order to meet the requirements outlined above, we first analyzed the deployment requirements for each hospital leveraging the Artemis cloud. After breaking down underlying hardware and software components, we concluded that each hospital's web service deployment constituted only four underlying components:

1. A VPN endpoint capable of creating a layer-2 (or higher) interface between a hospital's medical data hub (MDH—see Fig. 1) and the Artemis data captor software.
2. A data hander interface capable of collecting real time data from a hospital's MDH, isolating it, and passing it to the hospital's IBM Infosphere Streams software.
3. An isolated instance of IBM's Infosphere Streams software—known as the *Online Analysis* component—which can be used to extract relevant information from patient data.
4. A shared centralized DB2 database providing isolated and long-term data storage for each hospital.

This logical architecture can be seen in Fig. 3.

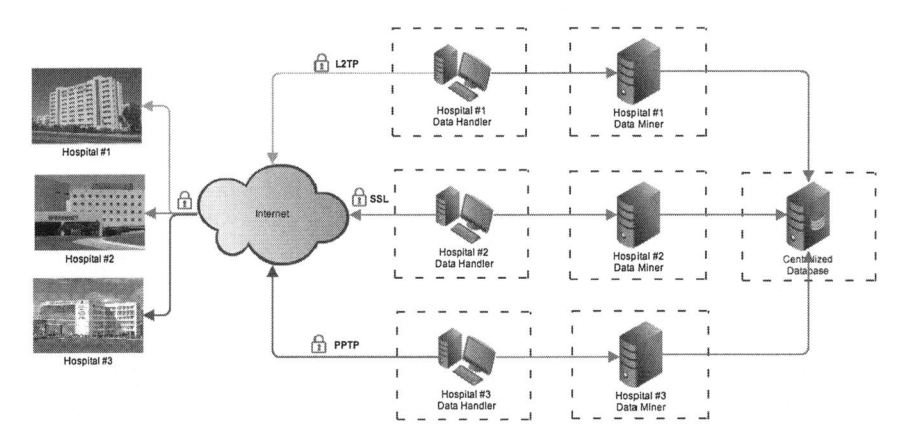

Fig. 3 Logical PaaS deployment model

As we can see, each hospital can be deployed via an underlying set of PaaS resources capable of fulfilling the software requirements needed by the Artemis web interface. Once underlying resources have been deployed, the Artemis cloud can interact with each hospital's resources and provide a high-level and standardized web service interface for each deployment.

Underlying PaaS Architecture

In order to provide the ability to deploy the logical PaaS model outlined in the previous section, we were tasked with selecting the underlying software and hardware capable of meeting the requirements outlined in Sect. 5.1 of this paper. Of upmost importance was ensuring the selected underlying software was capable of utilizing a plethora of hardware in multi-vendor environments while providing a standardized API interface capable of interacting securely with virtualized resources.

After much research, we decided to leverage VMware's vSphere hypervisor to meet our underlying software requirements. vSphere, a low level hypervisor software, allows us to pool any number of hardware resources into a logical cluster capable of providing high availability and disaster recovery [8]. This virtualized datacenter is easily managed and monitored via VMware's vCenter product and is able to provision Artemis cloud resources while providing a robust and mature API interface for interacting with said components [8].

The vSphere product, when coupled with vCenter can provide [8]:

1. High availability in the datacenter, capable of avoiding downtime even during hardware failure.
2. The ability to add or remove hardware resources at any time with zero downtime.
3. Deployment of virtualized Artemis assets extremely fast based on pre-defined service templates.
4. Access to Artemis resources using robust and well test API interfaces.
5. Automatic provisioning of hardware resources to ensure a single hospital does not utilize all computation resources.
6. Host isolation capable of reducing the attack surface of the Artemis cloud and ensuring hospitals are 100 % isolated from all other internal components.
7. Pooling of underlying hardware and storage resources to prevent unexpected data loss, even during hardware failure.
8. Snapshotting of instances in time to facilitate system roll-back, recovery, or knowledge extraction.
9. A simple and secure administrative web interface capable of safely manipulating and monitoring underlying hardware resources in real time.

As we can see, the vSphere and vCenter product exceed our requirements for an underlying PaaS platform that can be leveraged by the Artemis cloud framework.

Conclusions and Future Works

Thanks to recent advances in cloud technologies, like VMware's vSphere and vCenter products, it is possible to provide a robust and security underlying PaaS framework that can be leveraged by the Artemis cloud proposed by McGregor. Such a PaaS architecture can ensure high availability and seemingly infinite resource expansion capable of facilitating real time analysis of remote patient data. Furthermore such a robust architecture, when coupled with McGregor's Artemis cloud design, can be used to provide effective and secure web services that can be leveraged by hospitals worldwide.

Currently the Artemis cloud architecture, and the proposed underlying PaaS model, is being deployed at the University of Ontario Institute of Technology. Future work will focus on tightly integrating the proposed PaaS framework with McGregor's Artemis web service interface to ensure the effective provisioning and management of virtualized hospital resources.

By coupling Artemis with our proposed PaaS model, we can showcase the plausibility and effectiveness of leveraging cloud-based resources to provide real-time knowledge extraction from network-based patient monitoring devices.

References

1. McGregor C (2011) A cloud computing framework for real-time rural and remote service of critical care. Presented at computer-based medical systems (CBMS), 24th international symposium on 2011
2. McGregor C (2005) e-Baby web services to support local and remote neonatal intensive care. In: HIC 2005 and HINZ 2005: proceedings, health informatics society of Australia, p 344
3. McGregor C (2011) System, method and computer program for multi-dimensional temporal data mining. US Patient Office
4. McGregor C, Heath J, Wei M (2005) A web services based framework for the transmission of physiological data for local and remote neonatal intensive care. Data Management, Published by the IEEE Computer Society
5. Savolainen E Cloud service models
6. Finkelstein SM, Speedie SM, Potthoff S (2006) Home telehealth improves clinical outcomes at lower cost for home healthcare. Telemedicine J e-Health 12(2):128–136
7. CareCloud (2013) Carecloud, web-based medical practice management software, vol 2013, p 11
8. Lowe S (2011) Mastering VMware VSphere 5
9. Armbrust M, Fox A, Griffith R, Joseph AD, Katz R, Konwinski A, Lee G, Patterson D, Rabkin A, Stoica I (2010) A view of cloud computing. Commun ACM 53(4):50–58
10. Blount M, Ebling M, Eklund JM, James A, McGregor C, Percival N, Smith KP, Sow D (2010) Real-time analysis for intensive care. IEEE Eng Med Biol Mag 29(2):110–118
11. Mell P, Grance T (2011) The NIST definition of cloud computing (draft). NIST Special Publication 800, p 145

MABLIPS: Development of Agent-Based Vehicle Location Tracking System

Yun-Yao Chen, Yueh-Yun Wang and Chun-Wei Lin

Abstract Location positioning and information delivering techniques have recently become important research issues with the increasing of Internet Map service and GPS technology. Similarly, location-aware technologies for bike riders have also become a new issue for bike positioning. In this paper, a mobile agent technology system for locating bicycle position is thus developed. The proposed system can also automatically push and pull the message for location-aware. Combining multi-agent technology with push technique in the proposed system, it can thus enhance the intelligence of bike location tracking and route planning.

Keywords Mobile agent · Location positioning system · Pushing system

Introduction

Agent technology has shown great potential for solving problems in large scale distributed systems. Wireless communication and network services have substantially changed the landscape of communications. The associated facilities of

Y.-Y. Chen
Institute of Manufacturing Information and Systems, National Cheng Kung University, Tainan City, Taiwan, Republic of China
e-mail: yychenzx@gmail.com

Y.-Y. Wang
Department of Leisure and Sports Management, Far East University, Tainan City, Taiwan, Republic of China
e-mail: pamwang@cc.feu.edu.tw

C.-W. Lin (✉)
Innovative Information Industry Research Center (IIIRC), Shenzhen Key Laboratory of Internet Information Collaboration, School of Computer Science and Technology, Harbin Institute of Technology Shenzhen Graduate School, Harbin, China
e-mail: jerrylin@ieee.org

Y.-M. Huang et al. (eds.), *Advanced Technologies, Embedded and Multimedia for Human-centric Computing*, Lecture Notes in Electrical Engineering 260, DOI: 10.1007/978-94-007-7262-5_132, © Springer Science+Business Media Dordrecht 2014

the leisure bike industry have been developed rapidly in recent years due in part to the advocacy of the government in Taiwan. Therefore, information technologies used in the leisure bike industry have substantially increased. With the huge popularity of mobile devices and remarkable advances in mobile communications, mobile devices have gained much attention from bicycle riders. It allows bicycle riders and users to pinpoint their location and find relevant information on demand. Push technology comprises computers that regularly and actively transmit information via the Internet through certain standards and protocols according to the users' demands [1]. Combining push technology and agent technology with mobile devices, a more convenient interface for its users and improvements to the shortcomings of traditional push technologies can be produced. Many related studies have focused on the efficiency of delivering information to users [2–7]. Many agent standards [8, 9] have been also proclaimed by several organizations [10, 11]. The intelligent agents for message delivery and push/pull research issues are, however, still in urgent need of solutions.

In this paper, a mobile agent-based bicycle positioning system with a mobility agents to free travel amongst the hosts in the network is proposed. The states and codes can be transported to other execution environments in the network, thus reducing the time in a batch process with the resume mechanism [12, 13]. The proposed active push model can also reduce the resource requirements for delivering messages [14].

System Architecture

In the proposed architecture, bike riders are assumed to be equipped with mobile devices. Mobile devices with GPS modules continuously and automatically update the location of the rider to the server through a 3G network. Users who subscribe to the pushing services can receive location information from the bike rider's mobile device. The operation steps are shown in Fig. 1.

For the clients who wish to receive the bike riders' location information must subscribe to the pushing services from the push server by using their devices such as PC, TPC, or laptop. The server provides a firewall system and anti-hack techniques for preserving and securing user information. The Mobile devices carried by bike riders can keep receiving GPS signals continuously. The GPS signals may disappear in some locations but the system will automatically restart the signal searching functions to keep it working continuously. Applications on mobile devices can handle the data and display the latitude, longitude and speed information of the bike by using Universal Description Discovery and Integration UDDI to search for web services on the web server, which are stored with the Web Services Description Language WSDL message in the directory. Received messages can be parsed as readable information for users and displayed on the screen of the mobile device.

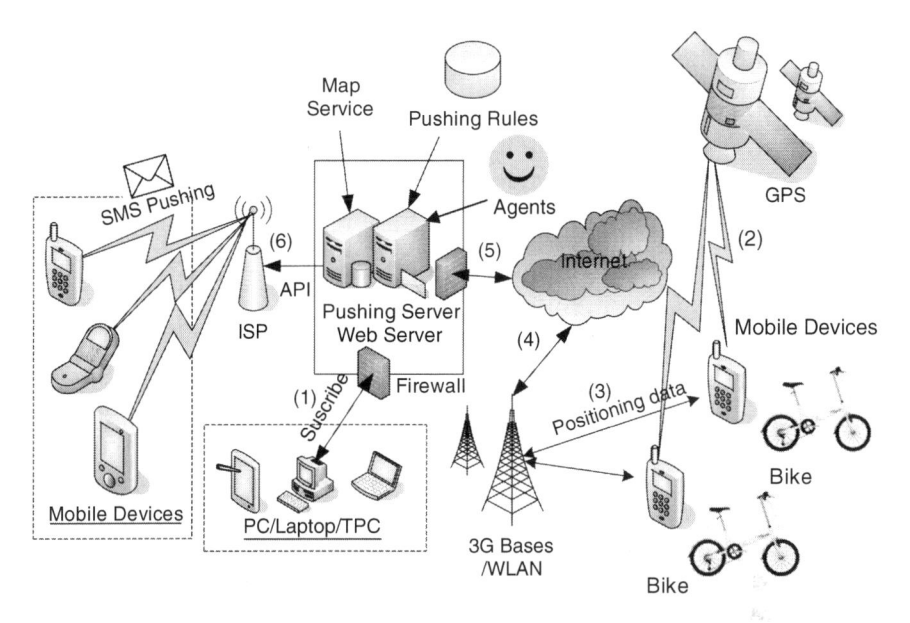

Fig. 1 The proposed architecture

All the information can be transferred by 3G network or Wireless Local Area Network LAN (WLAN). The devices carried by bike riders must able to connect with a 3G network and wireless LAN. When the network disconnects for some unpredictable reason, the system will reconnect to the network and finish any unfinished jobs dispatched by the user.

To access web services on a web server, all applications from a bike rider's mobile device must be identified and obtained an authorization. Some push rules such as certain time interval or certain date/time will trigger the SMS pushing service and mail pushing service. The listening service on the pushing server keeps listening to the contents and is triggered when the rules are satisfied the defined condition.

MABLIPS Framework

The framework of the proposed mobile agent-based bicycle location and information pushing system (MABLIPS) has multiple levels shown in Fig. 2.

The Information Aggregation Layer (IAL) level is composed of various modules to aggregate the proper information needed in the system. The lowest level also contains agencies for acquiring information from different modules. The Middleware Layer (ML) includes BTMC Stationary Agency (BSA), which is responsible for coordinating all of the agencies in IAL. The upper layer is

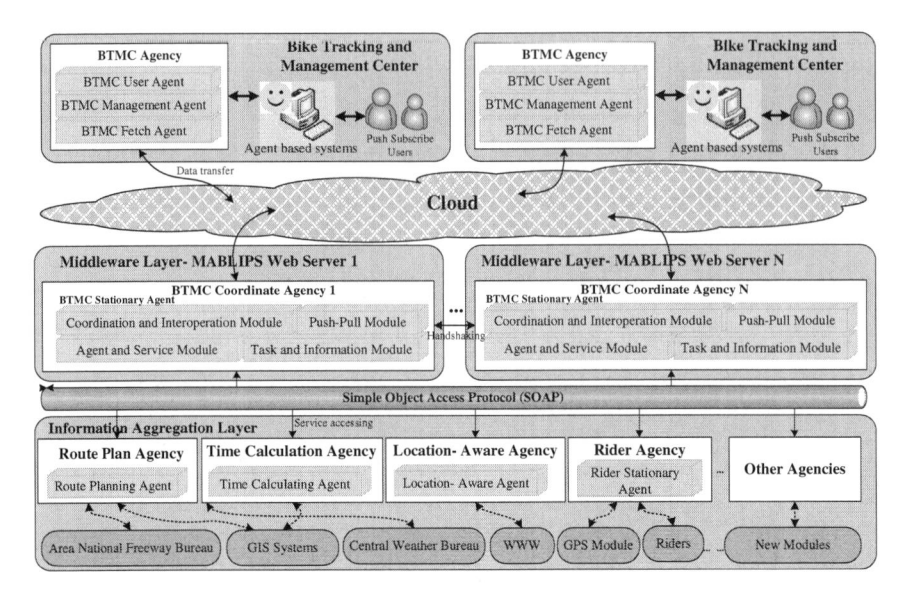

Fig. 2 MABLIPS architecture

described as the Bike Tracking and Management Center (BTMC), which includes the BTMC Agency responsible for delegating tasks to the lower level agents and analyzing the information from these agents. It can also dispatch mobile agents to lower level agencies to fulfill unforeseen tasks, and is a gateway between the Push Subscribe Users (PSU) and agent-based systems proposed in this research. All send/receive information is based on the standard of Simple Object Access Protocol (SOAP) with the Extensible Markup Language (XML) data type.

Architecture Design

Bike Tracking and Management Center

The BTMC consists of a Bike Tracking and Management Center Agency (BTMCA), the Agent-Based Systems (ABS) and the Push Subscribe Users (PSUs). The BTMCA comprises the BTMC User Agent (BUA), the BTMC Management Agent, and the BTMC Fetch Agent (BFA) for performing the tasks as follows:

- Create mobile agents and dispatch them to different agencies.
- Specialize the information from lower level agents and generate reports or control proposals to the BTMC users.
- Generate tasks dynamically and assign these tasks to lower level agents.
- Accept human commands from push subscribe users through agent based systems.

Middleware Layer: MABLIPS Web Server

The ML is based on BTMC Agencies (MTMCA) for performing the tasks as follows:

- Group lower level agents into a cluster according to the assigned task and coordinate these agents to dynamically accomplish the task.
- Serve as agent server and maintain the available services of agents.
- Decompose the assigned tasks by the BTMC to the IAL. Track and monitor those tasks and integrate the information from lower level agents and report to the BTMC agents.
- Interoperate with other agents to solve inter-network problems.

Information Aggregation Layer

The IAL mainly consists of the Route Plan Agency (RPA), the Time Calculation Agency (TCA), the Location-Aware Agency (LAA), and the Rider Agency (RA). The other agent models can be also integrated into the IAL for the future needs. The tasks can be performed as follows:

- Process real-time data from GIS or other data aggregation modules.
- Realize vehicle re-identification, estimate travel time, detect incidents, and forecast weather.
- Improve detection algorithms and select proper algorithms dynamically.
- Track real-time traffic conditions and alert predefined events to BTMC opera.

Agent Communication

In this research, the partial dynamic hierarchical control architecture [13] is adopted to dynamically organize different agents based on task decomposition. The numbers of agents are grouped into virtual clusters. The agent communication and control model are shown in Fig. 3, indicating the agent communication flow and the grouping clusters of agents.

When the RA receives a request from users, it automatically parses the request into certain information data types. The agents can be created to dispatch the information to BCAs. The BA is capable of receive the delay of the heavy network transmitting by the users. While the BTMC Coordinate Agency BCAs accept the data sent by BA, it automatically decomposes tasks assigned by the BA, and then starts to respectively aggregate the information provided by RPA, TCA, RA and RSA. The same data types of information can be recognized and adopted by BCAs.

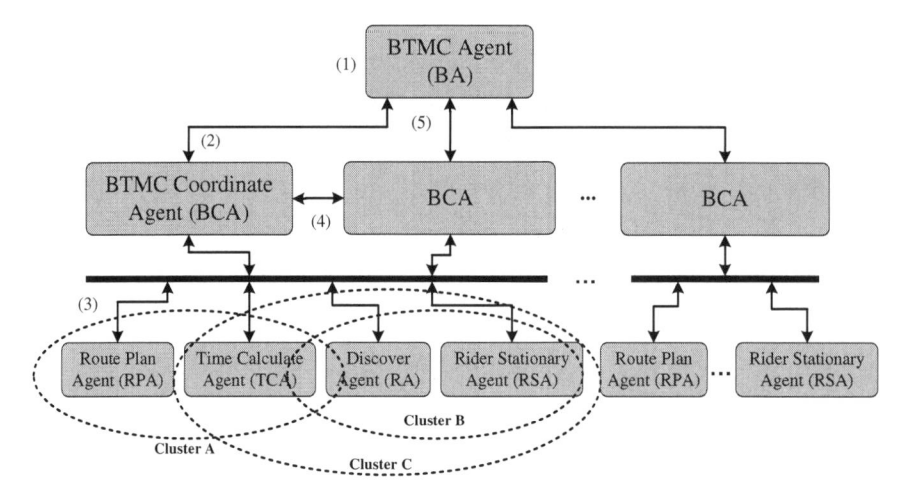

Fig. 3 Agent communication model

Meanwhile, a number of agents can be dynamically grouped into virtual groups and interact with each other to perform the same tasks. These virtual groups can interact through the same interface and same data types. A BCA can coordinate these agents in groups, and interact with other BCAs. The BCAs are able to manage different connections between each other and can handle and avoid the conflicts of reputational data transmitting. After that, the final suggested results return to BA, which displays the best processed results and the best descriptions for users. The proposed model can focus on interactions between different agents. Push subscribe users can receive the best results to locate themselves, obtaining the best routes for riding.

System Development

The system is developed by Apache 2.3 on Microsoft Windows XP Professional operating system with MySQL 5.0.51a database system. The mobile device application was implemented by Microsoft ASP.NET/C# and using .NET Framework on Windows Mobile 6 (HTC Touch Pro 2) mobile device operating system. Based on the defined pushing rule, information can be transmitted to end users in real-time. A few pushing rules are triggered to push SMS message or send the email with the biker's positioning information, which will be displayed on both a biker's mobile device and the installed application by the end user. The pre-defined rules may include the following conditions.

1. When bike riders ride through certain regions, i.e. the bike rider rides into a certain city (Taipei or ILan).

2. When mobile devices equipped by bike riders have no response for a certain time interval, i.e. the mobile device is unavailable to track for 30 min.
3. When the bike riders manually send out alarms, i.e. the bike rider uses their mobile device to send out emergency information.
4. When the bike rider's route distance exceeds a certain value, i.e. riding for 300 m.

Conclusion

In this paper, the concept of agent technology and an intelligent agent approach for the interaction with standard protocol web services are thus proposed. The framework of MABLIPS is also stated in details. The proposed architecture can be concluded with the following advantages: (1) Mobile agents can enhance the ability of push systems to deal with uncertainty in dynamic environments. (2) MABLIPS integrates multiple modules to enable the comprehensive support of the push system for bike positioning. (3) The open architecture with standard information transmission protocols allows new detection systems to be added by incorporating them into sub-agent systems.

Acknowledgments This research was partially supported both by Shenzhen peacock project, China, under contract No. KQC201109020055A and Shenzhen Strategic Emerging Industries Program under Grants No. ZDSY20120613125016389.

References

1. Li S, Li G, Chen Y (2008) A study on information push in personalized education system. In: International symposium on knowledge acquisition and modeling, pp 534–537
2. Carzaniga A, Wolf AL (2001) Content-based networking: a new communication infrastructure. In: NSF workshop on an infrastructure for mobile and wireless systems, pp 1–5
3. Cao J, Feng X, Lu J, Chan H, Das SK (2002) Reliable message delivery for mobile agents: push or pull. In: International conference on parallel and distributed systems, pp 314–320
4. Chen C, Ding J, Hua G, Chen Y (2009) Design and implementation of SMS employment agent based on ontology. In: World congress on software engineering, pp 489–492
5. Franklin M, Zdonik S (1998) Data in your face: push technology in perspective. In: ACM SIGMOD international conference on management of data, pp 516–519
6. Jacobsen HA (2001) Middleware services for selective and location-based information dissemination in mobile wireless networks. In: The advanced topic workshop on middleware for mobile computing, pp 154–169
7. Kozina B, Glavinic V, Kraljevic L (2006) Agent-based messaging system for m-learning. In: IEEE mediterranean electrotechnical conference, pp 1213–1216
8. Brewington B, Gray R, Moizumi K, Kotz D, Cybenko G, Rus D (1993) Mobile agents for distributed information retrieval. In: Intelligent information agents, chapter 15, pp 355–395
9. Zhang Y, Liu J (2003) Mobile agent technology. Tsinghua University Press, Beijing, pp 9–11
10. Mobile agent system interoperability facilities (MASIF) http://www.omg.org

11. The foundation for Intelligent Physical Agents http://www.fipa.org
12. Lange D, Oshima M (1998) Programming and developing java mobile agents with aglets. Addison Wesley Longman Inc., Botson
13. Lange DB (1998) Mobile objects and mobile agents: the future of distributed computing. In: European conference on object-oriented programming, pp 1–12
14. Sun T, Wang Y (2005) Implementation of a chat system with push and mobile agent technologies. In: International conference on intelligent agents, web technologies and internet commerce, international conference on computational intelligence for modeling, control and automation, vol. 2, pp 336–360
15. Du J, Tian Y, Zuo M, Zhou Y (2007) The realization of push and pull model in the agent-based electronic commerce platform. In: International conference on control, automation and systems, pp 17–20

Data Sensing and Communication Technology for the IoT-IMS Platform

Tin-Yu Wu and Wen-Kai Liou

Abstract When the Future Internet is defined as a dynamic global network infrastructure, the Internet of Things (IoT), an integrated part of it, is viewed as one of the most important technologies. Distinct from former definition and applications of IoT and the IoT based on *Radio Frequency IDentification* (RFID), the current IoT means the interconnection of devices and appliances via the Internet. However, because the development of the IoT technologies varies, there is neither an unified standard nor a standardized interoperability to provide an unified network software architecture for the IoT. In this paper, we propose a method for integrating the data detected by sensors in the IoT-IMS communication platform with specifications, including Zigbee, *Wireless Sensor Network* (WSN), RFID and IEEE 1451, and converting the data to *Sensor Be One Code* (SBO Code), a coding format that supports all kinds of sensors. Through this system, the data gathered by different sensors can be unified to meet the IoT requirement.

Keywords IoT · WSN · Intelligent sensor · IEEE 1451

Introduction

With the popularization of broadband network and the enhancement of transmission rate, users' requests for novel and diverse multimedia services and applications are never satisfied and the notion of ubiquitous services therefore

T.-Y. Wu (✉) · W.-K. Liou
Department of Computer Science and Information Engineering, National Ilan University, Ilan City, Taiwan, Republic of China
e-mail: tyw@niu.edu.tw

W.-K. Liou
e-mail: kevin1990115@gmail.com

Y.-M. Huang et al. (eds.), *Advanced Technologies, Embedded and Multimedia for Human-centric Computing*, Lecture Notes in Electrical Engineering 260, DOI: 10.1007/978-94-007-7262-5_133, © Springer Science+Business Media Dordrecht 2014

becomes very important. To meet user demand for network services, the *Internet Service Providers* (ISPs) keep making efforts in presenting a novel integrated infrastructure in response to future broadband services. Moreover, this demand has initiated the idea of the *Next Generation Network* (NGN) and the *IP Multimedia Subsystem* (IMS) under the NGN framework [1]. In addition to the solutions for the next generation core networks, the concept of the *Internet of Things* (IoT) takes shape: RFID tags, sensors or QR code embedded in the objects are linked via the Internet to endow the objects with the intelligence or ability to communicate with human beings and other objects. Therefore, the IoT enables not only machine/device/employee management/control, but also remote monitoring of home appliances and vehicles.

To cope with such a trend, this paper focuses on the IoT-IMS based data sensing and communication technology. Based on the IoT-IMS communication platform, the Smart Client Platform gathers the data detected by different sensors (QRCode, RFID, Barcode and RuBee, for example), converts the data to *Unique IDentifiers* (UID) by EPC coding scheme, and writes the UID on smart tags. By this method, objects and devices can be easily interconnected via the IoT. Furthermore, *EPC Information Services* (EPCIS) of EPC Global Network are built on the Smart Client Platform for information and data transmission while the All-IP heterogeneous network server is established to integrate different transmission standards and incompatible data to EPC Sensor Profile.

Based on EPCglobal [2], this paper presents *Sensor Be One code* (SBO Bode), a novel coding scheme to integrate data formats of heterogeneous sensors and allow data transmission in either Zigbee, RFID, Bluetooth or Wi-Fi. Two main research directions of this paper include:

- Establish IMS-URI to integrate IMS with the IoT.
- Present SBO Code to integrate data formats and transmission interfaces of IEEE 1451 standards and other standards.

Related Work

Introduction to the EPCglobal Network

For wireless radio system integration, the EPCglobal Network [3–11] was first developed by the Auto-ID Center, an unique partnership between almost 100 global companies and labs of seven leading research universities: the Massachusetts Institute of Technology in the US, the University of Cambridge in the UK, the University of St. Gallen in Switzerland, the University of Adelaide in Australia, Fudan University in China, Keio University in Japan, and Information and Communication University in South Korea; the labs were later renamed the Auto-ID Labs. Auto-ID solutions presented by the Auto-ID labs include *Electronic*

Table 1 EPC data structure

Header	EPC manager	Object class	Serial number

Product Code (EPC), proposal for network elements and passive tags and readers. In 2003, the research results produced by the Auto-ID Labs promoted the establishment of EPCglobal Inc., an international nonprofit joint venture between GS1 and GS1 US. As the commercial successor of the Auto-ID Labs, EPCglobal Inc. aims at architecting open and autonomous EPCglobal standards and promoting wider adoption of EPC technology.

Electronic Product Code

The EPC number is either a 64 bits or 96 bits identifier, a string of data in four partitions: header, EPC manager, object class and serial number (Table 1). The 96 bits version of the EPC, the EPC-96, can identify approximately 268 million manufacturers, each of which can have 16 million object classes and 68 billion serial numbers for every unique item. Table 2 displays the coding scheme of EPC-96 Type I based on GID.

- Header (assigned by EPCglobal): Identifies the length, type, structure, version, and generation of the EPC. GS1 assigns the keys to each class of object.
- EPC Manager (assigned by EPCglobal): Identifies the company that manages the data, usually the manufacturer of the product the EPC is attached to.
- Object Class (managed by the company): Identifies the class or type of product.
- Serial Number (managed by the company): Identifies every unique item of the object class.

Conversion Between EPC and Bar Code

GS1 has been recently making great efforts in integrating the EPC code with QR code and thus there is no clear explanation for the conversion between the EPC and QR code currently. Therefore, only the conversion between the EPC and bar code is mentioned here. To convert bar code to the EPC correctly, we need the following information: header, filter value, partition value and EPC manager number.

Table 2 EPC-96 I type coding scheme

Header	EPC manager	Object class	Serial number
8 bits	28 bits	24 bits	36 bits

The *Uniform Resource Identifier* (URI) coding scheme is usually adopted by RFID to enable data interchange in large-scale systems. To inform the operators who read and write about the tag format, the URI is a string of characters: urn : epc : tag : Enc Name: Encoding Specific Fields, where Enc Name refers to the coding format of the tag.

Internet of Things

The *Internet of Things* (IoT) means that through RFID, infrared sensors, *Global Positioning System* (GPS), laser scanners, etc., all independent devices can be connected via wireless links using specific protocols. Therefore, the devices are endowed with the ability to communicate with each other and intelligent identification, positioning, tracking, monitoring, controlling and management thus can be realized. Features of the IoT as below.

- In the IoT, the information about the objects is retrieved from RFID tags, sensors, QR code, or other means.
- The information about the objected is transmitted via the Internet because the IoT is connected to the Internet.
- By using various intelligent computing techniques, the IoT can process large amounts of data and accomplish intelligent control of objects.

Architecture of the Internet of Things

IoT is mainly divided into three layers: perception layer, network layer and application layer (Fig. 1). The main function of the perception layer is to identify objects and collect information. The key techniques in this layer include RFID and WSN (Wireless Sensor Network) [12–15], and the sensing devices include sensors, readers, IP Cams and *Micro Electro Mechanical Systems* (MEMS). The main function of the network layer is information transmission and processing, including integration of heterogeneous networks, center for intelligent information processing, information center, and management center. The primary techniques in the network layer include 2G/3G, Wi-Fi, WiMax and Zigbee. The main function of the application layer is to control input/output and provide services in cloud service platforms. The application support layer provides interfaces between the application layer and the network layer, integrates different networks and supports applications.

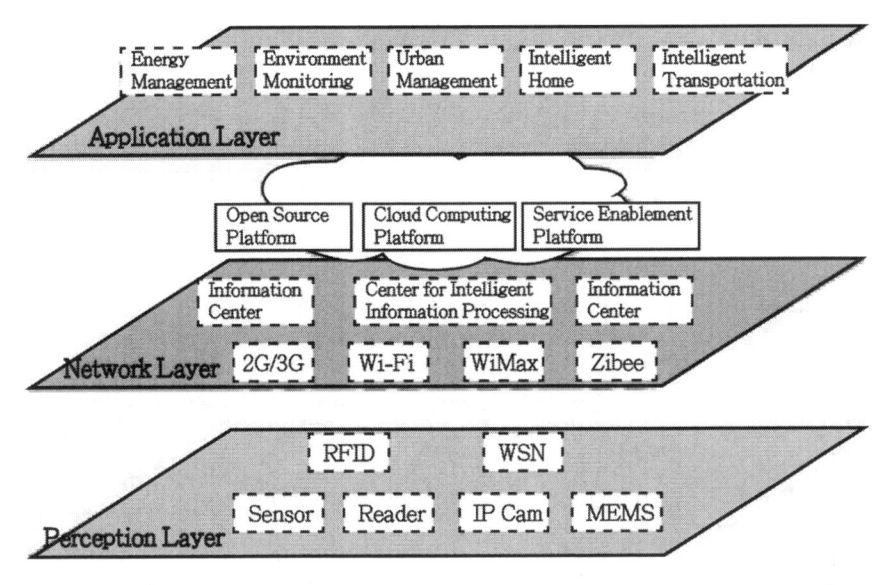

Fig. 1 Architecture of the internet of things

IP Multimedia Subsystem

As part of 3GPP Release 5, the *IP Multimedia Subsystem* (IMS) [1] is defined as a collection of core network functions in *Next Generation Networks* (NGN) that support multimedia services delivery on mobile devices. Based on the 3GPP Release 4 packet-switched domain, Release 5 adds an IP-based multimedia subsystem that uses the *Session Initiation Protocol* (SIP) to control calls or manage data via packet-switching functions. Therefore, the IMS provides support for mobile applications. Though independent of circuit-switched systems, the IMS can connect to circuit-switched networks and satisfy requirements for security, billing, roaming and QoS. In addition to the standardization of IMS by 3GPP, the IMS architecture is also adopted by 3GPP2 for IP multimedia services. Moreover, the IMS has been adopted, adapted and optimized by ITU-T.

Responsible for processing SIP signaling and responses, the IMS core comprises three kinds of *Call Session Control Functions* (CSCF): *Proxy-CSCF* (P-CSCF), *Serving-CSCF* (S-CSCF) and *Interrogating-CSCF* (I-CSCF). As the most important component in the IMS architecture, the CSCF is responsible for multimedia session processing: multimedia session control, address translation, service negotiation, and so on.

IEEE 1451 Smart Transducer Interface Standard

To unify sensor interface standards, *IEEE Technical Committee* 9 (TC-9) in 1993 discussed the possibility of developing a smart sensor communication interface standard and founded the IEEE 1451 smart transducer interface standards. A series of workshops have been established and defined a set of standards for intelligent sensors.

A Transducer Electronic Data Sheet, TEDS, refers to a set of electronic data in a standardized format stored in an EEPROM attached to the STIM. The TEDS includes the manufacturer ID, model number, serial number, measurement range, electrical output range, sensitivity, power requirements, and calibration data. The TEDS allows the self-identification, self-description, automatic setting of parameters (range, sensitivity, proportion, for example), self-diagnosis and self-calibration of the sensors (Table 3).

Proposed Architecture

The proposed architecture of this paper is based on the IoT-IMS communication platform, in which the data from different sensors is gathered by a *Network Capable Application Processor* (NCAP) and converted to SBO Code based on the EPC coding format. Objects therefore can be interconnected via the IoT. Here, we focus on how to transmit these messages and information to the cloud network.

Table 3 The IEEE 1451 family of standards

IEEE 1451.0 (2007) (in balloting)	Defines the common functions, communication protocols, and transducer electronic data sheet (TEDS) formats for IEEE 1451.X standards
IEEE 1451.1 (1999) (published standard)	Defines a common object model for smart transducers and interface specifications
IEEE 1451.2 (1997) (published standard)	Defines a TEDS and a point-to-point digital interface and communication protocols between the smart transducer interface module (STIM) and the network capable application processor (NCAP)
IEEE 1451.3 (2003) (published standard)	Defines a transducer interface standard for distributed multi drop systems: transducer bus interface module (TBIM) and transducer bus controller (TBC)
IEEE 1451.4 (2004) (published standard)	Defines a mixed-mode (analog/digital) transducer interface (MMI)
IEEE 1451.5 (2007) (in balloting)	Defines a transducer-to-NCAP interface and TEDS for wireless transducers
IEEE 1451.6 (in progress)	Defines a transducer-to-NCAP interface and TEDS using the CANopen protocol
IEEE 1451.7 (2010) (in progress)	Defines an interface and communication protocol between transducers and RFID systems

Fig. 2 IMS-IoT network
architecture

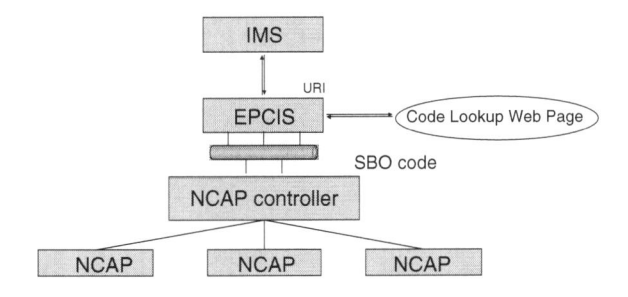

To integrate different standards of wireless communication and compatible data, we refer to *EPC Information Services* (EPCIS) in the EPCglobal Network. For heterogeneous sensor integration, our proposed architecture enables All-IP heterogeneous networks to acquire the names and addresses of smart tags via the IoT for sharing data in different standards through the platform. Finally, by using previous techniques together with the IoT-IMS communication platform, we can create associations between objects: the Family Tree-ID Relationship will be combined with IMS-URI to form associations between objects in the execution process of the research project. The project execution will achieve the integration of the identifiers under the IoT-IMS framework.

With the aim of heterogeneous sensor integration, this paper first builds a Smart Client Platform based on EPC SGTIN coding scheme. In the platform, we propose *Sensor Be One code* (SBO code) to integrate data formats of heterogeneous sensors.

SBO code is divided into several partitions to mark the transmission method, transmission content, sensor manufacturer, sensor serial number and so on. By using SBO coding scheme, we can integrate sensor data of various WSN as the basis of WSN for the IoT.

In this architecture, the NCAP gathers the data from different sensors and uploads it to the cloud database for remote users to access data. Figure 2 displays our proposed IMS-IoT network architecture. Based on IEEE 1451 architecture, the NCAP at the bottom are responsible for receiving data returned by sensors within a fixed range and uploading the data to the NCAP Controller, the large server in the range that is responsible for data management and relaying. The NCAP Controller converts the data uploaded by the NCAP to SBO code and stores it in EPCIS for users to access. At last, we integrate the IMS with EPCIS so that the IMS server can connect to EPCIS through URI.

Based on EPC SGTIN-96 coding scheme, the 256-bit code is designed as shown in Fig. 4. The code includes a Header for 14 bits, Manufacturer ID for 29 bits and Sensor Class for 11 bits, 40 bits in subtotal. The rest 202 bits are Sensor Serial Number for 24 bits and Transmission Content for 178 bits. The code thus is converted into the format: urn : sbo : id : transmission method : manufacturer code : sensor class : sensor serial number : transmission content. As shown in Fig. 3, we can convert ID to IMS-URI to complete the network deployment for the IoT.

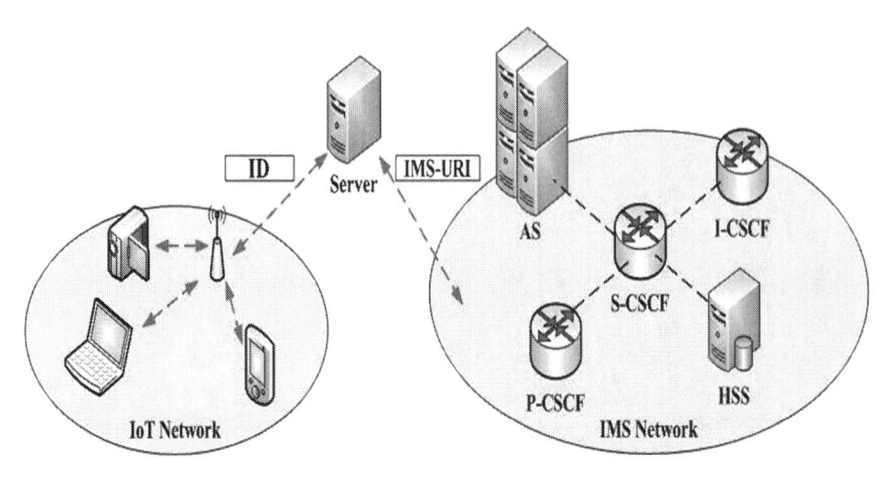

Fig. 3 Converting ID to IMS-URI

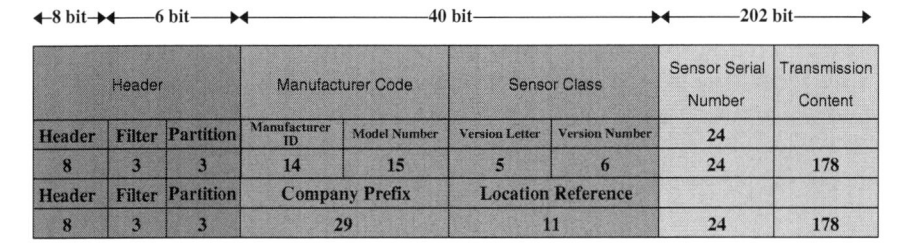

◄—8 bit—►◄——6 bit———►◄—————————————40 bit—————————————►◄———————202 bit———————►								
Header			Manufacturer Code		Sensor Class		Sensor Serial Number	Transmission Content
Header	Filter	Partition	Manufacturer ID	Model Number	Version Letter	Version Number	24	
8	3	3	14	15	5	6	24	178
Header	Filter	Partition	Company Prefix		Location Reference			
8	3	3	29		11		24	178

Fig. 4 SBO coding scheme

Achievements

The equipment adopted in this research project includes cc2530 Zigbee module, different types of sensors, NI PCI signal acquisition card to acquire Zigbee transmission, and NI LabVIEW 2011 software to convert data from Zigbee sensors to SBO code and relay data to remote SQL database for storage. As shown in Fig. 5, to construct and examine our proposed architecture and to test the differences between intelligent and traditional sensors, we use IEEE 1451-based intelligent sensors as the sensor nodes in the IoT. The major difference between intelligent and traditional sensors is that an intelligent sensor has a EEPROM to store the TEDS for plug and play data measurement.

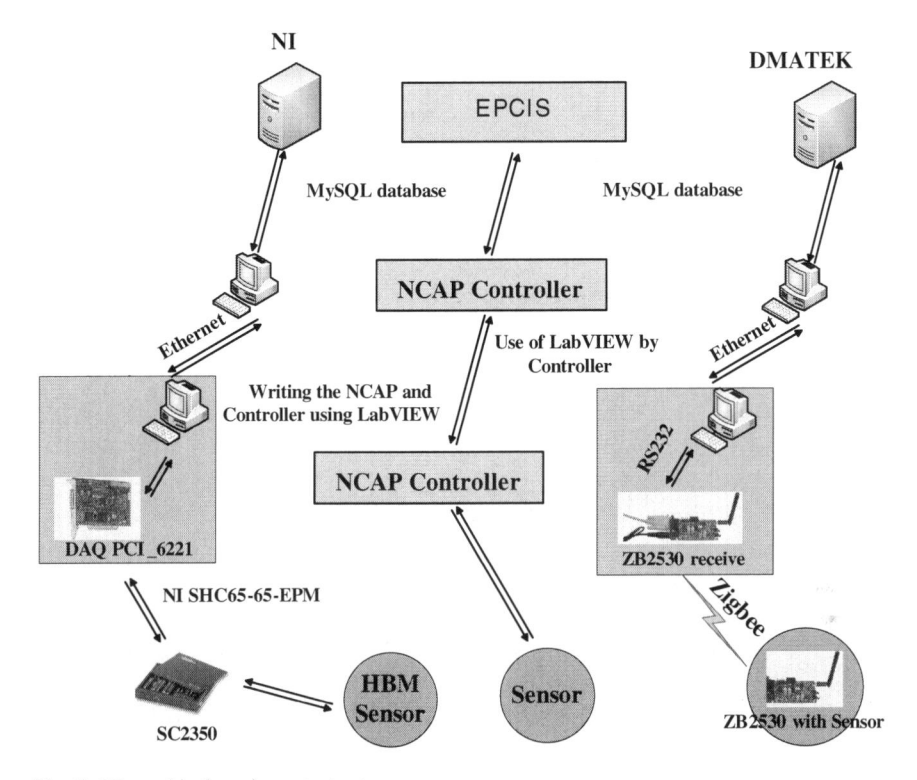

Fig. 5 Hierarchical equipment structure

Conclusion

This paper points out a possible direction for the IoT: collecting the data from Zigbee, Bluetooth, Wi-Fi, integrating the data into the same SBO Code and implementing an operable IoT system. Our future objective is to assign an independent IPv6 to each object. Due to hardware limitations, computing capability and power of sensors, the biggest challenge and also the ultimate goal for the future IoT is that every object has an IP address, which is waiting to be achieved by international standards and related equipment providers.

Acknowledgments This study was supported by the National Science Council, Taiwan, under grant No. NSC 101-2219-E-197-003.

References

1. Hwang J, Kim N, Kang S, Koh J (2008) A framework for IMS interworking networks with quality of service guarantee. In: The 7th international conference on networking 2008, ICN 2008, pp 454–459, 13–18 Apr 2008
2. Epcglogal Tag Data Standard v. 1.5 http://www.epcglobalinc.org/standards/tds/tds_1_5-standard-20100818.pdf
3. Zhang DY, Zhu YL, Chen HN (2008) An algorithm for deployment of RFID readers in EPC network. In: The 4th international conference on wireless communications, networking and mobile computing 2008, WiCOM '08, pp 1–4, 12–14 Oct 2008
4. Lee SD, Shin MK, Kim HJ (2007) EPC vs. IPv6 mapping mechanism. In: The 9th international conference on advanced communication technology, pp 1243–1245, 12–14 Feb 2007
5. Gao J, Liu F, Ning H, Wang B (2007) RFID coding, name and information service for internet of things. In: IET conference on wireless mobile and sensor networks 2007, (CCWMSN07), pp 36–39, 12–14 Dec 2007
6. Chang YC, Chen JL, Lin YS, Wang SM (2008) RFIPv6: a novel IPv6-EPC bridge mechanism. In: international conference on consumer electronics 2008, ICCE 2008. Digest of technical papers, pp 1–2, 9–13 Jan 2008
7. Thiesse F, Floerkemeier C, Harrison M, Michahelles F, Roduner C (2009) Technology, standards, and real-world deployments of the EPC Network. Internet computing, IEEE, pp 36–43, Mar–Apr 2009
8. Sanchez Lopez T, Kim D (2007) A context middleware based on sensor and RFID information. In: The 5th annual IEEE international conference on pervasive computing and communications workshops 2007, PerCom Workshops '07, pp 331–336, 19–23 Mar 2007
9. Huifang Deng, Haiyan Kang (2010) Research on high performance RFID code resolving technology. In: The 3rd international symposium on intelligent information technology and security informatics (IITSI) 2010, pp 677–681, 2–4 Apr 2010
10. Chalasani S, Boppana R (2007) Data architectures for RFID transactions. Industrial informatics, IEEE transactions, pp 246–257, Aug 2007
11. Pathak R, Joshi S (2009) Recent trends in RFID and a java based software framework for its integration in mobile phones. In: The first Asian himalayas international conference internet 2009, AH-ICI 2009, pp 1–5, 3–5 Nov 2009
12. Choi YS, Jeon YJ, Park SH (2010) A study on sensor nodes attestation protocol in a Wireless Sensor Network. In: The 12th international conference on advanced communication technology (ICACT) 2010, vol 1, pp 574–579
13. Yueqing R, Lixin X (2010) A study on topological characteristics of wireless sensor network based on complex network. In: The international conference on computer application and system modeling (ICCASM) 2010, vol 15, pp 486–489
14. Ilyas MU, Radha H (2006) End-to-end channel capacity of a wireless sensor network under reachback. In: The 40th annual conference on information sciences and systems 2006, pp 1713–1718
15. Choi SH, Kim BK, Park J, Kang CH, Eom DS (2004) An implementation of wireless sensor network. IEEE Trans Consum Electron 50(1):236–244

Adaptive Multi-Hopping MAC Mechanism for WSN Scheduling

Lin-Huang Chang, Shuo-Yao Chien, H. F. Chang and Tsung-Han Lee

Abstract In this paper, we extend the routing enhanced MAC (RMAC) protocol to combine wireless sensor network (WSN) technology with energy-efficiency and quality of service (QoS) guarantee transmission scheduling mechanisms. This energy-efficient QoS-aware scheduling over multi-hopping WSNs is called adaptive multi-hopping MAC (AMH-MAC). Based on the RMAC scheme, we further adjust the transmitted power and the number of multi-hopping nodes dynamically to reduce the transmission latency and energy consumption. We conduct the simulation using network simulator 2 (ns-2) to show the improvement of the performance in terms of end to end latency and energy consumption as compared with RMAC scheme.

Keywords RMAC · Adaptive transmitted power · Scheduling · Multi-hop

Introduction

With the characteristics of low cost and low power consumption, wireless sensor network (WSN) communication is suitable for real-time and emergency services, such as mountain emergency rescue, mining or farm monitoring.

L.-H. Chang · S.-Y. Chien · T.-H. Lee (✉)
Department of Computer Science, National Taichung University, Taichung,
Taiwan, Republic of China
e-mail: thlee@mail.ntcu.edu.tw

L.-H. Chang
e-mail: lchang@mail.ntcu.edu.tw

H. F. Chang
Department of Computer and Communication Engineering, Taipei Chengshih
University of Science and Technology, Taipei, Taiwan, Republic of China

Y.-M. Huang et al. (eds.), *Advanced Technologies, Embedded and Multimedia
for Human-centric Computing*, Lecture Notes in Electrical Engineering 260,
DOI: 10.1007/978-94-007-7262-5_134, © Springer Science+Business Media Dordrecht 2014

The idle listening for the well-known duty-cycling schemes in WSN MAC protocol is inefficient and wastes significant energy. To reduce this energy consumption of idle listening, some duty-cycle MAC protocols, such as S-MAC [1], T-MAC [2], and RMAC [3], have been introduced. The S-MAC and T-MAC mitigate the energy consumption of idle listening by using fairness contention to transmit packets in a periodic synchronized listen/sleep schedule. The increase of contention probability, due to the synchronized wake-up time at the sensor node's signal coverage during the same short period, and end to end delivery latency, due to the single hop characteristic in one duty cycle, would deteriorate the performance of S-MAC and T-MAC mechanisms. Therefore, the RMAC protocol, with contention-based and synchronization-based mechanisms, exploits cross-layer routing information to allow multi-hopping transmission within a single duty-cycle. Low power listening (LPL) is another well-known mechanism which employs a preamble in front of each outgoing packet from physical layer for transmission detection. These mechanisms, such as B-MAC [4] and X-MAC [5], could reduce the energy consumption, however, the latency issue due to the asynchronous mechanisms would be a problem.

In this paper, we take into account the end to end latency issue by employing the RMAC protocol with synchronization and multi-hopping scheduling mechanisms. Furthermore, with the LPL mechanism for reducing energy consumption during the RMAC idle time in mind, we dynamically control the transmitted power to reduce transmission energy consumption and dynamically adjust the number of multi-hopping in one duty cycle to properly increase the hopping numbers and consequently reduce the end to end latency. The proposed adaptive multi-hopping MAC (AMH-MAC) mechanism is expected to provide duty cycle scheduling in WSN MAC protocol with energy efficiency and low end to end latency.

Related Work

The S-MAC of well-known WSN MAC protocol, similar to IEEE 802.11 DCF mode, is designed with periodic synchronized listen/sleep schedule. The listening period, with sensor nodes' radio being set enabled, consists of SYNC period to synchronize the sensor nodes' clocks and DATA period to deliver packets with RTS/CTS mechanism for the reservation of transmission scheduling during SLEEP period. During the SLEEP period, every sensor node goes to sleep to save energy unless it is scheduled to send or receive data.

The duty cycle S-MAC protocol provides scheduling mechanism with energy efficiency as compared with traditional MAC protocols. However, the end to end delivery latency could be increased because a packet is forwarded over only a single hop for each duty cycle. The RMAC is therefore designed to forward a data packet with multiple hops in one duty cycle to reduce the end to end delivery latency.

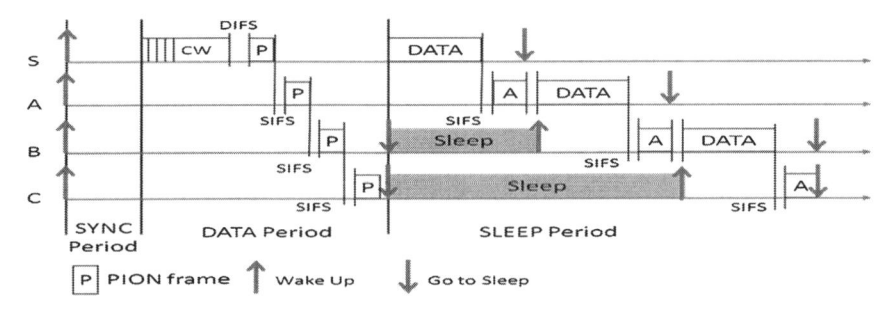

Fig. 1 The overview of RMAC operational and scheduling mechanism

The overview of RMAC operational and scheduling mechanism is shown in Fig. 1. RMAC uses a small control frame, called pioneer frames (PIONs), during the DATA period across multiple hops to allow all nodes along the path learn when to be awake. A PION control packet is employed to reserve and confirm communication, such as RTS and CTS respectively, for the corresponding data transmission schedules in SLEEP period within a single duty cycle. When a relaying node, such as node A or B which receives the data packet from the upstream nodes S or A and forwards it to the downstream nodes B or C respectively, receives PION frame, it sets its next wakeup time according to the hop count information in the PION so that it can wake up in its turn to receive or transmit data packet during SLEEP period.

The Hybrid MAC (HMAC) mechanism is proposed by Wang et al. [6] by combining the TDMA scheme to the multi-hopping scheduling of RMAC scheme to achieve energy saving and low delivery latency. However, HMAC mechanism uses fixed time slot for packet transmission which could limit the scheduling to certain number of nodes. Liu et al. [7] proposed the Q-MAC mechanism for traffic classification with the cut-off of contention window to achieve QoS service demand. But the packet transmission of Q-MAC did not support multi-hopping issue.

Some mechanisms and issues related to the QoS services of WSN MAC protocols are addressed above. Based on the RMAC protocol, in this paper, we propose an AMH-MAC mechanism to dynamically control the transmitted power to reduce transmission energy consumption and dynamically adjust the number of multi-hopping in one duty cycle to further reduce the end to end latency.

Design of Adaptive Multi-Hopping MAC Protocol

In RMAC mechanism, each node has to wake up in DATA period for listening even if it will not be scheduled to receive or forward data during this round of duty cycle. This results in the energy waste during this wake up for idle listening.

On the other hand, the number of hops, N, in a single duty cycle for RMAC multi-hopping scheme could be a factor affecting the transmission cycles for a packet travelling from a source node to the destination node. The research from Cho and Bahk [8] proposed a hop extended MAC (HE-MAC) scheme to optimal wake up duration with respect to packet latency along with power consumption. However, they did not provide any dynamic adjustment of N value for different transmission paths.

The operational mechanism and power transmission of AMH-MAC is shown in Fig. 2. In AMH-MAC mechanism, the SYNC packet, carrying N value and path ID, informs all nodes in the path during SYNC period. All nodes in the path will adjust the wakeup/sleep time according to the updated N value during the DATA period instead of waking up and idling for all DATA period. The dynamical adjustment of N value carried in SYNC packet on the one hand will relay PION frame to more nodes, and on the other hand, reserve the power consumption of nodes, which will not participate in the transmitting or receiving packets, by keeping them sleep during this duty cycle.

The path ID is used for nodes to determinate whether they will participate in this DATA period. For those with different path ID will sleep in DATA period till next SYNC period to save more energy.

Besides adjusting the number of multi-hopping nodes dynamically, the proposed AMH-MAC mechanism also dynamically adjusts the transmitting power to the receiving nodes during the DATA period. The research on the location-based routing protocol from He et al. [9] and Shang et al. [10] revealed that sensor nodes learn the position and distance of each other with RSSI signal. In AMH-MAC, the SYNC packet is broadcasted with maximum radio power to as many covered nodes as possible to reduce the number of broadcast while maintaining the synchronization functionality. However, each node uses point to point communication to its receiving neighbors with optimized transmitted power during the DATA period, as shown for the transmission of the unicast PION/data packets in Fig. 2. This dynamic adjustment of transmitted power will reduce the power consumption for nodes participating in DATA period.

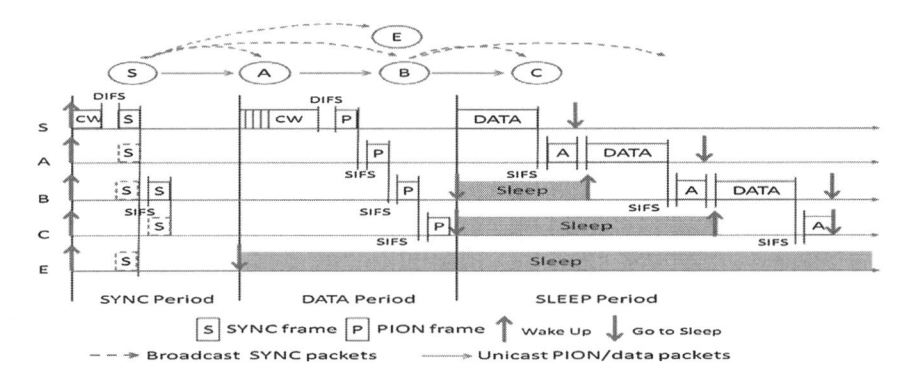

Fig. 2 The operational mechanism and power transmission of AMH-MAC

Performance Evaluation

We conduct the ns2 simulation with free space model to analyze the performance of AMH-MAC as compared with S-MAC and RMAC protocols. Due to the limit of the page length in this paper, we can only provide partial simulated results, such as the case with N equals to 4 in Table 1, conducted in the research.

In AMH-MAC, we add a long sleep time parameter, which equals to DATA period plus SLEEP period, in the SYNC packet to indicate long sleep without wakeup until the next duty cycle when a node does not participate any receiving or transmitting packets. The chain topology shown in Fig. 3 with nodes 1–24, separated from 100 m each, is used to simulate the experiment with constant bit rate (CBR) traffic loads of 100 packets at rate of 0.02 packet/sec.

The simulation results of energy consumption distribution for S-MAC, RMAC and AMH-MAC mechanisms are illustrated in Fig. 4a–c, respectively, where the total energy consumption is divided into receive (rx), transmit (tx), sleep and idle energy. As shown in Fig. 4a, b for S-MAC and RMAC schemes respectively, the idle listening, which contributes significant ratio of energy consumption, consumes much more energy as compared with our proposed AMH-MAC mechanism in Fig. 4c. The S-MAC scheme, with only one hop per single duty cycle, suffers from the idle listening power consumption significantly. On the other hand, the dynamic adjustment of transmitted power and modification of SYNC packet, such as path ID which keeps unnecessary wakeup for nodes not participating in transmission

Table 1 Simulation parameters

Parameters	S-MAC	RMAC	AMH-MAC
Bandwidth		20 Kbps	
Rx power		0.5 W	
Idle power		0.45 W	
Sleep power		0.05 W	
Sync range	100 m	100 m	200 m
Data/PION range	100 m	100 m	100 m
Transmitted power (200 m)	Null	Null	24.5 dBm
Transmitted power (100 m)		21.49 dBm	
SYNC period (ms)		55.2	
DATA period (ms)	104.0	168.0	168.0
SLEEP period (ms)	3025.8	4241.8	4241.8
Long SLEEP period (ms)	Null	Null	4409.8
Total cycle (ms)	3158.0	4465.0	4465.0
Duty cycle		5 %	
N value	Null	4	4

Fig. 3 The chain topology with multiple hops

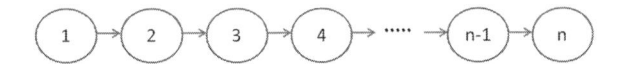

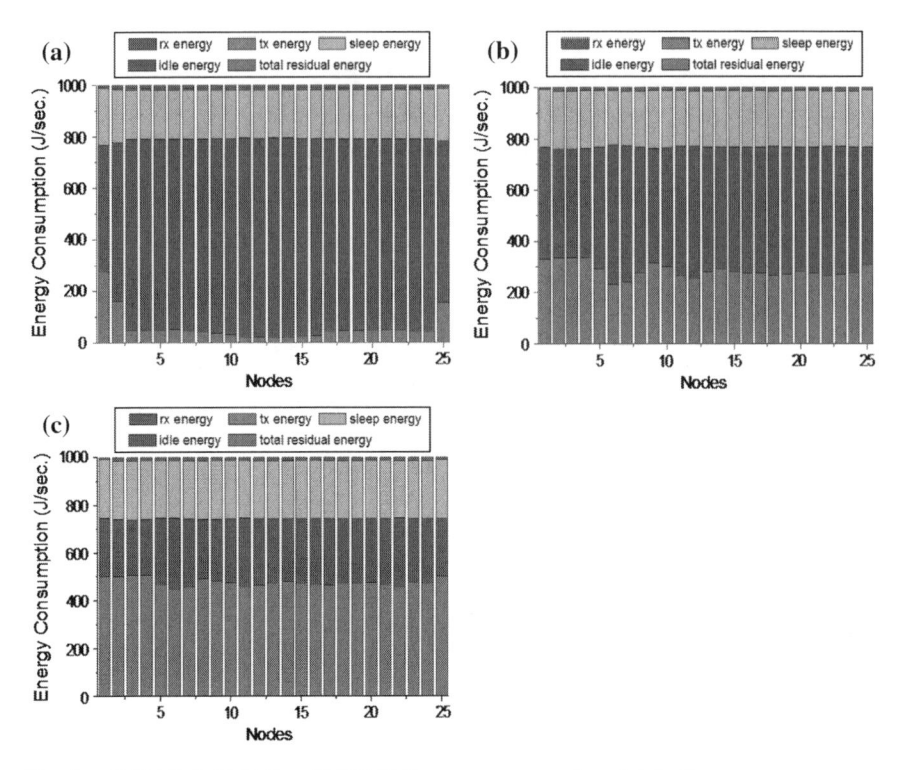

Fig. 4 **a** S-MAC, **b** RMAC, **c** AMH-MAC energy consumption distribution

during the duty cycle, in AMH-MAC further reduce the idle listening energy consumption as compared with RMAC scheme. The results reveal that significant improvement in total residual energy in our proposed AMH-MAC mechanism as compared with S-MAC and RMAC schemes.

The simulation result for end to end deliver latency is shown in Fig. 5 for three different schemes. As noted, the multi-hopping characteristic plays a significant role in latency. Therefore, due to the one hop only ability per single duty cycle for S-MAC, the latency is much higher than other schemes. The latency of AMH-MAC is very close to the RMAC result because we only present the result with the same N value but not the dynamic N cases in AMH-MAC due to the limit of page length. Although the proposed AMH-MAC mechanism with fixed N value shows about the same latency as compared with RMAC, however, it significantly improves the energy consumption.

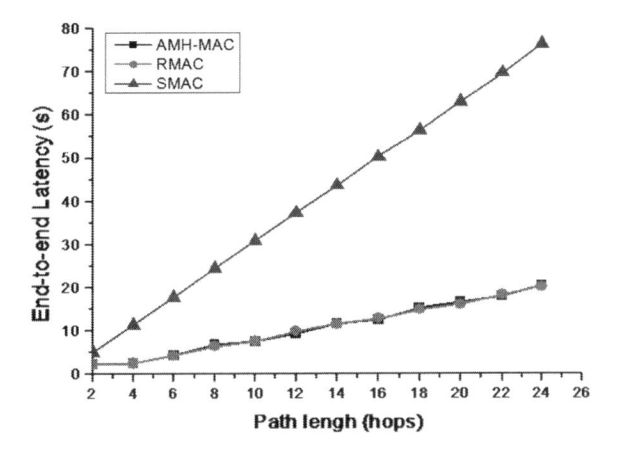

Fig. 5 End to end latency for different schemes

Conclusion

In this paper, we have proposed an adaptive multi-hopping MAC (AMH-MAC) for multi-hopping transmission scheduling mechanism, extended from RMAC protocol, with dynamical adjustment of transmitted power and number of multi-hopping nodes. The simulation results show that the proposed AMH-MAC mechanism provides significant improvement in terms of energy consumption and end to end deliver latency as compared with S-MAC and RMAC, though the results for the dynamic adjustment of N values were not shown in performance evaluation session due to the limitation of page length.

Acknowledgments The authors would like to acknowledge the support from the National Science Council of Taiwan (No. NSC 101-2221-E-142-003, NSC100-2221-E-142-008) and National Taichung University regarding the MoE project (No. 1020035480A).

References

1. Ye W, Heidemann J, Estrin D (2002) An energy-efficient MAC protocol for wireless sensor networks. In: Proceedings of INFOCOM, June 2002, pp 1567–1576
2. van Dam T, Langendoen K (2003) An adaptive energy-efficient MAC protocol for wireless sensor networks. In: Proceedings of first international conference on embedded networked sensor systems (SenSys 2003), Nov 2003, pp 171–180
3. Du S, Saha AK, Johnson DB (2007) RMAC: a routing-enhanced duty cycle MAC protocol for wireless sensor networks. Proceedings of INFOCOM, May 2007, pp 1478–1486
4. Polastre J, Hill J, Culler D (2004) Versatile low power media access for wireless sensor networks. In: Proceedings of the second ACM conference on embedded networked sensor systems (SenSys), Nov 2004, pp 95–107
5. Buettner M, Yee GV, Anderson E, Han R (2006) X-MAC: a short preamble MAC protocol for duty-cycled wireless sensor networks. In: Proceedings of the 4th international conference on embedded networked sensor systems (SenSys) 2006, pp 307–320

6. Wang H, Zhang X, Abdesselam FN, Khokhar A (2010) Cross-layer optimized MAC to support multihop QoS routing for wireless sensor networks. IEEE Trans Veh Technol 59(5):2556–2563
7. Liu Y, Elhanany I, Qi H (2005) An energy-efficient QoS-aware media access control protocol for wireless sensor networks. In: Proceeding of mobile adhoc and sensor systems conference 2005
8. Cho KT, Bahk S (2012) Optimal hop extended MAC protocol for wireless sensor networks. Comput Netw 56(4):1458–1469
9. He T, Huang C, Blum BM, Stankovic JA, Abdelzaher T (2003) Range-free localization schemes for large scale sensor networks. In: Proceeding of the 9th annual international conference on mobile computing and networking (MobiCom'03) 2003, pp 81–95
10. Shang Y, Ruml W, Zhang Y, Fromherz M (2004) Localization from connectivity in sensor networks. IEEE Trans Parallel Distrib Syst 15(11):961–974
11. McCanne S, Floyd S ns Network Simulator. http://www.isi.edu/nsnam/ns/

Avoiding Collisions Between IEEE 802.11 and IEEE 802.15.4 Using Coexistence Inter-Frame Space

Tsung-Han Lee, Ming-Chun Hsieh, Lin-Huang Chang,
Hung-Shiou Chiang, Chih-Hao Wen and Kian Meng Yap

Abstract Recently, more and more wireless networks have been deployed and start provide varies services to customers. Such as smart-phones integrate heterogeneous wireless devices, it may cause unintended interactions between multiple radios using difference radio access technologies. Thus, heterogeneous wireless network will be the trend of future network. The Co-channel interference problem in heterogeneous wireless networks is more and more important. This paper focus on the coexistence problem between IEEE 802.11b/g/n and IEEE 802.15.4 protocols in the ISM 2.4 GHz band. The performance impact on IEEE 802.15.4 network under IEEE 802.11 wireless network has been analysis, and results show that IEEE 802.15.4 has serious collision problem if there has no appropriate scheduling mechanisms among two wireless protocols. A CIFS (Coexistence Inter-Frame Space) has been proposed in this paper to effectively enhance the IEEE 802.15.4 transmission probability which is implemented in IEEE 802.11 nodes to

T.-H. Lee · M.-C. Hsieh · L.-H. Chang (✉) · H.-S. Chiang · C.-H. Wen
Department of Computer Science, National Taichung University of Education,
Taichung, Taiwan, Republic of China
e-mail: lchang@mail.ntcu.edu.tw

T.-H. Lee
e-mail: thlee@mail.ntcu.edu.tw

M.-C. Hsieh
e-mail: BCS100101@gm.ntcu.edu.tw

H.-S. Chiang
e-mail: bcs100110@gm.ntcu.edu.tw

C.-H. Wen
e-mail: BCS100102@gm.ntcu.edu.tw1

K. M. Yap
Department of Computer Science and Networked Systems, Sunway University,
Bandar Sunway, Malaysia
e-mail: kmyap@sunway.edu.my

Y.-M. Huang et al. (eds.), *Advanced Technologies, Embedded and Multimedia for Human-centric Computing*, Lecture Notes in Electrical Engineering 260, DOI: 10.1007/978-94-007-7262-5_135, © Springer Science+Business Media Dordrecht 2014

observe the transmission opportunity for IEEE 802.15.4 under wireless coexistence environment.

Keywords IEEE 802.11 · IEEE 802.15.4 · Coexistence · Co-channel interference

Introduction

Increasingly devices are equipped with multiple radio interfaces. This is due to the decreasing cost and size of radio embedded technology and the increasing demand from end-users. Thus, heterogeneous wireless networks are currently receiving a significant amount of interest in research. This paper focuses on the co-channel interference between IEEE 802.11 b/g/n [1–3] and IEEE 802.15.4 [4] protocols in the ISM 2.4 GHz band. In Fig. 1, the IEEE 802.11b/g has eleven channels between 2.412 and 2.462 GHz [5]. Each channel occupies 22 MHz and up to 3 separate channels can be simultaneously used without any mutual interference. In the other hand, the IEEE 802.15.4 has sixteen channels between 2.4 and 2.4835 GHz. Each channel occupies 3 MHz. The IEEE 802.15.4 will suffer serious interference problems from IEEE 802.11 if there is no proper arrangement for the mutual scheduling mechanism. Thus, the goal of this research is try to improve the performance of IEEE 802.15.4 in the heterogeneous wireless network environment.

Furthermore, it can explore to three different interference situations depend on the IEEE 802.15.4 transmission range. Thus, the interference circumstances can be divided into three ranges which are shown in Fig. 2.

- Range 1: a range in which IEEE 802.15.4 nodes and IEEE 802.11 nodes can sense each other. Such as the range between AP and Zigbee C.

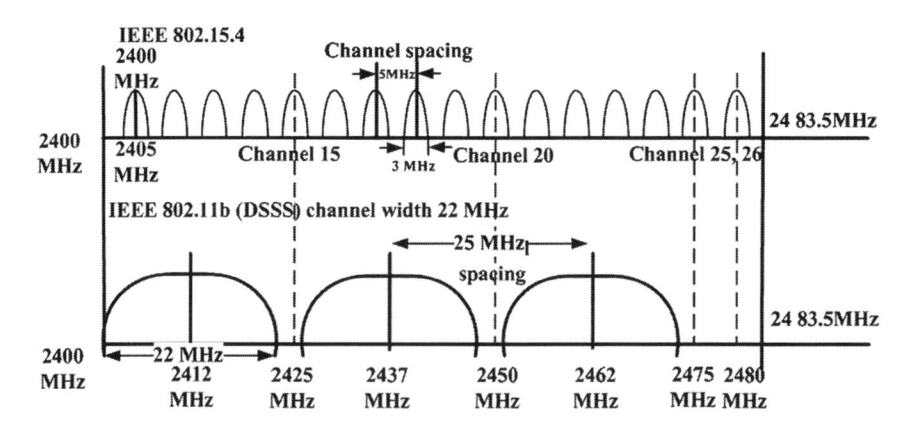

Fig. 1 The channel allocation in the 2.4 GHz band

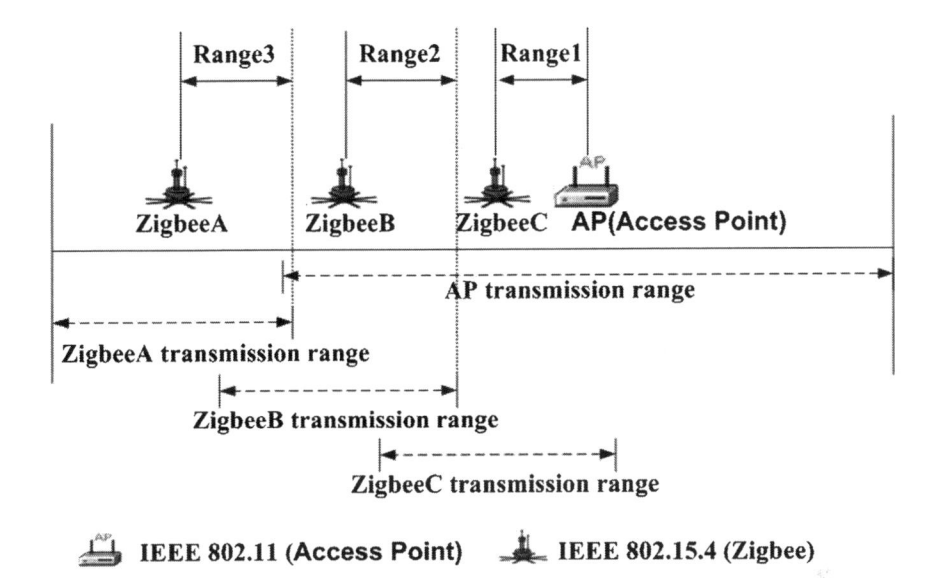

Fig. 2 The coexistence cases of IEEE 802.11 and IEEE 802.15.4

- Range 2: a range in which IEEE 802.15.4 nodes can sense IEEE 802.11 nodes, but IEEE 802.11 can't sense IEEE 802.15.4. Such as the range between AP and Zigbee B.
- Range 3: a range in which cannot sense each other. Such as the range between AP and Zigbee A.

The rest of this paper is organized as follows. Related Work will discuss related works for co-channel interference between IEEE 802.11. and IEEE 802.15.4. Packet Error Analysis of IEEE 802.15.4 Under IEEE 802.11 Coexistence describes the co-channel interference model in wireless IEEE 802.15.4 and IEEE 802.11 coexistence networks. The Coexistence Inter-Frame Space describes the proposed Coexistence Inter-Frame Space mechanism. Performance Evaluation evaluates the performance of the proposed CIFS by PRISM simulation. Finally, summarizes conclusions of this paper and give suggestions for future work in Conclusion.

Related Work

In [6–8], authors proposed different interference models. Simulation results can reduce the bit error rate (BER) and the packet error rate (PER) effectively. In [9, 10], interference avoidance is achieved by means of energy detection (ED); Yi et al. [9], both using "Safe Distance" and "Safe Offset Frequency" to reduce the BER and PER. In [10], author builds a neighbor table to allow nodes switch to relatively low-interference channel to reduce the packet loss rate. In [11] authors employs a separate signaler node between two IEEE 802.15.4 nodes to emit a carrier signal (busy-tone), thereby enhancing the IEEE 802.15.4's visibility to IEEE 802.11.

The reason of interference has been studied in [12, 13]. Authors try to avoid the interference by using packet scheduling mechanisms in the MAC layer. In [12], the authors use the channel switching and routing synchronization to avoid interference of IEEE 802.11. Yuan [13], the authors take into account difference interference from Range 1 to Range 3. However, all above literatures use packet scheduling mechanism to avoid interference in the MAC layer. In [14], MCSP (Multi-Canal Scheduling Protocol) has been proposed to avoid the lighthouse collision. Shin [15] was modified the state transition probabilities in IEEE 802.15.4 Markov process model. Simulation results show the proposed model match to the theoretical expressions. The longer distance between IEEE 802.11 and IEEE 802.15.4 gains higher throughput. However, IEEE 802.11 nodes are not able to affect the throughput of IEEE 802.15.4 when IEEE 802.15.4 nodes outside the IEEE 802.11's transmission range, such as Range 3 in Fig. 2.

In [12–14] was focus on the MAC layer arrangement for the mutual scheduling mechanism. In [6–10] was focus on the Range 1 which IEEE 802.15.4 nodes and IEEE 802.11 nodes can sense each other, but there were inconspicuous display problem. Most of literatures above focus on the Range 2 co-channel interference, and most of literatures ignore the impact on Range 1. To alleviate the interference problem, the IEEE 802.15.4 employs the Carrier Sense Multiple Access with Collision Avoidance (CSMA/CA) mechanism. However, IEEE 802.15.4 still has serious collision problem if there has no appropriate scheduling mechanisms for Range 1 co-channel interference. Therefore, we consider the performance enhancement of IEEE 802.15.4 for Range 1 co-channel interference in this paper.

Packet Error Analysis of IEEE 802.15.4 Under IEEE 802.11 Coexistence

In this section, we investigate the interference between IEEE 802.11 and IEEE 802.15.4 in the Range 1. A simulation environment was developed using the PRISM [16] simulator as shown in Fig. 3; Table 1 shows the related parameters for the simulation. In Table 1, the length of Slot-Time is different between IEEE 802.11 and IEEE 802.15.4. IEEE 802.15.4's Slot-Time is sixteen times to IEEE 802.11. IEEE 802.11 has shorter DIFS (DCF, Distributed Coordination Function IFS). It cause IEEE 802.11 always has higher transmission opportunity than IEEE 802.15.4. The unfairness transmission opportunity will resulting packet dropped in IEEE 802.15.4 due to reach the maximum number of back-off restriction.

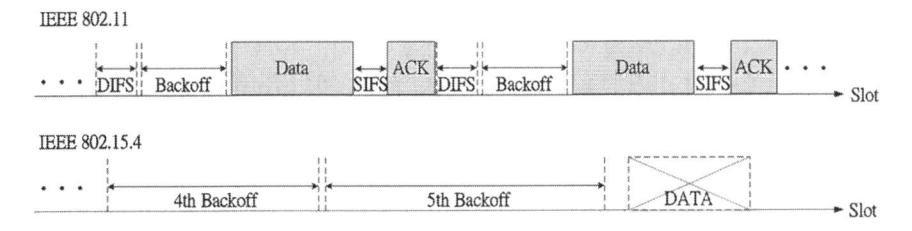

Fig. 3 Original IEEE 802.15.4 and IEEE 802.11 packet transmission in range 1

Table 1 Simulation configuration and parameters

	IEEE 802.15.4	IEEE 802.11b
Packet length (byte)	133	1,500
Data rate (bps)	250 k	2 M
Retry limit	3	7
CWmin	$7(2^3-1)$	$31(2^5-1)$
CWmax	$31(2^5-1)$	$1,023(2^{10}-1)$
Backoff-max	5	6
A slot time	320 μs	20 μs
DIFS		50 μs
SIFS	192 μs	10 μs

The Coexistence Inter-Frame Space

This section, a Coexistence Inter-Frame Space (CIFS) has been proposed to enhance the IEEE 802.15.4 transmission probability under the IEEE 802.11 network. In Fig. 4, the CIFS is implemented in IEEE 802.11 nodes, which try to observe the transmission opportunity to IEEE 802.15.4 under wireless coexistence environment. Figure 5 shows the IEEE 802.15.4 transmission probability in different CIFS. The result shows the shorter CIFS cause lower transmission probability in IEEE 802.15.4. The IEEE 802.15.4 transmission probability reaches to 100 % when the CIFS around 1.2 ms. Moreover, result shows the length of CIFS is proportional to the transmission probability under wireless coexistence environment.

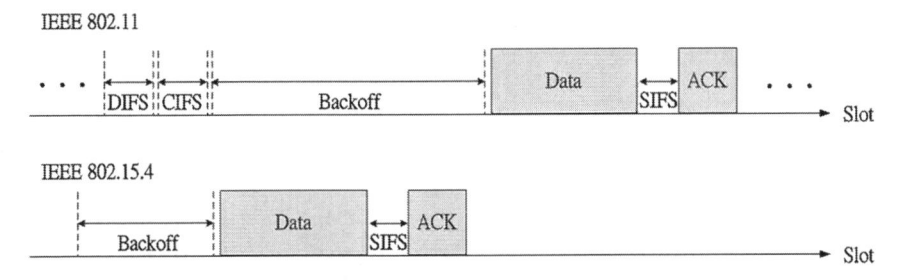

Fig. 4 Implement the CIFS to postpone the IEEE 802.11 packet transmission in range 1

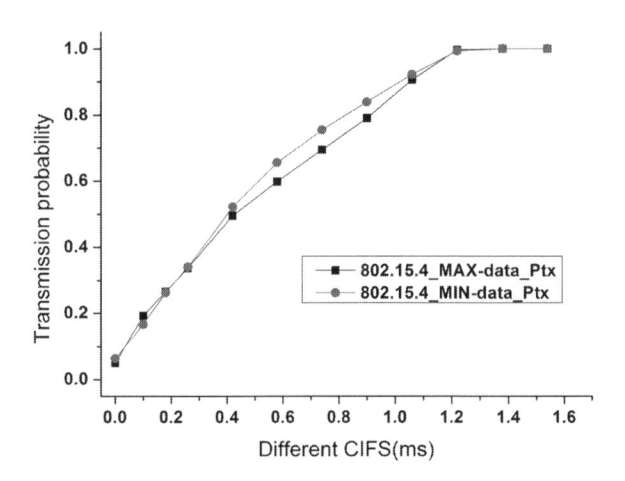

Fig. 5 The transmission probability of IEEE 802.15.4 in different CIFS

Performance Evaluation

In order to demonstrate the relationship between the payload size and transmission probability in IEEE 802.15.4 and IEEE 802.11 coexistence networks. Our probabilistic model was emulated by PRISM [16] simulator. The simulation environment is summarized in Table 1, which considers the operation of IEEE 802.15.4 with one pair of source and destination nodes in the multiple IEEE 802.11b WLANs coexistence environment.

We obtain transmission throughput from Eqs. (1) and (2) for IEEE 802.11 and IEEE 802.15.4 respectively.

$$S_{11} = \frac{L_{11}}{slot_{11} * IEEE\ 802.11\ A\ Slot\ Time} \tag{1}$$

$$S_{15.4} = \frac{L_{15.4}}{slot_{15.4} * IEEE\ 802.15.4\ A\ Slot\ Time} \tag{2}$$

The S_{11} presents the IEEE 802.11 throughput in coexistence environment, and L_{11} is the packet length of IEEE 802.11. $slot_{11}$ is the total slots to transmit a data packet in IEEE 802.11. $S_{15.4}$ is IEEE 802.15.4 throughput in coexistence environment. $L_{15.4}$ is IEEE 802.15.4 packet length, and $slot_{15.4}$ is total slots to transmit an IEEE 802.15.4 packet.

$$DRU_{11} = \frac{S_{11}}{IEEE\ 802.11\ Data\ Rate} \tag{3}$$

$$DRU_{15.4} = \frac{S_{15.4}}{IEEE\ 802.15.4\ Data\ Rate} \tag{4}$$

Finally, from Eqs. (3) and (4) will obtain the utilization on desire data rate for IEEE 802.11 and IEEE 802.15.4 respectively.

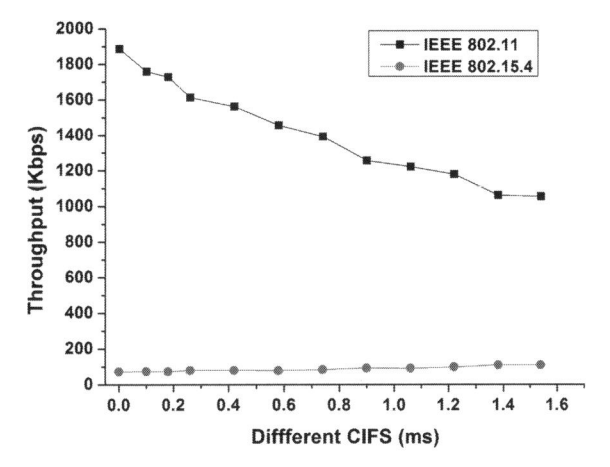

Fig. 6 IEEE 802.11 and IEEE 802.15.4 throughput in different CIFS

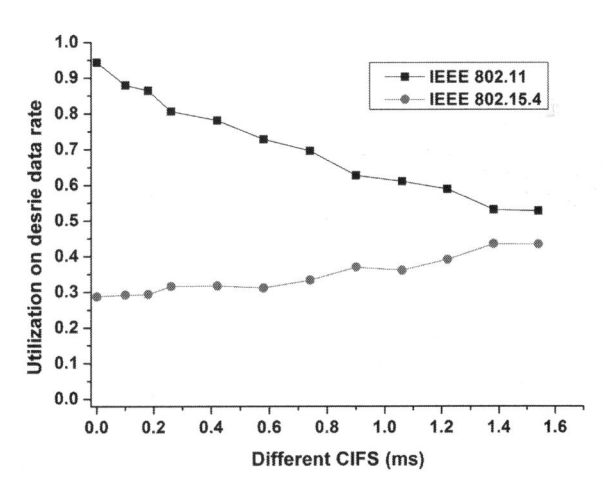

Fig. 7 The utilization on desire data rate in IEEE 802.11 and IEEE 802.15.4 data in different CIFS

Figure 6 shows IEEE 802.15.4 with CIFS has better probability against to original IEEE 802.15.4. It is because that the CIFS will postpone the data transmission in IEEE 802.11 to observe the transmission opportunity for IEEE 802.15.4. Thus, the throughput of IEEE 802.15.4 is proportional to the length of CIFS. However, the throughput of IEEE 802.11 is inversely proportional to the length of CIFS.

In Fig. 7, it shows a trade-off between the utilization on desire data rate in IEEE 802.11 and IEEE 802.15.4. However, coexistence among IEEE 802.11 and IEEE 802.15.4 in the 2.4 GHz band is an important issue in order to ensure that each wireless service maintains its desired performance requirements.

Conclusion

In this paper, the performance impacts on the IEEE 802.15.4 network under the IEEE 802.11 wireless network have been investigated. Simulation results show the IEEE 802.15.4 has serious interference problems in the Range 1 if there are no appropriate scheduling mechanisms among two protocols. The result shows, the proper CIFS (around 1.2 ms) in IEEE 802.11 nodes will increase IEEE 802.15.4 transmission probability. We have improved the performance of IEEE 802.15.4 networks from co-channel interference in the ISM band. In addition, the short IEEE 802.15.4 packet size may take relief from the co-channel interference from IEEE 802.11. Thus, a dynamic packet adjustment via a Channel ranking scheme is consider as the future work.

Acknowledgments The authors would like to acknowledge the support from the National Science Council of Taiwan (No. 101-2119-M-142-001, 100-2221-E-142 -002) and National Taichung University regarding the MoE project (No. 1020035480A).

References

1. IEEE 802.11 Work Group (1999) Part 11: wireless LAN medium access control (MAC) and physical layer (PHY) specification: high-speed physical layer extension in the 2.4 GHz Band, ANSI/IEEE Std 802.11b
2. IEEE 802.11 Work Group (1999) Part 11: wireless LAN medium access control (MAC) and physical layer (PHY) specifications: further higher data rate extension in the 2.4 GHz band, ANSI/IEEE Std 802.11g
3. IEEE 802.11 Work Group (1999) Part 11: wireless LAN medium access control (MAC) and physical layer (PHY) specifications: further higher data rate extension in the 2.4 GHz band, ANSI/IEEE Std 802.11n
4. IEEE 802.15 Work Group (2003) Part 15.4: wireless medium access control (MAC) and physical layer (PHY) specifications for low-rate wireless personal area networks (LR-WPANs), ANSI/IEEE Std 802.15.4
5. Musaloiu R, Terzis A (2007) Minimising the effect of wifi interference in 802.15.4 wireless sensor networks. Int J Sen Netw 3(1):43–54
6. Tamilselvan GM, Shanmugam A (2010) Inter and intra cluster scheduling for performance analysis of coexistence heterogeneous networks. Int J Comput Appl 1(8):0975–8887
7. Tamilselvan GM, Shanmugam A (2011) A cluster based interference mitigation scheme for performance enhancement in IEEE 802.15.4. J Sci Ind Res 70(9):756–761
8. Tamilselvan GM, Shanmugam A (2011) Multi hopping effect of ZigBee nodes coexisting with WLAN nodes in heterogeneous network environment. 978-1-4577-0183-2/09/$26.00 2011 IEEE
9. Yi P, Iwayemi A, Zhou C (2011) Developing ZigBee deployment guideline under WiFi interference for smart grid applications. IEEE Trans Smart Grid 2(1):110–120
10. Won C, Youn JH, Sharif HA, Deogun J (2005) Adaptive radio channel allocation for supporting coexistence of 802.15.4 and 802.11b. In VTC, 2005
11. Zhang X, Shin KG (2012) Cooperative carrier signaling: harmonizing coexisting WPAN and WLAN devices. IEEE Trans Netw, no. 99
12. Shin SY, Park HS, Choi S, Kwon WH (2007) Packet error rate analysis of ZigBee under WLAN and bluetooth interferences. IEEE Trans Wireless Commun 6(8):2825–2830

13. Yuan W (2011) Coexistence of IEEE 802.11b/g WLANs and IEEE 802.15.4 WSNs: modeling and protocol enhancements. Technische Universiteit Delft, PhD Thesis
14. Sahraoui M (2012) Collisions avoidance multi-channel scheme for the protocol IEEE 802.15.4. International conference on information technology and e-Services (ICITeS), pp 1–9
15. Shin SY (2013) Throughput analysis of IEEE 802.15.4 network under IEEE 802.11 network interference. AEU—Int J Electron Commun
16. Prsim http://www.prismmodelchecker.org/

Implementation and Evaluation of Multi-Hopping Voice Transmission over ZigBee Networks

Lin-Huang Chang, Chen-Hsun Chang, H. F. Chang and Tsung-Han Lee

Abstract The wireless sensor network (WSN) technology has been one of the most active research topics in the last several years. Besides sensing the environmental or physical data, there has been a growing need on supporting voice communication, particularly under emergency conditions, in WSNs. In this paper, we will conduct the implementation of voice codec on embedded system using multi-hopping ZigBee communication. The implementation issues, in terms of codec and transmission rate-dependence consideration, as well as the voice quality with multi-hopping are analyzed and evaluated for the designed test-bed.

Keywords WSN · ZigBee · Speex · Xbee · Bit rate

Introduction

The wireless sensor network (WSN) technology has been one of the most active research topics currently. The ZigBee [1] standard, based on IEEE 802.15.4 physical radio standard [2], is designed for low rate wireless personal area network (LR-WPAN) applications with the deployment of low-cost and low-power consumption wireless sensors or devices. The traditional operation to applications of

L.-H. Chang · C.-H. Chang · T.-H. Lee (✉)
Department of Computer Science, National Taichung University, Taichung, Taiwan, Republic of China
e-mail: thlee@mail.ntcu.edu.tw

L.-H. Chang
e-mail: lchang@mail.ntcu.edu.tw

H. F. Chang
Department of Computer and Communication Engineering,
Taipei Chengshih University of Science and Technology, Taipei,
Taiwan, Republic of China

Y.-M. Huang et al. (eds.), *Advanced Technologies, Embedded and Multimedia for Human-centric Computing*, Lecture Notes in Electrical Engineering 260, DOI: 10.1007/978-94-007-7262-5_136, © Springer Science+Business Media Dordrecht 2014

WSN is restricted on infrequently sensing environmental or physical data, such as temperature, pressure and light information with low duty cycle. Besides sensing and reporting data asynchronously, there has been a growing need on providing real time applications in WSNs. The transmission of voice, particularly under emergency conditions, has been proven to be feasible in WSNs [3]. However, there are still some issues, such as codec tuning and rate-dependent voice quality, need to be resolved for voice transmission using ZigBee communication.

The basic characteristic of ZigBee includes (1) low transmission rate of 250 Kbps, (2) short transmission distance of about 50–100 m (with different transmission power, it could be extended to 300 m), (3) low power consumption, and (4) operation at 2.4 GHz ISM band. It supports point-to-point communication with multi-hopping transmission.

In this paper we will conduct the implementation of Speex [4] voice codec on embedded system with Xbee [5] wireless sensor nodes. We will present the system architecture, voice codec consideration of Speex implementation, and transmission rate-dependence issues for the designed voice transmission test-bed over multi-hopping WSNs. The voice quality is always one important issue to achieve the realization of voice transmission using ZigBee communication. We will perform the voice quality analysis with multi-hopping WSNs in this research.

The rest of the paper is organized as follows. Section Related Work reviews the related works in voice transmission using WSN. The designed test-bed of voice transmission system using multi-hopping ZigBee communication is described in details in section System Design, followed by the performance analysis in section Performance Evaluation. Finally, section Conclusions addresses the conclusion of this research.

Related Work

Although the voice communication over ZigBee networks has attracted much attention currently, there are still some challenges to deliver timeliness guarantees or provide quality of service (QoS) with multi-hopping transmission. The research in [3] has presented the evaluation and implementation of a voice codec using Z-phone project to transmit voice in full-duplex mode over ZigBee networks. They investigated some criteria in selecting the most suitable codec and provided the performance analysis of the Z-phone. However, the system is based on single hop ZigBee scenario. The researches in [6, 7] have implemented the voice over sensor network with multi-hopping. They also provided the performance analyses, such as delay, packet loss and quality of service (QoS) on their test-beds. Unfortunately, the analyzed mean opinion score (MOS) values, which is a key factor to evaluate QoS of voice communication, in these researches are mostly below 3.0, corresponding to some users dissatisfied level.

In [8], the research group used the PCM codec, TLV320AIC1107 voice processing chip with ADPCM coding rates 16 kHz to 64 kHz, for the coding/decoding procedures to implement push-to-talk (PTT) functionality of voice communication across multi-hopping ZigBee networks based on CC2430 architecture. The packet loss of the designed test-bed with more than 3 hops was higher than 10 % or even as high as 40 % or more which may not provide acceptable voice quality.

The research in [9] provided some results on the time analysis in terms of a variety of delays for voice transmission over multi-hopping ZigBee networks. Their measurements revealed that the device probing delay is about 30 ms, including 15 ms for device channel link delay and 15 ms for trigger delay. The research in [10] on the other hand suggested the synchronization of relay nodes to the sending and receiving nodes to reduce the contention delay in the WSN chain topology. This design however, required additional synchronization signal overhead periodically.

In this paper we will conduct the implementation of open source Speex voice codec to transmit voice packets over multi-hopping ZigBee networks using Xbee wireless sensor nodes. The system performance, such as signal strength, delay, and rate-dependent voice quality, is carried out for the implemented test-bed.

System Design

Hardware Components and Design Issues

The system architecture with process flow of the designed test-bed is shown in Fig. 1, where the multi-hopping scenario is represented by the relay node besides sending and receiving nodes. The sending and receiving nodes are designed to be the ZigBee voice gateways which are implemented on the BeagleBone embedded systems with Xbee version Serial 1 wireless sensor nodes. At the sender side, when the analog signal is converted to digital data, it is encoded by Speex encoder, where the coding rates are adjusted to study the rate-dependence on voice quality. The Xbee version Serial 1 ZigBee node, with maximum transmission rate 250 Kbps, can carry 100 bytes data in payload. To avoid the issues and efforts from reassembling audio frames due to packet loss, each Xbee packet should carry complete audio frames instead of audio byte streams in the payload. The Xbee hardware structure and internal data flow diagram is illustrated in Fig. 2 which will be explained in details shortly.

Due to relatively small transmitting rate and limited buffer in Xbee hardware structure, it is unable to handle and process all audio streams from Speex encoder in corresponding time. This will result in significant packet loss and delay during the internal process. Therefore, we use leaky bucket mechanism to implement a flow control in ZigBee voice gateway to modulate the rate difference between Speex encoder and Xbee device. When the voice data is stored in traffic buffer of voice gateway, the flow control module with leaky bucket mechanism will be triggered to read 100 B voice data every 3 ms from traffic buffer to Xbee buffer. The triggered rate is tuned to adapt to the Xbee transmission rate and limited buffer

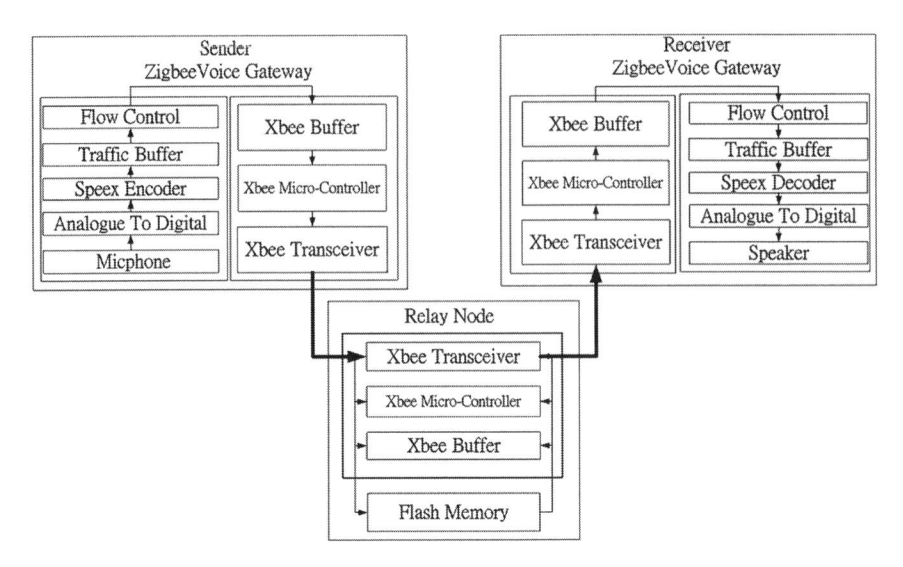

Fig. 1 The implemented system architecture with process flow

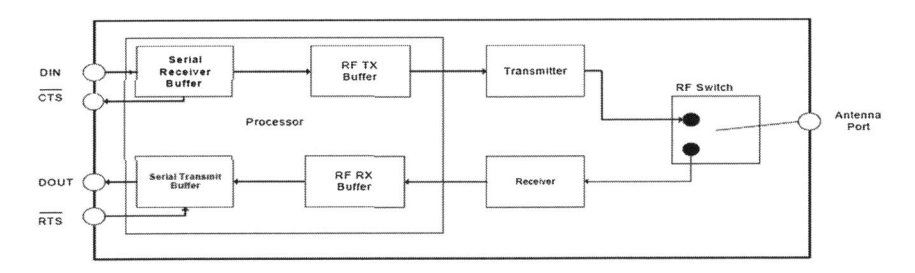

Fig. 2 The Xbee internal data flow diagram

space. On the other hand, the relay node, performing only the forwarding of voice packets, is simply a Xbee sensor node with flash memory. Because of the strict requirement in end-to-end delay for real time voice communication, the relay node, which does not conduct any additional operation or coding process but extend the transmission distance, aims to relay voice packets efficiently.

As shown in Fig. 2, there are two registers, RF TX buffer and RF RX buffer, in Xbee structure. When the RF TX buffer is full, the CTS pin will signal to stop receiving data from DI buffer. The resuming process of CTS signal will cause additional latency, therefore, it is important not to overwhelm the RF TX buffer. That is the significance to design a flow control in sender voice gateway and actually similar situation for the receiving case. The RF switch block provides the transceiver conversion within a very short time between the transmitting and

Fig. 3 The multi-hopping topology and transmission flow

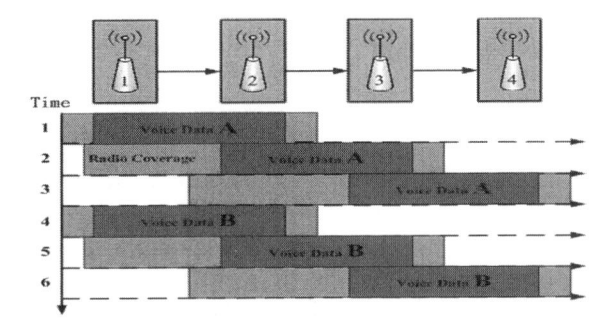

receiving antenna ports. The RF switch in relay node is implemented to attach the destination address of the relayed packets into the next hop address.

Figure 3 illustrates the multi-hopping topology as well as the voice data transmission flow in our implementation. Each sensor node, arranged in multi-hopping chain topology, is separated from 50 to 110 m. The measurements of the test-bed are conducted by sending voice data from node 1 relayed via nodes 2 and/or 3 and finally received by node 4.

Audio Codec Using Speex

The relatively low transmission rate, 250 Kbps ideally, of ZigBee wireless networks, and small payload, 100 bytes, using Xbee wireless sensor nodes is a challenge for voice data transmission. One of the top issues to face this challenge is to select a reliable and efficient voice encoder for codec implementation. The research in [3] proposed an implementation of open source Speex voice compression technology over ZigBee network environment. Speex, based on the code-excited linear prediction (CELP) speech coding algorithm, provides a good execution efficiency and voice quality. The fixed-point arithmetic instead of using floating point unit (FPU) for Speex compression will reduce the power consumption and computing time in WSN nodes with low processing ability and low power digital signal processor (DSP). The design goal of Speex is to optimize the voice data with high quality speech and low bit rate.

Speex employs multiple bit rates to support three different modes of operation, including ultra-wideband (32 kHz sampling rate), wideband (16 kHz sampling rate) and narrowband (telephone quality with 8 kHz sampling rate), and some features such as voice activity detection and variable bit rates. Since Speex is robust to loss packets, it is a good candidate for ZigBee communication environment. Due to the lower computational resource requirements and available transmission rate in WSN nodes, the narrowband operation is selected to be implemented in our design. In Speex narrowband mode, there are three major

Fig. 4 Speex 15 Kbps
encoding structure

Frame Size		Frame Size	
4 Bytes	38 Bytes	4 Bytes	38 Bytes

Table 1 Speex frame size

Bit-rate	15 Kbps	11 Kbps	8 Kbps
Sample	300	220	160
Frame size	38 B	28 B	20 B

tasks, including line spectrum pair (LSP) quantization, adaptive codebook search, and fixed codebook search. Depending on the difference of data rates, narrowband provides 8 different sub-modes to handle quantization and un-quantization of the pitch and its gain.

In this paper, we employ three different compression ratios of Speex narrowband mode, including 15, 11 and 8 Kbps, in our designed system. The frame size for Speex narrowband 15 Kbps is 20 ms, which corresponds to 300 samples in a frame. Speex 15 Kbps encoding structure is shown in Fig. 4 where each frame consists of 4 B header to record bit-rate parameters and 38 B voice data payload.

Table 1 lists the corresponding frame sizes for three different bit rates which will decide the number of frames embedded in each packet of Xbee 100 B payload. Figure 5 shows the design of Speex frames in Xbee packets. Without losing the encoding/decoding process and quality for our applications, we design and implement one Speex header only in a Xbee packet with 100 B payload to reduce the overhead of voice data. Therefore, each Xbee packet can carry two, three and four encoded voice frames for 15, 11, and 8 Kbps bit rates, respectively. As notice, the voice quality is affected significantly by the packet loss during transmission and the received signal strength indicator (RSSI) value could be an important and instant indicator related to packet loss. In this study, we will evaluate the relationship between encoding bit rates and RSSI values on voice quality to provide an optimized voice communication over ZigBee networks.

Fig. 5 Xbee voice packet
format

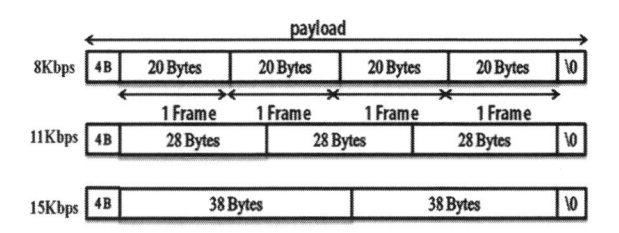

Table 2 Parameters setting

Xbee parameter	Value
Traffic loads	250 Kbps
Serial interface data rate	57,600
Topologies	Point-to-multipoint
Network mode	Beaconless
Data packet size	84, 88, 80 (byte)
Transmit power	1 mW(0 dBm)
Speex parameter	**value**
Mode	Narrowband
BitRate	8, 11, 15 (kb/s)
Quality	4, 6, 8

Table 3 RSSI/LQI measurements

Distance (m)	Relay node (RSSI/LQI)	Receive node (RSSI/LQI)
50 m	−38 dBm/119	−39 dBm/116
80 m	−50 dBm/86	−51 dBm/83
110 m	−56 dBm/68	−58 dBm/63

Performance Evaluation

The implemented parameters for Xbee and Speex setting of the designed test-bed are listed in Table 2.

Network Performance Analysis

In this sub-section, we conduct the RSSI and link quality indicator (LQI) measurements for different transmission distances. For simplicity in test-bed arrangement, the multi-hopping scenario is measured with 3 nodes topology, sending from node 1, relaying via node 2 and receiving at node 3. The RSSI/LQI values detected from relaying and receiving nodes with a series of different distances are measured and three of them, 50, 80 and 110 m, are listed in Table 3. As the transmission distance increase, the LQI value decreases.

The network delay measurements are conducted by processing original audio file using Speex encoding followed by constant packet inter-arrival time for voice data transmission. The measured delays from sending node to relay node, separated in 50, 80, and 110 m respectively, are plotted in Fig. 6. The corresponding delays are 6, 35 and 58 ms. As the transmission distance increases, the delay increases.

On the other hand, the results from sending node (via relay node) to receiving node, separated in 100, 160 and 220 m respectively, are illustrated in Fig. 7, which corresponding to 2-hop transmission. The corresponding measured delays are about 11, 64 and 120 ms.

Fig. 6 The delay from
sending node to relay node

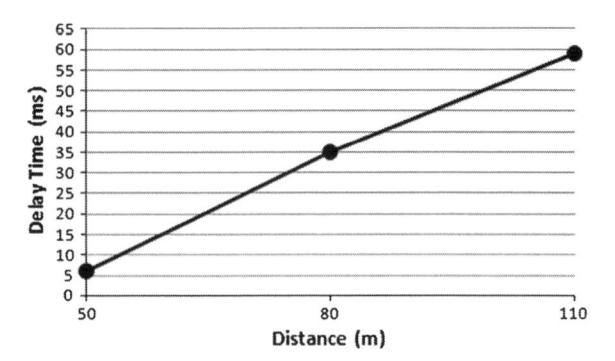

Fig. 7 The end-to-end delay
from sending node to
receiving node

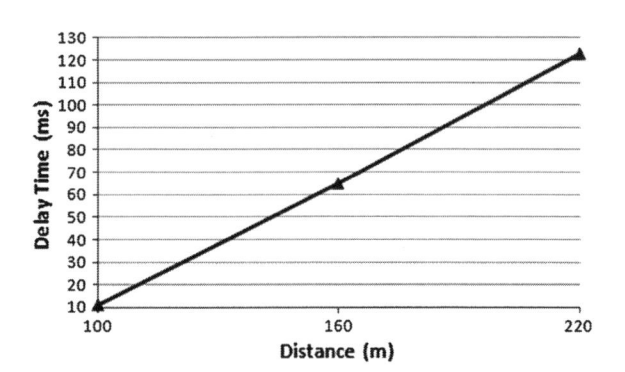

Voice Quality Analysis

The voice quality in terms of perceptual evaluation of speech quality (PESQ) could deteriorate when the packet loss increases during transmission. As notice, longer transmission distance may incur larger packet loss due to reduction in signal strength and longer delay. In this sub-section, we conduct the experiments on PESQ_MOS voice quality measurements with different LQI values. The voice signal as the input of PESQ measurements is sending from node 1 and extracted from node 3 as the output of PESQ measurements.

The voice quality results are illustrated in Fig. 8. When the LQI value is more than 100, the PESQ_MOS value for 15 Kbps bit rate case is better than those of 11 and 8 Kbps bit rates. The measured PESQ_MOS values are close to the referenced data from Speex project. However, when the LQI drops to 85, the voice quality of 15 Kbps case deteriorates significantly and its PESQ-MOS values is lower than those of 11 and 8 Kbps bit rates. When the LQI even drops to 80 or 70, the voice quality of both 15 and 11 Kbps cases decline dramatically.

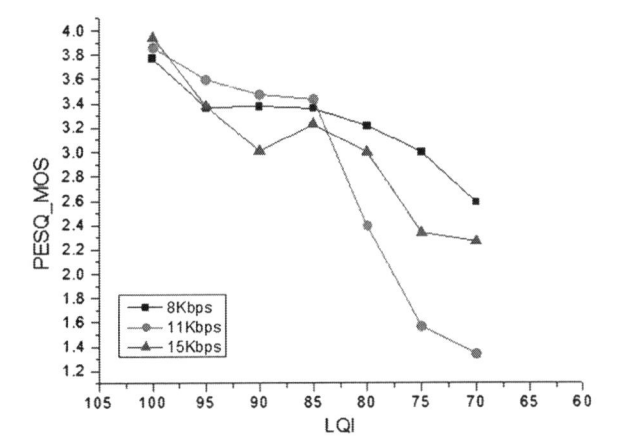

Fig. 8 Voice quality measurement

Conclusions

In this paper, we have designed and implemented the Speex voice codec on embedded system with Xbee WSN nodes using multi-hopping ZigBee communication. The design and implementation issues regarding the Speex codec and Xbee nodes have been addressed and discussed. From our design, we have implemented a multi-hopping ZigBee test-bed for performance analyses in terms of distance-signal strength dependence, network delay and PESQ-MOS voice quality. The design, implementation and evaluation result will provide a test-bed to realize voice transmission using ZigBee communication.

Acknowledgments The authors would like to acknowledge the support from the National Science Council of Taiwan (No. NSC 101-2221-E-142-003, NSC100-2221-E-142-008) and National Taichung University regarding the MoE project (No. 1020035480A).

References

1. Zigbee Alliance (2008) Zigbee specification. Zigbee Alliance. 053474r17 edn
2. IEEE 802.15.4 standard (2003) Wireless medium access control (MAC) and physical layer (PHY) specification for low-rate wireless personal area networks (LR-WPANs)
3. Touloupis E, Meliones A, Apostolacos S (2012) Speech codecs for high-quality voice over zigbee applications: evaluation and implementation challenges. IEEE Commun Mag
4. Speex (2006) a free codec for free speech. Available at http://www.speex.org
5. Xbee Data Sheet (2009). Available at http://www.digi.com
6. Mangharam R, Rowe A, Rajkumar R, Suzuki R (2006) Voice over sensor networks. 27th IEEE international of real-time systems symposium, pp 291–302
7. Song HY, Cho SH (2011) Performances of IEEE 802.15.4 unslotted CSMA-CA for voice communications. The 17th Asia-Pacific conference on communications
8. Brunelli D, Teodorani L (2008) Improving audio streaming over multi-hop Zigbee networks. Proc IEEE ISCC 2008:31–36

9. Brunelli D et al (2008) Analysis of audio streaming capability of zigbee networks. Proc EWSN 2008:189–204
10. Woon W, Wan T (2008) Performance evaluation of IEEE 802.15.4 wireless multi-hop networks: simulation and testbed approach. Int J Ad Hoc Ubiquitous Comput 3(1)

A Lightweight Intrusion Detection Scheme Based on Energy Consumption Analysis in 6LowPAN

Tsung-Han Lee, Chih-Hao Wen, Lin-Huang Chang, Hung-Shiou Chiang and Ming-Chun Hsieh

Abstract 6LoWPAN is one of Internet of Things standard, which allows IPv6 over the low-rate wireless personal area networks. All sensor nodes have their own IPv6 address to connect to Internet. Therefore, the challenge of implementing secure communication in the Internet of Things must be addressed. There are various attack in 6LoWPAN, such as Denial-of-service, wormhole and selective forwarding attack methods. And the Dos attack method is one of the major attacks in WSN and 6LoWPAN. The sensor node's energy will be exhausted by these attacks due to the battery power limitation. For this reason, security has become more important in 6LoWPAN. In this paper, we proposed a lightweight intrusion detection model based on analysis of node's consumed in 6LowPAN. The 6LoWPAN energy consumption models for mesh-under and route-over routing schemes are also concerned in this paper. The sensor nodes with irregular energy consumptions are identified as malicious attackers. Our simulation results show the proposed intrusion detection system provides the method to accurately and effectively recognize malicious attacks.

Keywords 6LoWPAN · Energy consumption · Intrusion detection

T.-H. Lee · C.-H. Wen · L.-H. Chang (✉) · H.-S. Chiang · M.-C. Hsieh
Department of Computer Science, National Taichung University of Education,
Taichung, Taiwan, Republic of China
e-mail: lchang@mail.ntcu.edu.tw

T.-H. Lee
e-mail: thlee@mail.ntcu.edu.tw

C.-H. Wen
e-mail: BCS100102@gm.ntcu.edu.tw

H.-S. Chiang
e-mail: BCS100110@gm.ntcu.edu.tw

M.-C. Hsieh
e-mail: BCS100101@gm.ntcu.edu.tw

Y.-M. Huang et al. (eds.), *Advanced Technologies, Embedded and Multimedia for Human-centric Computing*, Lecture Notes in Electrical Engineering 260,
DOI: 10.1007/978-94-007-7262-5_137, © Springer Science+Business Media Dordrecht 2014

Introduction

6LoWPAN is an acronym IPv6 over Low-Power wireless Area Networks. 6lowpan has defined encapsulation and header compression that allows IPv6 packets to be sent to and received from over IEEE 802.15.4 (Wireless sensor network) based networks. The packet size of IPv6 is much larger than the size of IEEE 802.15.4 MPDU. Thus, the adaptation layer is joining between the network layer and the data link layer to fragmentation and recombination the IPv6 packets into IEEE 802.15.4 radio.

There are two routing schemes in 6LoWPAN. And the characteristic of route-over is hop-by-hop. Mesh-under and route-over routing schemes can be considered as end-to-end and hop-by-hop transmission respectively. Hop-by-hop fragmentation and reassembly generates more delay but achieve better fragment arrival ratio. Whereas end-to-end scheme has less latency, but fragment loss has high probability.

In 6LoWPAN protocol stack, the last two layers are based on IEEE 802.15.4 physical and data-link layers. Thus, we investigate the energy consumption of IEEE 802.15.4 for the last two layers in the beginning, and then the 6LoWPAN performance analysis in route-over and mesh-under routing schemes are both consideration in this paper.

In order to construct the exact energy consumption model in 6LoWPAN, many study presented energy model to analysis the energy consumption on sensor node, and propose the method to improve the performance of network. We not only consider the reference of IEEE802.15.4 but also consider the reference of IEEE802.11. The work in [1], the author consider all the behavior and state of sensor to calculate the energy consumption in 802.11, and dynamic altered the routing path if the remain energy of sensor is low. The research in [2, 3] propose the energy consumption mode base on Markov chain. This model consider the multihop and error-prone channel condition to analysis the performance of 802.15.4. And the Fatma Bouabdallah [4] propose the energy consumption mode to calculate the energy consumption and transmission probability on sensor node, and use weight to altered the transmission path base on energy and quality of service.

In [5–9], authors introduced several types of attacks and the method against malicious attack. There has various attack in 6LoWPAN, such as Denial-of-service, wormhole and selective forwarding attack. Based on our research, the Dos attack is the major attack in both WSN and 6LowPAN. The Dos who launch we call the "jammer", the jammer may be is a simple device or a jamming station. And then, we will introduce four generic jammer modes:

- Constant jammer: This kind of jammer transmission large packets to occupies the channel, and make the channel busy cause the normal node can not transmission the packet.
- Deceptive jammer: Deceptive jammer transmission the packet with constant interval and without any gap.
- Random jammer: The random jammer change state between sleeping and jamming in random interval. It is good for the jammer that does not has unlimited power support. And this jammer is also difficult to be detected.

- Reactive Jammer: This jammer always sensing the channel. It always in idle mode when channel is free, and then transmission packet when it sensing there has someone transmission the packet.

The goal of above jammer is make the normal node always turn into receive mode and receive the packet, so the node will exhaust it energy.

There are few previous research works in detecting and against the malicious attack. The work in [9], the author use energy consumption mode to detect intrusion, and it use energy prediction model to identify the type of attacks. The concept of this paper is similar to our research. However, the proposed scheme in this paper based on an accurate energy consumption models in both route-over and mesh-under routing schemes to detect malicious nodes in 6LoWPAN. The goal of this research is to recognize malicious attacks by using energy consumption model compression which is not only based on the energy rising rate, but also take account the node's total energy consumption and traffic loads.

The research work [10] analysis the traffic load on sensor in network, and monitored the traffic load to detect whether is intrusion. Actually, only the sink node has the ability to monitoring all traffic loads, so the proposed detect scheme cannot put in the generic sensor node. Ponomarchuk et al. [11] analysis the energy consumption of sensor node attacked by Dos attack. From the result, it proves the energy consumption is increased when the node be attacked by Dos attack. The research work in [12], author analysis and compare the energy consumption with several Key cryptography in WSN, such as RSA-1024, SHA-1, AES. But actually in WSN and 6LoWPAN, put the key cryptography in node will reduce the data utilization. However, using the key cryptography will decrease the performance of sensor node.

The rest of the paper is organized as follows. Section Energy Prediction Model of 6LowPAN introduces energy prediction model for both route-over and mesh-under routing schemes in 6LoWPAN. Section The Lightweight Intrusion Detection Scheme Based on Energy Consumption Analysis describe the proposed lightweight energy prediction based intrusion detection scheme to recognize malicious attacks. Section Simulation Results shows the simulation results for the proposed scheme. Finally, Conclusion presents our conclusions and suggests the future work.

Energy Prediction Model of 6LowPAN

Energy Consumption Model of Node Transmission/Receive Process

Figure 1 shows the transmission process from source to destination nodes. First at all, the source node will go into a contention period T_C before the packet transmission. In the contention period, node will performed the CCA to check the channel condition. The node is going to transmission the packet when channel is

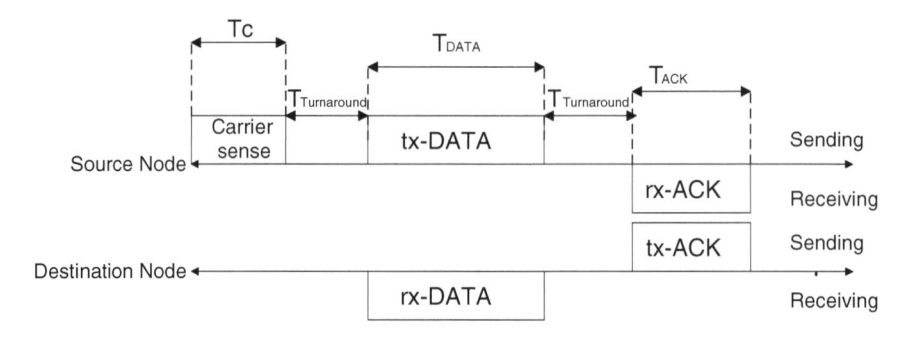

Fig. 1 Node transmission/receive process

free. Otherwise, it will turn to the backoff stage to wait for channel free. Node will wait for a turnaround time ($T_{turnaround}$) to change the mode from transmission to receive mode after the packet transmission. When the destination node receives a data frame correctly, it will send an ACK to indicate the successfly reception after $T_{turnaround}$.

The total energy consumption E_{total} is the total energy consumed in both source and destination nodes.

$$E_{total} = (1 - P_{tr}) \times E_{source} + E_{destination} \tag{1}$$

where, E_{source} is the energy consumption in source node, and the $E_{destination}$ is the energy consumption in destination node. $(1 - P_{tr})$ represents the packet success transmission probability. The E_{source} can be obtain from Eq. (2).

$$E_{source} = T_C \times E_{sensing} + T_{DATA} \times E_{tx} + 2 \times (T_{turnaround} \times E_{sensing}) \tag{2}$$

T_C, T_{DATA}, $T_{turnaround}$ and T_{ACK} are the contention time before node transmission, the time duration of transmission data, the time duration of node change state and the duration of node return ACK respectively. $E_{sensing}$, E_{tx} and E_{rx} are the energy consumption of sensing, the energy consumption of transmission the packets, the energy consumption of receive the packets. The $E_{destination}$ is given by:

$$E_{destination} = T_{DATA} \times E_{rx} + T_{turnaround} \times E_{sensing} + T_{ACK} \times E_{tx} \tag{3}$$

The $E_{destination}$ include the T_C, T_{DATA}, $T_{turnaround}$ and T_{ACK}. The T_{DATA} is obtained from:

$$T_{data} = \frac{Slots \times Slot_{DATA}}{250 \, \text{kbps}} \tag{4}$$

Slots is the number of slots in one packet. $Slot_{DATA}$ is the number of bits in one slot. One symbol period is 16 µs, and one slot is 20 symbols. Thus, one slot is carried 80 bits of data.

$$Slot_{DATA} = 20 \text{ symbol} \times 16 \text{ μs} \times 250 \text{ kbps} = 80 \text{ bits} \qquad (5)$$

The $T_{turnaround}$ is given by:

$$T_{turnaround} = (12 \text{ symbols} \times 4 \text{ bits})/250 \text{ kbps} = 192 \text{ μs} \qquad (6)$$

The contention time T_C is the duration of node performs CCA, it about 8 symbol which is around 128 μs.

Energy Consumption Model for Mesh-Under Routing Scheme

In the energy prediction model for mesh-under routing scheme, the energy prediction model for mesh-under is given by:

$$E_{mesh} = [(P_{tr}^{x+h}) \times x \times E_{total} + \delta \times E_{recombination}] + \sum_{K=1}^{3} K[(1 - P_{tr}^{x+h}) \times x \times \qquad (7)$$
$$E_{total} + \delta]$$

Equation (7) indicates the energy consumption of node when transmit x fragment packets for mesh-under routing scheme in 6LowPAN. P_{tr}^{x+h} is the probability of all fragments successful transmission in h hops count. h, x, k and δ represent the number hops count, the number of fragments, the time of retransmission, the process delay to reassemble all fragments respectively.

Energy Prediction Model for Route-over Routing Scheme

Equation (8) represents the energy consumption of the node for route-over routing scheme in 6LowPAN. Consider the characteristic of route-over, the energy model for route-over is given by:

$$h\{[(P_{trans}^{x}) \times x \times E_{tx-rx} + \delta] + \sum_{K=1}^{3} K[(1 - P_{trans}^{x}) \times x \times h \times E_{tx-rx} + \delta] + T_{frag}\}$$
$$(8)$$

P_{trans}^{x} is the probability of successful transmission of a fragment in single hop. T_{frag} is the delay time to process fragmentations. The major feature of route-over is hop-by hop fragmentation and reassembly. In each hop, all fragments will recover to a completed IPv6 packet. Thus, we can consider route-over scheme as hop by hop forwarding from source to destination.

The Lightweight Intrusion Detection Scheme Based on Energy Consumption Analysis

In the proposed intrusion detection scheme, all nodes have to morning its own total energy consumption by using our energy prediction models, and the sampling rate is 0.5 s. Base on our simulation results, we assume that node is regarded as attacked if the energy consumption rate is increased over 30 % (*Energy-Rise*$_{threshold}$) from pervious energy consumption record. The intrusion detection scheme will regard the node as malicious and remove the node from the route table in 6LowPAN.

The Intrusion Detection Scheme denote by Table 1.

Simulation Results

In this section, we present the energy consumption analysis for 6LoWPAN routing schemes. Our simulation was emulated by Qualnet [13]. The simulation parameters are shown in Table 2.

Table 1 Intrusion detection scheme

```
Computing Energy Consumption //Computing Energy Consumption of nodes
alarm = 0                          //initial alerm,count the times of attacked
if(Current Time − Pervious Sampling Time ≥0.5)
//0.5 s is the energy consumption sampling rate
{
    Calculate the Current Energy Consumption of node
    Energy consumption RiseRate = (Current Energy Consumption−
    Prediction Energy Consumption)/
    Prediction Energy Consumption × 100
    if (Energy consumption RiseRate > EnergyRise_threshold)
    //EnergyRise_threshold is 30 %,
    {
        alerm++
        if(alerm = = 2)
        //if there is two consecutive alarms, node is regarded as attacked and
        //remove the node from the route table in 6LowPAN.
    }
    else {
        alerm = 0
    }
}
```

Table 2 Simulation Parameters

IPv6 Packet Size	6,400 bytes
Number of Fragments	50
Hop Counts	2 hop
Routing	From node 1 to node 3 through node 2

The Energy Consumption of Normal Traffic Load and Dos Attack Traffic Load

The Fig. 2 is the energy consumption of node in normal and attack scenario from ideal channel. The node with normal traffic load transmits packet from 0.5 s, and the attack node attack from 2 s until normal node transmission end. The normal node and attack node transmission in interval 10 ms and 5 ms, respectively. We monitor the energy consumption each 0.5 s.

In Fig. 2, the energy consumption of node in normal traffic load is stable, and the node transmission end at 4 s. The node transmission end around at 18 s when node is under Dos attack. And the number of total received packets at receiver in normal and Dos attack scenarios are 50 and 0 respectively.

The Nodes Energy Consumption Sampling Rate in the Proposed Intrusion Detection

In this section, we detect the energy consumption every 0.5 s, and the detect threshold is 30 %. The intrusion detection scheme determined that the node has be attack when the energy increase rate over 30 %.

Figure 3 is the total energy consumption in normal and attack traffic loads. The Dos attack has the highest energy consumption 0.373 mJ per 0.5 s than normal

Fig. 2 Energy consumption between normal and Dos traffic loads

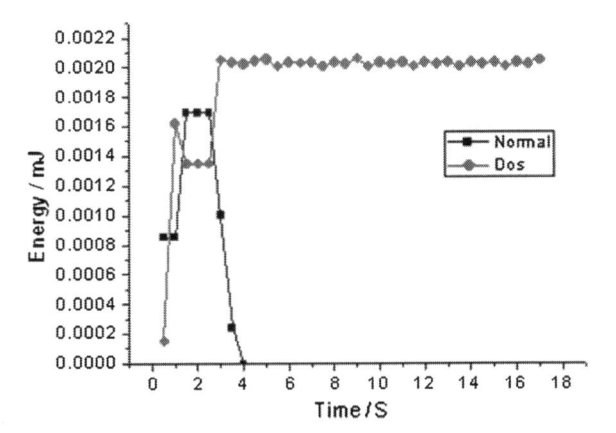

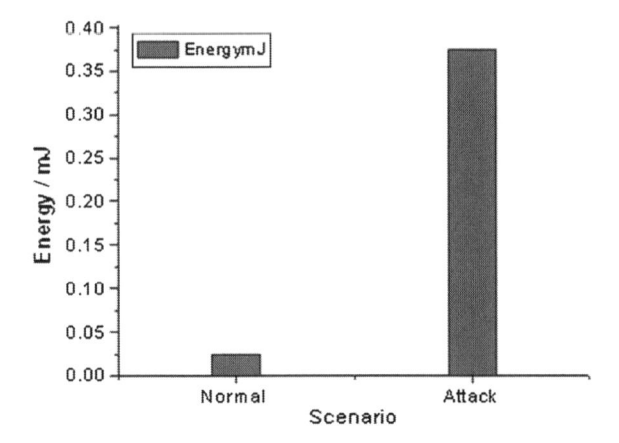

Fig. 3 The total energy consumption in normal and Dos attack traffic loads

traffic load 0.024 mJ/0.5 s due to the malicious node receives lots of attack packets. Furthermore, the probability of channel collision also increase make the retransmission times increase cause from the Dos attack in the malicious node.

Conclusion

The security issue has become more and more important in Internet of things. In this paper, a lightweight intrusion detection scheme for 6LoWPAN is proposed, which based on accurate energy consumption models to detect malicious nodes. The results show the proposed lightweight intrusion dection scheme is more efficient and accurate. From the smulation result, the energy consumption will receive large increasement from Dos attack, and the detection rate is 100 %.

Acknowledgments The authors would like to acknowledge the support from the National Science Council of Taiwan (No. 101-2119-M-142-001, 100-2221-E-142 -002) and National Taichung University regarding the MoE project (No. 1020035480A).

References

1. Lee TH, Marshall A, Zhou B (2005) A framework for cross-layer design of energy-conserving on-demand routing in multi-hop wireless networks. IEE Mobility
2. Performance Analysis of IEEE 802.15.4 MAC Protocol for WSNs in Burst Error Channels (ISCIT 2011)
3. Analytical Modeling of Multi-hop IEEE 802.15.4 Networks (2012). IEEE transactions
4. Bouabdallah F, Bouabdallah N, Boutaba R (2009) On balancing energy consumption in wireless sensor networks. IEEE Trans Veh Technol 58(6)

5. Manju VC, Kumar MS (2012) Detection of jamming style DoS attack in wireless sensor network. Parallel distributed and grid computing (PDGC). 2012 2nd IEEE international conference
6. Mpitziopoulos A et al (2009) A survey on jamming attacks and countermeasures in WSNs. Commun Surv Tutorials, IEEE 11.4(2009):42–56
7. Pelechrinis K, Iliofotou M, Krishnamurthy SV (2011) Denial of service attacks in wireless networks: the case of jammers. Commun Surv Tutorials, IEEE 13.2(2011):245–257
8. Modares H, Salleh R, Moravejosharieh A (2011) Overview of security issues in wireless sensor networks. Computational intelligence, modelling and simulation (CIMSiM), 2011 third international conference
9. Shen W et al (2012) A new energy prediction approach for intrusion detection in cluster-based wireless sensor networks. Green communications and networking. Springer, Berlin, pp 1–12
10. Ponomarchuk Y, Seo DW (2010) Intrusion detection based on traffic analysis in wireless sensor networks. 19th annual wireless and optical communications conference (WOCC)
11. Kim K, Hong J (2010) Analysis of power consumption of S-MAC protocol according to DoS attack. 4th IEEE international conference on new trends in information science and service science (NISS)
12. Wander AS et al (2005) Energy analysis of public-key cryptography for wireless sensor networks. Third IEEE International Conference on Pervasive Computing and Communications, PerCom 2005
13. QualNet simulator. http://www.qualnet.com/

A *k*-Cooperative Analysis in Game-Based WSN Environment

Hsin-Hung Cho, Fan-Hsun Tseng, Timothy K. Shih, Li-Der Chou, Han-Chieh Chao and Tin-Yu Wu

Abstract In wireless sensor networks (WSNs), the coverage problem usually accompanies the energy-saving issue, which ensures the fundamental functions are workable. Therefore, the sensor nodes must be deployed and survived in the target place for a long time, and sustain the trade-off between expected coverage and limited battery energy. The lesser sensor nodes are awake in an epoch, the longer network lifetime is achieved. It shows that the coverage problem must be solved based on the energy efficiency viewpoint. In this paper, we propose a game-based WSNs environment and solve the energy saving issue with the minimum number of competition players, which implies the most suitable duty-cycle for all sensor nodes. In the simulation results, the proposed approach achieves the lowest power consumption and longest network lifetime.

H.-H. Cho (✉) · F.-H. Tseng · T. K. Shih · L.-D. Chou
Department of Computer Science and Information Engineering, National Central University, Taoyuan, Taiwan, Republic of China
e-mail: hsin-hung@ieee.org

F.-H. Tseng
e-mail: fanhsuntseng@ieee.org

T. K. Shih
e-mail: timothykshih@gmail.com

L.-D. Chou
e-mail: cld@csie.ncu.edu.tw

H.-C. Chao · T.-Y. Wu
Institute of Computer Science and Information Engineering, National I-Lan University, I-Lan, Taiwan, Republic of China
e-mail: hcc@niu.edu.tw

T.-Y. Wu
e-mail: tyw@niu.edu.tw

H.-C. Chao
Department of Electrical Engineering, National Dong Hwa University, Hualien, Taiwan, Republic of China

Y.-M. Huang et al. (eds.), *Advanced Technologies, Embedded and Multimedia for Human-centric Computing*, Lecture Notes in Electrical Engineering 260, DOI: 10.1007/978-94-007-7262-5_138, © Springer Science+Business Media Dordrecht 2014

Keywords Wireless sensor network · Energy saving · Integer linear programming · k-cooperative game

Introduction

Since the battery capacity of sensor node is limited and finite, the power consumption issue is always a hot research topic in WSNs [1, 2]. Therefore, the power consumption is the most critical design factor. The duty-cycle is a well-known scheme for saving the energy consumption in entire network, because it limits the sleeping time of sensor nodes in each epoch. Owing to fractional duty-cycle values are too large, the setting of duty-cycle usually accompanies with the coverage issue. If a sleeping time of sensor node is too long, it may make the covered area decreasing due to the insufficiently awake sensor nodes in an epoch. Some researchers have proposed some mechanisms which have the ability to self-adjust duty-cycle value, such as Distance-based Duty Cycle Assignment (TDDCA) [3]. However, the network topology of WSN is usually randomly distributed that may occur various overlapping areas. In order to avoid the misjudgment, the Dynamic Duty-cycle Dynamic Scheduling Assignment (DDDSA) [4] is our afore-proposed scheme. In this work, we consider that the duty-cycle assignment in overlapping area includes the competed phenomenon, therefore the game theory is useful for such environment. Furthermore, the proposed game model is able to apply to any game in wireless networks. For details, we defined the number of links as the players in the game. Then, utilize the number of link to express the competed phenomenon with an unprecedented attempt. In the proposed k-cooperative algorithm, we successfully demonstrated the competed phenomenon, decreased more redundant links, and achieved energy efficiency without losing the communicated quality.

This paper is organized as follows. In Related Works, the related literatures are introduced, such as the TDDCA scheme, DDDSA scheme, and the game-based WSNs. One thing should be noticed that the DDDSA scheme and game-based WSNs is our fore-proposed methods. After that, we analyze the game-based WSNs and define a novel k-cooperative problem and solve it by the proposed k-cooperative algorithm in Game-Based WSNs Duty Cycle Assignment. The Simulation is the simulation result, and Conclusion and Future Work is the conclusion and future work of this work.

Related Works

Traffic-Adaptive Distance-Based Duty Cycle Assignment

The researchers [3] regarded the traffic relay is relative to the distance to the sink and they analyzed the performance of receiver-based mechanism through math models [5]. Receiver-based mechanism contains several elements such as an expected traffic rate and an assignable duty-cycle value. They assume the source sensor nodes are uniformly allocated around the central sink node with the transmission range r_T. The hierarchical concentric circles are composed of n rings, and the circle is $(n-1)r$ away from the sink node with width r_T, which is represented by the nth ring. The traffic generating by the source sensor nodes in the nth ring within the ring outside of the nth ring per unit time is expressed by Γ which is given in [3] as

$$\Gamma = \lambda_g \rho_s \pi (R^2 - [(n-1)r_T]^2), \tag{1}$$

The variable λ_g represents the average traffic generation rate, ρ represents the node density of source sensor nodes, and R represents the radius of the network area. A node in the $\frac{r}{r_T}$ ring with a distance r from the sink node and the number of sensor nodes in the nth ring is defined as

$$N_n = \rho_r \pi \left\{ (nr_T)^2 - [(n-1)r_T]^2 \right\}. \tag{2}$$

The average traffic rate [6] of a node is defined as

$$\lambda_r = \frac{\lambda_g \rho_s \pi \left\{ R^2 - \left[\left(\frac{r}{r_T} - 1 \right) r_T \right]^2 \right\}}{\rho_r \pi \left\{ \left[\left(\frac{r}{r_T} \right) r_T \right]^2 - \left[\left(\frac{r}{r_T} - 1 \right) r_T \right]^2 \right\}}. \tag{3}$$

In [3], the authors consider the transmission time is relative to the duty-cycle assignment. The higher duty-cycle can make the more sensor nodes available in an epoch and avoid the probability for relays to achieve a lower latency. However, the number of awake sensor nodes is increased and the number of idle sensor nodes also increased, which consume more energy. To overcome this problem, they improve the duty-cycle that minimizes the energy consumption for a given traffic rate. Based on CSMA protocol [7], a packet should be sent after it sends a request to send (RTS) packet and waits for the clear to send (CTS) packet from destination. Sensor nodes do not transmit or receive packets which called idle sensor nodes, and the power consumption of idle sensor nodes is approximately the same with active sensor nodes in WSNs [8]. Thus, given a constant value P to represent the idle listening, transmission, and reception. Assuming the topology is a Poisson distribution, and the expected time [5] before RTS/CTS handshake is

$$t_H = \left(e^{\xi dN} - 1 \right)^{-1} \left(T_{RTS} + N_p N_r T_{CTS} \right). \tag{4}$$

The T_{RTS} and T_{CTS} are the delay time for sending the RTS packet or CTS packet respectively.

The expected total time for a complete communication is defined as

$$t_C = T_{RTS} + \chi T_{CTS} + T_{DATA} + T_{ACK}, \tag{5}$$

and the total time consumption is $t_t = t_H + t_C$. The expected energy consumption $\bar{P}$ is defined as

$$\bar{P} \simeq \mathcal{P}(d + \lambda_r t_H) + \left(2 - e^{-\xi dN}\right)t_t. \tag{6}$$

The TDDCA is based on above-mentioned ideas that assign the duty-cycle value dynamically. In TDDCA, each node has a counter which records the numbers of the initial and retransmitted RTS packets. If received retransmitted RTS packets in the current epoch outweighs the total number of the received initial RTS packets that expresses the neighborhood of this node in a severe situation. At this moment, the duty-cycle of this node should be increased for reducing the traffic load. Otherwise, the duty-cycle should be subtracted 1 % to eliminate the power consumption.

Dynamic Duty-Cycle and Dynamic Scheduling Assignment

The researchers proposed the DDDSA scheme [4] to allocate the duty-cycle value and reduce the power consumption in WSNs. The proposed scheme improves the uncorrected information of sensor nodes within the overlapping area in TDDCA. They utilized the idea in TDDCA which using the ring to define the distance-based network, and modified the function of counter in each sensor nodes. The original counter records the number of received retransmission RTS packets. Unlike TDDCA, the difference between the previous and current received retransmission RTS packets is calculated in DDDSA. In order to deal the sensor nodes in the overlapping area, they defined the information from each neighbor as a priority. Then, they used this information as a priority to prioritize which sensor nodes need to be served first. If a sensor node must provide service for a low priority neighbor, it must postpone to a few seconds. If there is no higher priority sensor nodes need to be served, the sensor nodes in overlapping area start to work. Although the problem of overlapping area is solved, the duty-cycle value is not the most appropriate for each sensor nodes due to 1 % per stage. In other words, the issues of power consumption and traffic load are solved imperfectly.

Game-Based WSNs

The main concept of game theory is that the two or more players compete for the reward thus achieves a fair situation in whole environment [9]. In recent years, the researchers apply game theory to wireless networks [10]. In this paper, we model all of the duty-cycle values of sensor nodes as the players of game theory. The duty-cycle value is a decisive factor for the WSNs, because the power consumption is based on the number of awake sensor nodes. The sensor nodes whether enter into the sleep state must decide by the game, and each sensor nodes have to come up with their own information. It should be mentioned that the game in this work only happens in the overlapping area, and each sensor node owns an initial duty-cycle value. The sensor nodes cannot adjust their duty-cycle value arbitrarily in order to maintain the fairness. Since the summation of changed duty-cycle value is not zero, it belongs to the non-zero-sum game in game theoretic. In brief, the game-based WSNs model is responsible for coordinating the duty-cycle vale of sensor nodes within the overlapping area.

Game-Based WSNs Duty Cycle Assignment

The non-zero-sum game represents that the profits of players are not equal to zero, which could be a win–win result. In our environment, the main object is decreasing the power consumption and extending the network lifetime by controlling duty-cycle value. The influence of duty-cycle value covers the multi-hop situation, even the entire network. It implies that a player not only considers its cost but also the cost of its neighbors. According to the duty-cycle value is a regional variable, each player just concerns the current information from itself and its neighbors. If we want to set the adjusted duty-cycle value as a reward to the winner, we must consider its fairness. Because we assure the duty-cycle value will not affect the other player's game. Therefore, we must find out the number of players for each round of the game. The follow-up two subsections, we will introduce what is the *k*-cooperative, and clearly define the proposed problem. For better readability, we have listed all the symbols in Table 1 before next subsection.

Outline of k-cooperative

We assume the duty-cycle assignment problem is a *k*-cooperative problem, and the constant *k* represents the numbers of player who joined this subset of game. They are responsible for the most suitable duty-cycle value from the competition in this game subset. The proposed *k*-cooperative problem is similar to the well-known *k*-

Table 1 Definition of the improtant notations

Variable	Definition
h	The set of the received RTS from its neighbors
L	The set of the link between two sensors
C	The set of the traffic information which through the such sensor nodes
X	The set of the same link from its neighbors
λ_μ	The average traffic of a node in overlapping area u
T	The threshold for the traffic

connectivity problem [11]. The concept of k-connectivity problem is simply defined as follows. Given a graph $G = (V, E)$, N is a set of minimum connected k-neighbors, and $N \subseteq V$. In short, there are least k neighbors for a node in such network topology, and this node obtains information from these k neighbors only. In game-based WSNs, only one node and its neighbors joined a round game, and the entire game model is composed of several rounds. Depend on above definition, we may encounter some problems: (1) The node obtained duty-cycle value may imbalance for the global network topology; (2) For a high-density environment, the node has a large number of neighbors so that it leads the game model operation become inefficient. In order to solve these problems, we try to limit the constant k so that the imbalance of duty-cycle value will be decreased and maintained a certain level of performance. The problem should be abided by the principle that each node has a relation with at least one node. Try to decrease the constant k from a game, thus the sum of k of all game is the minimum that achieved by greedy method is our major goal.

Definition of k-Cooperative Problem

In this subsection, the clear definition of k-cooperative problem is stated and introduced. First of all, we consider the constant k is a crucial factor for the network performance. Because there may be two or more neighbor nodes in the coverage of one sensor node, this node is able to acquire the information from its neighbors. Moreover, we can use fewer sensor nodes to complete a round of the game. However, we have to find out the most appropriate value of k such that the game will be carried out smoothly and without losing fairness.

We have to define a correlation value $X_{i,j}$ to ensure each node has least one correlation with its each neighbor. The correlation value also affects the connection relation of whole topology directly. If it is well defined, the repeated computing sensor nodes will become less. However, every sensor nodes still maintain an indirect connection relation with each node. The indirect connection relation is that a node is the one hop and also n-th hops neighbor of another node in the same time. For example, when a node A has two neighbors B and C, and the neighbor B has a neighbor D which is also a neighbor of node A. Herein, the neighbor B is the one hop and aslo two hops neighbor of node A in the same time. Wherein above situation, the value $X_{A,B}$ is equals to 2. If the $X_{A,B} \geq 1$, it represents that there is a connection relationship between these two sensor nodes. In order to decrease the operation cost, we try to let the $X_{i,j} = 1$ for every two sensor nodes. If we achieve this goal, it means that every node joins one game only. However, a node has some neighbors which correlation value $X_{i,j} > 1$, this value will be an indicator filtering out a part of the sensor nodes which have larger value, because they have more connection chance to join other game. We can use the set theory to find out the correlation value as follow.

$$X_{i,j} = |N_i \bigcap N_j| \tag{7}$$

$N_{i,1} = \{n_1, \ldots, n_k\}$ is the set of one hop neighbors of *i*st sensor nodes, and $n_j \in N_{i,1}$. The $N_{j,1}$ is the set of one hop neighbors of *j*st sensor nodes, and they are also two hops of N_i, we denoted by $N_{i,2}$.

In order to solve this problem, we formulate the *k*-cooperative problem based on the Integer Linear Programming (ILP). The ILP for the *k*-cooperative problem is defined as follows.

Minimize

$$\sum_{i=1}^{n} k_i \tag{8}$$

Subject to

$$k_i \geq 1 \tag{9}$$

$$X_{i,j} > 0 \tag{10}$$

$$\lambda_\mu \geq 0, \text{ for } \forall u \in \{0, 1\} \tag{11}$$

$$0 \leq x_i \leq 1, \text{ for } \forall i \in V \tag{12}$$

k-Cooperative Algorithm

Before introducing the *k*-cooperative algorithm, we firstly design an initialization algorithm to create the basic network model and let each sensor nodes obtain the required information. Since we need both of the neighbor's and two hops neighbor's information, the two times broadcasts are unavoidable. Firstly, each sensor node broadcasts the own information to its neighbors, so that any sensor nodes have the required information of its neighbors in this moment. Secondly, each sensor node broadcasts again, but its contents have been substituted with the neighbors' information rather than the own information. Then, any sensor nodes not only have one hop neighbor's information but also have two hops neighbor's information via the two times broadcasts. In our proposed method, we consider that each link include a relationship with a neighbor. Hence, we use of these results to calculate the intersection of one hop neighbors and two hops neighbors. These results represent the cooperative relationship between each sensor nodes and such results can let the *k*-cooperative algorithm operate smoothly. The pseudo code is shown in Table 2.

Table 3 is the *k*-cooperative algorithm which also the main idea of our mechanism. Although our motivation and goal is the minimum cost, the maintenance of link quality is also an important point to ensure the execution of WSNs. Besides, this research is based on the distance-based WSNs and the game only happens in the

Table 2 Initialization of k-cooperative algorithm

k-cooperative *Initialization*
1. Broadcast RTS to its neighbors, and receive RTS from its neighbors
2. Broadcast h
3. Find out the intersection of h from each 1-hop neighbors then save them to $X_{u,n}$
4. Next we calculated the same neighbors for u that are based on the intersection of $X_{u,n}$ then save the traffic information which through the such sensor nodes to C
5. Run the k–cooperative algorithm
Run the following at each node u

overlapping area. We still consider the average traffic λ_μ of a node in overlapping area u as an initial condition for k-cooperative algorithm. If the λ_μ is higher than the threshold T, we compare the link situation $X_{\mu,n}$ by calculating intersection h. A link between two sensor nodes will be removed if these two sensor nodes have another path to communicate with each other. If a $L_{\mu,a}$ of $X_{\mu,n}$ that the sensor node a has maximal number of received RTS, the $L_{\mu,a}$ will be removed due to the enough neighbors of sensor node a. In the other case, if two arbitrary sensor nodes only have one same link, then we use of the traffic as a metric to decide which link must be removed. A simple assumption is that if the lower traffic through a sensor node, this sensor node does not need more link to relieve traffic, which shown as step 12. It represents that if more traffic through a sensor node, it needs more neighbors to play the game in order to coordinate out an optimal duty-cycle value.

Simulation

Parameters Setting

In order to check the performance of k-cooperative algorithm, we conduct extensive simulations in MATLAB [12]. The network size is 500 (meter) multiplies 500 (meter), which also equals to 250,000 (square meter). The number of deployed sensor nodes ranges from 100 to 1,000. Moreover, we compare the packet delivery ratio within the threshold T from Γ to $\Gamma/3$ in Fig. 3. The initial sensing radius of sensor node is 60 (meter) and initial power is 200 mAh. The energy consumption of a sensor by transmitting, receiving one byte data are 0.0144 and 0.00576 (mJ) [13].

Simulation Results

In Fig. 1, we compare the power consumption of all sensor nodes. It is clear that the higher number of sensor nodes leads to more power consumption, and the gap between original game and k-cooperative vice versa, because more sensor nodes

Table 3 *k*-cooperative algorithm

k-cooperative algorithm
01 *If* $\lambda_\mu > T$
02   *If* $\mid X_{u,n} \mid > 1$
03   *for* $i = 1{:}n$
04   *for* $u = 1{:}\mid X_{u,i} \mid$
05   *If* $\mid h_{index} \mid = Max(X_{u,i}(\mid h \mid))$
06   *Remove* $L_{u,\ index}$ *of node u*
07   *end if*
08   *end for*
09   *end for*
10   *else if* $\mid X_{u,n} \mid = 1$
11   *for* $u = 1{:}\mid X_{u,i} \mid$
12   *If* $\mid C_{index} \mid = Min(X_{u,i}(\mid C \mid))$
13   *Remove* $L_{u,\ index}$ *of node u*
14   *end if*
15   *end for*
16   *end if*
17 *end if*

Run the following at each node u

bring up more links. However, we can decrease the number of links without losing the availability of the game via *k*-cooperative algorithm. It is well known that the consumption of power is based on the initialization of link establishing and the process on those links. The result shows the effectiveness of our proposed algorithm.

The Fig. 2 shows the simulation result of the network lifetime between the original game-based WSNs and *k*-cooperative WSNs. According to the result of Fig. 1, the number of links affects the length of network lifetime directly. This result represents that the decreased links achieve lower power consumption, which saves more power average on every nodes.

Fig. 1 Comparison of power consumption between original game-based WSNs and *k*-cooperative WSNs

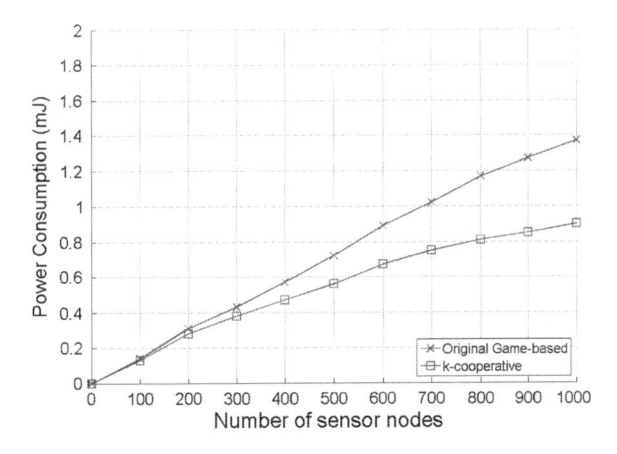

Fig. 2 Comparison of
network lifetime between
original game-based WSNs
and *k*-cooperative WSNs

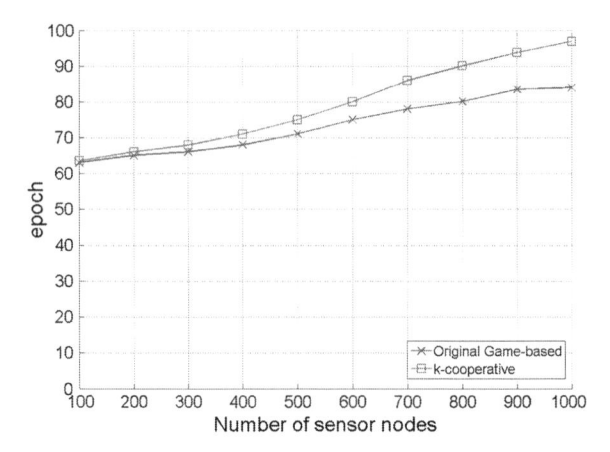

Fig. 3 Comparison of packet
delivery ratio between
original game-based WSNs
and *k*-cooperative WSNs

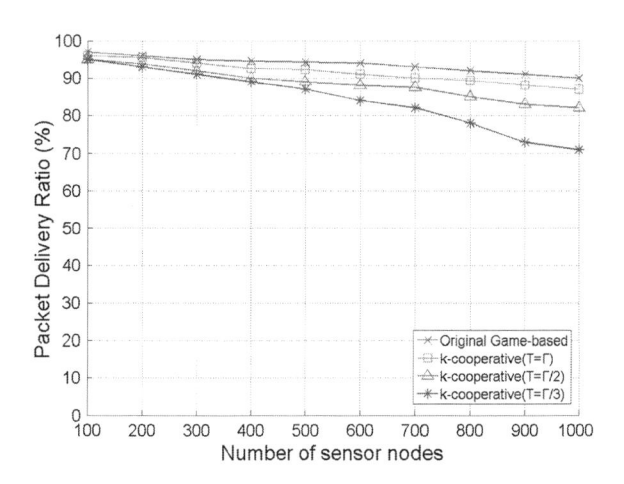

The Fig. 3 reflects the packet delivery ratio between original game-based
WSNs and *k*-cooperative WSNs. In this simulation setting, we investigate the
threshold T, because this parameter is a key factor for our proposed algorithm. If
the threshold T is well-defined, the result between original game-based WSNs and
k-cooperative WSNs are almost the same $(T = \Gamma)$. Obviously we see that the
smaller the threshold $(T = \Gamma/3)$, the more redundant links will be removed that
there are not enough links to relieve traffic. On the other hand, if we set a bigger
threshold T, some redundant links will not be removed so that the better packet
delivery ratio is acquired. Thankfully the result just equal to the original game-
based WSNs, because our proposed method is a subset of the original game-based
WSNs.

Conclusion and Future Work

In this paper, we proposed a novel method viz *k*-cooperative algorithm to decrease the redundant links and eliminate from energy exhaustion. We not only consider to the energy efficiency issue but also ensure the communication quality which only 2 % of the gap with original game-based WSNs. The simulation results show that the proposed algorithm achieves the lower power consumption and longer network lifetime, with slightly influence on packet delivery ratio only. In the future work, the exquisite analysis of threshold T will be investigated and studied to achieve the better performance.

Acknowledgments This research was also partly funded by the National Science Council (NSC) of the Taiwan under grants NSC 101-2221-E-197-008-MY.

References

1. Aziz AA, Sekercioglu YA, Fitzpatrick P, Ivanovich M (2012) A survey on distributed topology control techniques for extending the lifetime of battery powered wireless sensor networks. IEEE Commun Surv Tutorials 15(1):121–144
2. Suhonen J, Kohvakka M, Kaseva V, Hämäläinen TD, Hännikäinen M (2012) Low-power WSN technology," in Low-power wireless sensor networks, chap. 1, Springer US, pp 1–6
3. Zhang Y, Feng CH, Demirkol I, Heinzelman WB (2010) Energy-efficient duty cycle assignment for receiver-based convergecast in wireless sensor networks. In: Proceeding of the IEEE global telecommunications conference (GLOBECOM), Miamim, Florida, USA, pp 1–5
4. Cho H-H, Chang J-M, Chen C-Y, Huang S-Y, Chao H-C, Chen J-L (2011) An energy-efficient dynamic duty-cycle and dynamic schedule assignment scheme for WSNs. In: Proceeding of the IEEE Asia-Pacific services computing conference (APSCC), Jeju, Korea, pp 384–388
5. Zorzi M, Rao RR (2003) Geographic random forwarding (GeRaF) for Ad Hoc and sensor networks: multihop performance. IEEE Trans Mob Comput 2(4):337–348
6. Merlin CJ, Heinzelman WB (2010) Duty cycle control for low-power-listening MAC protocols. IEEE Trans Mob Comput 9(11):1508–1521
7. Ziouva E, Antonakopoulos T (2002) CSMA/CA performance under high traffic conditions: throughput and delay analysis. Comput Commun 25(3):313–321
8. Crossbow Technology. Available at http://www.xbow.com/. Accessed 2 July
9. Politis C, Dixit S, Lach HY, Uskela S (2004) Cooperative networks for the future wireless world. IEEE Commun Mag 42(9):70–79
10. Brown DR, Fazel F (2011) A game theoretic study of energy efficient cooperative wireless networks. J Commun Networks 13(3):266–276
11. Penrose MD (1999) On k-connectivity for a geometric random graph. Wiley Random Struct Algorithms 15(2):145–164
12. MathWorks. Available athttp://www.mathworks.com/. Accessed 2 July
13. Liang W, Chen B, Yu XJ (2008) Response time constrained Top-k query evaluation in sensor networks. In: Proceeding of 14th IEEE international conference on parallel and distributed systems (ICPADS), Melbourne, Australia, pp 575–582

An Enhanced Resource-Aware Query Based on RLS-based SIP Presence Information Service

Jenq-Muh Hsu and Yi-Han Lin

Abstract Many innovative services over the standardized SIP-based IMS signaling infrastructure have been widely deployed in the next-generation converged networks. The presence is a key feature for designing context-ware network service. It can efficiently reflect resource availability through subscription and notification mechanisms. Thus, it can apply the presence service to design a resource list server (RLS) to collect a set of resource information, such as resource availability and allocation in order to provide an efficient resource reporting for resource subscribers. In this paper, an improved RLS-based Resource-Aware Query Routing scheme is proposed, in which each server can obtain its neighbors' utilization states notified from RLS. Besides, the proposed scheme is capable of minimizing message traffic between local presence server and its neighboring presence servers while maintaining full responsiveness for the resource subscribing management.

Keywords Resource list server · Resource-aware query · Presence service

Introduction

With the rapid development of mobile services, mobile service providers will shortly deploy the SIP-based [1] mobile services based on IP Multimedia Subsystem (IMS) [2] framework for next-generation services. Such services can

J.-M. Hsu (✉)
Department of Computer Science and Information Engineering,
National Chiayi University, Chiayi, Taiwan, Republic of China
e-mail: hsujm@mail.ncyu.edu.tw

Y.-H. Lin
Department of Computer Science and Information Engineering,
National Chung Cheng University, Chiayi, Taiwan, Republic of China
e-mail: lyha94@cs.ccu.edu.tw

Y.-M. Huang et al. (eds.), *Advanced Technologies, Embedded and Multimedia for Human-centric Computing*, Lecture Notes in Electrical Engineering 260, DOI: 10.1007/978-94-007-7262-5_139, © Springer Science+Business Media Dordrecht 2014

exploit the joint presence information between a consumer and a vendor, such as a subscriber (i.e. SIP entity requesting a service) and a notifier (i.e. SIP resource providing a specific service). Presence information is a collection of contextual attributes within a discovery of available and suitable services. Service discovery has been a topic of research and standardization activities for quite many years. It also allows to providing a service by appropriately composing the service functionality based on the other available service functionality. Hence, service discovery is an essential piece within a service provisioning architecture.

Presence is a key feature of the context-aware applications. Presence-based applications typically leverage upon contextual information such as location, availability, schedule, and local information, such as "Gas Station Finder" or "Restaurant Finder" services, for providing the nearest available resource information to the requesters based on exploiting the presence attributes of requesters or of known end-points.

Presence information actually is stored across a distributed set of presence servers due to the issues of information locality. That is, the queries satisfying the requesting predicates on presence attributes are often routed to multiple servers finding suitable resources for requesters. Thus, an efficient context-aware query mechanism is needed to effectively find the resource information for satisfying the resource requirement from requesters. This paper adopts the features of Resource-Aware Query Routing (RAQR) [3] to proposed a LRS-based RAQR for efficiently query the resource information among distributed SIP presence services.

The rest of the paper is organized as follows. Related Work briefly introduces the related work. The proposed RLS-based resource-aware query routing is presented in The RLS-Based RAQR Scheme for LRF Applications. Experimental Result shows the experimental results and performance evaluation. Finally, a conclusion is made in Conclusions.

Related Work

A minimization of unnecessary notification traffic in IMS presence system is presented in [4] for reducing the SIP message traffic and enhancing the SIP-based service efficiency. In which, an optional resource list server (RLS) can manage all subscriptions of presentities on a resource list to package and forward the presence information of all presentities in bundles to all authorized watchers according to their subscription preferences. Therefore, it can efficiently reduce the SIP messages among user agents and application services through the packed bundle of the presence information. In addition, SIP event throttling [5] is also efficiently reduced the SIP traffic messages. It uses the RLS connecting a set of event resources and applies a throttle mechanism to limit the ratio of SIP event notification among the RLS and watcher applications. Therefore, event throttling mechanism can reduce the number of event notification messages without increasing the message sizes. However, a water application initiates the event throttling mechanism. It will lose the frequently-updated information due to the message throttling. A proper throttling rate of SIP message events will be acted as a trade-off between message reduction and information loss.

Presence-based application typically leverage upon context-aware attributes, such as location, availability, schedule, and local information. It can provide the nearest and available resources or services to the mobile users according to the contextual information of requesting users and responding context-aware information from presence application. They are generally called Live Resource Finder (LRF) applications [3]. The innovation of LRF application can increase the matching degree of service requests according to the location of resources and other contextual attributes.

Resource-Aware Query Routing scheme for LRF-based applications is proposed in [3]. It uses the spatial-temporal distribution and consumption of resources and forwards the LRF queries to alternate presence servers. It can efficiently avoid the flooding problem of query and presence information and uses the Quality-of-Response (QoR) metrics to choose the proper presence server to match the LRF query for responding the available resource according to the current contextual information. For absorbing the advantages of RAQR scheme, this paper adopts the features of RAQR scheme to propose an enhanced resource-aware query based on RLS-based SIP presence service for efficiently improve the query qualities of resource requesting for mobile users.

The RLS-Based RAQR Scheme for LRF Applications

In this section, the proposed RLS-based RAQR scheme is illustrated by using SIP events for LRF applications. The main feature of RLS-based RAQR scheme is the utilization of state notification on a Presence Server (PS) list that points to the presence servers having a surplus of available resources. Resource utilization state is indicated that it facts a crunch state in its local resources. A presence server also maintains additional information, such as information expiration for up-to-dating the newest presence information. In addition, each RLS is associated with the other presence server on a PS list and tracks the resource utilization states of presence servers located in the same domain for evaluating to select the proper available resources.

For obtaining the PS list, the available supplier spool of each presence server should be large enough. Thus, the proposed RLS-based RAQR employs a simple SIP event notification where the originating presence server (the one facing the crunch) delivers a SUBSCRIBE message (informing its need for a certain resource) with expiration time with zero value, which is sent to the local RLS (managing all subscriptions to presence servers on a PS list). The local RLS also subscribes to the state of the other non-local presence server, and thus SUB-SCRIBE message is forwarded to the remote RLSs in different domains. On receiving the SUBSCIBE responses, the local RLS will uses the QoR values of the presence servers to determine the proper presence servers that can satisfy the resource load of its request. The local RLS finally responses to the originating server with a NOTIFY message, offering a list of presence servers containing the information of available resources.

Figure 1 shows the SIP message flow used in RLS-based RAQR scheme. The RLS relieves the watcher from subscribing to and managing notifications for all addresses in its contact list. Upon the PS list stored on local RLS, the RESERVE message is issued to PS1 and the utilization state of PS2 is set to SURPLUS in the same domain. The local RLS also subscribes the states of the other non-local presence servers. And then SUBSCRIBE message with an asking rate denoting the additional resources required by the user is forwarded to the remote RLS in different domains. Since PS1 and PS2 fall into the surplus state with available resources, they respond with NOTIFY messages which responding rates offered by r1 and r2 respectively. Remote RLS provides the PS list in which the available supplier pool of each server is large and enough used by another NOTIFY message. In the local RLS, while receiving the SUBSCIBE responses, it exploits the QoR values of presence servers to determine the proper set of presence servers satisfying the load of resource requests. Finally, the subscriber will receive a NOTIFY message with a list of presence servers indicating the available resources.

Experimental Result

In this section, the experiments are performed to evaluate the performance of the proposed RLS-based RAQR scheme [6]. For evaluate the performance of RLS-based RAQR relative to alternative available resource discovery manners, it performs extensive simulation-based studies.

In the experiment, an overlay network of 25 presence server were connected another ones generated by BRITE (http://www.cs.bu.edu/brite/). Each RLS manages all subscription to five presence servers. The coverage area of the entire service provider is represented by a $(10,000 \times 10,000)$ grid and the presence server manages an identically-sized partition of the grid. If a particular service is matched to a customer's query, its state will be changed from "available" to "unavailable" during the service time. At the end of the duration, the service' state will be resumed to available at the presence server which originated the query.

Parameter settings in the experiment are depicted in Fig. 1. For the resources, it assumes that upper threshold T_H is $20T$, lower threshold T_L is $10T$ and resource distribution follows an exponential distribution with a mean of 25. All periods are expressed as a factor of T, where T corresponds to the length of a clock tick in the simulation.

Table 2 shows the experimental result of the simulation setting illustrated Table 1. Total number of query divided by number of presence servers equals to the average arrival query per presence server. Thus, average arrival query per

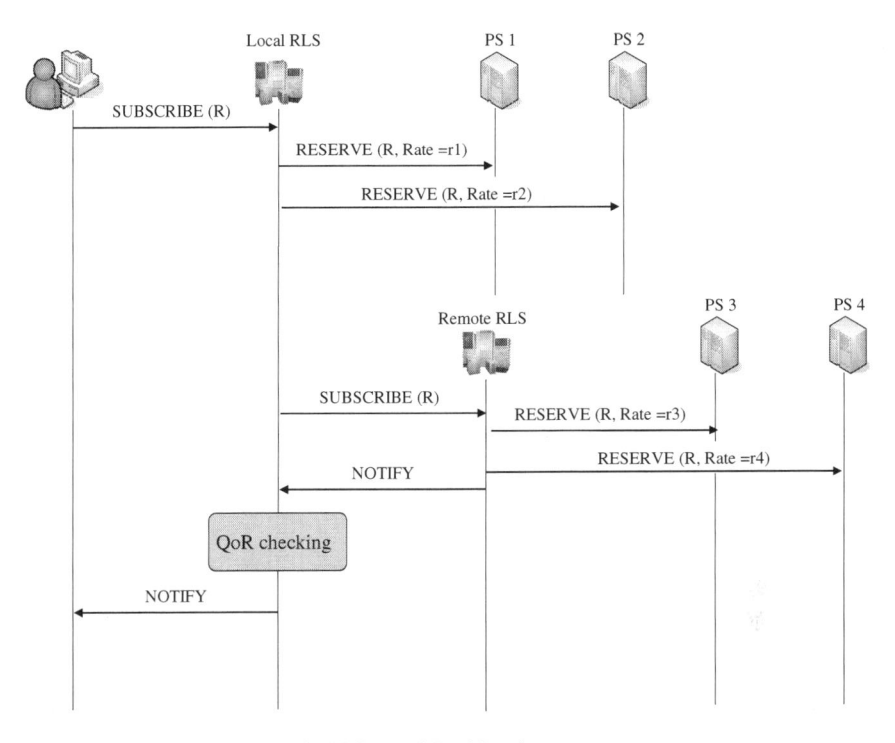

Fig. 1 SIP message flow used in RLS-based RAQR scheme

Table 1 Parameters setting of simulation

Parameter	Value
Number of presence server	25
Number of domain	5
Total number of resources	641
Mean of resource distribution	25
T_L	10
T_h	20
Mean of inter arrival time per query (λ_{qat})	1
Mean of service time per resource consumption (λ_{rs})	0.5
End time	60
Mean of the number of resource consumption per query (λ_{noorr})	10

presence server is 61.12. That is, arrival query per presence server is also equal to the value that λ_{qat} multiplied by end time is 60.

In the experiment, two discovery manners querying available resource are used to compare the difference of evaluating performance: RLS-based RAQR which LRF queries are forwarded from RLS to alternate presence server and RAQR without RLS which LRF queries are routed from originating PS to alternate presence server. The result indicates that RLS can be used for reducing signaling load since the

Table 2 Experimental result

Result	RAQR without RLS	RLS-based RAQR
Message number	1,731	636
Total number of query	1,528	1,528
Number of query miss	266	167
Query miss rate	0.17	0.11
Delay time	145.76	56.94

Presence User Agent (PUA) subscribes to the resource list on the RLS. Instead of subscribing to all members in a list, the PUA can subscribe to the PS list that conveys utilization states of individual PS. Hence, the reducing number of SIP messages in RLS-based RAQR will be better than which in RAQR without RLS. A missed query is occurred while there are no available resources locally and with a resource crunch. The result indicates that RLS-based RAQR leads to the more query matches better than RAQR without RLS. As the same result, the delay time of resource requesting in RLS-based RAQR is also much few better than RAQR without RLS.

Conclusions

In this paper, an enhanced resource-aware query based on RLS-based SIP presence service is proposed. The main feature of the proposed scheme utilizes the state notification in a PS list reflecting the resource state of PS server whether it has sufficient available resources or not.

The main advantage of adopting RLS is to package the presence information of all presentities in bundles and forward it to all authorized watchers. Thus, the RLS-based RAQR scheme is capable of minimizing message traffic between local presence server and its neighboring presence servers while maintaining full responsiveness for the resource subscription management. For roaming users, the number of messages between visited and home network will be effectively reduced for packing the various individual notification messages into a packaged notification message.

In the future, additional contextual attributes will be considered, such as localized information, resource popularity and user preferences, in our proposed scheme to enhance the computing of resource selection accurately.

References

1. Rosenberg J, Schulzrinne H, Camarillo G, Johnston A, Peterson J, Sparks R, Handley M, Schooler E (2002) SIP: session initiation protocol. IETF RFC 3261. Available at http://www.ietf.org/rfc/rfc3261.txt
2. Zhuang W, Gan YS, Loh KJ, Chua KC (2002) Multi-domain policy architecture for IP multimedia subsystem in UMTS. In: The IFIP TC6/WG6.2 and WG6.7 conference on network control and engineering for QoS, security and mobility, vol 235. pp 27–38

3. Chakraborty D, Dasgupta K, Misra A (2006) Efficient querying and resource management using distributed presence information in converged networks. In: The 7th international conference on mobile data management (MDM'06)
4. Wegscheider F (2006) Minimizing unnecessary notification traffic in the IMS presence system. In: The 1st international symposium on wireless pervasive computing
5. Niemi A (2007) Session initiation protocol (SIP) event notification extension for notification throttling, IETF Draft. Available at http://tools.ietf.org/id/draft-niemi-sipping-event-throttle-05.txt
6. Lin YH (2007) Efficient resource subscribing management using distributed presence information in converged networks. Mater Thesis, National Chung Cheng University

Part XIV
Networking and Applications

Authentication of Real-Time Communication System Using KIS Scheme

Binayak Kar and Eric Hsiao-kuang Wu

Abstract In global communication environment, signature computation will be frequently performed on a relatively insecure device that cannot be trusted all times to maintain the secrecy of the private key. To deal with this, Dodis et al. [1] proposed a strong key-insulated signature schemes whose goal is to minimize the damage caused by secret-key exposures. This environment will become more important when we focus on real time communication like telephony, TV shopping, electronic voting etc. Any flaws in the authentication system cause a critical damage to the real time environment. Considering this scenario we proposed a KIS scheme based on elliptic curve cryptography, which minimizes the damage of key exposer. Its security is based on elliptic curve discrete logarithm problem (ECDLP) assumption, and efficient in terms of computational cost and signature size.

Introduction

Real-time systems (RTS) [2] are systems, intended for the interaction (in the first turn, for control) with real physical objects and this interaction must run at a real physical time, in those temporary scales, in which live these objects. There are distinguished Soft-RTS (violation of timing restrictions in some range does not lead to the system failure), Hard-RTS (violation of timing restrictions leads to the system failure) and Firm-RTS (violation of timing restrictions in some range does not lead to the system failure with certain probability).But, communication is a

B. Kar (✉) · E. H. Wu
Department of Computer Science and Information Engineering, National Central University, Jhongli, Taiwan
e-mail: binayakar@wmlab.csie.ncu.edu.tw

E. H. Wu
e-mail: hsiao@csie.ncu.edu.tw

Y.-M. Huang et al. (eds.), *Advanced Technologies, Embedded and Multimedia for Human-centric Computing*, Lecture Notes in Electrical Engineering 260, DOI: 10.1007/978-94-007-7262-5_140, © Springer Science+Business Media Dordrecht 2014

very important issue for RTS. Consider any of the real time communication service like telephony, online shopping, TV shopping, electronic voting or any broadcasting services. We are handling all these communication by some insecure devices. Several Signature scheme has been proposed for the secure authenticated communication by using several cryptographic mechanisms. But these insecure devices cannot be trusted all times to maintain the secrecy of the private key of the scheme. Dodis et al. [1] proposed a Key-Insulated signature scheme whose main goal is to minimize the damage caused by the secret key exposes.

Considering various real-time communication services [3], we have designed a key-insulated signature scheme that can provide a secure authentication service. In which if the signing key of the signer is leaked still the adversary cannot generate the original signature as the signing key is valid for particular time period and after that it is get updated. The verifier can check the validity of the signature using the same verification key. The verification key is not necessary to update regularly as the signing key; this reduces the communication overhead. Because when the signer update the signing key, if it will update the verification key then it has to transmit the verification key to the receiver followed by the certificate revocation list. Then the receiver will access the certificate authority to check the validity of the verification key. Strong key-insulated signature scheme facilitate the mechanism to update the signing key without updating the verification key. We have used elliptic curve cryptosystem to design our algorithm; as an elliptic curve group could provide the same level of security afforded by an RSA-based system with a large modulus and correspondingly larger key; e.g. a 256-bit ECC public key should provide comparable security to a 3,072-bit RSA public key and the size of a DSA public key is at least 1,024 bits, whereas the size of an ECDSA public key would be 160 bits [4].

The remainder of this paper is organized as follows. Related works are presented in Related Works. Mathematical Primitives describes mathematical primitives for building our protocol. Basic Model of KIS Scheme presents the basic model of KIS scheme. We will describe the new KIS scheme in KIS Scheme for Real-Time System. We give the correctness proof of our proposed scheme and analyze the performance and security in Analysis of the New KIS Scheme. The paper is concluded in Conclusions.

Related Works

Key-insulated cryptosystems [5] focused on the case of public-key encryption. The goal of key-insulated security is to minimize the damage caused by secret-key exposures. Now in this electronic era digital signing is at the heart of Internet based transactions and e-commerce. In this global communication environment, signature computation will be frequently performed on a relatively insecure device (e.g., a mobile phone) that cannot be trusted to completely (and at all times) maintain the secrecy of the private key. In [1] Dodis et al. focused on key-leakage of digital signatures. They proposed a strong key-insulated signature scheme. In which they construct strong $(N - 1; N)$-key-insulated schemes based on any trapdoor signature scheme; that leads to very efficient solutions based on, e.g., the RSA assumption in the random oracle model. In [6] Deleito et al. proposed a new strong and perfectly key-insulated signature scheme, more efficient than previous proposals and whose key length is constant and independent of the number of insulated time periods. It is a forward-secure scheme in which an adversary needs to compromise an user at a second time period before being able to compute future secret keys.

Broadcasting and communications networks are used together to offer hybrid broadcasting services which incorporate a variety of personalized information from communications networks in TV programs. These services have many different applications that run on user terminals. Malicious service providers might distribute applications which may cause user terminals to take undesirable actions. An environment is necessary where any service provider can create applications and distribute them to users. In [7] proposed a protocol in which, a broadcaster distributes a signing key to a service provider that the broadcaster trusts. As a result, users can verify that an application is reliable. Another application of the KIS scheme is bidirectional broadcasting service [8, 9] in which a signer communicates with a huge number of receivers. In which, to renew the verification key, the signer has to send his new verification key to all receivers in an authentic manner. An efficient strong key-insulated signature scheme is proposed in [10]. The scheme is more efficient and is secure under the discrete logarithm assumption in the random oracle model. Traditional identity-based signature schemes typically rely on the assumption that secret keys are kept perfectly secure. But more and more cryptographic primitives are deployed on insecure devices such as mobile devices, key-exposure seems inevitable. It does not matter how strong the scheme is, once the secret key is exposed, it will be broken completely. To deal with this problem in [11, 12] use the key-insulation mechanism to minimize the damage of key-exposure in IBS schemes. Certificate-based key-insulated signature scheme [13] is proven to be existentially unforgeable against adaptive chosen message attacks in the random oracle model. Identity (ID)-based key-insulated cryptography has received much attention from cryptographic researchers. Reference [14] is a new and efficient ID-based key-insulated signature scheme with batch verifications.

Mathematical Primitives

In this section, we discuss the elliptic curve and its properties. We then discuss the rules for adding points on elliptic curve and the elliptic curve discrete logarithm problem.

Elliptic Curve over Finite Field

Let a and $b \in Z_p$, where $Z_p = \{0, 1, \ldots, p-1\}$ and $p > 3$ be a prime, such that $4a^3 + 27b^2 \neq 0 \pmod{p}$. A non-singular elliptic curve $y^2 = x^3 + ax + b$ over the finite field $GF(p)$ is the set $E_p(a, b)$ of solutions $(x, y) \in Z_p \times Z_p$ to the congruence $y^2 = x^3 + ax + b \pmod{p}$, where a and $b \in Z_p$ are constants such that $4a^3 + 27b^2 \ / = 0 \pmod{p}$, together with a special point O called the point at infinity or zero point.

The condition $4a^3 + 27b^2 \neq 0 \pmod{p}$ is the necessary and sufficient to ensure that the equation $x^3 + ax + b = 0$ has a non-singular solution [15]. If $4a^3 + 27b^2 = 0 \pmod{p}$, then the corresponding elliptic curve is called a singular elliptic curve. If $P = (x_P, y_P)$ and $Q = (x_Q, y_Q)$ be points in $E_p(a, b)$, then $P + Q = O$ implies that $x_Q = x_P$ and $y_Q = -y_P$. Also, $P + O = O + P = P$, for all $P \in E_p(a, b)$. Moreover, an elliptic curve $E_p(a, b)$ over Z_p has roughly p points on it. More precisely, a well-known theorem due to Hasse asserts that the number of points on $E_p(a, b)$, which is denoted by $\#E$, satisfies the following inequality [16]: $p + 1 - 2\sqrt{p} \leq \#E \leq p + 1 + 2\sqrt{p}$. In addition, $E_p(a, b)$ forms an abelian group or commutative group under modulo p operation.

Addition of Points on Elliptic Curve over Finite Field

The following parameters about the proposed proxy signature scheme over the elliptic curve domain are required. We take an elliptic curve over a finite filed $GF(p)$ as $E_p(a, b) : y^2 = x^3 + ax + b \pmod{p}$, where a and $b \in GF(p)$. The field size p is considered as a large prime. We take G as the base point on $E_p(a, b)$ whose order is n, that is, $nG = G + G + \cdots + G(n \text{times}) = O \pmod{p}$.

The elliptic curve addition differs from the general addition [17]. Let $P = (x_1, y_1)$ and $Q = (x_2, y_2)$ be two points on elliptic curve $y^2 = x^3 + ax + b \pmod{p}$, with $P \neq -Q$, then $R = (x_3, y_3) = P + Q$ is computed as follows: $x_3 = (\lambda^2 - x_1 - x_2) \pmod{p}$,

$$y_3 = (\lambda(x_1 - x_3) - y_1) \pmod{p}; \text{ where } \lambda = \begin{cases} \dfrac{y_2 - y_1}{x_2 - x1} \pmod{p}, \text{ if } P \neq Q \\ \dfrac{3x_1^2 + a}{2y_1} \pmod{p}, \text{ if } P = Q \end{cases}$$

In elliptic curve cryptography, multiplication is defined as repeated additions. i.e., if $P \in E_p(a, b)$, then $4P$ is computed as: $4P = P + P + P + P \pmod{p}$

Discrete Logarithm Problem

The discrete logarithm problem (DLP) is as follows: given an element g in a finite group G whose order is n, that is, $n = |G|$ and another element $h \in G$, find an integer x such that $g^x = h(\bmod n)$. It is relatively easy to calculate discrete exponentiation $g^x(\bmod n)$ given g, x and n, but it is computationally infeasible to determine x given h, g and n, when n is large.

Elliptic Curve Discrete Logarithm Problem

Let $E_p(a, b)$ be an elliptic curve modulo a prime p. Given two points $P \in E_p(a, b)$ and $Q = k \cdot P \in E_p(a, b)$, for some positive integer k. $Q = k \cdot P$ represents the point P on elliptic curve $E_p(a, b)$ is added to itself k times. The elliptic curve discrete logarithm problem (ECDLP) is to determine k given P and Q. It is relatively easy to calculate Q given k and P, but it is computationally infeasible to determine k given Q and P, when the prime p is very large.

Basic Model of KIS Scheme

In this section, we will discuss the basic model of KIS scheme [1]. This KIS scheme consists of five polynomial time algorithms $(Gen, Upd^*, Upd, Sign, Vrfy)$.

Gen: Key generation algorithm is a probabilistic algorithm that takes as input a security parameter 1^k and the total number of time periods N. It returns a master key SK^*, a verification key VK, and an initial signing key SK_0.

Upd*: Partial key generation algorithm is a probabilistic algorithm that takes as input indices i and j for time periods and a master key SK^*. It returns a partial key $SK'_{i,j}$.

Udp: Key update algorithm is a deterministic algorithm that takes as input indices i, j, a signing key SK_i, and a partial key $SK'_{i,j}$. It returns a signing key SK_j for time period j.

Sign: Signing algorithm is a probabilistic algorithm that takes as input an index i for a time period, a message M, and a signing key SK_i. It returns a pair $\langle i, s \rangle$ consisting of a time period i and a signature s

Vrfy: Verification algorithm is a deterministic algorithm that takes as input a verification key VK, a message M, and a pair $\langle i, s \rangle$. It returns a bit b. If b is $TRUE$, then accept the signature, otherwise reject

KIS Scheme for Real-Time System

In this section, we describe the different phases of our proposed Key-Insulated Signature Scheme.

Setup Phase

In this section, we generate a large prime p. We take an elliptic curve over the finite field $GF(p)$ as $E_P(a, b) : y^2 = x^3 + ax + b \pmod{p}$ such that the elliptic curve discrete logarithm problem (ECDLP) becomes intractable. We take G as the base point on $E_p(a, b)$ whose order is q. Then randomly generate a private key d in the range $0 < d < q$ and computes the public key Q, as $Q = d \cdot G$. $h(\cdot)$ represents a secure one-way hash function (for example, SHA-1). In summary, we list the generated domain parameters in Table 1.

Key Generation Algorithm

Randomly select d' in the range $0 < d' < q$ which is kept by the signer. Compute the master key d_0 as $d_0 = d - d'$, stored in the secure device. Calculate $Q_0 = d_0 \cdot G$ and $Q' = d' \cdot G$.

Set the verification key as $VK : <q, G, Q, Q_0, Q', h_1, h_2>$

Partial Key Generation Algorithm

From the secure device randomly select k_A, $0 < k_A < q$ and compute $R_A = k_A \cdot G$. Set $(x_A, y_A) = R_A$ and compute $r_1 = x_A \pmod{q}$. Calculate $v_1 = h_1(r_1, T)$ using time period T. Compute the partial key $s_1 = v_1 \cdot d + r_1 \cdot d_0 \pmod{q}$.

Send (s_1, r_1, T) to the signer.

Table 1 Parameters generated in setup phase

m	Message to be signed
p	A large prime number
$E_p(a, b)$	Elliptic curve over $GF(p)$
G	A base point on $E_p(a, b)$
q	Order of G
$h_1(\cdot, \cdot), h_2(\cdot, \cdot, \cdot, \cdot)$	Secure one-way hash functions
d	A private key, $0 < d < q$
Q	A public key, $Q = d \cdot G \pmod{q}$

Key Update Algorithm

The signer calculates $v_1 = h_1(r_1, T)$ and verify the partial key as follows:

$$s_1 \cdot G = v_1 \cdot G + r_1 \cdot Q_0$$

if it is successful, then the signer compute the signing key for the time period T as:

$$sk_T = (s_1 + d')(\bmod q).$$

Signing Algorithm

Now the signer randomly select k_S, $0 < k_S < q$ and compute $R_S = k_S \cdot G$. Set $(x_S, y_S) = R_S$ and compute $r_S = x_S(\bmod q)$. Calculate $v_S = h_2(r_1, r_S, T, m)$ by using the input massage m and time period T. Then compute $\delta_S = (v_S \cdot k_S - sk_T)(\bmod q)$. Finally the signer transmits the signature tuple with the message $\langle m, (\delta_S, v_S, r_1), T \rangle$ to the verifier for verification.

Verification Algorithm

Using $Q, Q_0, Q', m, (\delta_S, v_S, r_1)$ and T the verifier compute $v_1 = h_1(r_1, T)$ and $W = v_S^{-1}(\bmod q)$. Compute $X = (x'_S, y'_S) = (G \cdot \delta_S + v_1 \cdot Q + r_1 \cdot Q_0 + Q')W$ if $X == 0$, then reject the signature. Otherwise compute $r'_S = x'_S(\bmod q)$.
 Accept the signature if and only if $r'_S == r_S$.

Analysis of the New KIS Scheme

Correctness Proof

Proof 1 We show that the equation $s_1 \cdot G = v_1 \cdot G + r_1 \cdot Q_0$ holds good in order to check the validity of the partial key as follows:

$$L.H.S = s_1 \cdot G = (v_1 \cdot d + r_1 \cdot d_0)G$$
$$= v_1 \cdot d \cdot G + r_1 \cdot d_0 \cdot G = v_1 \cdot Q + r_1 \cdot Q_0 = R.H.S$$

Proof 2 We have the equation $\delta_S = v_S \cdot k_S - sk_T$ to check whether the signature is valid or not. Multiplying G on both side of the equation, we will get:

$$G \cdot \delta_S - G \cdot (v_S \cdot k_S - sk_T) = v_S \cdot k_S \cdot G - sk_T \cdot G$$
$$= v_S \cdot k_S \cdot G - (s_1 + d') \cdot G = v_S \cdot k_S \cdot G - (s_1 \cdot G + d' \cdot G)$$
$$= v_S \cdot k_S \cdot G - (v_1 \cdot Q + r_1 \cdot Q_0 + Q')$$

Rearranging the equation we will gate:

$$k_S \cdot G = (G \cdot \delta_S + v_1 \cdot Q + r_1 \cdot Q_0 + Q') \cdot v_S^{-1}$$

i.e. $(x'_S, y'_S) = (G \cdot \delta_S + v_1 \cdot Q + r_1 \cdot Q_0 + Q')W$

As a result, we have $x_S == x'_S$, that is, $x_S(\bmod q) == x'_S(\bmod q)$ and hence, $r_S == r'_S$.

Performance Analysis

To achieve reasonable security, RSA and DSA should employ 1024-bit moduli, while a 160-bit modulus should be sufficient for ECC [4]. Moreover, the security gap between the systems increases dramatically as the moduli sizes increases. For example, 300-bit ECC is dramatically more secure than 2,048-bit RSA or DSA [18]. Since we constructed our scheme over elliptic curves, so we can get better security with small key size. Due to small signing and verification key, the signature size is reduced and the computational cost for signing and verification is efficient.

For analysis of computational costs required for different phases in our scheme, we use the following notations shown in Table 2.

It is clear to note that the computational costs of different phases of KIS Scheme are as follows: Key Generation Algorithm: $T_{OA} + 2T_{ECM}$;

Partial Key Generation Algorithm: $T_{OA} + 2T_{OM} + T_{ECM} + 2T_{\bmod} + T_H$;

Key Update Algorithm: $2T_{OA} + 2T_{ECM} + T_H + T_{\bmod}$;

Signing Algorithm: $T_{OA} + T_{ECM} + T_{OM} + T_H + 2T_{\bmod}$;

Verification Algorithm: $3T_{OA} + 4T_{ECM} + T_{INV} + T_H + T_{\bmod}$

Table 2 Notations used in computation of computational cost

T_E	Time taken for an exponential operation
T_H	Time taken for a one-way hash function $h(\cdot)$
T_{INV}	Time taken for a modular inverse operation
T_{EMC}	Time taken for a scalar (elliptic curve) multiplication
T_{OM}	Time taken for an ordinary multiplication
$T_{\bmod}$	Time taken for a modular operation
T_{OA}	Time taken for an ordinary addition

Security Analysis

In the standard framework of identification protocols the signer sends an initial message, the verifier sends a random challenge and the signer respond with some answer. Here we have a master key which is stored on a physically-secure device. At the beginning of time period T, the signer interacts with the secure device in order to obtain a key sk_T valid for current time period. The signer (who may be operating on an insecure device) proves his identity to a verifier in period T using key sk_T. Let us consider an adversary, who interacts with the signer in an execution of the identification protocol during various time periods, and may additionally compromise the insecure device and obtain the temporary keys. As the validity of the key is for limited time period, so the adversary will not be able to successfully impersonate the signer during any time period other than those in which it compromised.

The security of a KIS scheme means a KIS scheme is secure against a signing key leakage. i.e. the signing key sk_T was leaked by a signer and used by a third party(adversary). Still the KIS scheme is considered secure as the adversary cannot generate valid signature using the leaked signing key sk_T. Since signing key sk_T is valid for that T period of time in which the compromise occurred.

The security of a Strong KIS scheme means a KIS scheme is secure against a signing key or master key leakage. If signing key or a master key was leaked by a signer and used by a third party, still the adversary cannot generate valid signature, because the master key d_0 and d' are managed by two different servers. Even if the master key d_0 is leaked, the signing key sk_T cannot be updated without d'.

Conclusions

In this paper, we have considered different aspect of real-time communication system and proposed a secure and reliable authentication mechanism. We have designed a key-Insulated Signature scheme that minimized the damage caused by key-exposure. The main feature of this algorithm is it is based on elliptic curve cryptography. This reduces the key size as well as the signature size and increase the security compared to DLP and RSA as the security is based on elliptic curve discrete logarithmic problem. We have computed the computational cost; in addition performance of the scheme is analyzed.

Acknowledgments The paper is supported partially by the "Cross-platform smart phone-enable stroke prevention and rehabilitation health system" project and partially by the NSC-101-2811-E-008-013 project.

References

1. Dodis Y, Katz J, Xu S, Yung M (2003) Strong key-insulated signature scheme. In: Public Key Cryptography PCK, pp 130–144
2. Kopetz H (2011) Real-time system: design principles for distributed embedded applications, vol 25. Springer
3. Lin L, Lin P (2007) Orchestration in web services and real-time communications. Commun Mag, IEEE 45(7):44–50
4. Barker E, Barker W, Burr W, Polk W, Smid M (2011) Recommendation for key-management-part 1: general (revision 3). NIST Spec Publ 800:57
5. Dodis Y, Katz J, Xu S, Yung M (2002) Key-insulated public key cryptosystems. In: Advances in cryptology Eurocrypt, Springer pp 65–82
6. Gonzalez-Deleito N, Markoitch O, DallOlio E (2004) A new key-insulated signature scheme. In: Information and communication security, pp 9–12
7. Ohtake G, Ogawa K (2012) Application authentication for hybrid services of broadcasting and communication networks. In: Information security applications, pp 171–186
8. Ohtake G, Hanaoka G, Ogawa K (2010) Efficient provider authentication for bidirectional broadcasting services. IEICE Trans Fundam Electron, Commun Comput Sci 93(6):1039–1051
9. Matsuda T, Hanaoka G, Ogawa K (2007) A practical provider authentication system for bidirectional broadcast service. In: Knowledge based intelligent information and engineering systems. Springer, pp 967–974
10. Ohtake G, Hanaoka G, Ogawa K (2008) An efficient strong key-insulated signature scheme and its application. In: Public key infrastructure, PKI, pp 150–165
11. Weng J, Liu S, Chen K, Li X (2006) Identity based key-insulated signature with secure key-updates. In: Information security and cryptology. Springer, pp 13–26
12. Weng J, Liu S, Chen K, Ma C (2007) Identity based key-insulated signature without random oracles. In: Computational intelligence and security, pp 470–480
13. Du H, Li J, Zhang Y, Li T, Zhang Y (2012) Certificate based key-insulated signature. In: Data and knowledge engineering. Springer, pp 206–220
14. Wu TY, Tseng YM, Yu CW (2012) Id based key-insulated signature scheme with batch verifications and its novel applications. In: Int J Innovative Comput, Inf Control 8(7(A)):4797–4810
15. Nickalls R (1993) A new approach to solving the cubic: Cardan's solution revealed. In: The mathematical Gazette. pp 354–359
16. Stallings W (2003) Cryptography and network security, principles and practices. Practice Hall
17. Koblitz N (1987) Elliptic curve cryptosystem. Math Comput 48(177):203–209
18. Remarks on the security of elliptic curve cryptosystem: a Certicom white paper, CTEC Cryptosystem, updated July (2000). http://www.certicom.com/

Innovative Wireless Dedicated Network for e-Bus

Eric Hsiao-Kuang Wu, Chung-Yu Chen, Ming-Hui Jin and Shu-Hui Lin

Abstract In recent years, several kinds of wireless network have been widely deployed. In this work, we focus on the e-bus-system for movements of buses. Passengers can get the bus information through the intelligent bus stop when they are waiting for the bus. In today's environment of Taiwan, the transportation information is provided through some specific Mobile network operators. In this case, it costs a huge expenditure every month for the charge of data transmission. Therefore, we propose a new network system with Digital Mobile Radio (DMR) and Super Wi-Fi to replace the mobile network operator. Furthermore, when the catastrophe happened, the proposed system can also change to be used for the rescues. The implementation of the system architecture, the application of DMR and super WiFi in the proposed system, and the operation scenario are described in this paper.

Keywords Digital mobile radio · Intelligent transportation systems

E. H.-K. Wu (✉) · C.-Y. Chen
Department of Computer Science and Information Engineering, National Central University, Jhongli, Taiwan
e-mail: hsiao@csie.ncu.edu.tw

C.-Y. Chen
e-mail: giawoba@wmlab.csie.ncu.edu.tw

M.-H. Jin · S.-H. Lin
Department of Smart Network System Institute, SNSI Institute for Information Industry, No. 133, Minsheng East Road, Taipei 10574, Taiwan
e-mail: Jinmh@iii.org.tw

S.-H. Lin
e-mail: vickylin@iii.org.tw

Y.-M. Huang et al. (eds.), *Advanced Technologies, Embedded and Multimedia for Human-centric Computing*, Lecture Notes in Electrical Engineering 260, DOI: 10.1007/978-94-007-7262-5_141, © Springer Science+Business Media Dordrecht 2014

Introduction

In the recent years, many public and private organizations, such as police forces, fire brigades and health emergency associations [1], have built their own trunking or simulcast digital radio solutions to replace the old analog radio networks. In this paper we focus on the application of communication techniques in e-bus-system. If the bus stop provides the bus information, passengers can grasp it immediately, and can appropriately adjust the schedule. In Taiwan, in order to enhance the service performance of the bus transport system, Ministry of transportation and communications (MOTC) has developed Intelligent Transportation Systems (ITS) [2] plan. The combination of electronics, communications, information and the management of transportation technology have strengthened the supervision and management of the bus. Thereby, the proposed plan is to attract people to take the bus and, therefore, to increase the usage of public transportation. Its main objectives are to save the energy consumed by the transportation and to avoid the traffic congestion. The e-bus-system in Taiwan is used to transmitting the information through some specific Mobile networks. In this case, it needs to pay a lot of monthly expenditure for data transmission. And it always introduces burden to the government and the bus companies. In paper [3], it compared some common professional mobile radio (PMR) standards. The result shows that Digital Mobile Radio (DMR) [4] has been identified as the best solution, which grants cost saving, high coverage, spectral efficiency and simplicity in network configuration and is well suitable in wide area with a low/medium density of traffic. And this characteristic is suitable to be applied for the e-bus-system.

In this paper, we propose a new network system with DMR and Super Wi-Fi to replace the current system, which is supported by the mobile network operator. The proposed system will use for the e-bus-system as usual, however, it can also be used for emergency communication in rescue mission when the catastrophe happens. Thus the proposed system provides a compensative solution for the public network that is usually unable to communicate when the occurrence of disaster.

The organization of this paper is as follows. The architecture of the proposed system is provided in the following section. Experiments presents the implementation and experiments of the proposed system. The experiment is given in Discussions. And, finally, we provide some conclusions in the last section.

Innovative e-Bus System

System Architecture

Our e-bus-system is divided into three parts, as shown in Fig. 1. First, we use the DMR technique to implement the communication links between BUS and the central control server. We will set the DMR based GPS device in the bus

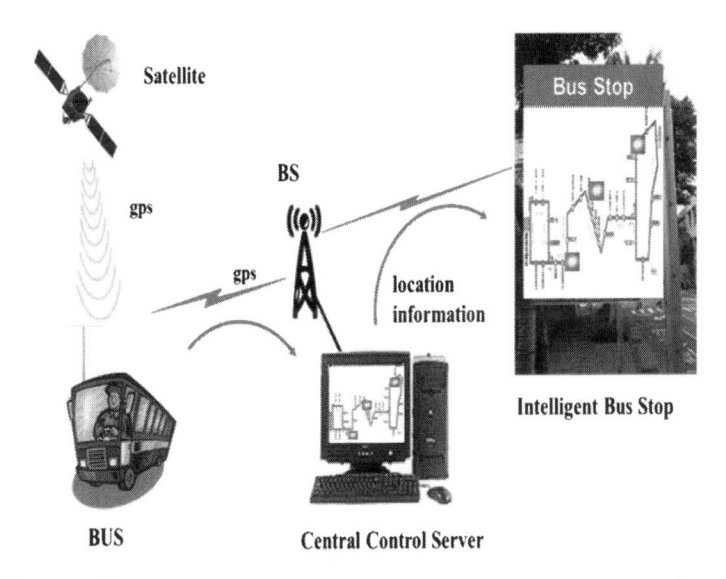

Fig. 1 System architecture

afterwards. It will transmit the GPS information back to the server side with the fixed time intervals, and the format is following the [5]. Second, we build the central control server to record and process the GPS information. We transformed the GPS information into simply bit string, and then transmit to the bus stop. The bus stop will show up the LED to let public know where is the bus. The bus stop we designed is the dynamo bus stop, so that it can't handle too complexity information. Third, we use super Wi-Fi to send data to the bus stop. The reason why we use super Wi-Fi will be explain later. Finally, our system replace the network system which provided by the network operator.

Digital Mobile Radio

DMR is one of Digital PMR solution proposed by the European Telecommunications Standards Institute (ETSI) and it is diffused digital technologies in Europe. This technology was developed by ETSI to grant gradual migration from the analogical conventional system to the digital mode without new licenses and without changing the existing network architecture. DMR uses 2 time slots on a 12.5 kHz bandwidth carrier and adopts time division multiplexed accessing (TDMA) with 4-FSK modulation. Since the modulated signal has constant envelope, a transmitter can work in saturation (clipping) mode (C class or superior) with very low consumption. It is noted that DMR has the maximum bit rate of 9.6 kbps and can works in simulcast mode to provide a wider coverage area (until 80 km), using a frequency pair only. Network and terminals can be dual mode,

Fig. 2 DMR communication device

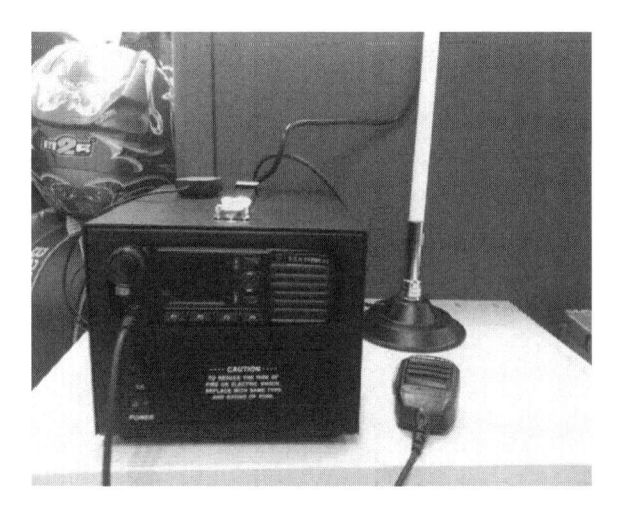

thus granting the coexistence of analog and digital devices. It provides for voice and data, in which tier 1 is direct mode, tier 2 is conventional mode, and tier 3 is trunking mode. Figure 2 shows the example of DMR communication equipment. It will be set in the bus for experiment, and the work is to send the GPS information back to the central control side periodically. Every device has the unique ID so that we can identify the bus. Furthermore, each device is equipped with a digital two-way radio. Both bus driver and dispatcher can communication with each other to deal with emergencies or temporary dispatching immediately. It enhances the ability of being adaptable.

Super Wi-Fi

The proposed system adopts the super Wi-Fi as the communication technique. The super Wi-Fi, or IEEE 802.22 [6] and IEEE 802.11af as it is technically known, is a term coined by the United States Federal Communications Commission (FCC) to describe a wireless networking proposal. FCC plans to apply it for the creation of longer-distance wireless internet. The use of the trademark "Wi-Fi" in the name has been criticized by people because it is not based on common known traditional Wi-Fi technology or endorsed by the Wi-Fi Alliance. Instead of using the 2.4 GHz radio frequency of Wi-Fi, the 'Super Wi-Fi' proposal uses the lower-frequency white spaces between television channel frequencies. These lower frequencies allow the signal to travel further and penetrate walls better than the higher frequencies previously used. The FCC's plan is to allow those white space frequencies to be used for free, as happens with Wi-Fi and Bluetooth. Here we have noticed that the data is send from the base station to all the bus stops. It can be expected that the data flow will be much bigger than the traffic between bus and central control side, and DMR may not deal with it. In our design, we use super Wi-Fi to handle this part to send the corresponding bus position to the bus stop. Compared to DMR, super Wi-Fi has farther transmission range and higher transmission rates, and the transmission rate is the main reason why we use the super Wi-Fi.

Central Control Server

In the central control server, its main functions are to process the received location information from buses by referring to the geographical information and bus routes information base. Here we need to associate the received GPS information to the position of the bus stop, and expressed as a bit string of the bus routes. The software process of the proposed system in the central control server is illustrated in Fig. 3. The bus will send its bus ID with location information, which includes longitude and latitude, to the central control server periodically. As shown in Fig. 3, when received the location information sent by the bus, the central control server associates the longitude and latitude with the geographic information base to determine its location in the map. Then the server determines the route of the reporting bus by referring to its bus ID. It is noted that the bus route could be helpful for the location identification of the bus. Finally, the central control server sends the bus location to the bus stops, which are associated with this bus route, and transformed it into web page for query.

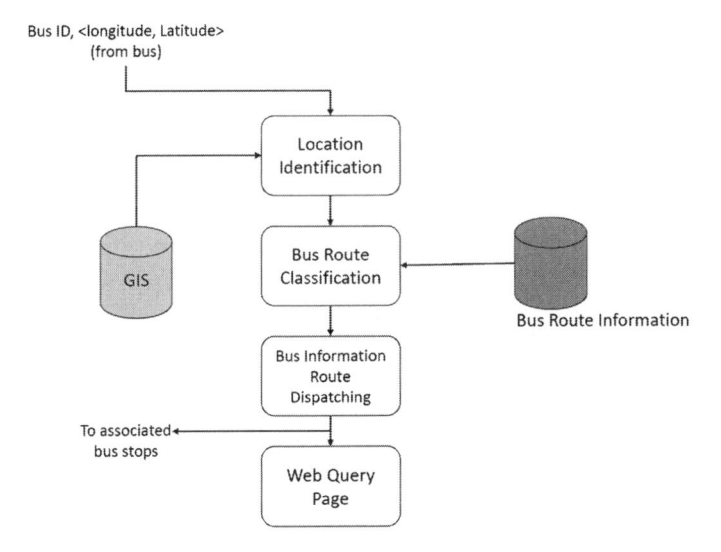

Fig. 3 Information process of central control server

One bus stop is corresponding to one bit, in which the bit set to one represents the bus stop that the bus is going to arrive. When the bus stop receives the bit string, it will light up the LED to show the bus current position. Here we have set up the web site to show the records and bus information, furthermore we also build our own GIS (Geographic Information System) [7] to show the bus currently position on the map. The record and the map will be dynamic update periodically, so that public can also grasp the latest BUS information on the web site. Combine these three parts; our e-bus-system has been completed.

Experiments

In this section, an experiment is performed by using our system to test in the real environment. We performed our experiment in the Taipei City. Here we set the DMR equipment in our personal vehicle, and then we follow the bus to imitate the real situation. And we took the travel of Bus 518 as the experimental example. That is Bus 518 equipped with the DMR communication device to periodically report its location information during the experiment.

In Fig. 4, the lines represent the road, and triangle represents the current location of the bus. It is noted that the map used in Fig. 4 can be constructed in layer approach. Thus we can selectively define which kind of information to be visible on the map. This function is very useful for user interface. For example, as there are many bus routes in the city, we can choose to show some specific routes only. It is noted that the relationship between the bus stops and the current bus location or the bus arrival time prediction [8] is more meaningful for the passengers. Thus the determined location information of the bus shall be mapped to longitude and latitude of the nearest bus stop. Figure 5 shows the bus position being related to the bus stops associated with the route of this bus, and arrived time during the experiment is also provided. It is noted that, in addition to the route selection, the system also provides the capabilities of car tracing, map view, history information, and the transformation of GPS. Thus users can access versatile information from the system.

Discussions

In the past, traditional analog radio system offers some great benefits, such as lower cost of ownership, to meet the requirements of users of radio communication range and voice communications, simple and reliable operation capability. However, analog radio technology has reached its cap because the radio spectrum interval is the main limitation to achieve the accurate and reliable communications. Thus spectral efficiency and channels cannot meet the requirements of digital communications due to the problems of congestion and interference.

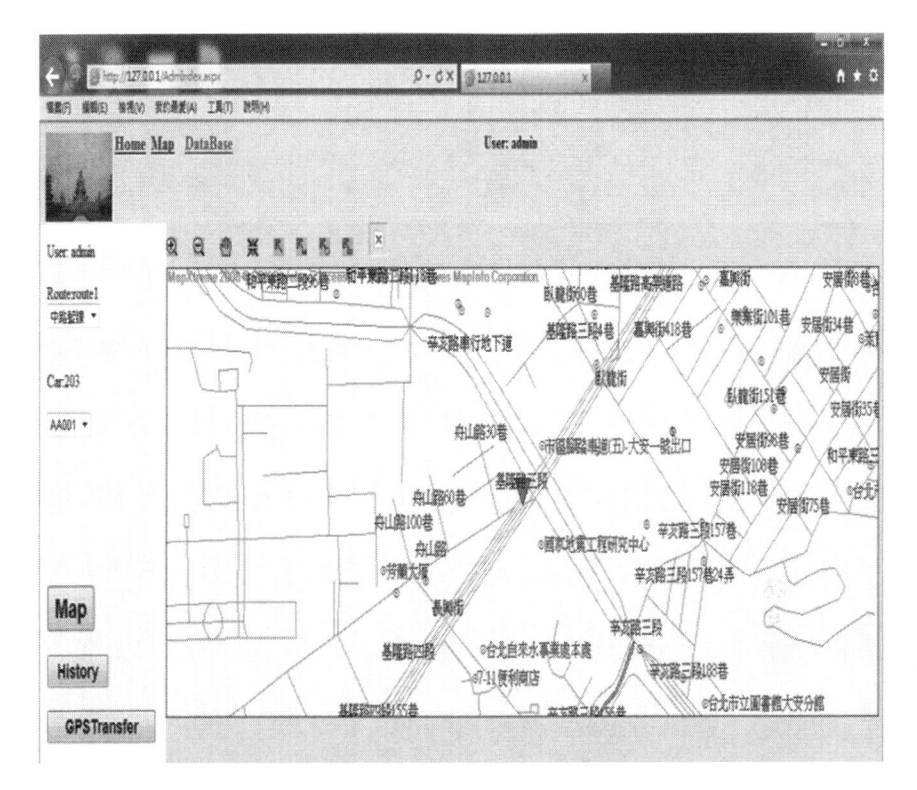

Fig. 4 Geographic information of the experiment (*Filled triangle* the bus location)

Our system can divide into three parts. First part is the connection between bus and the central control server. In this part the DMR communication device is use for sending the GPS information back to the central control side. However, our system has only experiments with one bus currently, so when the bus increased, system capacity is need to be concerned. Second part is the central control server. This part processes the GPS data. It correspond GPS information to the position of the bus stop. When the bus stop receives the information, it will lights up the LED to show the bus current position. Final, the third part is the connection between central control server and the bus stop. Our system use Super Wi-Fi to send the data. Because we have noticed that the data is send from the base station to all the bus stop, Can be expected that the data flow will much bigger than the traffic between bus and central control side, and DMR may not deal with it.

In our experiment, the proposed e-bus-system demonstrates that it can provide accurate information for query. However, after analysis the GPS records we have noticed that the GPS single is unstable in real environment. Because the bus may pass through the tunnel or the viaduct, GPS device may be unable to receive the signal from satellites. In such locations, the system will not receive the information sent by the bus. We also notice the GPS accuracy has a little deviation, these

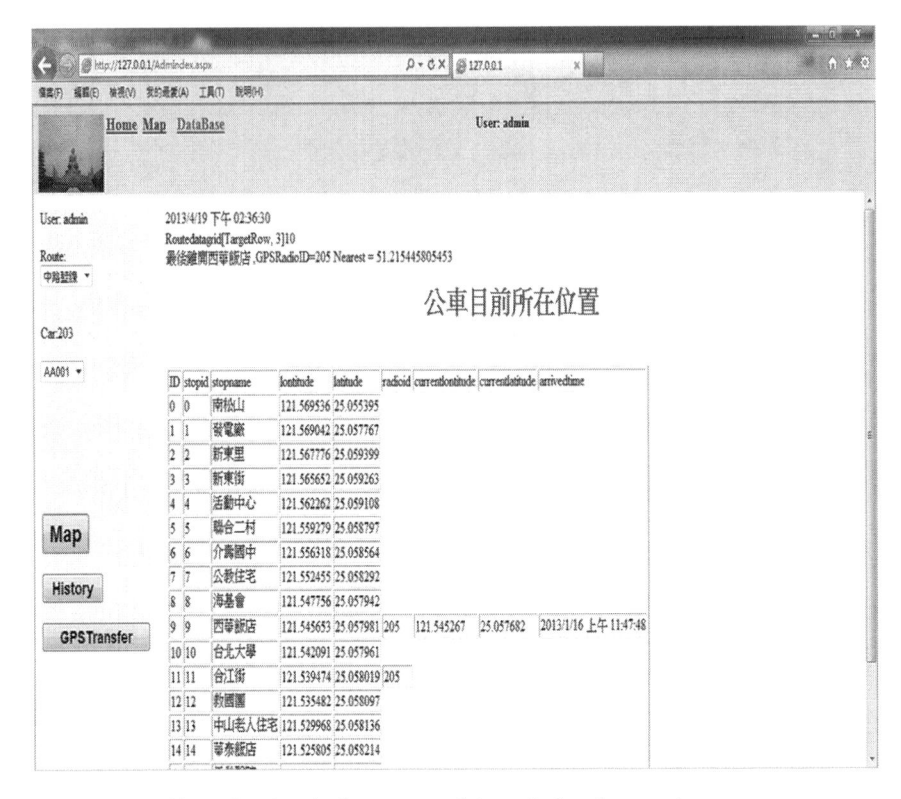

Fig. 5 Bus position related to the bus stops and times during the experiment

problems are necessary to be concerned. Furthermore, in our experiments, the GPS information is submit every twenty seconds. However, we noticed that during these twenty seconds, the bus may drive through two or more bus stop. If we increase the frequency of the bus information, it will be more accurate. But the server will face the big data issues. How to balance the tradeoff is also a big challenge.

In the geographic information system, we have built our own map. The reason why we didn't use the Google map is that we have to get the fully control to add additional layer into the map, such as the bus stop information, bus route information, etc. This is not possible to be implemented with Google map. In addition, as we can fully obtain and control the geographic information of the map and assign the bus stops at specific locations, it provides the capability to tune accuracy of the location identification especially when the bus information is missed. Furthermore, based on the current architecture, we can develop more attractive applications in the future.

Our system will use for the e-bus-system as usual, however, when the catastrophe happened, DMR can also use for the communication in rescue mission, seize the prime time of rescue. It solve the problem that public network is usually unable to communicate when disaster happened. Some similar system such as P25 [9], TETRA [10] are already used widely in police forces, fire brigades in other countries. In this paper, we demonstrate the feasibility of the proposed system for e-bus service from the real experiment. And we need to develop some rescue-based scenarios to verify the usability and to measure the system performance.

Conclusions

Due to the environment in Taiwan, most of the network resources (i.e. spectrum) have been allocated for some specific usages such as private mobile networks, digital TV, radio broadcasting, etc. It is necessary to carefully reuse the existing spectrum for the development of some valuable information distribution systems. In this paper, we implement a new network system with DMR and Super Wi-Fi to be used as the communication in e-bus-system to replace the current system that utilizes the spectrum band of mobile networks. Currently, the DMR is used the transmission from bus to the control server and the super Wi-Fi is adopted for transmission from the control server to the bus stops. In order to simplify the communication architecture, the simpler architecture may be possible after further examination on the performance and interference of both communication spectrums.

In the proposed system, passengers can easily get the bus information through the intelligent bus stop or on our web site immediately. However, as mentioned in previous section, we have learned that the GPS single is not stable in real environment. Fortunately, we can access the map information directly. Therefore, we plan to develop some location refinement techniques according to the speed of the bus and geographic information to solve this problem in the near future. Our work is just at the beginning toward the wide deployment. In addition to the location refinement, we are expecting to work on the digital surveillance system with the intersection monitors. There still a lot of applications worth to be developed.

Acknowledgments The paper is supported partially by the "Cross-platform smart phone-enable stroke prevention and rehabilitation health system" project and partially by the NSC-101-2811-E-008-013 project.

References

1. Pawelczak P, Prasad RV, Xia L, Niemegeers IG (2005) Cognitive radio emergency networks—requirements and design. In: Proceedings of IEEE international symposium dynamic spectrum access networks (DySPAN), pp 601–606
2. Weiland RJ, Purser LB (2000) Intelligent transportation systems. Transportation in the new millennium
3. Onali T, Sole M, Giusto DD (2011) DMR networks for health emergency management: a case study. In: 7th international wireless communications and mobile computing conference (IWCMC)

4. ETSI, TS-102-361, electromagnetic compatibility and, radio spectrum matters (ERM): digital mobile radio, (DMR) systems (2007)
5. Langley RB (1995) NMEA 0183: a GPS receiver interface standard. GPS world 6.7, pp 54–57
6. Stevenson C, Chouinard G, Lei Z, Hu W, Shellhammer S, Caldwell W (2009) IEEE 802.22: the first cognitive radio wireless regional area network standard. IEEE Commun Mag 47:130–138
7. Burrough PA et al (1998) Principles of geographical information systems, vol 333. Oxford university press, Oxford
8. Lin W-H, Zeng J (1999) Experimental study of real-time bus arrival time prediction with GPS data. Trans Res Rec 1666:101–109
9. TIA, TIA-102: project P.25 (2004)
10. ETSI, EN-300-392, terrestrial trunked radio (TETRA) (2009)

Cross-Platform Mobile Personal Health Assistant APP Development for Health Check

Eric Hsiao-Kuang Wu, S. S. Yen, W. T. Hsiao, C. H. Tsai, Y. J. Chen, W. C. Lee and Yu-Wei Chen

Abstract Our team proposes a concept allowing patients taking health check anytime everywhere, which can increase patients' attention to their own health condition. To improve the user experience and convenience, the system must be designed to simply operate and easily connect with the medical devices. Moreover, the system must have the ability to communicate between the patients and doctor or medical personnel. In this paper, we illustrate our system, such as user interface, storage, display and cloud system. The user interface is designed with standards-based Web technologies. We use PhoneGap to build cross-platform mobile apps with HTML, JavaScript, and CSS. Because patients need to keep their record of health check, we use SQLite database for storage. Moreover, for the patients'

E. H.-K. Wu (✉) · S. S. Yen · W. T. Hsiao · C. H. Tsai · Y. J. Chen · W. C. Lee
Department of Computer Science and Information Engineering, National Central University, Jhongli, Taiwan
e-mail: hsiao@csie.ncu.edu.tw

S. S. Yen
e-mail: freehourse2sh@gmail.com

W. T. Hsiao
e-mail: wayne12345.168@gmail.com

C. H. Tsai
e-mail: a968574123@hotmail.com

Y. J. Chen
e-mail: denny923@livemail.tw

W. C. Lee
e-mail: win60615@gmail.com

Y.-W. Chen
Department of Neurology, Landseed Hospital, No. 77, Kwang-Tai Road, Ping-Jen City 32449 Tao-Yuan County, Taiwan
e-mail: yuwchen@gmail.com

Y.-M. Huang et al. (eds.), *Advanced Technologies, Embedded and Multimedia for Human-centric Computing*, Lecture Notes in Electrical Engineering 260, DOI: 10.1007/978-94-007-7262-5_142, © Springer Science+Business Media Dordrecht 2014

health check report which shall be easily understood, we design line charts to display the data. In this paper, we implement the Personal Health Assistant system.

Keywords Health check · PhoneGap · Cross-platform

Introduction

Nowadays, smartphones are in widespread use and enough to replace the feature phones [1]. With the smartphones growth, the applications are explosive increasing. The field of smartphones applications appeals lots of developers, because the operating systems of smartphones allow users to install unofficial applications on the OS, moreover providing the application market for developers to sell their own applications. There are several operating systems, Google Android, Apple iOS, Windows Phone which have its own application store, Google Play Store, Apple App Store, Windows Phone Apps store respectively.

The Population aged 65 and above has increased notably over time. According to Taiwan Ministry of the Interior, in the end of 2012 the Population aged 65 and above has occupied 11.15 % of Taiwan population [2]. Taiwan has become aging society. Furthermore, the elder need more care, especially in health care. Therefore health check becomes increasingly important. On the other hand, people in Taiwan are suffering from chronic disease for a long time. Hypertension, hyperlipidemia, diabetes, cancer, for example, are some of the main chronic diseases. In the 2011 the Statistics of Causes of Death show that chronic diseases occupy up to 75 % of total causes of death [3]. Besides, the percent of elder having chronic diseases is higher than younger. It infers health care for elder is considerably important. To prevent chronic diseases or as therapy, regular health check is a useful method. Moreover, regular health check is not only for this way. There are lots of benefits of regular health check, such as detection of disease symptom, earlier treatment, and so on.

Android is the smartphone market leader [4]. To integrate our medical devices with smartphone applications. We have developed the health check application for Android. Nonetheless, it is not enough, there are other smartphone operating systems, iOS, Windows Phone for example. In order to fulfill the usage form other operating systems, we have to develop other versions for them. Therefore we use PhoneGap as the tool to develop the cross-platform application with Web Language, HTML, JavaScript, and CSS. We do not have to know the programming language for android, iOS, and Windows Phone, yet we can fulfill the application for all operation systems. The advantage is decreasing the development cost, and simpler management.

In order to storing the data of our application, we use database to achieve our goal. PhoneGap storage API is based on the W3C Web SQL Database Specification and W3C Web Storage API Specification. Some devices already provide an implementation of this specification. This is why we using SQLite to implement our application [5].

We can view the health check records in the table by loading data from SQLite database, but sometimes the data is considerable that are impossible to view specifically. Instead, line chart is better to show the overview. The line chart in our app is written in JavaScript and uses the new canvas element to load graph data from SQLite database.

The mobile devices and the networks are inseparable. We cannot leave our application working on its own, because we should make differences between other applications. What we should do is that the devices can always exchange data with a remote server where can store data permanently. In this way, we integrate distributed data of the same person in a remote server, and then the users do not need to rely on a single device and they could access their data anytime everywhere.

Facebook, for example, provides users to share their statuses, photos, and videos on their post wall. Their friends can also like or reply the statuses. Therefore we integrate the feature with ours application. Patients can upload their photos, videos to cloud servers, sharing with their friends. Moreover adding patients' family or doctors as friend and uploading patients' health check report can let patients' family or doctors know their health condition, instead of sending email. Personal Health Assistant becomes not only health check recorder but also a social platform sharing photos, video, and even their health check report.

Ultimately, we fulfill our proposal. Patients could take health check anytime everywhere by using medical device and view the result in their smartphones, including body analysis, body temperature, blood glucose, and blood pressure, even sharing them with their doctors, family or friends to let them know patients' condition. We think the application can provide integration of the four medical detection devices, which are body analysis, body temperature, blood glucose, and blood pressure respectively. Also, it has the advantages of providing portable, convenient. Moreover, chart view is a feature providing patients or doctor to view obviously. The personal health assistant aims to play a necessary role in elders' and patients' daily life.

Related Work

A healthcare application should apply to everyone including patients and common people. In other words, we believe that monitoring health status by self in the daily life is the responsibility for everyone, so our application design to make more motive power for users to take care of their health through social networks. Some market-available healthcare products or applications did not offer to common users. For instance, there is an application [6] of observing and analyzing ECG (Electrocardiography) waveforms on Android devices. The application tends to give medical staff a convenient and portable platform to see the heart status of patients since traditional devices such as PCs which are too heavy to carry. ECG waveforms analyzing is unusual to common people, therefore, they could not recognize what happened on his status. Our application does not aim to specific people but common people who wonder their health status.

Another paper [7] introduces their product how to measure elderly's blood pressure (BP) and pulse rate. Here we have more measurement devices such as blood pressure (BP), Blood Glucose (BG), Body Temperature (BT) and user can record data about the Body Analysis Scale (BAS) by inputting BMI, weight and etcetera. Most important of all, the Daily Check App can send the application data via any useable network (3G, Wi-Fi) to the server, and the collected statics in server can represent to others via "Social Network". "Social Network" is a private platform allows you to post your healthy situation or chat with friend online, sharing every special moment to your old friends or families.

Ultimately, we want to view our health check on the go with diversity data of our health, so our team provides an application to collect and analyze the data and convert it into line chart. Above of functions available on the market in some application, but we want to create more interaction during the people and be attentive to others. For this purpose that everyone can share their health check and some photos or videos to their friend, nurses, and doctors to tell the user's condition and health recent, so we combine the social network.

Personal Health Assistant System

System Architecture

The entire system is based on PhoneGap which is a solution of the development of cross-platform mobile applications. PhoneGap allows us build user interface with web technologies such as HTML5, CSS3, and JavaScript. But there is a little difference between mobile applications design and website design. For example, mobile applications have to consider its screen size and its work flow which is different from web design. Consequently, our team imported a JavaScript framework—jQuery Mobile which is a unified, HTML5-based user interface system. JQuery Mobile solves the problems of responsive pages [8] to apply many distinct kinds of screen sizes. Besides, jQuery Mobile used AJAX technology on switching between pages. With combination of PhoneGap and jQuery Mobile, we could briefly build the entire architecture and the user interface. Moreover, it also keeps a good user experience [9].

Another feature of jQuery Mobile is that it provides many essential components of user interface which has been styled, and then we can simply design buttons, header, and footer and so on.

In the start page of the application, we directly place six large buttons lead to each measure pages of health status and setting page (Fig. 1). At each measure

Fig. 1 Start page

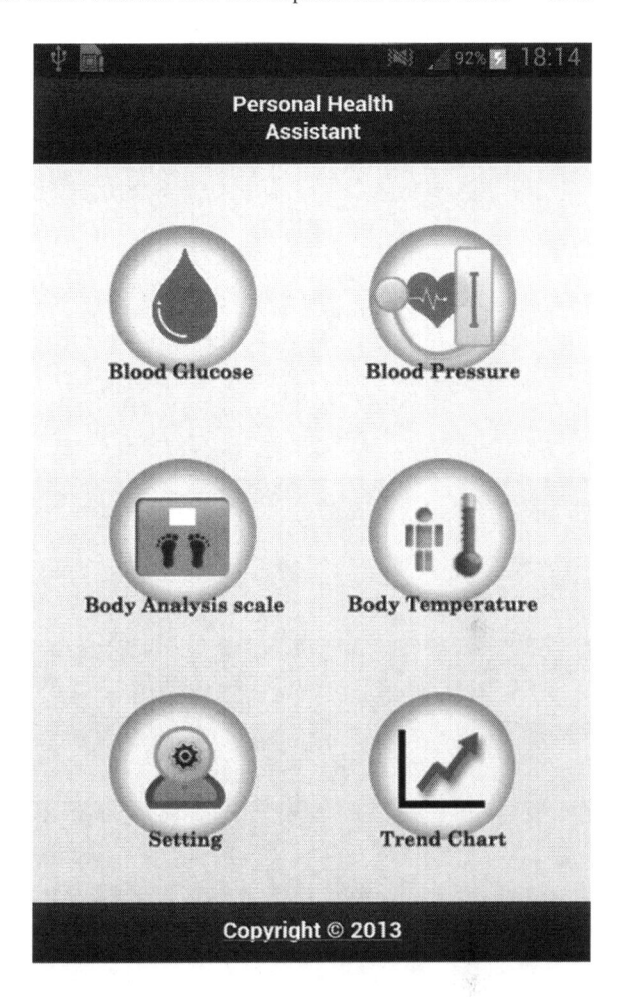

page, we divide it into four parts: Input, Data, Chart and Email, and these parts separately provide distinct functions. For this reason, we design four tabs on the top to switching these functions (Figs. 2 and 3). At the last, we also have a setting page for configurations, and a Trend Chart page to integrate all charts of measures in one page to see an entire situation of health about self.

For the input, we provide three ways to obtain measured value:

1. Users manually key in data.
2. Receive the data via Bluetooth protocol from the measure device
3. Receive the data via NFC protocol from the measure device

See Fig. 3, users can simply tap one of three buttons to obtain input and store into a database. For data display, the Data function will list all records from the database. For a run chart, the Chart functions provide a clear chart composed of all

Fig. 2 Blood glucose page

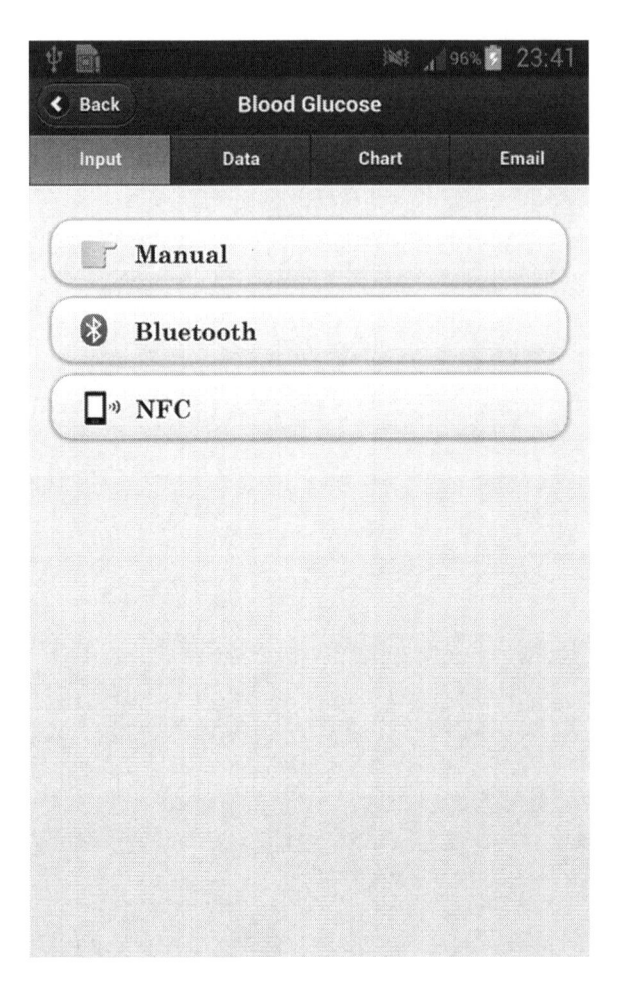

records. In the end, the Email function could send an E-mail contains weekly or monthly records to a doctor or someone by keying in an E-mail address.

Consequently the Fig. 4 show the hierarchy chart of our user interface design.

Storage on Smart Phones

SQLite is a relational database management system contained in a small C programming library. In contrast with other database management systems, SQLite is not a separate process that is accessed from the client application, but an integral part of it.

Fig. 3 Data tab-page

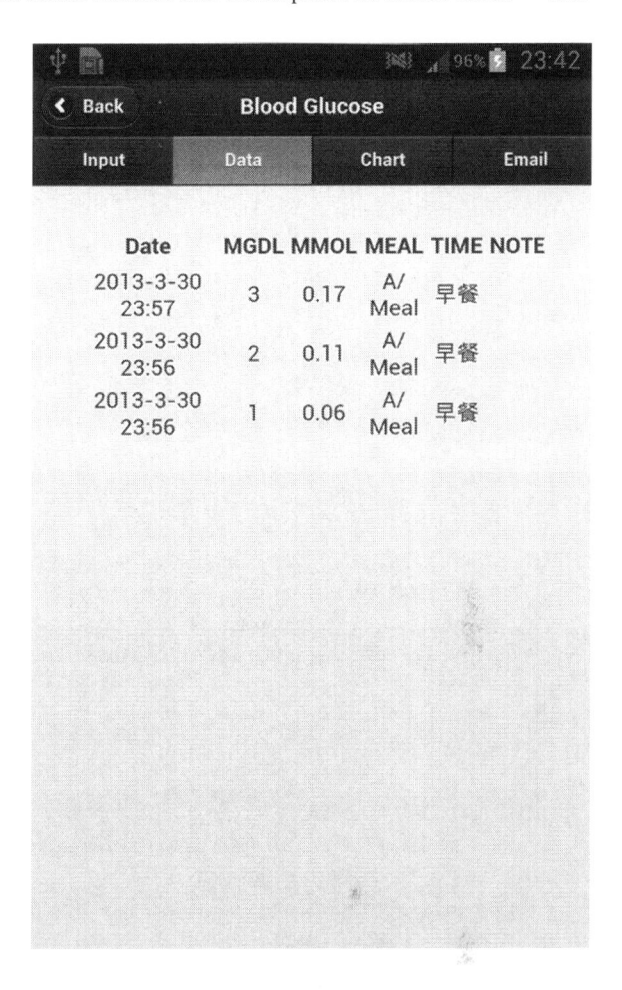

Fig. 4 Hierarchy chart of UI design

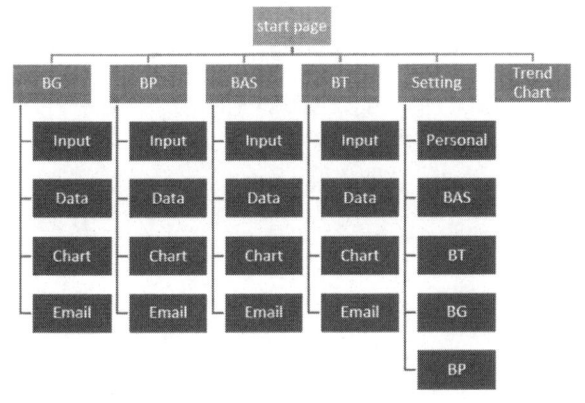

SQLite is ACID-compliant and implements most of the SQL standard, using a dynamically and weakly typed SQL syntax that does not guarantee the domain integrity.

SQLite is a popular choice as embedded database for local/client storage in application software such as web browsers. It is arguably the most widely deployed database engine, as it is used today by several widespread browsers, operating systems, and embedded systems, among others. SQLite has many bindings to programming languages. The source code for SQLite is in the public domain.

Display Chart on Smart Phones

For the purpose of data visualization on different platforms, we use up-to-date JavaScript and HTML5 technologies to draw the graph. JavaScript and now HTML5 canvas allow for quick and easy 2D drawing and are built into all modern browsers. HTML5 Canvas is also supported by all modern browsers and mobile devices meaning our charts and graphs will be seen by the widest possible number of users. Let's briefly introduce HTML5 canvas and JavaScript.

HTML5 Canvas

HTML5 canvas is a new HTML tag. It allows bitmap drawing and is controlled by JavaScript. In other word, it likes a piece of paper which is a part of page and we can draw on it by using JavaScript [10, 11].

Canvas uses a "fire and forget" drawing methodology to renders its graphics directly. If we want to change something, usually we must redraw the entire canvas and this is important when providing animated or interactive charts to users. To conquer this problem, canvas need fast to draw on and very responsive.

When we build a canvas, we have to define a drawable region in HTML code with height and width attributes. JavaScript code will access the area through a set of drawing functions like other common 2D APIs, thus allowing for dynamically generated graphics. Now HTML5 canvas usually is used to build animations, graphs, image composition, and games.

JavaScript

JavaScript is one of interpreted computer programming language. It is also the world's most popular programming language and implemented as part of web browsers so that client-side script could interact with the user, control the browser, communicate asynchronously and alter the document content that was displayed. It

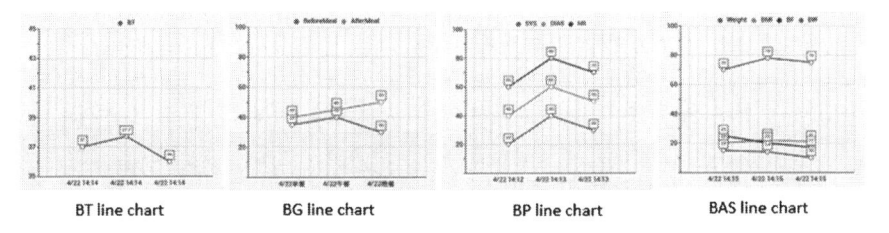

| BT line chart | BG line chart | BP line chart | BAS line chart |

Fig. 5 Line charts of BT, BG, BP, and BAS

is a multi-paradigm language, supporting object-oriented, imperative, and functional programming styles [12].

By using JavaScript and HTML5 technologies, our application can be cross-platform between iOS and Android and even other mobile operating systems.

In our application, Fig. 5 respectively shows distribution of BT, BG, BP and BAS data over time.

Cloud and Sociality

What does connecting with could actually do? When the data was sent to a remote server, the users can watch their records and charts directly. On the other hand, users do not even have anything but a browser; they could access their records anywhere. The result is that promotes the users' data portability.

On the implementation of the client side, we used PhoneGap Connection API to detect whether the Wi-Fi or 3G/4G enabled. If there are data which are not been uploaded, the client will send a HTTP request to the web server and upload remain parts of data in the SQLite database.

On the server side, we built the environment by LAMP [13] which is a web service bundle to process all requests from clients. We write a processor in PHP to receive requests and send back responses. Later, the processor will store the uploaded data to the MySQL database. Finally, the data will be presented with records or charts on the web pages with the help of a JavaScript library of charts. With charts, users could easily understand their health condition that is consistent with the mobile application.

IMS [14], IP Multimedia Subsystem, is an architectural framework for delivering IP multimedia services. The users can connect to IMS in various ways, such as via WLAN or 3G/4G, so does the IMS terminals may be mobile phones, PDAs, or computers. Therefore, we establish an IMS server in a Linux system. The users can have voice communications or video communications with their friends through our IMS server, following the SIP (Session Initiation Protocol) [15]. It provides the function like Skype allowing patients to chat together or share the experience to each other and help patients feel less lonely.

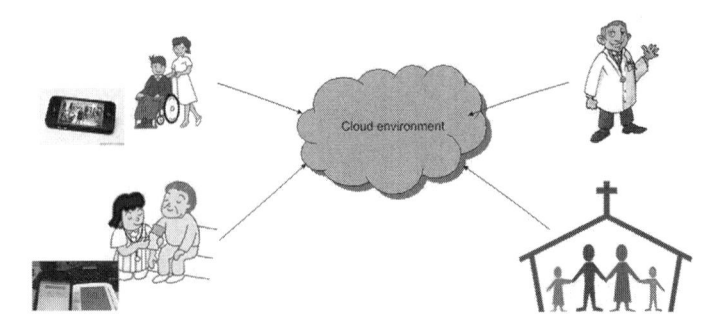

Fig. 6 Scenario of system

Scenario of System

Due to the implementation mentioned above, we can have a scenario for our system (Fig. 6). As far as the patients who need health care concerned, our system provide that not only patients can take health checks at home and uploading the result to the cloud server, but also the doctors can review the patients' reports. Furthermore, patients can use the social part of our application to know the situation of their family or friends. Considering the patients who are elder are not capable to read amount of words on the screen of smart phones, our social application are designed to only allow users sharing the photos and the videos which was taken recently. Even if the patients' bad health condition, the patients' do not lose the chance to know the situation of their family or friends.

Discussion

As far as elder is concerned, they were suffer from disease of every variety. Health condition needs to be considered anytime. Body analysis, body temperature, blood glucose, and blood pressure are the common health condition. By observing the value of them, we can discover the difference from previous data, even detecting the chronic diseases, to achieve the effect of prevention and treatment.

Although, elder can take health check at home, it is difficult to operate the complicated medical device. What they need is the simple and user-friendly way to use. Therefore, using Personal Health Assistant can simplify the complicated operation.

However, there are some difficulties to develop cross-platform application. First, PhoneGap is framework like wrapper that wraps the website with PhoneGap and deploy to the mobile platforms. It means the JavaScript functions are mapped to native function with intermediary. Actually PhoneGap is a web view to view the website which we write with HTML, JavaScript, and CSS. Therefore, comparing with the native application, PhoneGap application has a fatal disadvantage that it is inefficiency and slow. Second, there are some function are not be supported, because website app is not the same as mobile app. Take android for example; Android API use java as standard programing language, but web app is made with HTML not suitable for Android API. The solution is to write plugins that let JavaScript function call native code function. Nonetheless, it is not a good solution, writing plugins means you have to write with native code, against the original intention of cross-platform.

Conclusion

Integrating the smartphone application technology with medical service, and the universality of smartphone make taking health check more convenient and simpler. Therefore, elder or patients are willing to take health check more frequently. It can effectively make people pay close attention to own health.

Acknowledgments The paper is supported partially by the "Cross-platform smart phone-enable stroke prevention and rehabilitation health system" project and partially by the NSC-101-2811-E-008-013 project.

References

1. Zheng P, Ni LM (2006) Spotlight: the rise of the smart phone. Distributed systems online, IEEE 7.3
2. 2013 4th week 內政統計通報(101年底人口統計分析), 2013.01.26, department od stastics, ministry od the interior, R.O.C. (TAIWAN)
3. Stastics of Causes of Death (2011) Dec 2012, Department of health, executive Yuan, R.O.C. (TAIWAN)
4. Butler M (2011) Android: changing the mobile landscape. Pervasive Comput, IEEE 10(1):4–7
5. Junyan LV, Xu S, Li Y (2009) Application research of embedded database SQLite. In: IEEE international forum on information technology and applications, IFITA'09, vol 2
6. Lorenz A, Oppermann R (2009) Mobile health monitoring for the elderly: designing for diversity. Pervasive Mob Comput 5(5):478–495
7. Hii P-C, Chung W-Y (2011) A comprehensive ubiquitous healthcare solution on an Android™ mobile device. Sensors 11(7):6799–6815
8. Darie C (2006) AJAX and PHP: building responsive web applications. Packt Publishing Ltd
9. Hassenzahl M, Tractinsky N (2006) User experience-a research agenda. Behav Inf Technol 25(2):91–97
10. Allen S et al (2010) Pro Smartphone cross-platform development: iPhone, Blackberry, windows mobile and android development and distribution. Apress
11. Kessin Z (2011) Programming HTML5 applications: building powerful cross-platform environments in JavaScript. O'Reilly Media, ISBN: 978-1-449-39908-5
12. Negrino T (1998) JavaScript for the World Wide Web. Peachpit Press, San Francisco
13. Rosebrock E, Filson E (2006) Setting up LAMP: getting Linux, Apache, MySQL, and PHP working together. Sybex

14. Camarillo G, Garcia-Martin M-A (2007) The 3G IP multimedia subsystem (IMS): merging the Internet and the cellular worlds. Wiley
15. Rosenberg J, Schulzrinne H, Camarillo G, Johnston AR, Peterson J, Sparks R, Handley M, Schooler E (2002) SIP: session initiation protocol, RFC 3261, Internet engineering task force, June 2002

Cross-Platform and Light-Weight Stroke Rehabilitation System for New Generation Pervasive Healthcare

Eric Hsiao-Kuang Wu, C. C. Tseng, Y. Y. Yang, P. Y. Cai, S. S. Yen and Yu-Wei Chen

Abstract We propose a concept allowing disabled patient taking their rehabilitation treatment without going to the hospital. We design a portable device of rehabilitation and it can be lightweight in an advanced version. Particularly, we select PhoneGap to be a mobile development framework and to build the specific application for mobile devices using JavaScript, HTML5, and CSS3 instead of using device-specific language in our system. The great advantage is cross-platform that represent write once and run anywhere. In our software, we lead the patient to do the exercise of rehabilitation step-by-step and then store recovery status in database. Finally, the system will play restful music when patient do exercise. According to the mentioned above, we developed the Mobility Rehabilitation that combines mobility, entertainment, and storing ability.

Keywords Rehabilitation · PhoneGap · Lightweight · Cross-platform

E. H.-K. Wu (✉) · C. C. Tseng · Y. Y. Yang · P. Y. Cai · S. S. Yen
Computer Science and Information Engineering, National Central University,
Jhongli, Taiwan
e-mail: hsiao@csie.ncu.edu.tw

C. C. Tseng
e-mail: ben482922@gmail.com

Y. Y. Yang
e-mail: q0955895260@yahoo.com.tw

P. Y. Cai
e-mail: tigebrp6@hotmail.com

S. S. Yen
e-mail: freehourse2sh@gmail.com

Y.-W. Chen
Department of Neurology, Landseed Hospital, No. 77, Kwang-Tai Road, Ping-Jen 32449,
Tao-Yuan County, Taiwan
e-mail: yuwchen@gmail.com

Y.-M. Huang et al. (eds.), *Advanced Technologies, Embedded and Multimedia for Human-centric Computing*, Lecture Notes in Electrical Engineering 260, DOI: 10.1007/978-94-007-7262-5_143, © Springer Science+Business Media Dordrecht 2014

Introduction

Recently, Smart Phone has being widely used and the number of applications also increases. It is not only because of the prevalence of Smart Phone, but also the strong operating systems that provide the users to install software and games easily. Compared with all the existing operating systems, Android seems to be more developer-friendly since it releases it's source code and the kernel and file system that allow programmers to modify and fix the bugs. For example, other systems don't release the open source including Windows Mobile and Apple iOS. Due to the convenience for both users and developers, Android-based Smart Phone is now most popular in the market.

The percentage of the population of elders is increasing in Taiwan for decades. Stroke and dementia are key problems of the functional impairment among the old people, lead great impact to the whole society. Stroke is one of the major causes of dementia. Several risk factors might cause this cerebrovascular attack including diabetes, hypertension, hyperlipidemia and smoking. Besides the acute onset of neurological deficit, the pathological changes of small cerebral vessels have been associated chronic deterioration of mentality. Modified Framingham Stroke Risk Profile (FSRP) is proposed to predict the risk of stroke in 10 years and had been proven to be an independent predictor for the declining cognition over time. To prevent and cure the cerebrovascular and degenerative disorders, we need to develop novel strategies for their prevention, treatment and rehabilitation.

Android market, an online store could download the Android applications, provides variety of applications; nonetheless, the rehabilitation applications haven't caught people's attention. In the past, physical therapists have been gradually decreased, but patients who suffer from work injuries, sports injuries and stroke increased need the system. Moreover, in our current rehabilitation system, it is inconvenient for individuals with mobility problems to receive the treatment. Without the extra help, patients are difficult going to hospitals or rehabilitation clinics. These statements address the importance of rehabilitation applications, providing some simple methods for patients to take treatment at home. This may decrease the frequency for patients to go to the hospitals, make the best use of hospitals' facilities and handle issue of therapists' shortage. Thus, our goal is to design and implement a pervasive mobility rehabilitation system [1].

First, we use standardized web APIs to tap into device's motion sensor called accelerometer. The accelerometer can capture device motion in the x, y, and z direction. Then we can make the design of the corresponding sensing to meet the demand for rehabilitation motion, according to the different requirements of each patient.

Second, we choose the PhoneGap to develop the program that can run on the heterogeneous platform. Third, when the sensing procedures are completed and detected correctly, we need a mechanism to record the achievement of rehabilitation. We use a database to record date, time, frequency, and other related information of the rehabilitation by using the standardized web APIs.

Until now, almost functions have been completed, but we think the design of the system is not perfect enough to make the mobile rehabilitation device light and handy. In the later section, we will make a lighter device connect to our Smart Phone. This improvement makes our system be much closer to the "mobile" characteristic.

Related Work

In recently years, there are more and more papers explore in improving the rehabilitation training [2] uses android-based language a specified language to achieve a novel ubiquitous rehabilitation system (Ubi-REHAB) based on Augmented Reality (AR) technology, which is designed to enhance the recovery of upper extremity functions in patients with stroke. Although their system seems good, their development tool limits the usage of the program. On the contrary, we use PhoneGap to be the development platform in our rehabilitation system. This makes our system more attractive because of the cross-platform characteristic of Phonegap. It reduce the burden on the developer because they don't need to worry about the problem brought by the diversity of the platform. Most importantly, it reduces the threshold while the application is applied. It benefits patients. Another Ref. [2] shows a prototype for a new generation of game therapies based on Smartphone to achieve the purpose of rehabilitation of articulations. For some patients with serious condition, they can't exert their power. According to this situation, we develop another motion sensor which is much lighter than Smartphone to help more patients. This can address another advantage of our system. Patients not only can play game on Smartphone but also can do exercise on peripheral device in our system. Finally, "cross-platform" and "lighter" differentiate the difference from other paper.

Mobility Rehabilitation System

System Architecture

We proposed four mainly characteristics of our system development. First, we developed a mobile phone app for rehabilitation training. Patients can do the rehabilitation through the smart phone anytime at any place. Second, we use PhoneGap, a mobile development framework package Web application, as a native application to develop our application. We take advantage of its cross-platform characteristics to solve problems due to the variety of the platform. Third, we use a database to store patient's information. And the last one point is we develop another external embedded mobile device to reduce the full application weight comparing with using Smartphone to taking the rehabilitation training. According to the mentioned above, our system make patients to do rehabilitation simply, conveniently, easily, and efficiently.

Rehabilitation on Smartphone

With an increasing of mobile device usage in our society and the worldwide deployment of mobile application, the vision of Pervasive Healthcare or healthcare to anyone, anytime, and anywhere could be possible. In our stroke rehabilitation application, we design a suit of rehabilitation procedures especially for patients with stroke. In order to make sure that users could enjoy the rehabilitation without complicated instructions and misunderstanding, we make our system to be more user friendly for patients and we also make our user interface present with facilitation and the brisk style.

This application chiefly uses two sensors. One is originally built-in a smart phone, called accelerometer. The other sensor is Gyro-Sensor. Both of them capture device motion in the x, y, and z direction. Furthermore, our application uses database to preserve the data. So that doctors can know about patients' condition. We also play some music while patients take the rehabilitation training by our application.

In addition to achieve the exquisite design in our application, we also cooperate with the local hospital for further studying in the procedure of rehabilitation and the related movements. After we visit the rehabilitation center, we realize how the patients feel when they take the rehabilitation training. And due to this precious experience, we adjust the degree of difficulty of the program to an appropriate level. The hospital staff gave us many diverse professional suggestions. Through bilateral communication and coordination, we improve our application with proficiency as well as completeness. And following is the operative instructions to the application (Fig. 1).

When patients entering the application, patients would see two big buttons. One of them records the latest progress, the other lets user enter the main menu. In the primary stage, there are four types of the rehabilitation activities (Comb, Arms raised, Arms declined, Wrist flip) available for the patients. After choosing one of the buttons, there is a demonstrative picture to guide user how to do the motion. And just press the "start" button, set the game level and then start enjoying it. If the mission is finished, they could get reward as well. Besides, even if the game is forced to interrupt, the program could record the latest progress, so the previous score and the reward are still recorded in the database.

Rehabilitation on Bracelet

In the beginning, we use two embedded boards to develop a lighter sensor roughly (Fig. 2). One is Bluetooth module playing role as a transmission tool transmits

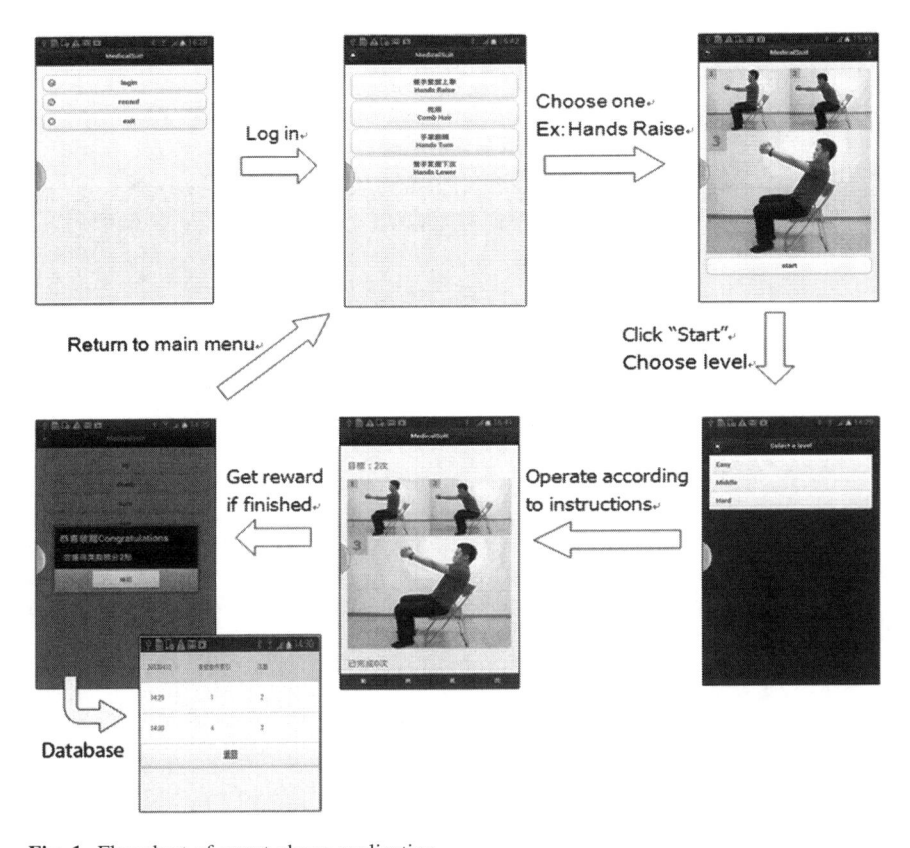

Fig. 1 Flowchart of smart phone application

statics about the angular velocity of the three-axis every moment from the other circuit board called Gyro-Sensor to smart phone. And the work for the Gyro-Sensor is motion sensing precisely. When phone receives data, it will determine whether the motion is correct or not. And this sensing motion will not be done until patients make the right motion. In the end, when all of the components are completed, we combine these two circuit board to a much perfectly lighter sensor we called bracelet. With this bracelet, rehabilitation process is much easier [3].

PhoneGap

We can say that there is a trend that more and more programmers or teams join the mobile developments of application. PhoneGap is just the platform that can produce cross-platform mobile application based on HTML, CSS, and JAVASCRIPT (Fig. 3). It has some great features worth mentioning. The first one is compatibility that make application write once and run anywhere (Fig. 4). Second, PhoneGap

Fig. 2 Bluetooth and Gyro-Sensor

use W3C standard and combine with jQuery Mobile make development more powerful. But it has the problem of lower performance compared to native code. So according to different request, choose suitable platform to develop applications to meet their needs. For example, high-level games need high-speed requirement, so developer use native code to develop. But in our system, we pay most of our attention on the utilization of the rehabilitation application, so we use PhoneGap to be our development platform [4]. In conclusion, the characteristic of "cross-platform" will benefit more patients [5].

G-Sensor and Gyro-Sensor

In our system, we use two kinds of sensor. And the difference between them is their sensing way. One is G-Sensor called accelerometer responses for detecting "triaxial acceleration". This sensor is also in the general smart phone. The other is Gyro-sensor which detects "angular velocity" in the x, y, and z direction. Giving a more precise description, these "3-axis" gyroscopes we used have a single sensing structure for motion measurement along all three orthogonal axes different from other gyroscopes that use two or more structures. This sensing way can eliminates

Fig. 3 Schematic diagram for PhoneGap

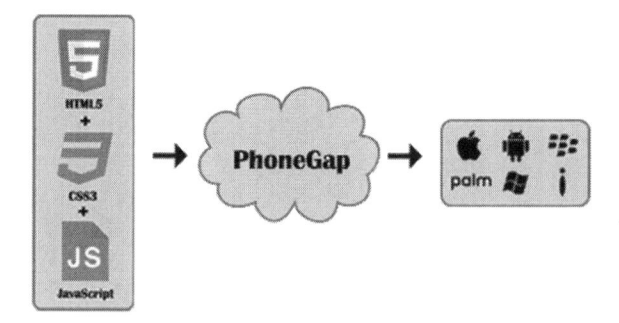

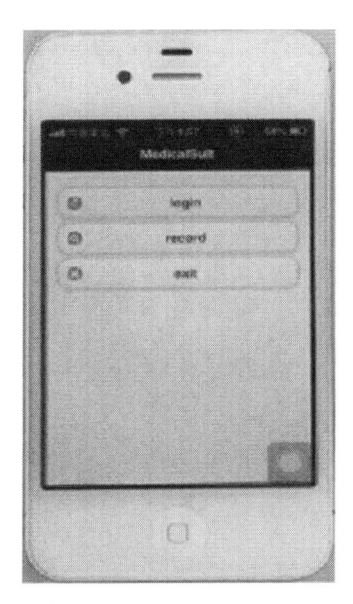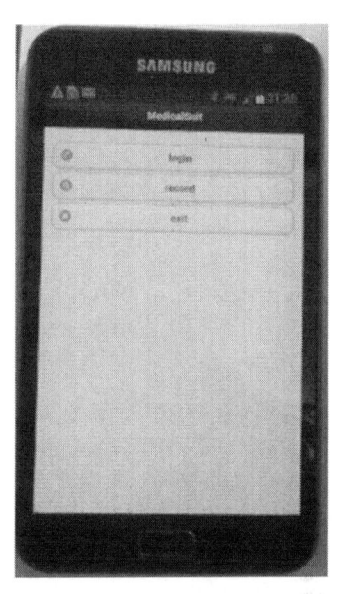

Fig. 4 Cross-platform characteristics of PhoneGap

any interference between the axes that inherently degrades the output signal, increasing accuracy and reliability of motion-controlled functionalities. From the above mentioned, we also can use gyroscope to be our powerful motion sensor [6].

Scenario of System

Due to the implementation mentioned above, we can have a scenario for our system (Fig. 5). For a patient who needs to do rehabilitation training, it is difficult to take a long trek to rehabilitation clinic. And even the patient arrives at the hospital, he can only do the boring training as usual. At this moment, our rehabilitation APP has an opportunity to show. No matter patient is traveling or going to some place lacking of rehabilitation clinics, he or she can still do rehabilitation.

After the treatment of our system to the patient, our system will record the patient's rehabilitation result and upload this record to database. Thus, doctor can check the database to learn about the condition of rehabilitation for patients.

Discussions

Most of physical disable person need to take some training for recovery. Rehabilitation can make significant improvement on the recovery of patients' body.

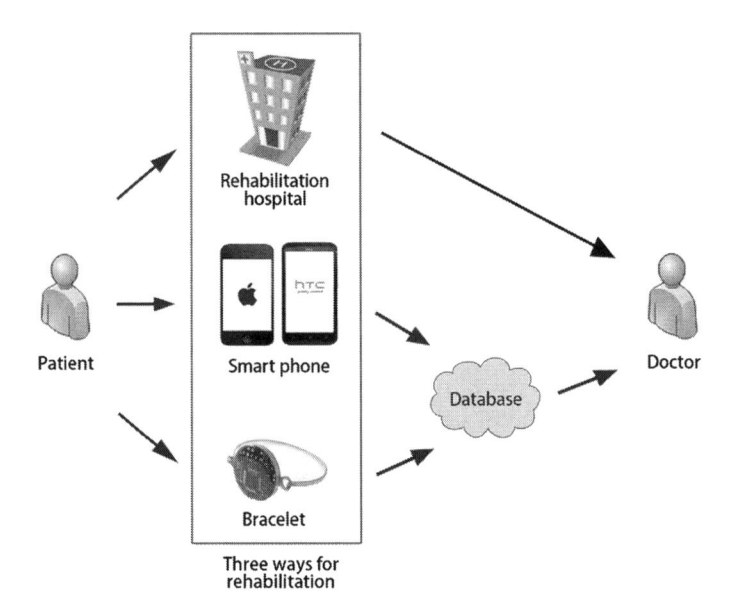

Fig. 5 Scenario of system

However, the patients need long-term rehabilitation to recovery after the operation. This makes tremendous changes in their life. Not only the patient's daily-life, but also their families'. Recently, the demand of rehabilitation is dramatically increasing in medical resource. Also, as the aging society approaching, the long-term care for the elder is becoming more and more important.

In the past, patients need to go to the hospital in person to use the rehabilitation instruments which are large and expensive. After getting home, they might not continue their rehabilitation because of the lack of the instruments. Our project is aim to dig out a way to let the patient of stroke to rehabilitate continually at their home in an easier and more comfortable way.

Conclusion

Due to the popularity of smart phones, there is a new developing area in medical care. Convenience and efficiency in the further medical care is going to be more important in the future. Therefore, we implement a cross-platform rehabilitation application on the smart phone., and we also realize an embedded bracelet for our mobile application.

In the future, we will pay attention to improve our accuracy of the motion detection by sensors. We can also implement a feedback mechanism in our mobile application, so we can provide a different rehabilitation training course by considering the different recovery situation of patients. Due to pursuing the passive health care, there are some undeveloped medical area we can explore and work on. And patients who suffer from stroke will be more and more cheerful when they take rehabilitation training day by day.

Acknowledgments The paper is supported partially by the "Cross-platform smart phone-enable stroke prevention and rehabilitation health system" project and partially by the NSC-101-2811-E-008-013 project.

References

1. Varshney U (2003) Pervasive healthcare. IEEE Comput Mag 36(12):138–140
2. Choi Y (2011) Ubi-REHAB: an android-based portable augmented reality stroke rehabilitation system using the eGlove for multiple participants. In: International conference on virtual rehabilitation (ICVR), 27–29 June 2011, pp 1–2
3. Deponti D, Maggiorini D, Palazzi CE (2011) Smartphone's physiatric serious game. In: IEEE 1st international conference on serious games and applications for health (SeGAH), 16–18 Nov 2011, pp 1–8
4. Motoi K (2005) Development of a wearable device capable of monitoring human activity for use in rehabilitation and certification of eligibility for long-term care. In: 27th annual international conference of engineering in medicine and biology society. IEEE-EMBS 2005, 17–18 Jan 2006, pp 1004–1007
5. Sin D (2012) Mobile web apps—the non-programmer's alternative to native applications. In: 5th international conference on human system interactions (HSI), 6–8 June 2012, pp 8–15
6. Smutny P (2012) Mobile development tools and cross-platform solutions. In: 13th international carpathian control conference (ICCC), 28–31 May 2012, pp 653–656
7. Barthold C (2011) Evaluation of gyroscope-embedded mobile phones. In: IEEE international conference on systems, man, and cybernetics (SMC), 9–12 Oct 2011, pp 1632–1638

Off-line Automatic Virtual Director for Lecture Video

Di-Wei Huang, Yu-Tzu Lin and Greg C. Lee

Abstract This research proposed an automatic mechanism to refine the lecture video by composing meaningful video clips from multiple cameras. In order to maximize the captured video information and produce a suitable lecture video for learners, video content should be analysed by considering both visual and audio information firstly. Meaningful events were then detected by extracting lecturer's and learners' behaviours according to teaching and learning principles in class. An event-driven camera switching strategy was derived to change the camera view to a meaningful one based on the finite state machine. The final lecture video was then produced by composing all meaningful video clips. The experiment results show that learners felt interested and comfortable while watching the lecture video, and also agreed with the meaningfulness of the selected video clips.

Keywords Video content analysis · Video composition · Virtual director · Lecture video analysis

D.-W. Huang · G. C. Lee
Department of Computer Science and Information Engineering, National Taiwan Normal University, Taipei, Taiwan, Republic of China

Y.-T. Lin (✉)
Graduate Institute of Information and Computer Education, National Taiwan Normal University, Taipei, Taiwan, Republic of China

Y.-M. Huang et al. (eds.), *Advanced Technologies, Embedded and Multimedia for Human-centric Computing*, Lecture Notes in Electrical Engineering 260, DOI: 10.1007/978-94-007-7262-5_144, © Springer Science+Business Media Dordrecht 2014

Cycle Embedding in Alternating Group Graphs with Faulty Elements

Ping-Ying Tsai and Yu-Tzu Lin

Abstract The alternating group graph, which belongs to the class of Cayley graphs, is one of the most versatile interconnection networks for parallel and distributed computing. Cycle embedding is an important issue in evaluating the efficiency of interconnection networks. In this paper, we show that an n-dimensional alternating group graph AG_n has the following results, where F is the set of faulty vertices and/or faulty edges in AG_n: (1) For $n \geq 4$, AG_n -F is edge 4-pancyclic if $|F| \leq n - 4$; and (2) For $n \geq 3$, AG_n-F is vertex-pancyclic if $|F| \leq n - 3$. All the results are optimal with respect to the number of faulty elements tolerated, and they are improvements over the cycle embedding properties of alternating group graphs proposed previously in several articles.

Keywords Alternating group graph · Pancyclicity · Fault-tolerant · Cayley graph · Cycle embedding · Interconnection network

P.-Y. Tsai (✉)
Taiwan Geographic Information System Center, Taipei, Taiwan, Republic of China
e-mail: bytsai0808@gmail.com

Y.-T. Lin
Graduate Institute of Information and Computer Education,
National Taiwan Normal University, Taipei,
Taiwan, Republic of China

Y.-M. Huang et al. (eds.), *Advanced Technologies, Embedded and Multimedia for Human-centric Computing*, Lecture Notes in Electrical Engineering 260, DOI: 10.1007/978-94-007-7262-5_145, © Springer Science+Business Media Dordrecht 2014

Printed by Publishers' Graphics LLC